AF874146

High Pressure Chemistry

NATO ADVANCED STUDY INSTITUTES SERIES

Proceedings of the Advanced Study Institute Programme, which aims at the dissemination of advanced knowledge and the formation of contacts among scientists from different countries

The series is published by an international board of publishers in conjunction with NATO Scientific Affairs Division

A	Life Sciences	Plenum Publishing Corporation
B	Physics	London and New York
C	Mathematical and Physical Sciences	D. Reidel Publishing Company Dordrecht, Boston, and London
D	Behavioral and Social Sciences	Sijthoff International Publishing Company Leiden
E	Applied Sciences	Noordhoff International Publishing Leiden

Series C – Mathematical and Physical Sciences

Volume 41 - High Pressure Chemistry

High Pressure Chemistry

Proceedings of the NATO Advanced Study Institute held in Corfu, Greece, September 24 - October 8, 1977

edited by

H. KELM

University of Frankfurt, Federal Republic of Germany

D. Reidel Publishing Company

Dordrecht : Holland / Boston : U.S.A. / London : England

Published in cooperation with NATO Scientific Affairs Division

Library of Congress Cataloging in Publication Data

CIP

Main entry under title:

High pressure Chemistry.

(NATO advanced study institutes series : Series C, Mathematical and physical sciences ; v. 41)

Includes index.

1. Chemistry, Physical and theoretical–Addresses, essays, lectures. 2. High pressure (Science)–Addresses, essays, lectures. I. Kelm, H., 1933- II. Series.

QD455.H48 541'.3 78-15304

ISBN 90-277-0935-1

Published by D. Reidel Publishing Company
P.O. Box 17, Dordrecht, Holland

Sold and distributed in the U.S.A., Canada, and Mexico
by D. Reidel Publishing Company, Inc.
Lincoln Building, 160 Old Derby Street, Hingham, Mass. 02043, U.S.A.

All Rights Reserved
Copyright © 1978 by D. Reidel Publishing Company, Dordrecht, Holland
No part of the material protected by this copyright notice may be reproduced or utilized in any form or by any means, electronic or mechanical, including photocopying, recording or by any informational storage and retrieval system, without written permission from the copyright owner.

Printed in The Netherlands

TABLE OF CONTENTS

PREFACE vii

P. L. M. Heydemann
GENERATION AND MEASUREMENT OF PRESSURE 1

B. J. Alder
COMPUTER SIMULATION AT HIGH PRESSURE 51

J. Jonas
MAGNETIC RESONANCE SPECTROSCOPY AT HIGH PRESSURE 65

D. Bloch and J. Paureau
NEUTRON SCATTERING AT HIGH PRESSURE 111

E. Whalley
HIGH-PRESSURE INFRARED AND RAMAN SPECTROSCOPY 127

W. B. Holzapfel
MÖSSBAUER SPECTROSCOPY ON SOLIDS UNDER PRESSURE 159

W. B. Holzapfel
X-RAY DIFFRACTION ON SOLIDS UNDER PRESSURE 177

D. Bloch
SOLIDS UNDER PRESSURE 199

E. U. Franck
EXPERIMENTAL INVESTIGATIONS OF FLUIDS AT HIGH PRESSURES AND ELEVATED TEMPERATURES 221

F. Hensel
THE EFFECT OF PRESSURE ON THE ELECTRONIC STRUCTURE OF FLUIDS 259

H. Kelm and D. A. Palmer
DETERMINATION AND INTERPRETATION OF VOLUMES OF ACTIVATION 281

K. Heremans
FAST REACTIONS IN SOLUTION UNDER HIGH PRESSURE 311

W. J. le Noble
ORGANIC MODEL REACTIONS UNDER PRESSURE 325

W. J. le Noble
ORGANIC PROBLEM REACTIONS PRESSURE 345

G. Jenner
EFFECTS OF PRESSURE ON THE FORMATION AND PROPERTIES OF POLYMERS 365

D. A. Palmer and H. Kelm
PRESSURE EFFECTS ON RAPID EQUILIBRIA IN SOLUTION 421

D. A. Palmer and H. Kelm
THE EFFECTS OF PRESSURE ON THE RATES OF REACTIONS OF COORDINATION COMPOUNDS IN SOLUTION 435

K. Heremans
PRESSURE EFFECTS ON BIOCHEMICAL SYSTEMS 467

J. Troe
KINETIC PHENOMENA IN GASES AT HIGH PRESSURE 489

H.-D. Brauer, R. Schmidt, and H. Kelm
HIGH PRESSURE EFFECTS ON PHOTOPHYSICAL AND PHOTOCHEMICAL PRIMARY PROCESSES 521

LIST OF PARTICIPANTS 569

INDEX OF NAMES 573

INDEX OF CHEMICAL COMPOUNDS 584

INDEX OF SUBJECTS 589

Preface

Recent advances in the field of high pressure techniques influenced me to propose an Advanced Study Institute in High Pressure Chemistry. It was intended that the summer school should devote itself exclusively to the description and discussion of the effects of pressure in chemistry. Besides typical effects on matter, the application of high pressure techniques to existing research methods were to be treated, as well as pressure effects on reaction rates and equilibria.

According to the concept of the Advanced Study Institute Program, the Summer School proceedings were meant to be a high level teaching activity. It was emphasized that the contributions should have the character of surveys rather than of highly specialized reports on recent research results.

Now, following the successful completion of the summer school, which involved very close cooperation with my colleagues during its preparation, it is my sincere wish to thank all the lecturers and contributors to this volume for the extreme care they used in preparing the lectures and manuscripts. I am especially grateful to the members of the organizing committee for their valuable assistance. Finally, the financial support of the Scientific Affairs Division of the North Atlantic Treaty Organization is equally appreciated by participants and organizers of the Advanced Study Institute.

Frankfurt am Main, May 1978 H. Kelm

GENERATION AND MEASUREMENT OF PRESSURE

Peter L.M. Heydemann

National Bureau of Standards
Washington, D.C. 20234, USA

ABSTRACT. This is a review of methods used to establish the primary pressure scale, of transfer gages and of the most common fixed points on the pressure scale. The major types of pressure generators and some important components like pumps, electrical feedthroughs and windows will be discussed. The review is generally limited to the hydrostatic environment.

I. DEFINITIONS

Over much of the pressure range to be discussed here pressure is homogeneous in space, hydrostatic and independent of time. It can therefore be defined as force per unit area exerted by a fluid on its containing walls

$$(1) \qquad p = \frac{F}{A}$$

or as energy density

$$(2) \qquad p = \left(\frac{\partial F}{\partial V}\right)_T$$

where F is the Helmholtz free energy with F = U - TS and dF = pdV - SdT.

At very high pressures when viscous liquids or even solids are used to transmit pressure one has to be more specific in the definition of hydrostatic pressure. This was pointed out by Decker et al. in their excellent review of high pressure calibrations [1] from which we will quote here extensively.

In order to specify precisely the terminology and describe the non-ideal features of an arbitrarily stressed system in a particular non-ideal laboratory situation, we introduce the symmetric stress tensor,

Kelm, H. (Ed.) High Pressure Chemistry, 1–49. Copyright © 1978, D. Reidel Publishing Company.
All Rights Reserved.

T, which is defined at each point in space. Measurable stress over finite areas can then be calculated by averaging. We note that each component of the stress tensor, in general, depends upon position and time, $T_{ij}(r,t)$, where r is the position vector. The pressure, which is also a function of position and time, is defined as the negative of the average of the three normal stress components

(3) $$p(r,t) = -\frac{1}{3}\ (T_{11} + T_{22} + T_{33}).$$

The shear stresses are given by the deviatoric stress tensor

(4) $$T_{ij}(r,t) = T_{ij}(r,t) - \delta_{ij}\ p(r,t).$$

We are now in a position to define unambiguously the meaning of hydrostatic pressure, which is characterized by zero deviatoric shear stress and isotropic normal stresses. Mathematically we write

(5) $$T'_{ij}(r,t) = 0$$

and

(6) $$-\ p(r,t) = T_{11}\ (r,t) = T_{22}\ (r,t) = T_{33}\ (r,t).$$

We note that this definition does not require static, i.e. time-independent, conditions although time-dependent changes generally involve shear. The equations define hydrostaticity at a point. If a region of space is to be hydrostatic, each point in the volume must satisfy the hydrostatic condition. This does not necessarily imply homogeneous, i.e. constant in space conditions. According to the above definition of pressure, nonhomogeneity in a hydrostatic medium at equilibrium can arise only from volume type forces such as gravitational, magnetic, or electric forces, which in practice are often quite small compared to the applied pressures. It is important to note that neither homogeneous pressure nor homogeneous stress implies hydrostaticity, but simply constancy in space.

In laboratory pressure systems the time dependence in T_{ij} generally arises from a change of the system from one equilibrium state to another. In a practical manner we define equilibrium as the state, which a system approaches asymptotically within a laboratory time scale. All systems under pressure will support time-dependent shear stress components with some characteristic relaxation time, τ, when momentarily disturbed from the equilibrium state. In many cases, especially with gases and liquids at low pressures, τ, may be a small fraction of a second and is usually negligible compared with measuring times. In highly viscous liquids and solids the relaxation times may be of the order of many hours or even years. In the formalism discussed here one can clearly distinguish between plastic solid and viscous liquid behaviour during this stress relaxation. For the liquid case the deviatoric stress T' will in time approach a zero value, but for solids T'_{ij} will approach some non-zero finite deviatoric stress state.

The approach to equilibrium will be highly dependent upon the details of the system. In solid-media systems stress relaxation will be very complicated. For systems containing only fluids in which the viscosity of the fluid can be assumed constant throughout the system, the approach to equilibrium is characterized by the stress components decreasing with time approximately exponentially. Spatial pressure differences within the system will obey the expression

$$(7) \qquad \Delta p \propto e^{-t/\tau}$$

where the characteristic time, τ, is directly proportional to the viscosity of the fluid but also highly dependent upon the geometry of the pressure chamber. One of the important consequences associated with this discussion is the fact that time-dependent shear stresses are generally associated with pressure changes, and adequate time must be allowed for the system to approach equilibrium before reliable measurements can be made.

To measure pressure experimentally one must measure the normal stress over a finite area. If p is homogeneous over that area, as it very nearly is in a fluid chamber, the simple force per unit area relationship is valid, and pressures can be determined with high accuracy. Pressures determined by force per unit area in solid-media systems measure some average stress over the specified area which may differ from the true average normal stresses on the surface as well as from pressures at points inside the bulk of the chamber. In general such errors will be of the order of the shear strength of the solid materials. The errors associated with measurements of material properties under these non-hydrostatic stress conditions may be more serious than the errors associated with the pressure measurement since the property measurement errors are dependent upon the stress sensitivity of the parameter being considered.

In this text we will predominantly deal with generation and measurement of hydrostatic pressure covering the range from a few hundred megapascal to about 10 GPa.

II. PRESSURE SCALE

In any branch of metrology the establishment and universal acceptance of a primary scale upon which all interpolation and extrapolation functions and devices can be based and to which they can be referred is of fundamental importance. In general the establishment of such a scale will involve specific procedures, apparatus, and precautions associated with the measurement in question and will be as closely related to the fundamental definition of the measured quantity as apparatus will permit.

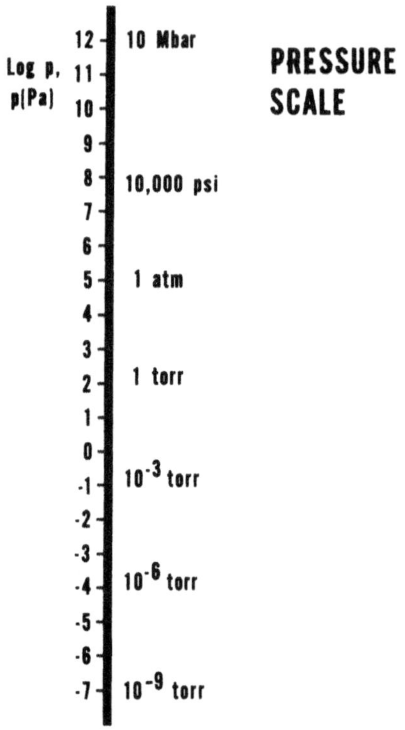

Fig. 1 The range of pressures available to science and industry covers more than 20 decades. Only about 3 of these fall into the "high pressure" range. It is no easier to measure pressure around 100 kPa than it is to measure 10 GPa, if state-of-the-art accuracy is to be attained.

The techniques and apparatus associated with the primary scale should be as simple as possible and the process should be a direct measurement of the quantity itself. It is preferable not to involve averaging, differentiation, integration, or other mathematical manipulations implied by a theoretical treatment. It is not expected that specific procedures and details will be permanent but, rather, that they will be temporarily accepted by the scientific community until a more direct and reliable or more accurate technique can be demonstrated. The fundamental nature, the reliability, and the accuracy will be of prime importance in contrast to sensitivity, convenience, and availability. It would be desirable to have but one technique or apparatus extend over all ranges of the measured quantity, but such condition is generally not possible.

Based on criteria of this type two basic measuring systems have developed and are used as the basis of the high pressure scale: the piston gage, and the measurement of shock wave and free surface velocities. The latter is used predominantly in solids but, lacking other more reliable methods of measurement at very high pressures, shock waves are nevertheless important for establishing part of the pressure scale.

III. PRESSURE MEASUREMENT

The range of pressures used in industry and science (Fig. 1) covers more than 20 decades. Within this range a large variety of gages and devices are used to define the pressure scale. Regardless of the design of these instruments two classes of gages must be distinguished: primary, and secondary or transfer gages. Primary pressure gages establish the pressure scale in terms of the basic unit of force and length, or energy and volume using basic thermodynamic relations without resorting to the calibration against other pressure gages of higher accuracy. A controlled clearance piston gage or a gas obeying the ideal gas law are examples of primary gages. Secondary or transfer gages measure pressure in terms of some other suitable quantity whose relation to pressure is determined by a calibration, that is a comparison with a primary gage or standard.

Deadweight loaded piston gages, piston and cylinder or piston and die devices, equations of state calculated from first principles, and the Rankine-Hugoniot relationships used for the measurement of pressure in shock waves are all examples of primary pressure gages. Piezo-resistance and piezo-capacitance gages, the ruby fluorescence line shift or the NaCl-equation of state with adjustable parameters are examples of transfer gages.

In practice the pressure scale in fluids above a few hundred kilopascal is defined by deadweight loaded piston gages to 2.6 GPa, by piston and cylinder devices to about 6 GPa and by transfer gages or other indirect methods to about 10 GPa, the end of the range of existence of fluids [2]. In shock wave experiments in solids the pressure scale can be defined by the Rankine-Hugoniot relationships.

To facilitate the realization of the pressure scale in the laboratory a large number of fixed points have been determined. From time to time interested groups review the available fixed point determinations [1, 3, 4] and recommend critically evaluated data for use by the high pressure community. There does not as yet exist a well defined pressure scale with a set of fixed points with assigned values that is universally accepted. However, work on such a scale [5] is now in progress and, since the need for an International Practical Pressure Scale is widely recognized, it may before long obtain the sanction of the appropriate international organizations.

In the following sections we will in turn discuss first the primary, then the secondary devices primarily as used with fluids. We will discuss the major fixed points and, in a cursory manner, review the application of the Rankine-Hugoniot relations to shock waves. There will undoubtedly be some overlap with later sections on the generation and containment of pressure, and on various apparatus and components.

IV. DEAD-WEIGHT LOADED PISTON GAGES

Piston gages measure pressure in terms of force per unit area using a piston loaded with weight and fitted into a cylinder containing a suitable fluid medium under pressure (Fig. 2). Force and area are readily measured at standard pressure and temperature (SPT: $p = 100$ kPa, $T = 296$ K), and methods have been developed to determine and account for corrections for elastic distortion, temperature, friction, surface tension etc. Several different types of gages have evolved from the efforts to minimize one or more of the large elastic distortion corrections. For pressures to 2.6 GPa only the controlled clearance piston gage [6] has been used successfully, but simple piston gages have been used for very accurate measurements up to about .8 GPa [7].

The pressure generated by a simple piston gage at its reference level is given by [8]

$$p = \frac{Mg(1 - \rho_{air}/\rho_M) + \gamma C}{A_{eff}(1 + bp)[1 + (\alpha_c + \alpha_p)(T - T_r)]} \tag{8}$$

where M is the mass of the load applied to the piston
g is the local acceleration of gravity,
ρ_{air}, ρ_M are the densities of ambient air and of the weights respectively,
γ is the surface tension of the fluid,
C is the circumference of the piston where it emerges from the fluid,
A_{eff} is the effective area of the piston-cylinder assembly at SPT
b is the pressure coefficient of the area,
α_c, α_p are the thermal expansivities of the cylinder and the piston respectively, and

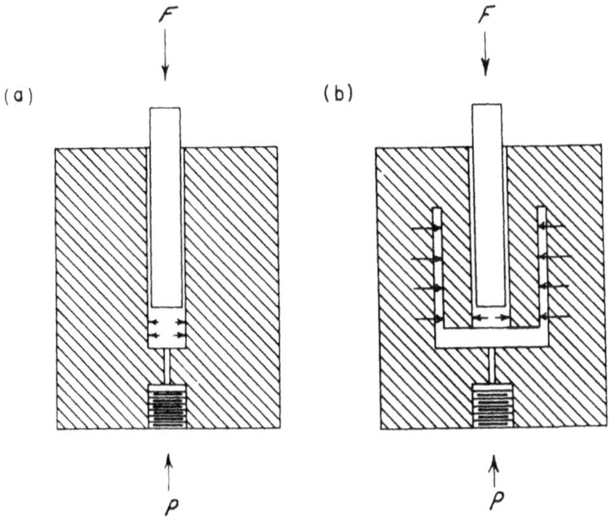

Fig. 2 Simple (a) and reentrant (b) piston gages establish the pressure scale by measuring pressure in terms of force per unit area; P pressure, F applied force.

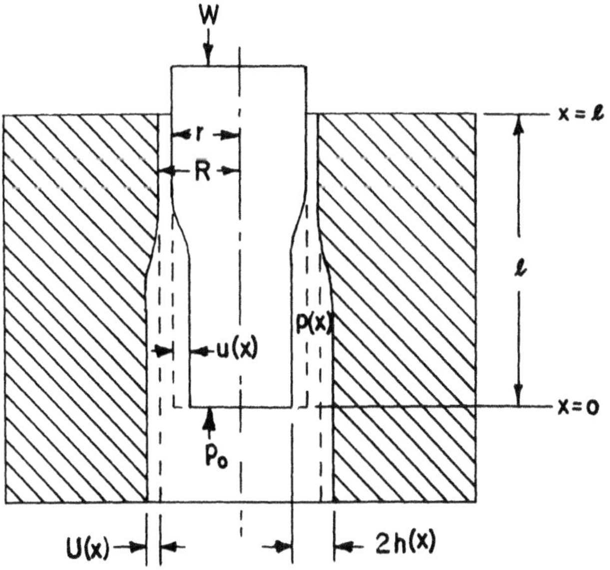

Fig. 3 The largest source of uncertainty in high pressure piston gages is the elastic distortion of the cylinder.

T, T_r are the temperature of the assembly and the reference temperature (standard temperature).

The area of a cylinder subjected to internal pressure varies as

$$(9) \qquad A_c(p) = A_c(o) \left[1 + 2p \frac{(1 + \mu_c) R^2 + (1 - \mu_c) r^2}{E_c (R^2 - r^2)} \right] ,$$

and the area of a piston subjected to external pressure varies as

$$(10) \qquad A_p(p) = A_p(o) \left[1 - 2p \frac{(1 - 3\mu_p)}{Ep} \right]$$

where $A_c(o)$, $A_p(o)$ are the respective areas at SPT,
and μ, E are the Poisson's ratios and elastic constants of the piston and cylinder material.

In a piston gage the pressure in the crevice between piston and cylinder drops from p at the bottom of the piston to zero at the top of the cylinder. Consequently the distortion is largest near the lower end of piston and cylinder as indicated in Fig. 3. Johnson [9] has shown that the change in the effective piston area is only half as large as indicated by equ. (10) and in analogy the same applies to the change in area of the cylinder. The effective area of a piston-cylinder combination is given by

$$(11) \qquad A_{eff} = \frac{A_c + A_p}{2}$$

and therefore the pressure coefficient of such an assembly is

$$(12) \qquad b = \frac{(1 + \mu_c) R^2 + (1 - \mu_c) r^2}{2 E_c (R^2 - r^2)} - \frac{(1 - 3\mu_p)}{2 E_p} .$$

With a cemented tungsten carbide piston and a steel cylinder the pressure coefficient is for R/r = 3

$$(13) \qquad b = 38 \times 10^{-13} - 4 \times 10^{-13} \frac{m^2}{N}$$

and at 0.7 GPa (7 kbar)

$$(14) \qquad bp = 2.7 \times 10^{-3} - 0.28 \times 10^{-3} .$$

This is a very large effect that is difficult to correct for and normal piston gages must therefore be calibrated against other more accurate standards as for example a controlled clearance piston gage.

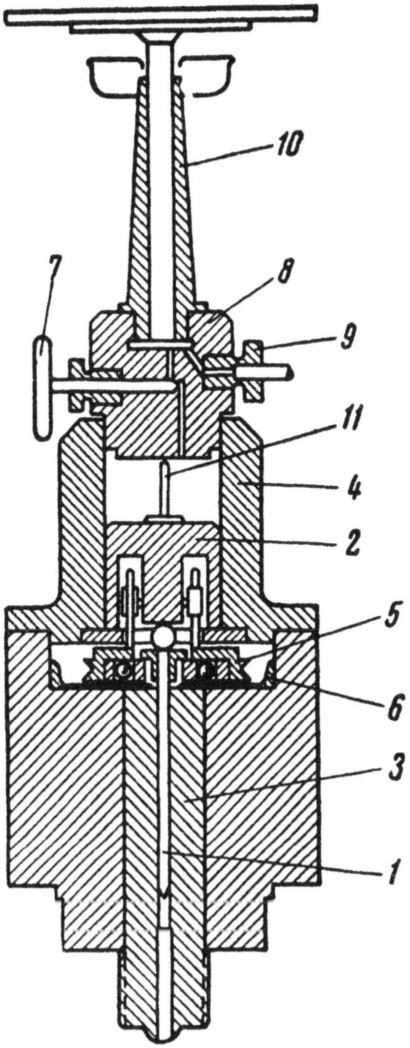

Fig. 4 This piston gage by Konyaev operated to beyond 2 GPa. The high pressure piston (1) is pushed by ram (2) into cylinder (3). Piston gage (10) generates the ram pressure.

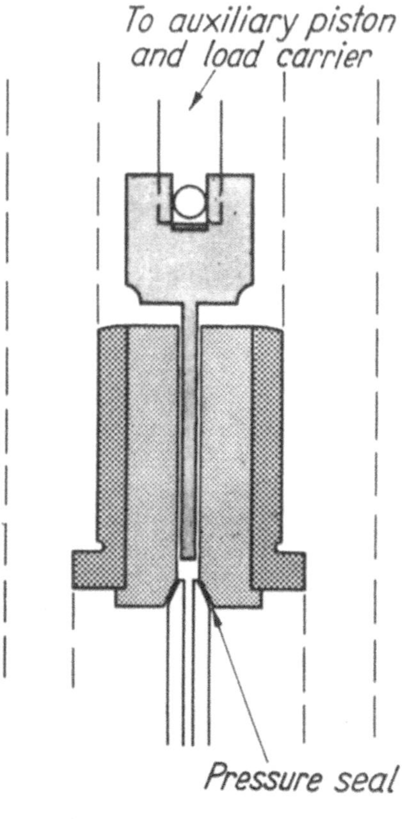

Fig. 5 Dadson of the National Physical Laboratory used this simple piston gage to determine the mercury melting pressure at 0°C. The long, slender piston is an important feature of the gage.

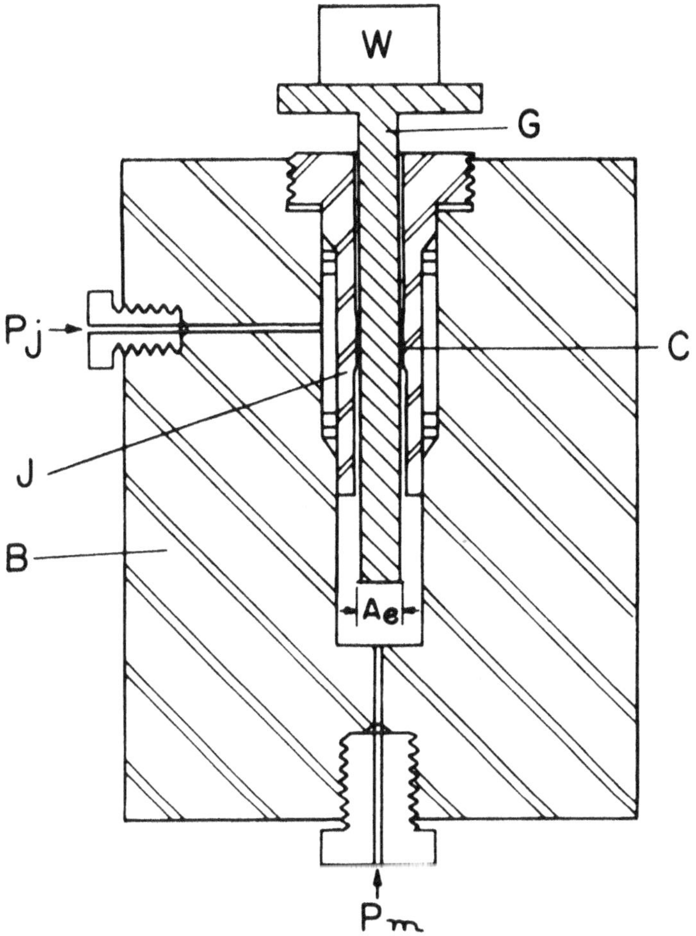

Fig. 6 The controlled clearance principle is widely used to establish the high pressure scale since it helps to reduce significantly the elastic distortion error. P_j is the external pressure that controls the clearance C between piston G and cylinder J. P_m is the measured pressure.

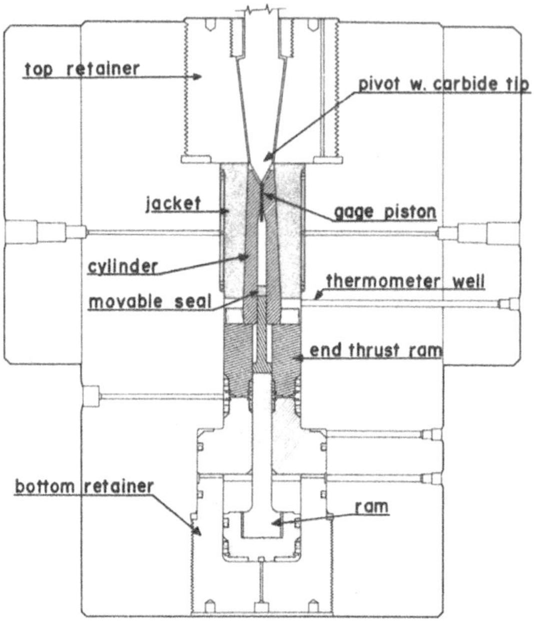

Fig. 7. NBS 2.5 GPa controlled clearance gage. 2 mm diameter cemented tungsten carbide gage piston. Pressure generated by hydraulic ram. Longitudinal support of cylinder from end thrust ram. Radial cylinder support from jacket. This gage was used to establish the Bi I-II transition pressure.

In spite of the large correction for elastic distortion errors simple piston gages have been designed for very high pressure. One example is the hydraulically loaded gage designed by Konyaev [10] and shown in Fig. 4. The hydraulic ram pressure is generated by a simple piston gage. Low pressure piston, ram piston and high pressure piston are all rotated to relieve friction. The use of hydraulic loading allows the use of larger diameter pistons with reasonable amounts of weights. This improves the accuracy of the area determination for the high pressure piston. It appears however that the instrument designed by Dadson of the National Physical Laboratory in England and shown in Fig. 5 was the only simple piston gage that was operated successfully at such high pressures.

In the controlled clearance piston gage (Fig. 6) a separate, external control pressure is applied to the outside of the cylinder to counteract the effect of the internal pressure. In fact, the cylinder can virtually be forced to conform to the piston at all pressures and only the distortion of the piston needs then to be considered. The pressure generated by a controlled clearance piston gage at its re-erence level is

$$p = \frac{Mg\,(1 - \rho_{air}/\rho_M) + \gamma C}{A_p(o)(1 + bp)[1 + (\alpha_c+\alpha_p)(T-T_r)][1+d(p_z-p_j)]} \tag{15}$$

where $A_p(o)$ is the area of the piston at SPT,
b is the pressure coefficient of the piston (see equ.(12), second term),
d is the jacket pressure coefficient,
p_j is the jacket or control pressure, and
p_z is the jacket pressure at which the clearance is reduced to zero.

The coefficents d, p_z and p_j are determined experimentally [8] but without need for another calibrated pressure gage.

Fig. 7 shows the controlled clearance piston gage used at the U.S. National Bureau of Standards for pressures up to 2.6 GPa. [6]. The pressure is generated inside the high pressure cylinder with a built-in hydraulic intensifier and measured with a cemented tungsten carbide gage piston of about 2 mm diameter. Up to 10 kN (1000 kg) are applied to the piston in the form of weights to generate the top pressure. Controlled clearance piston gages are commercially available [11] for pressures between 70 kPa and 2.6 GPa. Their potential accuracy ranges from about 30 ppm at the low end to about 2400 ppm (0.24%) at 2.6 GPa. Fig. 8 is a photograph of a gage for 1.4 GPa showing the high pressure cylinder on the top with the load applied through a yoke , the jacket pressure inlet at the top left and a manganin gage on the right for monitoring the pressure. The piston is, of course, rotated to relieve friction. The weights are handled by a hydraulic weight lifter.

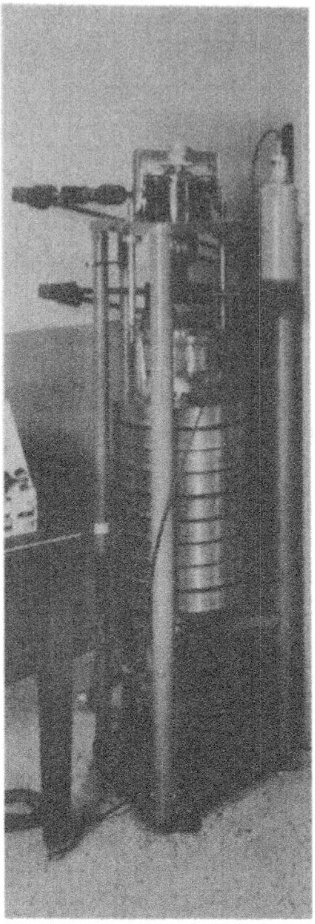

Fig. 8 Commercial version of a controlled clearance piston gage for 1.4 GPa with hydraulic weight lifter.

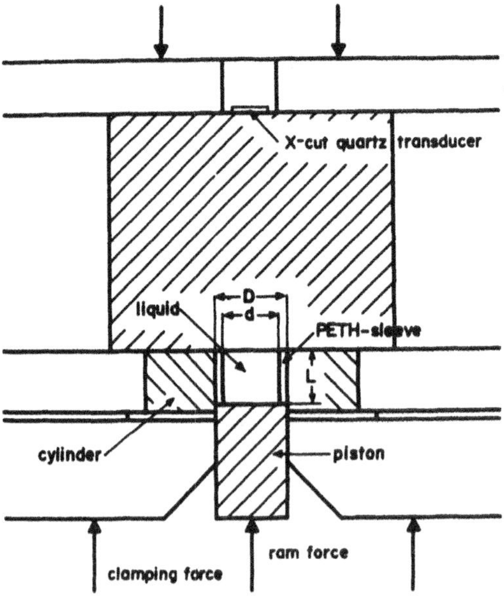

Fig. 9 5 GPa piston and die with polyethylene liquid container.

The use of piston gages, deadweight loaded, or hydraulically loaded, is not likely to extend much beyond 2.6 GPa due to need for long and slender pistons which are stressed close to their compressive strength.

V. HYDRAULICALLY LOADED PISTON AND CYLINDER DEVICES

Two major constraints appear to limit dead-weight loaded piston gages to pressures below about 2.6 GPa: the compressive strength of the rather long pistons and the large load to be applied to the piston. Higher pressures can be obtained by switching to rather short pistons with hydraulic loading. A variety of such devices have been used with solid media relieving friction by rotation or osillation of the piston [12]. Jayaraman et al. [13] were the first to suggest the use of a polytetrafluorethylene capsule inside the cylinder to contain a fluid pressure medium. Heydemann and Houck [14] suggested the use of a polyethylene liner for the same purpose with several obvious advantages. In the piston and die assembly with polyethylene liner shown in Fig. 9 the pressure inside the chamber can be determined in terms of the effective area of the assembly and the force applied. It constitutes a primary pressure gage.

The high pressure die consists of a cemented tungsten carbide cylinder (6% Co) of 12 mm internal and about 37 mm external diameter. The cylinder is 10 mm high. The outside cylindrical surface is tapered and press-fitted into a hardened steel cylinder of the same height and an outside diameter of about 140 mm. A mild steel ring of about 150 mm surrounds the steel cylinder. The interference at the steel - carbide interface generates a static radial support stress of about 1.1 GPa. A longitudinal clamping stress of up to 2.5 GPa is applied to the endfaces of the carbide cylinder and to a small area of the steel cylinder. A hydraulic ram applies sufficient force to the piston to generate pressures approaching 5 GPa in the high pressure fluid.

The effective area of the high pressure assembly is

$$A_{eff} = A_{co}(1 + \beta_{cyl}\bar{p}) \tag{16}$$

where A_{co} is the effective cylinder area at atmospheric pressure,

β_{cyl} is the pressure coefficient of this area, and

$\bar{p}$ is the internal pressure during an upstroke.

If β_{cyl} is calculated from simple elastic theory we obtain $\beta_{cyl} = 4.25 \times 10^{-12}\ Pa^{-1}$. A direct measurement [14] yields $\beta_{cyl} = 8 \times 10^{-12}\ Pa^{-1}$ but with a very large uncertainty. We have generally assumed $\beta_{cyl} = 5 \times 10^{-12}\ Pa^{-1}$.

The force supplied by the hydraulic ram is

$$F = A_{ro}(1 + \beta_r\bar{p}_r)\ \bar{p}_r \tag{17}$$

where A_{ro} is the effective ram area at atmospheric pressure,
β_r is the pressure coefficient of this area, and
$\bar{p}_r$ is the ram pressure during an upstroke.

A typical value for β_r is $7.5 \times 10^{-12} Pa^{-1}$.

The ram pressure, p_r, must be corrected for friction. The latter is determined by running many friction loops as shown in Fig. 10 during up and down strokes. The friction, p_{fr}, is generally a linear function of the pressure and can be expressed as

(18) $$\bar{p}_{fr} = a + b\bar{p}_r$$

with typical values for the constants being $a = 2.6 \times 10^5 Pa$ and $b = 0.035$.

The polyethylene sleeve further reduces the pressure inside the cavity by absorbing some of the applied force. This can easily be evaluated using a phase transition whose transition pressure must not be known but must be reproducible. The transition is induced using sleeves with different wall thicknesses. The ram pressure at transition is plotted vs. sleeve wall thickness and an extrapolation to zero wall thickness is made. This is illustrated in Fig. 11 for the Bi I-II transition. A sleeve wall thickness of 1.5 mm reduces the internal pressure 1.1% below the calculated pressure.

The internal pressure can now be calculated from

(19) $$\bar{p} = R(1 + \beta_r\bar{p}_r)(1 - \beta_{cyl}R\bar{p}_r)[\bar{p}_r - (a + b\bar{p}_r)]\ B$$

where R is the ratio of ram area to piston area (R = 146.12), and
B is the sleeve effect (B = 0.989 for 1.5 mm wall thickness).

The following table gives estimates of the contribution to the total uncertainty from various effects:

Ram area to piston area ratio	0.1 %
Ram cylinder expansion	<0.01 %
High pressure cyl. exp.	0.3 %
Ram pressure	0.1 %
Constant friction	0.1 %
Variable friction	0.2 %
Sleeve effect	0.2 %
Total uncertainty	1 %

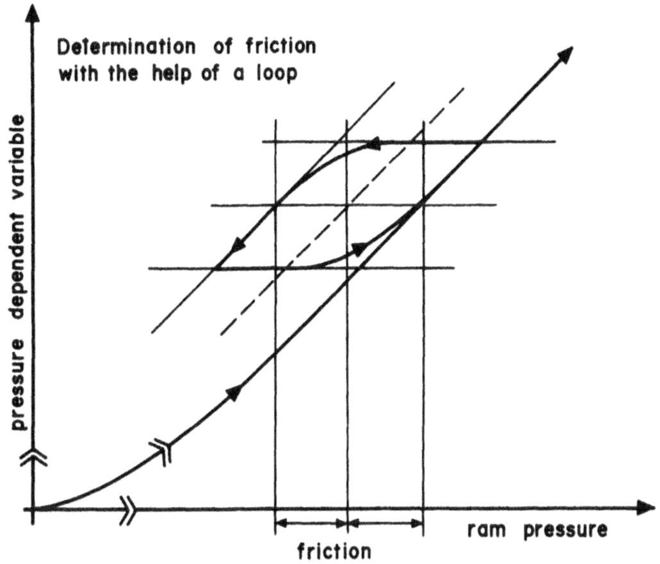

Fig. 10 Several such friction loops are taken during a run to determine the friction correction.

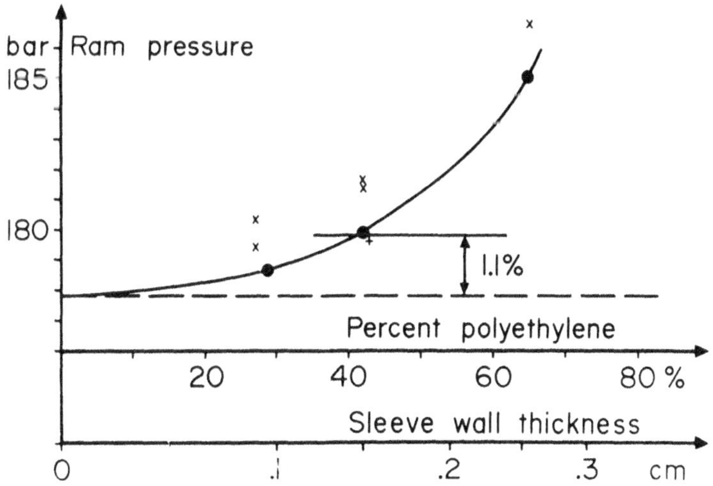

Fig. 11 The amount of force absorbed by the PE sleeve is determined by making phase transition measurements (here for Bi I-II) with different sleeve thicknesses and extrapolating this data to zero.

The very close agreement between the ram pressure extrapolated in Fig. 11 with the ram pressure expected for the Bi I-II transition indicates that the error in the pressure determination for this assembly is probably considerably less than 1%.

Like the 2.5 GPa dead-weight loaded piston gage the 5% GPa piston and die does not connect to other apparatus. However, we will show later that it can easily be used for ultrasonic, electrical and volumetric measurements on samples contained within the sleeve.

VI. TRANSFER GAGES

Many properties of materials are dependent upon pressure and can, in principle, be used to convert pressure into another suitable signal for ready measurement. There are a number of desirable features though: the measured parameter should depend upon pressure not on othe variables, as for example temperature; the parameter must be a unique function of pressure, preferably linear; the parameter should not exhib hysteresis; the sensor material should be well characterized and stable a single crystal would be preferred over an amorphous material. The piezoresistive manganin gage is used extensively as pressure transducer Semiconductor "band-gap", ultrasonic and dielectric transducers are beginning to be used. The NaCl and CsCl equations of state and the Rub fluorescence line shift enjoy widespread use particularly at pressures above 5 or 10 GPa. We shall discuss all of them here briefly.

Manganin Gages

The temperature coefficient of the electrical resistance of manganin is zero at some temperature between 20 and 40°C depending on composition and state of annealing. Manganin is an alloy of copper wit about 12% manganese and 4% nickel. It was in wide-spread use as a material for electrical resistors when it was first suggested as a pressure gage. Manganin has a pressure coefficient of the resistance o about 2.4×10^{-11} Pa^{-1}. This coefficient varies not only from lot to lot but also along the wire from a single spool. The resistance is quite sensitive to strains other than the volume strain generated by hydrostatic pressure and it depends on the way the coil was annealed. Babb [15], Cross [16] and later Houck and Bean at NBS have achieved ver reproducible measurements with manganin gages. Atanov [17] has investi gated and used large numbers of manganin gages. It may be advisable to follow the suggestions from these laboratories.

For high accuracy strain-free winding is of great importance. At the National Bureau of Standards (Cross [16], et al.) the wire is wound in a double helix on a ceramic mandrel with deep grooves to accept the helix (Fig. 12). Babb et al. have investigated different methods of winding manganin coils on fibre bobbins [18] and found no dependence upon the winding technique. Babb therefore has the coils wound on fibre bobbins laying each turn right next to the preceding one. These are then tied and coated with coil varnish. The wire used by these two groups is 0.127 mm in diameter covered with two layers of silk. Atanov uses 0.08 mm wire wound on grooved ceramic cores.

For precise resistance measurements four lead wires are attached to the manganin resistor. At NBS gold wires are silver brazed to the free ends of the manganin coil using a gas-oxygen flame. The borax flux can be removed by immersing the coil in boiling water for several minutes and then rinsing it in boiling water. Copper leads can be used instead of gold since its thermoelectrical voltage against manganin is small. When the gage is mounted on the pressure vessel plug or closure care must be taken to ensure that no intermittend shorts can develop between wires coming from the same or even the opposite end of the coil.

Cross, at NBS, annealed the manganin coils before they were connected to the header or plug by heating mandrel and coil to 140°C for 48 hours. The connections were then completed and the assembly was finally seasoned by several excursions to a pressure of not less than 800 MPa. Other laboratories use the more elaborate Bridgman cycle which consists of long periods of baking the coils at 140°C interspersed with cooling to -80°C with dry ice keeping the coil meticulously dry The coils can be left in the furnance overnight, cooled during the day, and the process could last for a full week. Atanov anneals manganin coils by discharging a capacitor bank through the gages raising the temperature of the wire to about 500°C for a very short time [19].

In most cases the resistance change of a manganin gage is measured with a suitable bridge. Because of the temperature dependence of the resistance it is often advantageous to place another manganin resistor in the bridge which is not pressurized but is in good thermal contact with the active bridge.

At NBS a seven decade, direct current bridge is used to measure the resistance of manganin gages. This gives a resolution of 0.0001 ohms corresponding to about 40 kPa. For less demanding measurements a 10-turn potentiometer will often suffice. Measurements are generally made with direct current but recent experience with the use of alternating current resistance bridges using ratio transformers indicate that a resolution comparable to that of the best conventional bridges can be attained with alternating current.

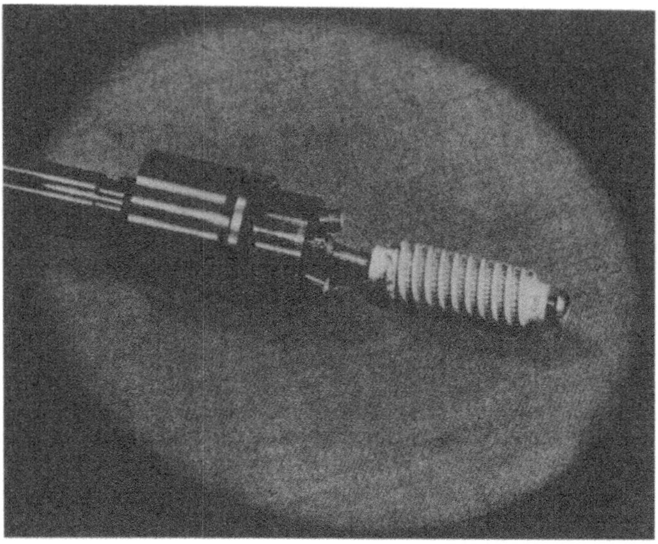

Fig. 12 Manganin pressure gage with dual double helix for strain-free mounting. Header with supported O-ring seal on the left.

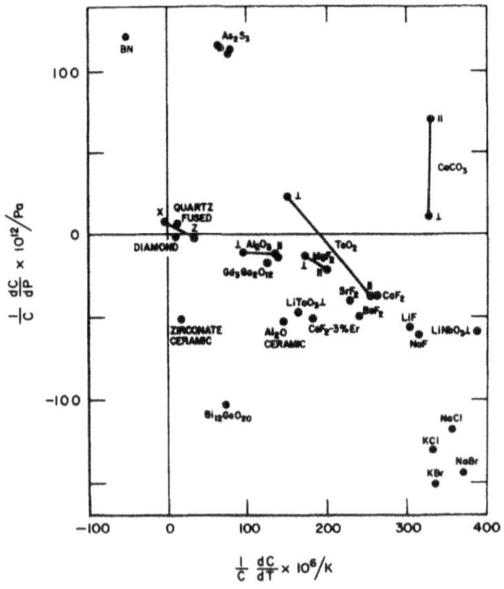

Fig. 13 Pressure and temperature coefficients of the capacitance of solid di-electric capacitors. Preferred combinations ar $CaCO_3$(||-⊥) and As_2S_3-$Bi_{12}GeO_{20}$.

Manganin gages can be contained in separate pressure vessels or be built directly into the apparatus where the measurement of pressure is needed. Care must be taken not to expose them to any rapid pressure changes. The fluid used should be a good insulator, free from moisture. Likewise the electrical leads should have high electrical leakage resistance. For the most accurate measurements it is necessary to calibrate and use the gage at the same temperature.

Manganin pressure gages can be calibrated over their entire range with controlled clearance piston gages or at certain pressures against fixed points. The resistance of manganin is a slightly nonlinear function of pressure and at least three points are therefore required. Depending on the range considered these could be atmospheric pressure, the melting pressure of mercury at 0°C (759 MPa), and the Bi I-II transition at 25°C (2.55 GPa) or several pressures on the mercury melting line. All of these will be discussed later. Since it is the resistance change rather than the absolute resistance that must be measured the pressure-resistance change relation is commonly expressed as

$$p = \frac{\Delta R}{R_o} \left[\alpha(1 - \beta\Delta R)\right]^{-1} \tag{20}$$

or as

$$p = A\Delta R + B(\Delta R)^2 \ . \tag{21}$$

Manganin wire is not a suitable pressure gage for solid or even very viscous media [20,21]. It has been used in shock waves [22] but it is difficult to see how it can provide reliable measurements under such conditions.

Manganin pressure gages complete with pressure vessels and read-out devices are commercially available [11]. Several of the National Standards Laboratories can provide calibrations of such gages against their primary standards.

The Solid Dielectric Capacitive Transducer

Andeen, Fontanella and Schuele [23] were the first to demonstrate the feasibility of using the capacitance change caused by hydrostatic pressure in solid dielectric capacitors as a transducer. Fig. 13 is a plot of the pressure and temperature coefficients of a number of solid dielectrics. The capacitance of three terminal capacitors containing such materials as dielectric can be measured with modern ratio transformer bridges with an accuracy of $1:10^7$ and a resolution of better than $1:10^8$. With CaF_2, the material that Andeen et al. preferred, a resolution of 1 kPa should be possible if the capacitor could be thermostated to better than 0.14 mK. In view of the large adiabatic effects in pressure vessels this is impossible, but the authors have nevertheless maintained a set of CaF_2 pressure gages over several years with a precision approaching the accuracy of the piston gage used to calibrate them. Colwell [24] has recently shown that the temperature problem can be significantly reduced if one measures the ratio of two capacitors with similar temperature coefficients but dissimilar pressure coefficients.

The capacitors must be inside the pressure vessel and in close thermal contact with each other. The measured capacitance ratio is then a function of temperature and pressure

$$\frac{1}{R}\frac{dR}{dT} = \frac{1}{C_1}\frac{dC_1}{dT} - \frac{dC_2}{dT} \tag{22}$$

and

$$\frac{1}{R}\frac{dR}{dp} = \frac{1}{C_1}\frac{dC_1}{dp} - \frac{1}{C_2}\frac{dC_2}{dp}\ . \tag{23}$$

Two pairs of dielectrics that satisfy the above condition and their coefficients are

	$\frac{1}{R}\frac{dR}{dT}$ [K^{-1}]	$\frac{1}{R}\frac{dR}{dp}$ [Pa^{-1}]
$CaCO_3$-a/$CaCO_3$-c	3.6×10^{-6}	59×10^{-12}
$Bi_{12}GeO_{20}$/As_2S_3	3.0×10^{-6}	213×10^{-12}.

With these pairs a pressure resolution of 1 kPa requires thermostating to only than 16 mK and 71 mK respectively. Colwell has further reduced the adiabatic effects by equating the heat adiabatically produced inside the pressure vessels against that abosrbed in the vessel walls by careful choice of materials. The temperatures of the vessels change by only about 4 mK after full-scale pressure changes of 140 or 700 MPa respectively. Thermal equilibrium within the vessel itself is re-established in 6 to 7 minutes after a full scale pressure change.

Capacitive transducers for the 700 MPa range are stable to better than 4 kPa, is less than the uncertainty of primary standards for this range. The dielectrics are well characterized materials with excellent long term stability. An automatic ratio transformer bridge with sufficient accuracy and resolution is available. The U.S. National Bureau of Standards will use a capacitance pressure gage as working standard for pressures up to 700 MPa.

Ultrasonic Pressure Gage

The elastic constants of solids are functions of pressure and temperature. In some cases, as for example in fuzed or single crystal quartz, they exhibit extremely small hysteresis with pressure and temperature. Instead of measuring the elastic constants it is more convenient to measure the change of a suitable resonance frequency using ultrasonic interferometry. For a quartz single crystal the change in shear wave resonance frequency for any angle of cut rotated about the X-axis is given by

$$\frac{1}{f}\frac{df}{dT} = \alpha + \frac{1}{2C}\frac{dC}{dT} \tag{24}$$

and

$$\frac{1}{f}\frac{df}{dp} = -\frac{K}{6} + \frac{1}{2C}\frac{dC}{dp} \qquad (25)$$

where $C = m^2 C_{66} + n^2 C_{44} + 2mnC_{14}$,
C_{ij} are the elastic constants,
α is the linear thermal expansivity; the small anisotropy was neglected, and
K is the compressibility; the small anisotropy was neglected, and m, n are the directional cosines.

Using the data of McSkimin et al. [25] and Bechmann et al. [26] the pressure temperature coefficients were calculated [27] as a function of the orientation about the X-axis (Fig. 14). There are two orientations with zero pressure coefficients: the AT-cut with a pressure coefficient of 1.12×10^{-11} Pa^{-1} and the BT-cut with a pressure coefficient of -1.57×10^{-11} Pa^{-1}. With quartz crystals of 75 mm length and a tracking, pulsed ultrasonic interferometer working around 10 MHz [28] resonance frequencies can be tracked and measured with a precision of better than 2 Hz corresponding to a pressure resolution of 18 kPa and 13 kPa respectively. Smaller transducers (8.9 mm diameter, 15.2 mm long) have a resolution of about 30 kPa.

The ultrasonic pressure transducers are much more rugged than either manganin gages or capacitance gages. They require only one or two leads which do not have to be very well insulated and they can, in fact, share these leads with other devices working with direct current. Their main use is anticipated for pressures above about 700 MPa and up to 3 GPa in apparatus such as the Bridgman/Birch pressure generator.

The NaCl Pressure Gage

The equation of state for a suitable solid calculated from first principles would be the ideal thermodynamic basis for a pressure scale. The equation of state for NaCl developed by Decker [29,30] is not a first-principles calculation. It contains some simplifying assumptions and a parameter which is adjusted using atmospheric pressure thermal expansivity data. The results of the calculation are, however, in such good agreement with experimental results over a wide pressure and temperature range that one is tempted to regard Decker's NaCl - pressure gage as thermodynamic pressure scale. In fact, many researchers have increasingly used it as such during the past ten years.

Decker chose NaCl because the major terms in the lattice energy of ionic crystals are well known, because sufficient experimental data were available to check the results of the calculation, and because it did not appear to have any pressure-induced polymorphic phase transitions. A transition to the CsCl-structure has since been identified at about 30 GPa [31] and for pressures exceeding this value the equation of state of CsCl might serve instead.

Decker assumes that the ionic crystal is composed of a set of harmonic oscillators with parameters

(26) $$\gamma_i = -\frac{d(\ln w_i)}{d(\ln V)}$$

where w_i is the eigenfrequency of the i-th mode, and V is the volume. Assuming that $\gamma_i=\gamma$ for all i, one arrives at the Mie-Gruneisen equation of state

(27) $$p = -\frac{d\,\Phi\,(V)}{dV} + \frac{\gamma}{V} E_v(V,T)$$

where p is the pressure,
$\Phi(V)$ is the lattice potential energy,
T is the absolute temperature, and
E_v is the vibrational contribution to the energy given by

(28) $$E_v = 2.25R\Theta + 6RTD(\Theta/T)$$

where R is the universal gas constant,
Θ is the Debye temperature,
$D(\Theta/T)$ is the Debye function.

The volume dependence of γ, the Gruneisen parameter, is approximated by

(29) $$\gamma = \gamma_o\left(\frac{V}{V_o}\right)^A ,$$

where V_o is the volume at atmospheric pressure and T = 25°C,
γ_o is the Gruneisen parameter at the same conditions, and
A is an adjustable parameter.

The potential energy function

(30) $$\frac{\Phi}{N_o} = -\frac{aq^2}{r} - \frac{C}{r^6} - \frac{D}{r^8} + Qbe^{-\frac{r}{\rho}} + Q^1b_-e^{-\frac{r\delta}{\rho_-}} + Q^1b_+e^{-\frac{r\delta}{\rho_+}}$$

contains attractive Madelung and van der Waals terms, and repulsive terms due to ion core wavefunction overlap. Here

q is the electron charge,
α is the appropriate Madelung constant,
C is the coefficient of the dipole-dipole interaction,
D is the coefficient of the dipole-quadrupole interaction,
Q is the number of nearest neighbors,
Q^1 is the number of next-nearest neighbors,
δ, is the ratio of next-nearest neighbor to nearest neighbor distance,
$b, b_+, b_-, \rho, \rho_+, \rho_-$ are constants in the repulsion terms.

The ion-like repulsion terms are only a small part of the total repulsion. It is therefore possible to reduce the number of constants by relating the constants in the ion-like repulsion terms to b and ρ, and further expressing ρ_+/ρ and ρ_-/ρ in terms of the ionic radii, and b_+/b and b_-/b in terms of the polarizabilities.

To determine the parameters ρ, b and A equ. (27) is evaluated for T_o, p_o and r_o. The equation is then multipled with V, differentiated with respect to V, and evaluated again at p_o. The resulting two equations can be solved simultaneously for b and ρ after choosing an initial value for A. An equation of state is then calculated from which lattice parameters are derived and compared with atmospheric pressure thermal expansion data taken from the literature. The parameter A is varied until a best simultaneous fit of the calculated adiabatic bulk modulus and thermal expansion data with literature data is achieved.

Decker [30] lists calculated pressures at selected temperatures and compressions for NaCl and CsCl with a grid close enough to permit linear interpolation with reasonably small errors. The agreement of the calculated data with data from the literature is very satisfactory. Decker estimates the uncertainty in the pressure calculated for a given compression to be 1.1% below 5GPa, 1.7% below 10 GPa and 2.4% below 20GPa. This, as he points out quite properly, is of the same order of magnitude as the best possible primary pressure measurements in those ranges. Close agreement is also found between the directly determined transition pressures for a number of polymorphic phase transitions with the pressures calculated from the measured compression of the NaCl-lattice at these points. The equation of state of NaCl was also calculated by Perez-Albuerne [32] and Weaver et al [33] with quite similar results. In the following table we compare for some polymorphic transitions the directly determined pressures with those calculated from NaCl compression data taken at the transitions.

Phase Transition Pressures. Direct determination and calculation from NaCl compression data.

	Bi I-II	Ba	Bi III-V
	all pressures in GPA		
Direct determination	2.549±.006 [36]	5.50±.05 [37]	7.75±.10 [38]
Decker [30], Jeffery's data[34]	2.532±.077	5.47±.13	7.60±.18
Inoue's data [35]	2.521±.033	5.41±.06	7.51±.06
Weaver[33], Jeffery's data[34]	2.54±.08	5.44±.11	7.55±.14

The NaCl gage can, of course, be used in any apparatus with provision for X-ray lattice constant measurements. This requires access with sufficiently wide angles through windows or transparent anvils (diamonds) or through the gasket areas in multianvil apparatus. The NaCl probe is often intimately mixed with the sample under investigation or is even used as the pressure transmitting medium. A serious drawback of the NaCl gage are the often excessively long exposure times caused by the small apertures of anvil presses.

The Ruby Fluorescence Gage

Barnett, Block and Piermarini [39] developed the ruby fluorescence pressure gage for high pressure apparatus with optical access to the pressure cell. Their system is expecially well suited for use with the diamond anvil cell at pressures up to at least 50 GPa and temperatures of several hundred degrees. The shift of the R_1 line with pressure is about 270 MPa/Å, the line width is about 7.5 Å near room temperature and the temperature coefficient is about 15 °C/Å. A resolution of 50 MPa can be achieved.

In operation (see Fig. 15) a small piece of ruby is placed with the sample under investigation inside the gasket of the diamond anvil cell. The cell is firmly attached (Fig. 16) to the movable stage of a microscope for coarse horizontal adjustment. A movable objective is used for fine adjustment of the sample in the diamond anvil cell. The ruby crystal is illuminated with light from a high pressure mercury lamp. A filter between light source and sample eliminates all light with wavelengths in the fluorescence range. A second filter following the diamond anvil cell filters out all light with wavelengths outside the fluorescence range. The light then enters a 25 cm monochromator whose output is measured with a sensitive photodetector. Since the resolution of the system depends to a great extent on mechanical rigidity of the monochromator and other parts of the optical system careful attention must be paid to rugged mechanical design. The output signal from the photodetector is plotted as a function of wavelength on a strip chart recorder. For this purpose the wavelength control on the monochromator can be driven with a synchronous motor. A fluorescence line recording allowing a precision of better than 200 MPa lasts about 5 minutes.

Piermarini et al [40] have calibrated the ruby fluorescence line shift against the NaCl equation of state of Decker [30]. These measurements were made in a diamond anvil cell. 4 parts NaCl and 1 part ruby (by volume), finely powdered and passed through a #325 mesh sieve, were mixed and placed in the gasket filling about half the volume. The rest of the space was filled with a 4:1 mixture of methanol and ethanol. The measurements were started at the highest load, corresponding to 19.5 GPa. After each set of measurements the pressure was reduced in steps of 1 to 2 GPa. At least 5 hours were allowed each time to let the system come to the new equilibrium. At each point an x-ray lattice measurement was made, preceded and followed by a measurement of the ruby line shift. The x-ray measurements required approximately 15 hours of exposure.

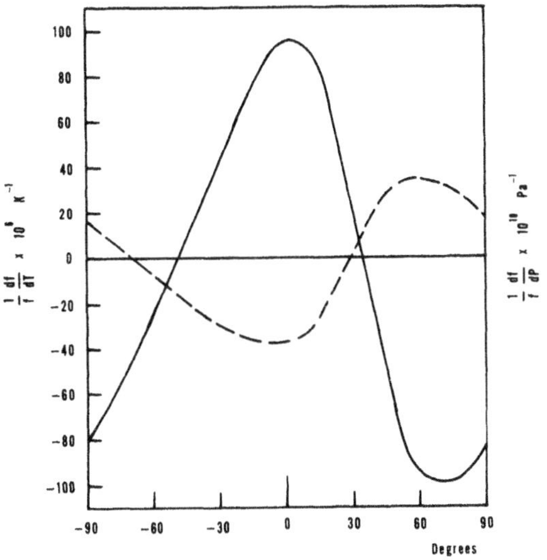

Fig. 14 Temperature (solid line) and pressure (dashed line) coefficients of the resonance frequency of a quartz crystal as function of the angle of rotation about the X-axis.

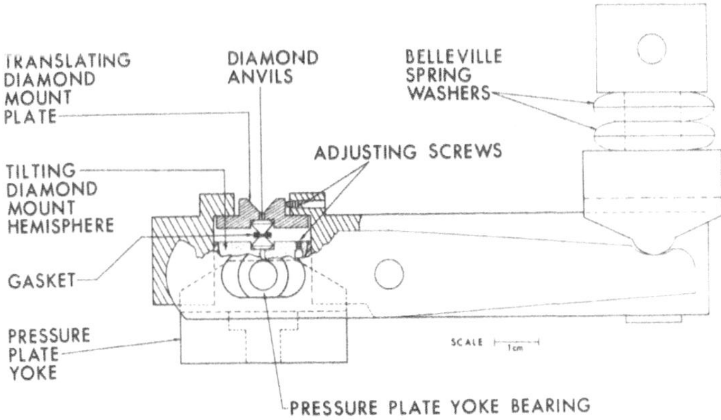

Fig. 15 NBS diamond anvil cell for pressures beyond 50 GPa.

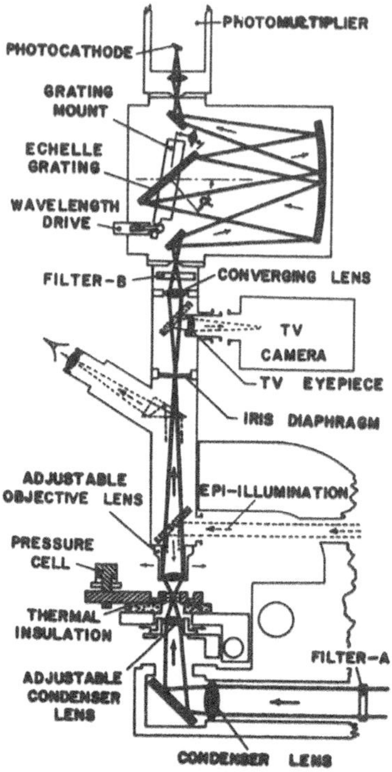

Fig. 16 Optical path for the ruby fluorescence pressure gage. Light from the high pressure mercury arc lamp enters from lower right. Filter A absorbs all radiation in the fluorescence band. A sample of ruby in the pressure cell fluoresces in the infrared. Filter B absorbs all light outside the fluorescence band. The light then enters the monochromator whose output is measured with a photomultiplier. A TV camera can be switched into the light path to permit sample observation on a TV-screen.

Least squares fits of the functions

(31) $$p_{NaCl} = A(\Delta\lambda)$$

and

(32) $$p_{NaCl} = A(\Delta\lambda) + B(\Delta\lambda)^2$$

to three sets of data were made. These were: data to 10.4 GPa, hydrostatic pressure; data above 10.4 GPa, non-hydrostatic pressure; and all data. Each point was weighted according to the uncertainty associated with the x-ray measurement. A second set of least squares fits of the same equation was made weighting each point according to the combined uncertainties from the equation of state of NaCl and the x-ray lattice constant measurement. The error due to the $\Delta\lambda$ measurement is considerably smaller than the two errors considered here. The authors suggest the relation

(33) $$p = (274.0 \pm 1.6)\ (\Delta\lambda)\ \text{MPa}$$

with 95% confidence for pressure determinations with the ruby R_1 line. The uncertainty stated above includes only contributions from random sources of error. The uncertainties suggested by Decker for the NaCl gage must be added to this.

Piermarini et al. [40] provide an interesting and detailed discussion of various sources of error.

It is interesting and important for the purpose of extrapolation that the quadratic terms were statistically insignificant in all six least squares fits. The ruby fluorescence gage has since been used to pressures as high as 100 GPa under static conditions [41].

III.4 FIXED POINTS ON THE PRESSURE SCALE

Fixed points greatly facilitate the in situ calibration of transfer gages particularly in high pressure equipment. Most fixed points are either solid-solid or liquid-solid phase transformations of pure substances. Apart from occurring in the right pressure range fixed points must satisfy a number of other requirements:

(a) they must be easily detectable with one or more standard types of measurement, as for example electrical resistance, volume, heat of fusion, ultrasonic propagation or magnetic properties;

(b) the transformation should be sharp and reversible;

(c) the temperature coefficient of the transition pressure should be small or well known.

(d) the substance should be easy to characterize or standardize;

A number of fixed points covering the hydrostatic range (to 10 GPa) hav evolved over the years and the following set of transition pressures wa suggested for general use at an international conference [42].

Transition	Pressure [GPa]	Est. Uncertainty [GPa]
Hg, freezing at 0°C	0.7569	0.0002
Bi I-II at 25°C	2.55	0.006
Th I-II at 25°C	3.67	0.03
Ba I-II at 25°C	5.5	0.2
Bi III-V at 25°C	7.7	0.3

A short discussion of some of these points follows.

Mercury

At the time of the conference [42] more than a dozen determinations of the mercury freezing pressure at 0°C had been made. From these a recommended, best value of 756.9 ± 2 MPa at 0°C was extracted. In the mean time two further determinations have been made at 757.1 MPa [43] and 756.9 [44] which both fall within the range of uncertainty of the recommended value.

Since the 1968 conference also the melting line of mercury between 0°C, 756.9 MPa and -38.841 °C, 0 MPa was redetermined by Houck [45]. According to Houck the melting line is best described by

$$p = 19.33114\ d + 0.0014055d^2 + 0.000067028d^3 \tag{34}$$

where p is in MPa, and d = T - 234.309 K. p is the absolute pressure.

The deviations of the measured data from this curve ranged from -0.096 to + 0.104 MPa with a residual standard deviation from the curve of 0.055 MPa.

Houck has developed a rather simple apparatus for the ready realization of the mercury melting line in the laboratory. Fig. 17 shows his small pressure vessel in vertical cross section and top view. The vessel is constructed of maraging steel with a well for a standard platinum resistance thermometer, a well for a sheathed thermocouple and a through-hole for the mercury. The lower end of the through hole is plugged. The upper end is connected through high pressure tubing to a tee (Fig. 18) and from there to the pressure generating system. A sheathed thermocouple is introduced through a fitting in the tee down into the mercury contained in the pressure vessel. The mercury is in direct contact with the pressure vessel. A mixture of n-pentane and i-pentane serves as pressure fluid. The pressure vessel is suspend

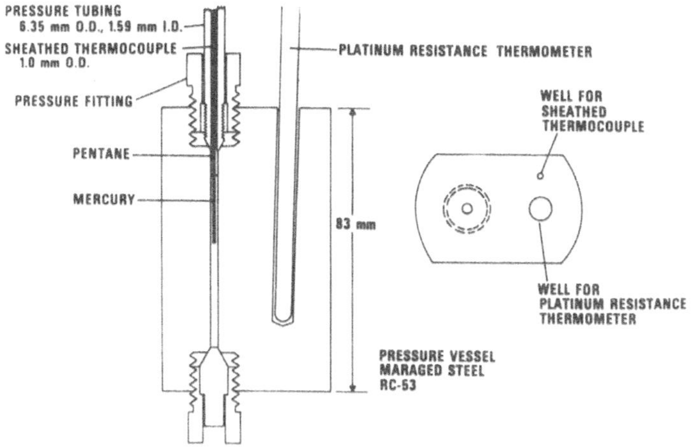

Fig. 17 Mercury Melting Line Apparatus.
Pressure vessel with thermometers.

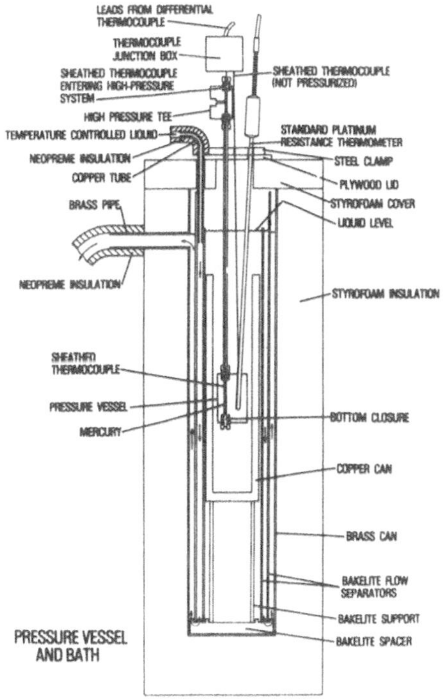

Fig. 18 Mercury Melting Line Apparatus Thermostated bath.

from the high pressure tubing in the bath shown in Fig. 18. The two thermocouples are combined in series - opposed to form a differential thermocouple pair. Their output is indicated on a strip chart recorder or a voltmeter with a resolution of 2 mK. The platinum resistance thermometer is read on a bridge with an accuracy corresponding to 1 mK.

Freezing and melting induced by small variations of the applied pressure can easily be detected by observing the differential temperatures between mercury and pressure vessel caused by heat flowing into or out of the mercury sample. The corresponding pressure can then be determined by calculating p from equ. (34) having measured the temperature of the vessel.

The imprecision of establishing the pressure scale with the help of the mercury line is very close to that expected from a dead-weight loaded piston gage in this range. Using the apparatus of Houck the mercury melting line can be realized easily and inexpensively in most laboratories.

Bismuth I-II and III-V Transitions.

Fig. 19 shows the phase diagram of bismuth. Of the many phase boundaries in the diagram two have found widespread use as fixed points on the pressure scale: the boundary between phases I and II, and the boundary between phases III and V.

The most accurate determination of any point on the Bi I-II phase boundary was the determination of its intercept with the 25°C isotherm by Heydemann [36]. His determination was based on measurements with a controlled clearance piston gage and in 1968 [42] his value of 2.55 ± 0.006 GPa was adopted as a fixed point. In the meantime Houck has also determined the slope of this phase boundary in the range 20 to 50°C. A value of -4.06 MPa/°C with a standard deviation of 0.05 MPa/°C, an estimated systematic uncertainty due to temperature of 0.08 MPa/°C, and an estimated uncertainty due to friction of 0.07 MPa/°C was found [46].

The Bi I-II transition is very sharp with a region of indifference of less than 2 MPa [47]. It can be detected in several ways, for example using the 4% volume change, the very large change in electrical resistance, or changes in ultrasonic velocity or absorption.

Haygarth et al. [48] determined the Bi III-V transition pressure as 7.75 GPa in a piston and cylinder device with solid pressure transmitting medium using the change of electrical resistance to detect the transition. A value of 7.7 ± 0.3 GPa was adopted in 1968 [42] as the current best value. This is consistent with values determined by Jeffer et al. [34] and Weaver et al. [35] using the NaCl gage.

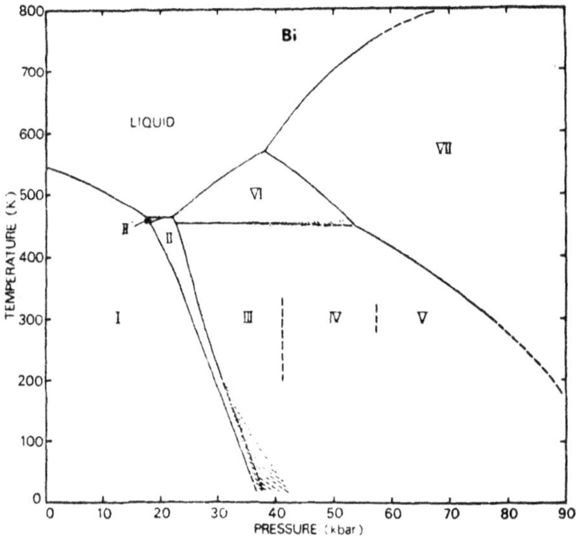

Fig. 19a Phase diagram for Bi after Shusuke Yomo. Note the different designation of the phases. V and VII in this diagram are normally referred to as III and V.

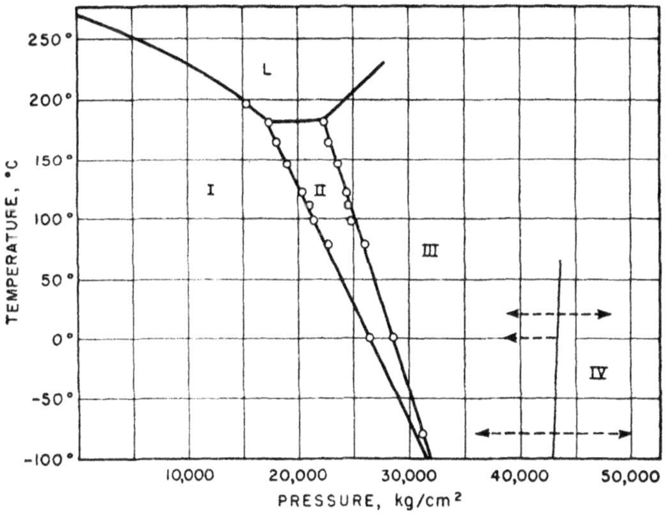

Fig. 19b Phase diagram for Bi after Bridgman

VII. SHOCK WAVES

In most materials, solid and fluid, the velocity of sound increases with pressure. A strong compression wave travelling in such a material will experience dispersion: those parts of the wave carrying the highest stress will travel faster overtaking the areas of lower stress. This results in a continuous steepening of the wave front into a shock wave. Only the dissipation of energy from the wave through various loss mechanisms prevents the formation of an infinitely sharp shock front.

Shock waves can be generated in fluids and solids with explosives, light gas guns, electrodynamic generators, laser pulses and many other means. Most frequently the shock front velocity and the particle velocity behind the shock front are measured and the thermodynamic p-V relations are obtained using the Rankine-Hugoniot relationships. These express the conservation of mass, momentum and energy across the shock wave. One assumes that the material on both sides of the shock front is in thermodynamic equilibrium.

With density ρ, pressure p, particle velocity u, internal energy E, and shock wave velocity U_s, and subscript zero indicating conditions in front of the shock wave, we have

conservation of mass

$$\rho_o(u_o-U_s) = \rho\,(u-U_s), \tag{35}$$

conservation of momentum

$$p_o+\rho_o(u_o-U_s)^2 = p + \rho(u-U_s)^2, \tag{36}$$

and conservation of energy

$$\rho_o(u_o-U_s)\,E_o+\frac{1}{2}\,\rho_o(u_o-U_s)^3 + p_o(u_o-U_s) = \rho(u-U_s)\,E + \frac{1}{2}\,\rho(u-U_s)^3 + p(u-U_s). \tag{37}$$

Equs. (36) and (37) can be further simplified and if we assume for the initial conditions p_o=0 and u_o=0 we have for the conservation of mass

$$\rho_oU_s = \rho(U_s-u), \tag{38}$$

for the conservation of momentum

$$p = \rho_oU_su \tag{39}$$

and for the conservation of energy

$$E-E_o = \frac{1}{2} p \left(\frac{1}{\rho_o} - \frac{1}{\rho}\right) \tag{40}$$

The locus of all states in the p-ρ plane assumed by the material in the shock wave is called the Hugoniot (curve). p and ρ are obtained from equs. (38) and (39). The shock wave velocity U_s can be measured in a number of ways. To measure the particle velocity u is much more difficult and one usually measures instead the free surface velocity, u_f, when the shock wave is reflected at the rear surface of the shocked material assuming that

$$u_f = 2u. \tag{41}$$

Several corrections must be applied to the shock wave data before it can be compared with static data. The most important of these is that due to the finite yield strength of the material. The pressure p in equs. (38) and (39) is not the hydrostatic pressure. It is essentially a uniaxial stress and the hydrostatic pressure is calculated from considering the pressure that would lead to the same density as that given by equs. (38) or (39).

The data obtained along a Hugoniot curve can be converted into data on isotherms using a suitable equation of state. The evaluation of phase change data requires furthermore the determination of temperature.

Shock wave measurements have been made in numerous solids at pressures beyond 1 TPa, but few measurements were made in fluids at pressures much exceeding the capability of static apparatus.

VIII. HIGH PRESSURE VESSELS, GENERATORS AND COMPONENTS.

It is, of course, impossible to give here an exhaustive account of all the various types of high pressure equipment in use today in the range to 10 GPa. Only a few instruments and components will be discussed to give an indication of the principles involved in the design of high pressure equipment.

Pumps

For pressures up to about 300 MPa reciprocating piston or membrane pumps are commercially available. These can be manually operated, motor driven or pneumatically operated and will compress liquids or gases. Electrically or pneumatically operated pumps can easily be converted to automatic operation establishing and holding a set pressure to better than 1% of reading over a wide range.

For fine control of pressures in a closed volume screw pumps are used (Fig. 20). These reduce the available volume and thereby raise th pressure by driving a piston into the pump cylinder. At pressures abov 300 MPa high pressure valves with drilled-out seat can serve as screw pumps. Their smaller displacement is usually adequate to fine adjust pressures in the 1.4 GPa range.

In the range to 1.4 GPa hydraulically driven intensifiers (Fig. 21) are used. A primary pressure (up to 100 MPa), for example from a pneumatically driven pump, is applied to the large diameter primary piston. The resulting force drives the high pressure piston. The intensifier piston stack can be driven back by applying a sufficien high pressure to the high pressure piston or a much lower pressure to the rear of the low pressure piston. Some intensifiers are rigged with check valves etc. to operate automatically like reciprocating pumps. More often the operator will be required to throw the valves in the proper sequence. It is very important to have pressure gages measu both the primary and the secondary pressure in order to be able to diagnose any malfunction promptly.

Pressures above about 1.4 GPa are mostly generated at the point of use by built-in intensifiers or rams. The NBS 2.6 GPa piston gage is a case in point. Another example is the so called Bridgman-Birch apparatus shown in Fig.22. In its present commercial version it generates pressures up to 3 GPa inside a usable cylindrical volume of 10 cm length and 1.9 cm diameter. A 100 to ram (top) pushes an unsupported area seal (see section on closures) into the steel cylinder. The other end of the cylinder is closed with a plug carrying a number o leads into the cylinder. The unsupported area seal shown in Fig. 23 ha performed markedly better than the bottom closure provided by the manufacturer. The cylinder is tapered and is pushed into a tapered jacked with the help of a 250 to ram (bottom). This generates a support press on the circumference of the cylinder and a longitudinal support pressur sufficient to sustain the high internal pressure. To lubricate the interface between cylinder and jacket we apply in this order: petroleu jelly, graphite powder, lead foil (0.1 mm), petroleum jelly and graphit This results in a friction coefficient under load of only 0.005. A precharging port is provided near the top of the cylinder just under th seal in its uppermost position. The seal will ride over the port witho loosing its ability to seal but a notch in the polytetrafluorethylene packing can be observed where it passed the port. The packing should b rotated a few degrees before it is reused.

For pressures beyond about 2.5 GPa the piston and die assembly shown in Fig. 9 has proved itself to be a very reliable work horse. The cylinder is radially supported by being press-fitted into a hardened steel ring and longitudinally by force applied through a bridge plate and washer from a large ram. At NBS an inexpensive, commercial 1000 to press is used for this purpose. A 300 to ram sits on the platen of the large press and provides the force driving the high pressure piston. The seal is made with the help of the polyethylene sleeve inside the high pressure cylinder. Fig. 24 shows a variety of sample arrangements:

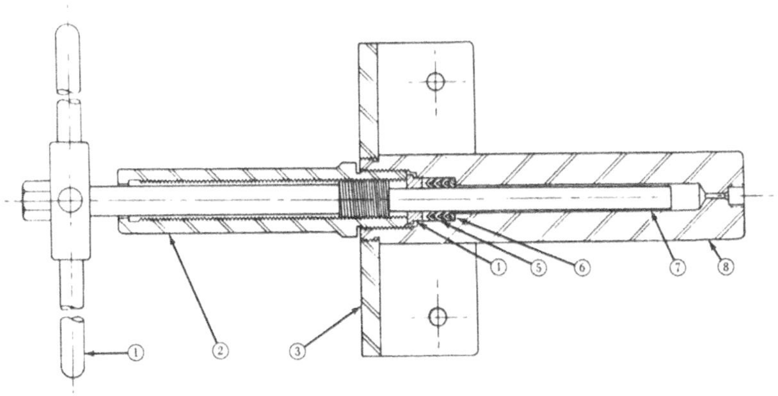

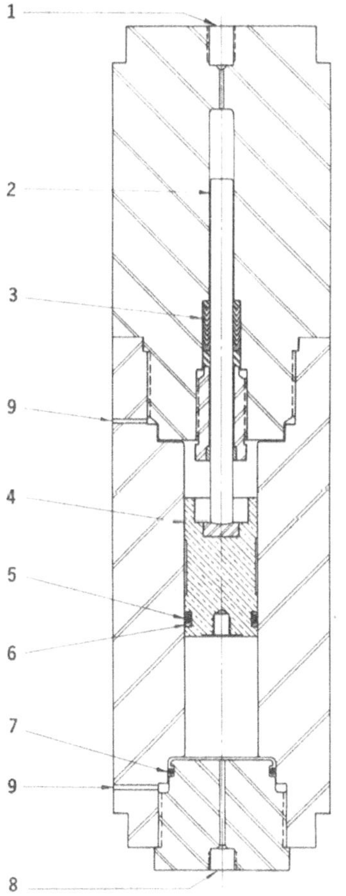

Fig. 20 Screw press used for the fine adjustment of pressure in high pressure installations.

Fig. 21 Intensifier for boosting low pneumatic or hydraulic pressure to high hydraulic pressure.

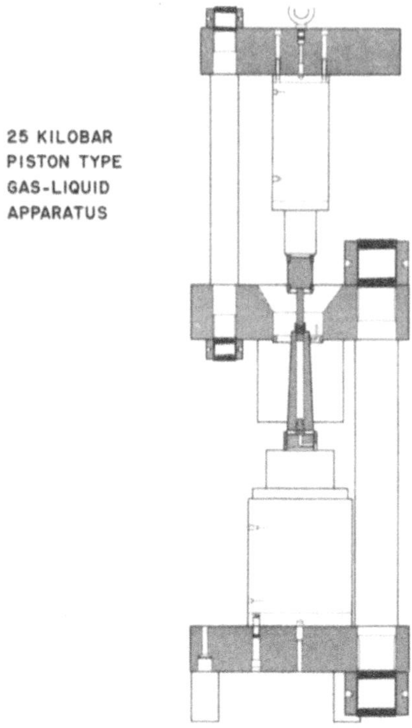

Fig. 22 Bridgman-Birch hydrostatic pressure generator to 3 GPa. An unsupported area seal is pushed into the cylinder by the upper ram to generate the pressure inside the cylinder. The tapered cylinder is pushed into a tapered jacket to generate outside support pressure. The usable volume at pressure is 10 cm long and 1.9 cm in diameter. Electrical leads are fed through the bottom closure (see Fig. 23).

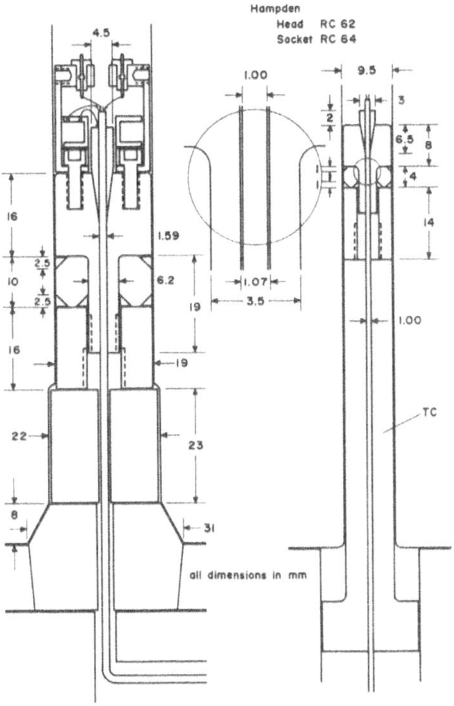

Fig. 23 Closures with electrical feedthroughs for 2.5 GPa service. The left plug is the bottom closure of the Bridgman-Birch apparatus. It consists of a Bridgman type unsupported area head and socket seal with NBS electrical feedthrough. The head carries a manganin gage and a socket for electrical connections. The plug on the right is the high pressure piston of the NBS 2.6 GPa piston gage. Its Bridgman head carries an electrical feedthrough with two leads.

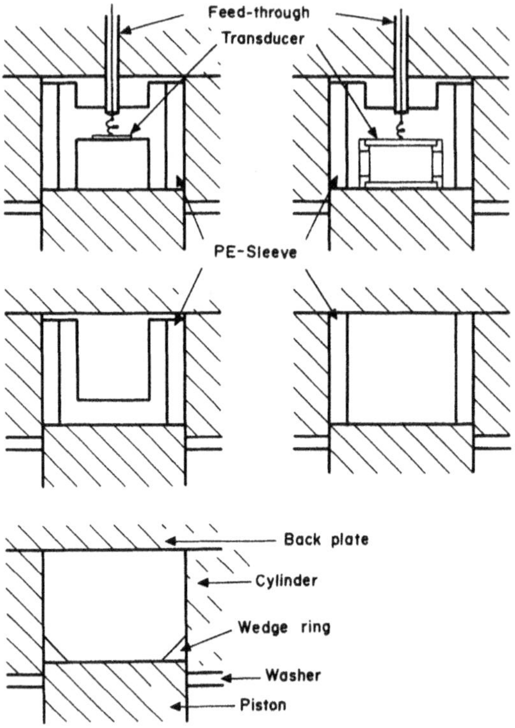

Fig. 24 Sample arrangement for the 5 GPa piston and die apparatus.
bottom left: cylinder filled with solid sample; wedge rings to prevent extrusion for ultrasonic and isothermal compression measurements.
center left: polycrystalline or amorphous samples; seal is made on rim machined on samples; pressure assures ultrasonic contact with back plate.
top left: single crystal sample with transducer inside the cell; electrical feedthrough operates to at least 3 GPa.
top right: ultrasonic measurements on liquids with high absorption made in fixed path cell.
center right: cell completely filled with liquid; ultrasonic and isothermal compression measurements possible.

(1) Lower left: the entire cylinder is filled with solid material. Wedge rings are used to prevent extrusion. Friction can be reduced by rotating (oscillating) the piston. Ultrasonic measurements can be made using the transducer on top of the back-up plate with the back-up plate serving as buffer.

(2) Center left: a sample of solid material is sealed under the PE-sleeve with the rim machined on the sample. The pressure presses the sample against the back-up plate. This is a good arrangement for ultrasonic measurement on solid, polycrystalline materials.

(3) Top left: For single crystals the transducer is placed directly on the sample and the signal is fed in and out through a high pressure electrical feed-through to be discussed in detail in the section on "Electrical Feedthroughs".

(4) Top right: For ultrasonic measurements particularly in liquids with high attenuation the fixed path cell is used.

(5) Center right: Most investigations on fluids are made in this cell. Ultrasonic measurements are made with the transducer on the back-up plate. Simultaneous isothermal compression measurements can be made.

Closures

Sealing a plug, window or feedthrough into a pressure vessel generally requires an elastically deforming sealant. At low pressures up to about 300 or 400 MPa a simple O-ring seal as shown in Fig. 25a works satisfactorily as static or dynamic seal. At higher pressures or when the clearance between the surfaces to be sealed is large metal wedge rings are often used with the O-ring making only the initial seal (Fig. 25b). The wedge ring constitutes an unsupported area seal. The wedge ring should be made of a material that deforms sufficiently to fill the clearance between the surfaces to be sealed. Seals that deform plastically are difficult to open or reuse. Nickel, beryllium-bronze, and stainless steel are frequently used for wedge rings. Double wedge rings (Fig. 25c) are occasionally used to cover large clearances.

Static and moving seals in the range to 3 GPa can be made most efficiently using the Bridgman unsupported area head-and-socket. Fig. 23 shows two such seals, the left one used as stationary bottom closure in a Bridgman-Birch 3 GPa generator and the right one used as movable seal on the piston generating the pressure in the NBS 2.6 GPa controlled clearance piston gage (Fig. 7). In these seals piston heads are partially supported by the elastic-plastic packing The area of the stem, which is unsupported, can be as little as 8% of the total cross sectional area without affecting the quality of the seal or causing pinch-off of the stem. The packing in these seals is polytetrafluorethylene

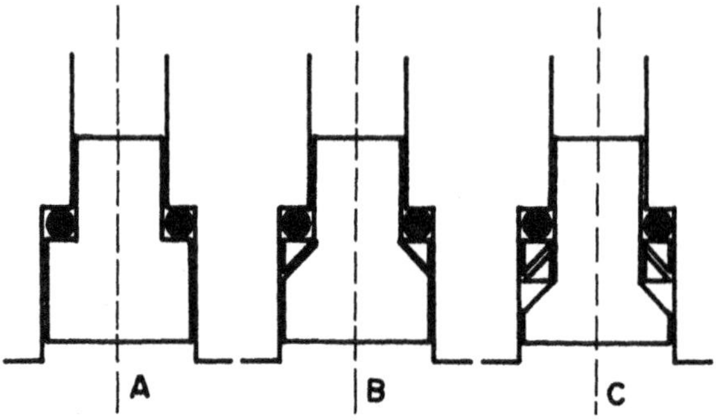

Fig. 25 O-Ring seals : (A) simple low pressure seal; (B) O-ring backed with wedge ring; (C) O-ring backed with two wedge rings to seal large clearance at high pressure.

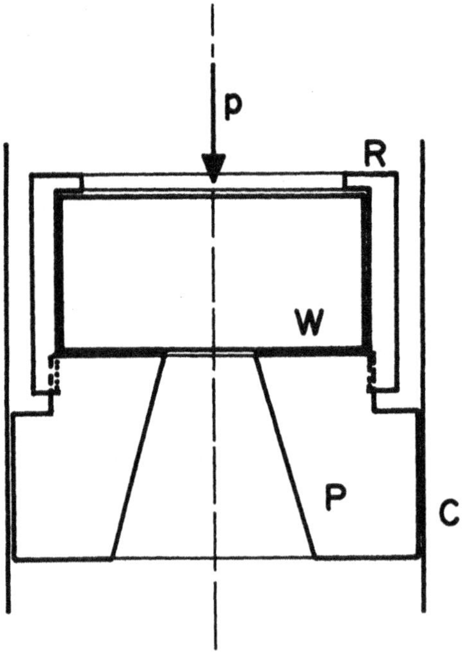

Fig. 26 Optical window (W) using Poulter seal; (C) cylinder, (P) plug, (R) retaining ring, (p) pressure.

supported by beryllium bronze wedge rings. The polytetrafluorethylene packing is machined to be slightly oversize to help make the initial seal. Head and socket are threaded to facilitate removal from the cylinder. The predominant failure mode are small circumferential and radial cracks in the socket. It appears, however, that these do not effect the performance of the seal. Splitting of the socket is rare.

The Bridgman seals shown in Fig. 23 are also provided with electrical feedthroughs and the larger one of the two carries a manganin gage and a socket arrangement for the connection of a furnace or other instrumentation.

For moving (dynamic) seals friction is often important. Bridgman unsupported area seals with polytetrafluorethylene packings as shown and an unsupported area of 8 to 10% show friction of less than 5% as measured by the full width of friction loops (see Fig. 10).

Windows

The most popular type of window for high pressure apparatus uses the socalled Poulter seal. The plug or other part on which the window is mounted (see Fig. 26) is lapped optically flat. The window, also optically flat, is then wrung onto this surface. The Poulter seal is a variation of the unsupported area seal, since part of the window is obviously not supported. Single crystal alumina, sapphire, quartz and other glasses have been used. Whalley [49] has recently extended the use of such windows to 4.5 GPa.

The diamonds in the diamond anvil cell (Fig. 27) are another type of window. Here the seal is made either with a special metal gasket containing the sample immersed in fluid or with a suitable gasket containing the solid pressure medium and sample.

Electrical Feedthroughs

Many types of electrical feedthroughs have been developed for different applications, pressure ranges and pressure vessels. A group of very simple to make feedthroughs [50,51] is based on the sheathed thermocouple wires available from several manufacturers. The sheathed thermocouple wires consist of one to four wires embedded in alumina, magnesia or zirconia with a stainless steel sheath swaged around it mostly by a drawing process. The wires could be any of the standard thermocouple wires or copper.

Fig. 28 shows an electrical feedthrough for pressures to 700 MPa and, with a different gland nut to 1.4 GPa, built into a standard NBS high pressure fitting. A sheathed thermocouple wire is silver-soldered into a piece of threaded and coned high pressure tubing. The solder joint is made in the first 6 to 8 mm from the coned end of the tubing. Sheathed wires with up to four copper conductors and an outside diameter of 3mm have been used. To seal and preserve the high insulation resistance the following procedure is usually followed: After assembling and cleaning the feedthrough it is baked at about 140°C for 8 to 12 hours. The assembly is then dropped into a dish with silicon oil (DC200/10 or similar) and left there for several hours. Afterwards

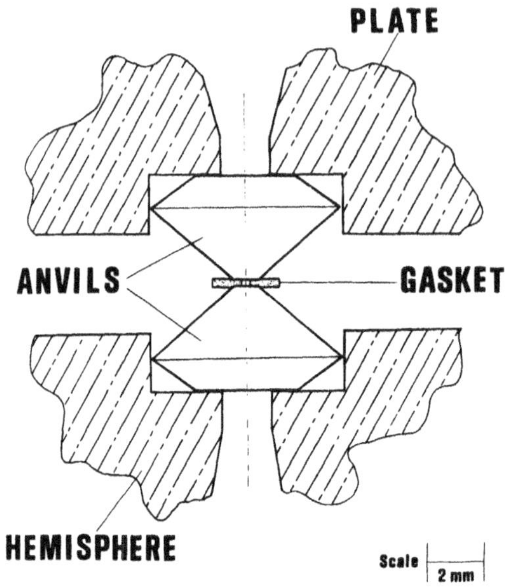

Fig. 27 Diamond anvils with metal gasket.

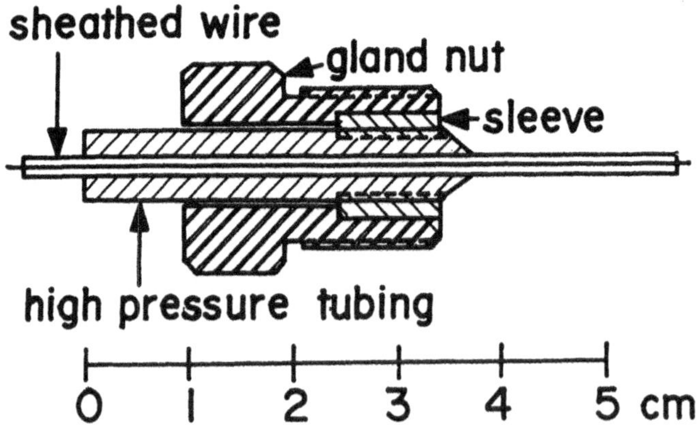

Fig. 28 Electrical feedthrough for 700 MPa. The sheathed thermocouple wire is silver-soldered into the piece of high pressure tubing. The wire sealed with silicone oil and varnish.

Fig. 29 Photograph of 4-wire feedthrough for 700 MPa. Sealed with silicone oil and varnish.

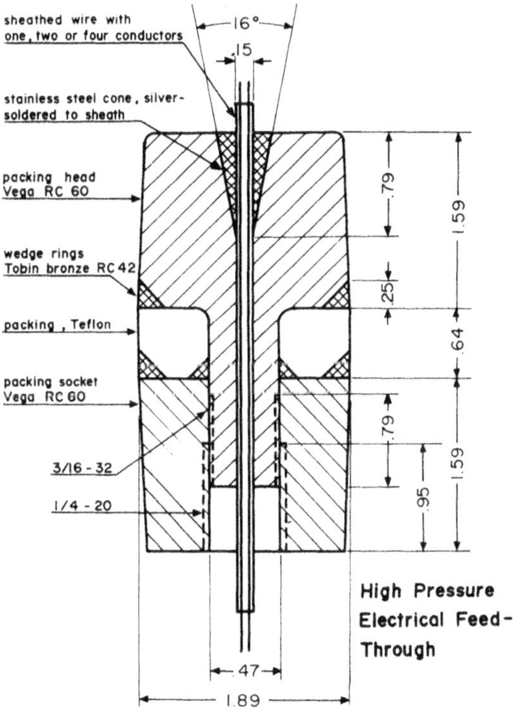

Fig. 30 Details of a 2.5 GPa plug with electrical feedthrough. All dimensions in centimeters.

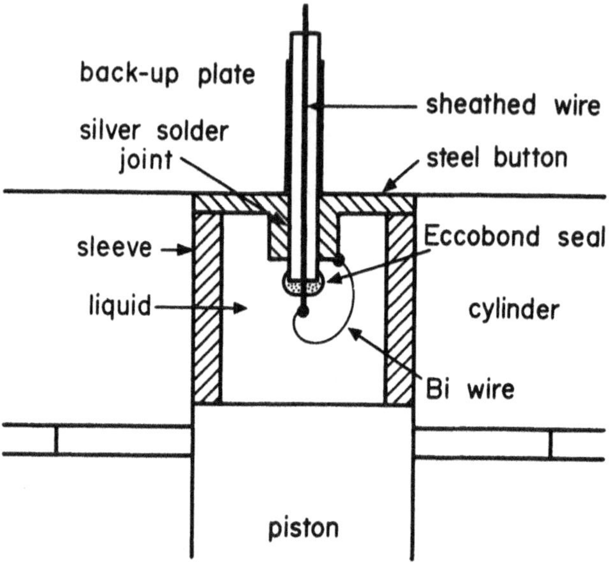

Fig. 31 Electrical feed-through for pressures to 3 GPa, DC to 30 MHz. The sheathed wire is silver-soldered into the steel button. Rim thickness is about 0.5mm; total height of the center part of the button is about 3.5 mm.

the baking process is repeated. This process of baking and soaking is repeated four or five times. After the last baking the excess oil is wiped off and the ends of the feedthrough are sealed with silicone varnish which is then baked in. Typically after this treatment a four-wire feedthrough for resistance measurements will have a resistance between any two wires or between the sheath and one wire of greather than 10^9 ohms. This insulation resistance can usually be maintained for years. Fig. 29 shows a photograph of a finished feed-through.

A feedthrough for service up to 2.5 GPa is shown in Fig. 30. It is also based on sheathed thermocouple wires. A cone of stainless steel is silver - soldered over the sheathed wire. This cone is seated with a dab of vaseline in a conical set with the same solid angle of 16 to 20 degrees. The conical seat terminates in a straight hole whose diameter is slightly larger than the outside diameter of the sheath. This is an unsupported area seal. Even after onehundred applications of pressures to 2.5 GPa the cone can usually be lifted out of the seat with two fingers. The silicone oil/varnish seal is used with this feedthrough.

The feedthrough shown in Fig. 31 has been used to beyond 3 GPa without failure. It consists of a sheathed wire with an outside sheath diameter of about 1.5 mm silver - soldered into a steel button as shown. The rim of the button is sealed with the polyethylene sleeve. The sheath is held in place by the solder and by the swaging action of the protruding button. This part must protrude 2.5 to 3 mm over the 0.5 mm thick rim to prevent blow-outs. Epoxy seals or the silicone oil/varnish seal described above work well to seal the feed-through.

REFERENCES

1. D.L. Decker, W.A. Bassett, L. Merrill, H.T. Hall, J.D. Barnett, J. Phys. Chem. Ref. Data 1, 773 (1972)
2. G.J. Piermarini, S. Block, J.D. Barnett, J. Appl. Phys. 44, 5377 (1973)
3. H.G. Drickamer, Rev. Sci. Instr. 41, 1667 (1970)
4. B. LeNeindre, K. Suito, N. Kawai, High Temperatures-High Pressures 8, 1 (1976)
5. AIRAPT Task Force on the International Practical Pressure Scale, 1977
6. D.P. Johnson, P.L.M. Heydemann, Rev. Sci. Inst. 38, 1294 (1967)
7. R.S. Dadson, R.C.P. Greig, Brit. J. Appl. Phys. 16, 1711 (1965)
8. P.L.M. Heydemann, B.E. Welch, in Experimental Thermodynamics Vol. II., B. LeNeindre and B. Vodar, editors, Butterworths 1975
9. D.P. Johnson, J.L. Cross, J.D. Hill, H.A. Bowman, Industr. Eng. Chemistry 49, 2046 (1957)
10. Yu. S. Konyaev, Prib. Tekh. Eksp. 4, 107 (1961)
11. Harwood Engineering Company, High Pressure Equipment Catalog
12. G.C. Kennedy, P.N. LaMori, J. Geophys. Res. 67, 851 (1962)
13. A. Jayaraman, R.G. Maines, Proc. Symp. Accurate Characterization of the High Pressure Environment, NBS Spec. Publ. 326, U.S. Gov. Printing Office, Washington, D.C. 1971,
14. P.L.M. Heydemann, J.C. Houck, J. Res. N.B.S. 71C, 11 (1967)
15. S.E. Babb in Mechanical Behaviour of Materials under Pressure, H. Ll.D. Pugh, editor; Elsevier Publ. Comp., 1970.
16. H.A. Bowman, D.P. Johnson, N.B.S. Report No. 5381 (1957).
17. Y.A. Atanov, E.M. Ivanova, Proc. Symp. Accurate Characterization of the High Pressure Environment, NBS Spec. Publ. 326, U.S. Govt. Printing Office, Washington, D.C. 1971
18. M.D. Boren, S.E. Babb, G.J. Scott, Rev. Sci. Instr. 36, 1456 (1965)
19. K.A. Alexeyev, Y.A. Atanov, L.L. Burova, Proc. Inst. Comm. Standards (USSR) 75, 44 (1964)
20. G.A. Samara, A.A. Giardini, Rev. Sci. Instr. 35, 989 (1964)
21. T. Kozuba, T. Yamamoto, Proc. 4th Intern. Conf. on High Pressure Kyoto 1974
22. P.J.A. Fuller, J.H. Price, Nature 193, 262 (1962)
23. C. Andeen, J. Fontanella, D. Schuele, Rev. Sci. Instr. 42, 495 (1971)
24. J.H. Colwell, Proc. 6th AIRAPT International High Pressure Conference, Boulder, CO. 1977
25. H.J. McSkimin, P. Andreatch, R.N. Thurston, J. Appl. Phys. 36, 1624 (1965)
26. R. Bechmann, A.D. Ballato, T.J. Lakaszek, Proc. I.R.E. 50, 1812 (1962)
27. V.E. Bean, Proc. 2nd Int. Conf. on High Pressure Engineering, Brighton, England (1975)

28. W.T. Angel, V.E. Bean, Rev. Sci. Instr. Rev. Sci. Instr. 46, 533 (1975)
29. D.D. Decker, J. Appl. Phys. 36, 157 (1965)
30. D.C. Decker, J. Appl. Phys. 42, 3239 (1971)
31. D.B. Larson, R.N. Keeler, A. Kusubov, B.L. Hord, J. Chem. Phys. Solids 27, 476 (1966)
32. E.A. Perez-Albuerne, H.G. Drickamer, J. Chem. Phys. 43, 1381 (1965)
33. J. Scott Weaver, T. Takahashi, W.A. Bassett, Proc. Symp. Accurate Characterization of the High Pressure Environment, NBS Spec. Publ. 326, U.S. Gov. Printing Office, Washington, D.C. 1971
34. R.N. Jeffery, J.D. Barnett, H.B. Vanfleet, H.T. Hall, J. Appl. Phys. 37, 3172 (1966)
35. see Table V in ref. [30]
36. P.L.M. Heydemann, J. Appl. Phys. 38, 2640 (1967)
37. J.C. Haygarth, I.C. Getting, G.C. Kennedy, J. Appl. Phys. 38, 4557 (1967)
38. J.C. Haygarth, H.D. Luedemann, I.C. Getting, G.C. Kennedy, J. Chem. Phys. Solids 30, 1417 (1969)
39. J.D. Barnett, S. Block, G.J. Piermarini, Rev. Sci. Instr. 44, 1 (1973)
40. G.J. Piermarini, S. Block, J.D. Barnett, R.A. Forman, J. Appl. Phys. 46, 2774 (1975)
41. P.M. Bell, H.K. Mao, Proc. 6th AIRAPT International High Pressure Conference, Boulder, Co., 1977
42. Symposium on the Accurate Characterization of the High Pressure Environment, Washington, D.C. 1968.
43. A.L. Ruoff, R.C. Lincoln, Y.C. Chen, J. Phys. D6, 1295 (1973)
44. S. Yamamoto, Proc. 4th International Conference on High Pressure, Kyoto, Japan, 1975
45. J.C. Houck, J. Appl. Phys. 48, 605 (1977)
46. J.C. Houck, J. Res. N.B.S. 74, 51 (1970)
47. R.J. Zeto, H.B. Vanfleet, E. Hryckowian, C.D. Bosco, Proc. Symp. Accurate Characterization of the High Pressure Environment, NBS Spec. Publ. 326, U.S. Govt. Printing Office, Washington, D.C., 1971
48. J.C. Haygarth, H.D. Luedemann, I.C. Getting, G.C. Kennedy, Proc. Symp. Accurate Characterization of the High Pressure Environment NBS Spec. Publ. 326, U.S. Govt. Printing Office, Washington, D.C., 1971
49. E. Whalley, A. Lavergne, Proc. 6th AIRAPT International High Pressure Conference, Boulder, CO., 1977
50. P.L.M. Heydemann, Rev. Sci. Instr. 38, 558 (1967)
51. P.L.M. Heydemann, Rev. Sci. Instr. 41, 1896 (1970)

COMPUTER SIMULATION AT HIGH PRESSURE

B. J. Alder

Lawrence Livermore Laboratory, University
of California, Livermore, California 94550

ABSTRACT. The use of either the Monte Carlo or molecular dynamics method to generate equations-of-state data for various materials at high pressure is discussed. Particular emphasis is given to phase diagrams such as the generation of various types of critical lines for mixtures, melting, structural and electronic transitions in solids, two phase ionic fluid systems of astrophysical interest as well as a brief aside of possible eutectic behavior in the interior of the earth. Then the application of the molecular dynamics method to predict transport coefficients and the neutron scattering function will be discussed with a view as to what special features high pressure brings out. Lastly, an analysis by these computational methods of the measured intensity and frequency spectrum of depolarized light and also of the deviation of the dielectric measurements from the constancy of the Clausius-Mosotti function is given that leads to predictions of how the electronic structure of an atom distorts with pressure.

1. INTRODUCTION

The computer simulation methods have no difficulty generating any pressure, however high. They only require an intermolecular potential, preferably pair-wise additive, although that is not necessary in principle. If that, however, is not the case, the computations are vastly more complex and no practical calculation exists as yet that overcomes this difficulty. Since all materials under sufficient pressure become metallic, where the pair-wise additive approximation is inadequate, one has to resort to approximate schemes such as effective pair-potentials; for example, the

Kelm, H. (Ed.) High Pressure Chemistry. 51–63. *Copyright* © 1978. *D. Reidel Publishing Company. All Rights Reserved.*

pseudopotential. These potentials, however, are now strongly density and temperature dependent. If one goes to still much higher pressures where the electrons are all stripped off the nuclei, the description of matter becomes simple again, namely by a pair-wise additive Coulomb potential between the nuclei neutralized by a uniform electron gas. In all these situations the computer simulation method can make a useful contribution. In the experimentally accessible range the calculations coupled to experiments can back-out an effective pair-potential that can be used to predict not as yet measured other properties or extend the range of measured ones. It is, however, in the experimentally inaccessible region that the computer method likely makes its most valuable contribution.

So far the above discussion had computer simulation methods in mind in which the particles obeyed classical mechanics. It is, of course, this limitation that led to the introduction of pair-potentials that in metals can be considered a stop-gap measure. To deal with at least the unbound electrons in a metal more accurately one needs to deal with them quantum mechanically and a few attempts of calculating the properties of an electron gas by computer simulation methods are under way but as yet too incomplete and complex to discuss here. Nevertheless they may ultimately provide a rather accurate description of matter under all pressures. Here we will confine ourselves to classical calculations but within that framework, in interpreting light scattering experiments, discuss in some detail how the electronic structure of an atom distorts as pressure is applied. This, after all, is the fundamental effect of all high pressure experiments.

2. PHASE DIAGRAMS

The computer methods can generate equations-of-state quite accurately with a few hundred particles, however phase transitions are distorted because the large fluctuations necessarily associated with these transitions are limited by the finite system. A particular difficulty also occurs with reproducing crystal transformations because of the symmetry imposed on the system by the shape of the cell to which the few hundred particles are confined. Thus with solid phases one is generally restricted to try out various crystal structures and determine the very small free energy difference between the various possibilities, a not very fruitful procedure. The distortion in fluid phase transitions is not serious unless one asks questions which are connected with long range correlations which is of not particular concern here.

2.1 Critical Line in Gas Mixtures

The computer results for the equation-of-state have shown the remarkable validity of the van der Waals model above the critical density.[1] That model consists of determining the thermodynamic properties and pair distribution function of a reference system, usually taken to be a purely repulsive potential. The attractive potential is then added as a perturbation utilizing the pair distribution of the unperturbed system. This theory depends on the observation that the structure of a fluid is primarily determined by the repulsive forces and one would guess increasingly so the higher the pressure. The only modification of the present procedure, which is utilized near normal conditions, that needs to be made involves the use of a hard sphere reference system. Under normal conditions the steep repulsive potential is usually replaced by an equivalent hard sphere potential to simplify matters. The choice of that hard sphere diameter is given by several recipes which makes that diameter temperature dependent.[2] This is likely to become an inappropriate procedure at higher pressure where the softness of the repulsive potential plays an increasingly dominant role. What needs to be done and has already been done to a considerable degree is to generate properties for soft repulsive potentials of the form r^{-n}, for example, or of the exponential type and then the task will be to find the power of n, which best reproduces experimental results after some reasonable attractive potential has been added on. That procedure is not significantly more difficult than the use of a hard sphere reference system since still a single reduced variable characterize the properties of a r^{-n} system.

Near critical densities, however, the choice of the form of the repulsive potential is unimportant since the density is low. Thus, one can, for example, easily predict the behavior of the critical lines in a mixture of gases by the van der Waals model. Indeed, van der Waals himself already used his theory to make such predictions. The only difference now is that it is possible to be more quantitative, since the properties of the reference system can be much more accurately represented than by means of the covolume, b. Furthermore, the attractive forces need no longer be represented by a constant value of the cohesive energy density but can be calculated accurately by the perturbation theory. This modified or augmented van der Waals theory can then be used to predict quantitatively the behavior of critical lines in mixtures provided the various pair interaction potentials are known.[3] It can thus be established under what conditions of differences in repulsive potentials and in strengths of attractive potentials the two critical points of the pure components are connected. As the two pure components differ more in their interaction potentials the two critical points not only become disconnected, but the critical temperature of the mixture becomes higher than that of either component; circumstances leading to so-called two-gas phase equilibrium.

2.2 Melting

The single most important result of the computer simulation method for equilibrium properties is that melting occurs in purely repulsive systems such as hard spheres.[4] Since all systems have rather steep repulsive potentials one can understand the universality of the phenomenon. At the same time there is still no *a priori* theory that can deal with melting and it remains the outstanding theoretical problem in equilibrium statistical mechanics. Nevertheless, there exists an old empirical observation based on geometric considerations and consistent with the fact that only a repulsive potential is required, that predicts melting very well in all known cases. This is Lindemann's rule which says that melting occurs when the root mean square excursion of a particle from its lattice site (the Debye-Waller factor) is about ten percent of its lattice spacing or roughly the substance is thirty percent expanded from close-packing.[5] Given the universality of this rule, it is hard to understand why geophysicists predicting melting under pressure widely use some other empirical procedure based on extrapolating low pressure data that leads to rather different predictions than Lindemann's rule.

The question of what ultimately happens to the melting line as pressure increases can be answered quite definitively.[6] These considerations must take into account the argument that there cannot be a solid-fluid critical point because it is impossible to go continuously from an ordered system to a disordered one or from some symmetry to no symmetry. As was previously pointed out compressing materials turns them metallic, weakening the repulsive forces till ultimately the potential becomes coulombic between the bare nuclei neutralized by an electron gas background. However, it was found by computer simulation that such a coulombic repulsive system called a one component plasma also has a solid-fluid phase transition which incidentally conforms quite well to Lindemann's criterion.[7] This state of affairs lasts till enormous pressures are reached where the density is so high that the nuclei have to be treated quantum mechanically. It is then easy to show that the lattice must melt even at the absolute zero because the zero point energy rises faster, namely as the reciprocal square of the lattice spacing then the potential energy which increases for coulombic potentials linearly with reciprocal lattice spacing as the density increases. These considerations, nevertheless, do not contradict the impossibility of a critical point because they predict a closed region in pressure and temperature below which a solid exists. Melting is always first order resulting from the combined effect of zero point energy and thermal energy. The surprising prediction[6] is that for protons the maximum temperature in this closed region is about 3000°K above which a solid cannot exist at any pressure. Thus most protonic stars can be predicted to have fluid interiors.

2.3 Electronic Transitions

Certain substances, in particular cerium, have a solid-solid critical point in apparent contradiction to the above general statement which does not allow continuous transitions from one symmetry to another. That difficulty is easily resolved since it was found that both crystal structures were the same on either side of the phase line below the critical point. Closely associated with this critical point is a melting extremum in cerium. Similarly in cesium a solid-solid phase transition has been observed between two identical crystal structures which however does not end in a critical point but instead intersects the melting line and again at a melting minimum. These phenomena can be explained in terms of an electronic transition that the atom undergoes as it is compressed. The effect of pressure is to stabilize electronic energy levels of higher angular momentum but lower electronic quantum number. At the density at which this electronic shift occurs the size of the atom suddenly shrinks and one might think that this effect could be well modelled by a repulsive step potential by which is meant that to a hard sphere core a repulsive mound or repulsive square-well of certain height and width has been added on. These systems can be investigated by computer simulation and indeed some of the qualitative features of these phase diagrams are reproduced.[8]

A solid-solid critical point is obtained if the width of the added mound is not too large. Of course, after the electronic phase transition has taken place the new phase does not have to have the same crystal structure as the previous one and if not, no critical point would be observed. The computer calculation, for reasons previously mentioned, has excluded that possibility. Also observed is a melting maximum with this square-mound potential that can be qualitatively understood in terms of larger freedom of motions in a fluid which allows fluid configuration to sample the smaller diameter before the solid can. The fluid then becomes denser in the mean than the solid for some density range because the solid is constrained to maintain its lattice structure, and, so to speak, all the lattice spacings have to shrink simultaneously when the appropriate pressure is reached. Unrealistic in these calculations is that the solid-solid phase transformation does not intersect the melting curve near its extremum and that the width of the mound is unrealistically large before these two phase lines intersect. The mound model seems to be too simplistic to represent these details well.

One would expect these electronic phase transitions to be quite a common phenomena, since many heavy elements, for example, all the rare earths, have their 4 f orbitals empty and some 5 shells filled under normal conditions. Indeed there is evidence from shock wave data on rare earth elements that there is a sharp break in the shock versus particle velocity plot at or below the melting point that leads into a more compressible phase.[9] This would be consistent with an electronic phase transition at a melting maximum or, if below melting with crossing a critical line in a solid that leads to a more compressible liquid phase. The sharpness of the break indicates that the fluid is denser than the solid. Such a first order electronic phase transition cannot occur in a fluid phase basically because a fluid is so flexible that it can sample or average over at least all ranges of distances less than a diameter of a particle. In the solid, as Lindemann's rule indicates, it is only possible by thermal fluctuation to average over structures in the potential which are less than 10% of the size of an atom. One would expect many of the effective size changes accompanying electronic transition to be of that order or less.

2.4 Interior of Jupiter

It has been suggested that under the pressure and temperature conditions in the interior of Jupiter, helium may not be completely soluble in hydrogen. In the interior the pressure is estimated to be about 40 Mbars and the temperature about 10,000°K or 1 e.v. Under those conditions hydrogen is certainly completely stripped of electrons, but for helium an estimated 100 Mbars is required so that it is only partially pressure and temperature ionized. Nevertheless, as our first attempt to predict the possibility of phase separation it shall be assumed that all electrons form a uniform background in which α particles and protons interact via their unscreened coulomb potential. Since these nuclei under these conditions are still well described by classical mechanics, the problem is easily studied by computer simulation.[10]

These calculations indeed show phase separation below a critical temperature of about 1 e.v. over a wide region of pressure. The qualitative reason for phase separation comes from potential energy considerations, namely that local charge neutralization is more effective in the pure phases than the mixture, given the constraint of uniform electron charge distribution. So to speak, in the mixture of uniform charge distribution the different species cannot each be surrounded by a charge neutralizing sphere radius that is as effective as in the pure species separately.

The mixture becomes eventually stable, that is the critical temperature goes to 0°K, at the enormous pressure of 10^8 Mbars because then the electron gas pressure dominates and the mixture becomes nearly ideal.

More realistic calculations applicable to the center of Jupiter would allow for the electron gas to distort, that is to increase its density in the neighborhood of the nuclei. This can be done through linear response by perturbation theory if this distortion is small. Since this is not the case one crude approximation would be to place effective partial charges on the helium nuclei. As can be readily seen, this reduces the charge difference between hydrogen and helium and since in the unscreened model the critical temperature was already close to center of Jupiter conditions any lowering would predict miscibility.

2.5 Interior of the Sun

The phase separation of stripped helium and hydrogen opens up the possibility of much higher temperature phase separation for highly charged nuclei in a hydrogen plasma. For example, completely stripped iron of charge Z=26 might separate in hydrogen below a temperature of 10^7 degrees or 1 kev since the potential energy scales as Z^2. Thus, the possibility exists that the small amount of iron which is believed to be distributed in the sun according to cosmic abundance values might be insoluble. If iron were less soluble at temperatures of about 1.5 kev than one part in 10^6, the cosmic abundance value being one part in 10^5, one could very well resolve the neutrino dilemma. That dilemma consists of not finding experimentally any neutrinos while stellar evolution models demand that they should be observed. These neutrinos are essentially a measurement of the temperature in the interior of the sun, because they are generated only by an extremely energetic reaction of two α particles so that they fuse into another nuclei that decays by giving off the searched for neutrinos. At 1.5 kev, the fusion will not take place unless the kinetic energy of collision corresponds to about 10 times the mean since otherwise the coulombic repulsion barrier cannot be overcome. Thus a small overestimate of the temperature would cut out the reaction since the number of particles in the high energy tail of the Maxwell-Boltzmann distribution would be drastically affected.

Such a lowering of the temperature could be achieved by removing the small amount of iron, since it contributes about 30% to the opacity in the interior of the sun. The reason for this small amount contributing so much is that this element is the only one of high Z value present in significant amount. Because this element at the temperature in the interior of the sun still has bound electrons it contributes through the bound-free transition significantly to the opacity. Iron has two inner electrons which are bound by 8 kev and the next 6 ones would be partially bound because they require about 2 kev to be ionized.

Calculations using dilute electrolyte theory, since iron is present in such small concentrations and at high temperature, show indeed that iron even if only of charge 20 would be soluble by less than one tenth cosmic abundance.[11] The proposal is hence that iron precipitates out into the center of the sun. The calculations cannot, however, be entirely trusted as yet. First of all, though the iron poor phase is indeed very dilute, the iron-rich phase that precipitates can by no means be adequately described by the dilute Debye-Hückel theory. One thus needs to introduce higher density corrections and once these were made they led to significant changes in the phase diagram. The next model to be investigated is to calculate for an effective Z of 24 the phase diagram over a wide range of iron concentration through an approximate integral equation which, however, has proved reliable for long range potentials, the so-called hypernetted-chain equations. These can be used at all densities of iron and the validity of the results can be checked at selected points by machine simulation methods. Once the reliability of that equation has been established, the further screening of electrons can be taken care of by perturbation theory as discussed in the case of Jupiter. By this procedure, it might thus be made theoretically plausible that the solubility of iron is significantly less than cosmic abundances.

2.6 Interior of the Earth

There appears to be another possible type of phase transition in the core of the earth that is, however, not well simulated by computer methods. It stems from the observation of sound propagation in the deep interior of the earth from which it can be concluded that there appears to be a region of some 500 km sandwiched between the solid and liquid core where the sound speed is nearly independent of pressure over some 500 Kbars. Since no known pure substance is expected to behave this way, it seems reasonable to associate this observation with the phase transition, in particular eutectic behavior.[12] The argument would be that going into that region as pressure and temperature increases, the iron would precipitate out and settle eventually in the core while the fluid follows the liquidus branch of the eutectic. The consequence would be that the concentration of presumably a lighter and hence smaller sound speed component would increase. The concentration increase would then compensate the usual pressure increase of the sound speed leading to a nearly constant value in this region.

This suggestion is supported, however weakly, by both static and dynamic data. If the component is either iron sulfide or iron silicide the possibility of eutectic behavior is large since they do so under normal conditions and under as high a static pressure (100 Kbars) as has been reached. Furthermore, the concentration required to keep the sound speed constant is in reasonable agreement with the observed nearly pressure independent liquidus. The proposal has two further consequences. One is not that only is the sound speed nearly constant in that region, but, as might be expected, the density is also nearly constant. This could lead to hydrodynamic instabilities in that region. The other consequence is that the eutectic melting temperature is considerable (about 30%) lower than that of pure iron. As a result the application of Lindemann's melting law to the eutectic would lower the temperature in the center of the earth by about 30% and these temperatures would then be in better agreement with other temperature estimates that have been proposed.

3. TRANSPORT PROPERTIES

The computer method of molecular dynamics is capable of generating any transport property under any desirable conditions given again an intermolecular pair-potential. There are, however, a few differences with respect to equilibrium properties that are worth emphasizing. Furthermore, it is interesting to point out what special features of transport behavior is brought out in high pressure studies.

The van der Waals model for transport properties reduces to that of hard spheres, since a mean attractive field does not exert any force on the particles. Indeed hard spheres give a very good account of transport properties, if again a suitable choice of the temperature dependent hard sphere diameter is made.[13] Should that be inadequate at very high density again a soft repulsive potential must be investigated directly. However, the perturbation methods, so successful in equilibrium, fail for non-equilibrium and that is because a small atractive potential can vastly change the trajectory of a particle.[14] That occurs when a perturbation potential can change a glancing collision to a bound trajectory. In fact, there exists as yet no fundamental theory which can deal practically with anything but a hard sphere system. For the hard sphere system, the Boltzmann equation is used at low density, and the collision rate scaling suggested by Enskog is used to boost the low density results to high density. The molecular dynamic calculations have shown this to be a good approximation except at densities near the melting point.

The correction near the normal melting point boosts the viscosity by a factor of about 2 above the Enskog value and similarly the diffusion coefficient is reduced by about a factor of 2.[15] The origin of this discrepancy is that the Enskog theory leaves out the effects of backscattering, that is the high probability of the reversal of a particle's velocity in dense media by the neighboring particles. In hydrodynamic language the medium acts viscoelastically on a short time and distance scale. It is clear that at higher pressures this effect will become even more important. It is proposed that this effect will probably be adequately dealt with by introducing a space and time dependent viscosity in the hydrodynamic or Navier-Stokes equation. This generalization of hydrodynamics is fairly easy to execute and will make the predictions of transport coefficients at high pressure more reliable. Since this theory is not confined to hard spheres, it is hoped that it will also be useful for other systems. In the meantime, the hard sphere model has been successfully used to predict the experimental viscosities and diffusion coefficients of various materials including metals under pressure. It has, for example, been used to predict the viscosity of liquid iron in the core of the earth.

The hard sphere calculations of the viscosity and the diffusion coefficient have shown that even when there are deviations from the Enskog theory, that the product of these two transport coefficients is remarkably constant since they deviate in opposite directions. The product of the viscosity and the diffusion coefficient is, furthermore, very well given by the macroscopic hydrodynamic calculations of Stokes, if slip boundary conditions are employed between the moving sphere and the viscous medium.[15] The surprising nearly quantitative validity of hydrodynamics on a microscopic scale is the central result of the computer simulation studies. It is this observation that leads to the above suggestion of generalizing hydrodynamics by introducing non-constant transport coefficients to make hydrodynamics even more quantitative on a molecular scale. As far as the validity of the Stokes-Einstein relation is concerned, it can be made use of under high pressure conditions by measuring only one of the transport coefficients, namely the one that is easier to measure, to predict the other.

The remarkable validity of hydrodynamics is also evident in its prediction of the neutron scattering function. A recent calculation of the neutron scattering function for hard spheres by computer simulation shows that both kinetic theory and hydrodynamics can predict the function very well.[16] This shows, first of all, that the neutron scattering function is very insensitive to the detailed molecular motions that go on in a fluid. Significant differences between the two theories and the calculation occur only at the highest densities, and these are, of course, of interest at high pressure. Both of the theories fail to take into account the backscattering effect, spoken of earlier. In the hydrodynamic theory, this is, however, easily taken care of by the viscoelastic theory, and indeed a crude introduction of that effect led to quantitative predictions of the neutron scattering function also at high density.

4. LIGHT SCATTERING AND DIELECTRIC BEHAVIOR

Much can be learned from the theoretical interpretation of spectral line shapes and intensities about the distortion of atoms and molecules under pressure. The theory is rather complex and still being worked on and so we shall here confine ourselves to the simplest possible situation, namely depolarized light scattering from rare gases at various pressures. Since the theory for deviations of the dielectric behavior from the constancy of the Clausius-Mosotti function is very similar to that of the depolarized light scattering we shall also discuss the complementary information obtained from these experiments.

For spherical atoms depolarization of light can only come about because of a temporary asymmetry in the polarization of a collection of rare gas atoms. Such an asymmetry results from asymmetric density fluctuation in which in some arbitrary region of the fluid more atoms are temporarily packed. This higher density region is necessary accompanied by another region that is less packed, since the average density must remain constant. On the average, of course, each separate region must also be of the mean density, but the temporary density fluctuation leads to an asymmetry in the polarization merely because more polarizable material is located in one region and less in another, the two regions defining a direction that can rotate the polarization of the laser light. Computer simulation can readily calculate this depolarization as the autocorrelation function of the induced-dipole tensor. The Fourier transform of this autocorrelation function gives the frequency spectrum of the depolarized light, while the initial value of the autocorrelation function gives the intensity of the line.

Comparing the theoretical results with experimental observations showed that the frequency spectrum under all conditions of temperature and pressure and a number of different rare gases was well reproduced while the intensity showed increasing deviation from experimental results at higher density.[17] At liquid densities the experimental intensity was a factor of about 8 less than the calculated one. What this meant is that accompanying the density fluctuation is an electronic distortion of the atom that nearly instantaneously responds to the density fluctuation, and that reduces the polarization caused by the density asymmetry. If the slow process in the polarization were the relaxation of the density asymmetry, then it is clear that the calculation which neglected the electronic distortions would still give the correct spectra. The electronic distortion would surely oppose the density fluctuation in that the electrons would, so to speak, flow from the high density region to the low density region in order to minimize their mutual repulsive potential energy, and increasingly so at higher density. Thus, the intensity would be very much reduced at the higher density due to the opposing effect of the electronic polarization. The intensity thus represents a very sensitive measure of the distortion of the wave function of an atom with pressure.

To quantitatively account for this electronic distortion a polarizability of a pair of atoms (over and above that of the two isolated atoms) as a function of their separation is introduced. This pairwise electronic distortion polarizability is as important to optical and electrical properties as the pair potential of interaction is to thermodynamic properties. It is also as difficult to calculate from quantum mechanical considerations as the pair potential. Hence one must resort to similar procedures in backing it out empirically from experiments as for the pair potential, leading to an inverted Lennard-Jones form for this function. One can also ask similar questions about the pair distortion polarizability as the pair potential, namely is it pair-wise additive and does it obey corresponding states. Preliminary investigations lead to positive answers to the last two questions.

A very similar induced dipole tensor autocorrelation function is involved in calculating the deviations from the Clausius-Mosotti dielectric function for the rare gases.[18] If the electronic distortion is again ignored, it can be shown that these deviations must necessarily be positive, contradicting experiments at high density. Introducing electronic distortions, but this time of the diagonal elements of the induced dipole tensor, rectifies that situation. Combining these results with the light scattering results which involve the difference in the pair polarizability distortion parallel and perpendicular to the axis connecting two rare gas atoms, the electronic distortion in the two different directions can be separately extracted. Thus much can be learned from these experiments, coupled with an interpretation that requires computer simulation, about electronic distortion under pressure.

Work performed under the auspices of the U.S. Energy Research & Development Administration, contract No. W-7405-Eng-48.

1. B. J. Alder, D. Young and M. Mark, J. Chem. Phys. 56, 3013 (1971).
2. J. A. Barker and D. Henderson, Ann. Rev. Phys. Chem. 23, 439 (1972).
3. M. Rigby, B. J. Alder, A. M. Sapse and C. E. Hecht, J. Chem. Phys. 52, 3665 (1970).
4. B. J. Alder and T. E. Wainwright, Phys. Rev. 127, 359 (1962).
5. F. A. Lindemann, Z. Physik. 11, 609 (1910).
6. J. P. Hansen, B. Jancovici and D. Schiff, Phys. Rev. Lett. 29, 991 (1972).
7. S. Brush, H. Sahlin and E. Teller, J. Chem. Phys. 45, 2102 (1966).
8. D. A. Young and B. J. Alder, Phys. Rev. Lett. 38, 1213 (1977).
9. R. Grover and B. J. Alder, J. Phys. Chem. Solids 35, 753 (1974).
10. E. L. Pollock and B. J. Alder, Phys. Rev. 15A, 1263 (1977).
11. E. L. Pollock and B. J. Alder, to be published.
12. B. J. Alder and M. Trigueros, J. Geophys. Res. 82, 2535 (1977).
13. B. J. Alder and J. Dymond, J. Chem. Phys. 45, 2061 (1966).
14. B. J. Alder, W. E. Alley and M. Rigby, Physica 73, 143 (1974).
15. B. J. Alder, D. Gass and T. Wainwright, J. Chem. Phys. 53, 3813 (1970).
16. B. J. Alder, W. E. Alley and S. Yip, to be published.
17. B. J. Alder, J. J. Weis and H. L. Strauss, Phys. Rev. 7A, 281 (1973), J. Chem. Phys. 59, 1002 (1973).
18. B. J. Alder, H. L. Strauss, and J. J. Weis, J. Chem. Phys. 62, 2328 (1975).

MAGNETIC RESONANCE SPECTROSCOPY AT HIGH PRESSURE

J. Jonas

Department of Chemistry, School of Chemical Sciences
and Materials Research Laboratory,
University of Illinois, Urbana, Illinois 61801 U.S.A.

ABSTRACT. The field of magnetic resonance spectroscopy at high pressure is discussed with the main emphasis on nuclear magnetic resonance. The lectures include the following main sections: 1. Introduction; 2. Experimental high pressure techniques in NMR; 3. Applications of NMR at high pressure; 4. Pure nuclear quadrupole resonance and electron spin resonance spectroscopy at high pressure.

1. INTRODUCTION

Nuclear magnetic resonance (NMR) spectroscopy represents a well established and powerful tool for studying the dynamic structure of matter. NMR relaxation experiments provide detailed information about molecular motions in gases, liquids and solids as the analysis of the relaxation data yields the zero-frequency Fourier transform of the time correlation function. One can determine the reorientational correlation time, τ_θ, defined in the standard form,

$$\tau_\theta = \int_0^\infty < P_2 \; (\vec{u}(0)\cdot\vec{u}(t) > dt/< {P_2}^2 > \; , \qquad (1)$$

where P_2 is the second Legendre polynomial. When spin-rotation interactions dominate the relaxation mechanism the measurements of the NMR relaxation times yield the angular momentum correlation time, τ_J, defined as

$$\tau_J = \int_0^\infty < \vec{J}(0)\cdot\vec{J}(t) > dt/< J^2 > \; , \qquad (2)$$

Kelm, H. (Ed.) High Pressure Chemistry. 65–110. *Copyright © 1978. D. Reidel Publishing Company. All Rights Reserved.*

where J is the molecular angular momentum. The NMR spin-echo technique provides a convenient way to determine the self-diffusion coefficient, D, in a liquid or gas

$$D = \frac{kT}{m} \int_0^\infty < \vec{v}(0) \cdot \vec{v}(t) > dt / < v^2 > , \qquad (3)$$

where v is linear velocity. It is not necessary at this point to discuss the nature of high resolution NMR spectroscopy which is a familiar field to all chemists.

A great majority of the NMR studies have used temperature as the only experimental variable, while pressure was left constant, usually at atmospheric pressure. The pioneering NMR study at high pressure was done by Benedek and Purcell[1]; the early studies have been reviewed by Benedek[2], and the more current activity has been discussed by Jonas[3,4].

In the following presentation of the magnetic resonance spectrsocopy at high pressure, emphasis has been placed on the NMR studies of liquids, with some comments on work on gases and solids. After a discussion of the experimental aspects of NMR spectroscopy at high pressure, the results of several specific studies will be used to illustrate various applications of this high pressure field. The lectures will be concluded by a brief discussion of electron spin resonance spectroscopy and pure nuclear quadrupole resonance spectroscopy performed at high pressure.

Before going into a detailed description of the experimental aspects, let us try to answer a basic question - why do we do NMR experiments at high pressure? The NMR studies of liquids have used temperature as the only experimental variable, while pressure was left constant, usually at atmospheric pressure. Such studies probe only a narrow region of the liquid state since, for any given temperature, a liquid can exist over a range of densities. The use of pressure provides another dimension over which to investigate the liquid system. It is important to realize that changing the temperature at constant pressure affects molecular motion in two distinct ways. Not only is the average kinetic energy of the molecule changed but there is always an accompanying change in the average volume available for the motion of the molecule. If one uses both pressure and temperature as experimental variables in an NMR experiment, only then can one separate the effects of density and temperature on molecular motion.

At first it would appear that the change in density for a liquid over the typical experimental range ($\sim 100^\circ C$) is insignificant; however, for a molecular liquid, a $30^\circ C$ temperature increase requires the pressure to be increased approximately 400-600 bar to maintain a constant density. Very often the temperature effect on the dynamics of molecular motions is much smaller than the combined density-temperature effects, and in many cases volume effects play the dominant role in influencing motions in the liquid state. This is physically quite reasonable because the molecules are so closely packed in the dense liquid that the forces between particles arise almost exclusively from the harsh short-ranged repulsive forces between nearest neighbors. Therefore, even small changes in density can produce a considerable change in the molecular dynamics of the system.

An illustrative example of the relative effect of density and temperature on molecular motions is given in Figure 1, which shows the reorientational correlation function obtained from the observed Raman spectra in liquid methyl iodide[5].

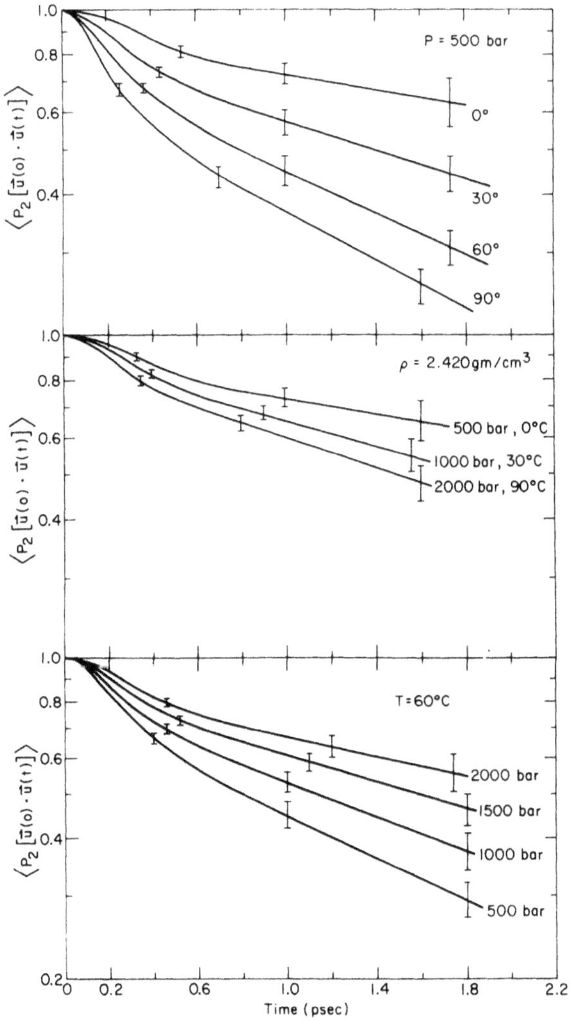

Fig. 1. The reorientational correlation function $< P_2[\vec{u}(t)\cdot\vec{u}(0)] >$ in liquid CH_3I.

Another example of the importance of separation of the effects of density and temperature is given in Figure 2 which shows the temperature dependence of the rotational correlation time, τ_θ, for overall reorientation in toluene-d_8 at constant pressure and at constant density[6]. It is clear that the temperature dependence

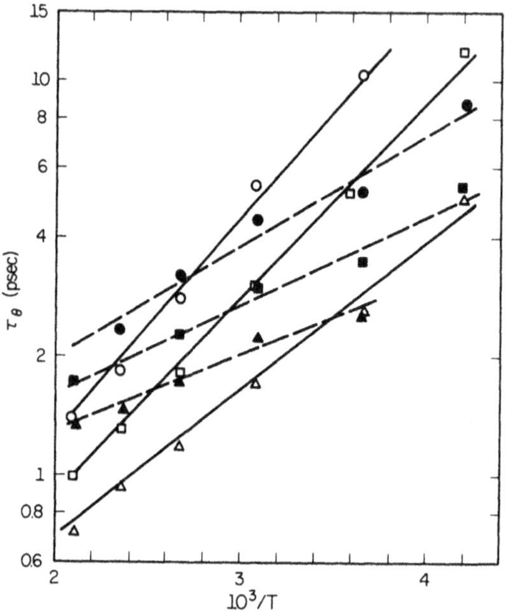

Fig. 2. The temperature dependence of the rotational correlation time, τ_θ, for overall reorientation of toluene-d_8 at constant pressure (full lines; △ P = 1 bar; □ P = 1.5 kbar; ○ P = 3 kbar) and at constant density (dashed lines; ▲ ρ = 0.96 g/cm^3; ■ ρ = 1.00 g/cm^3; ● ρ = 1.04 g/cm^3).

of molecular reorientation observed at constant pressure is due just as much to changes in density as to changes in the average energy of the molecules. This ability to separate density and energy effects is one of the most important results of the high pressure NMR experiments. Since most theoretical calculations of liquids are usually done at either constant density or constant temperature, the ability to present experimental data at constant density is vital if we are to improve our understanding of the liquid state.

Another reason for doing a high pressure experiment is to extend the measurements well above the normal boiling point and to permit study of supercritical dense fluids. A very good example of an additional advantage of the high pressure experiments is illustrated in Figure 3 which gives the phase diagram of heavy

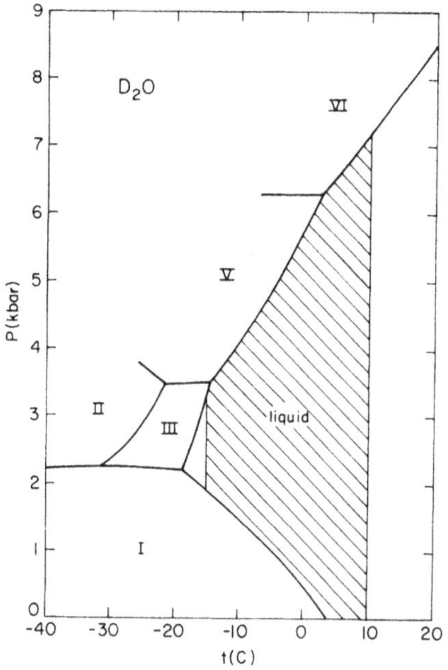

Fig. 3. Phase diagram for liquid D_2O. The ranges of pressures and temperatures of our measurements denoted by shaded area.

water. Using this phase diagram for D_2O, Defries and Jonas[7] were able to study relaxation and transport properties of D_2O at temperatures below $0^{\circ}C$; the range of their measurements is described by the shaded area. Without using pressure as an experimental variable, one could not perform these measurements.

2. EXPERIMENTAL HIGH PRESSURE TECHNIQUES IN NMR

i) General instrumentation

Except for the single coil geometry, there are no special requirements for the NMR pulsed spectrometer for the measurements[8] on liquids at high pressure. In view of the limited space in the high pressure probe, the single coil probe is a logical choice. Since changes in temperature and pressure lead to subsequent changes of electrical parameters in the NMR sample probe, which in turn cause variations in the effective radiofrequency field and its phase. A frequent reoptimization of pulse length and pulse phase is necessary and therefore it is advantageous to have a computer directed NMR spectrometer[9] which automatically carries all optimizations and also all relaxation measurements. A general schematic diagram of the spectrometer is shown in Figure 4.

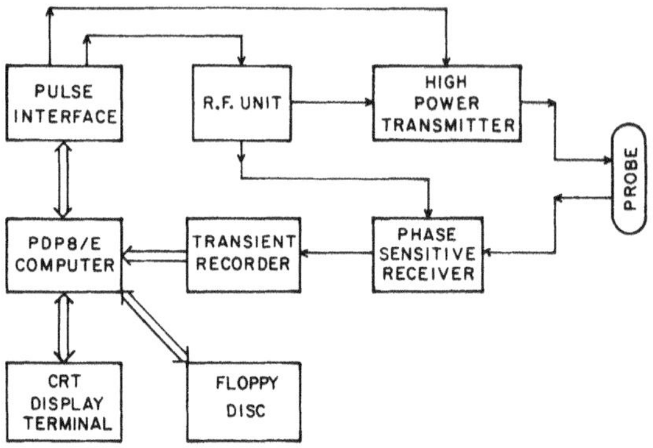

Fig. 4. Schematic diagram of the NMR pulsed spectrometer.

Due to the bulky high pressure probes and additional thermostating equipment, one should try to acquire an electromagnet with the widest gap possible, of course, capable of producing a magnetic field of at least 10 kG to 14 kG. A special wide gap (9.5 cm) electromagnet Varian model V-3800-1 producing a magnetic field of 14,092 G represents a very important part of our experimental system. The wide gap provided us with enough working space so that we could optimize the design of high pressure probes and sample cells. The electromagnet is equipped with field homogeneity coils and a solid state flux stabilizer in order to maintain the necessary homogeneity and stability of the magnetic field. A Varian model V-7900 current regulated 20 kW power supply and a Varian model V-3524 heat exchanger for temperature control of the magnet cooling water are used in our system. The homogeneity of the magnetic field will be discussed later when we deal with the Fourier transform method. The short term stability of the magnetic field is better than one part in 10^7.

In our laboratory we are using the high pressure generating system depicted schematically in Figure 5. It consists of Enerpac hand pumps, a pressure intensifier, a liquid separator, Heise Bourdon gauges, and necessary high pressure tubing and valves. The system can generate hydrostatic pressures up to 5 kbar and in a modified version up to 10 kbar. Carbon disulfide was used as a pressure transmitting fluid for proton relaxation measurements at low temperatures, and tetrachlorethylene at higher temperatures. For other nuclei, a mixture of 50% SAE 10 W motor oil and 50% kerosens was used for temperatures up to 150°C; at higher temperatures, above 300°C, even the motor oil deteriorates rapidly and silicone oils are recommended for temperatures up to 500°C.

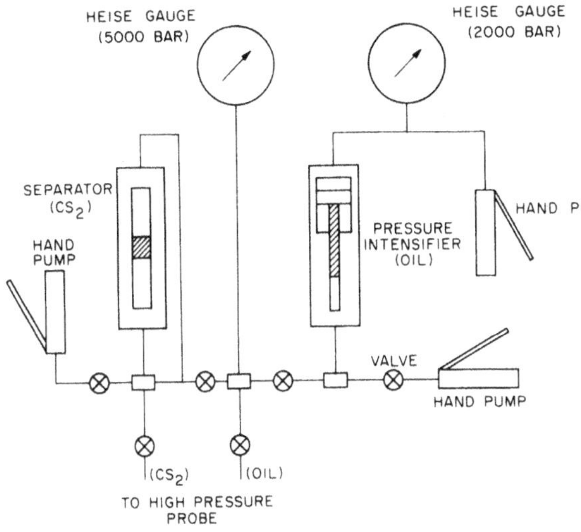

Fig. 5. Schematic drawing of the high pressure generating system.

For low temperatures and temperatures higher than 500°C inert gases are the obvious choice for the pressure medium. However, gas systems are much harder to make leak-free than are systems using a liquid pressure medium and more attention must be paid to safety precautions. The gas pressurizing system which we used in several recent studies[10,11] is schematically given in Figure 6.

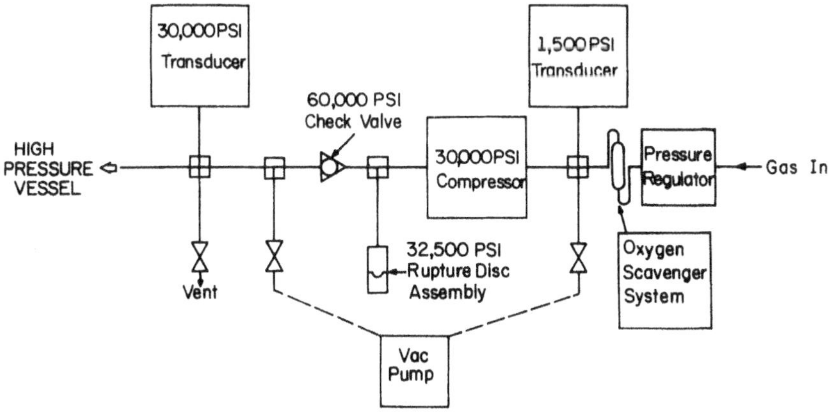

Fig. 6. Gas pressurizing system.

A material which is used to construct high pressure NMR probes must be nonmagnetic and of high mechanical strength. Table I gives the most common materials used for high pressure NMR. The main advantage of titanium alloys is their mechanical strength at high temperatures. Our own experience with titanium alloys is excellent.

Table I. Nonmagnetic materials for high pressure NMR.

Material	Tensile Strength (kpsi)
Berylco (Beryllium copper alloy)	175-200
Titanium alloys	
IMI 680[a]	180
6Aℓ-2Sn-4Zr-6Mo[b]	170
Stainless steel (type 316)	150

[a]Made in U. K. by IMI.

[b]Made in U. S. A.

Figure 7a shows schematically the high temperature titanium pressure vessel. This vessel was externally heated by a Nichrome wire noninductively wound on asbestos insulation. Only the middle part of the vessel was heated. On the contrary, the top of the vessel which contains the closure plug with rf and thermocouple terminals was cooled by a cooling jacket. The temperature on the surface of the titanium vessel near the top and near the bottom of the external heating coil was continuously monitored by a digital temperature measuring system using constantan/copper thermocouples. The fact that the closure plug was always maintained at temperatures below 100°C enabled us to use simple rf and thermocouple terminals (see Fig. 7b) even when working at temperatures well above 200°C. It was surprising to find that even when operating at high temperatures, the heating coils did not impair the magnetic field homogeneity over the sample volume. A readjustment of the field homogeneity controls was necessary for each setting of the power supply. There was no dissipation of heat to the magnet even at temperatures above 300°C since cooling plates were used around the high temperature vessel.

ii) High pressure NMR sample cell

Figure 8a is a schematic drawing of the high pressure sample cell. A glass (Pyrex 7740) or quartz to metal (SS 316) seal

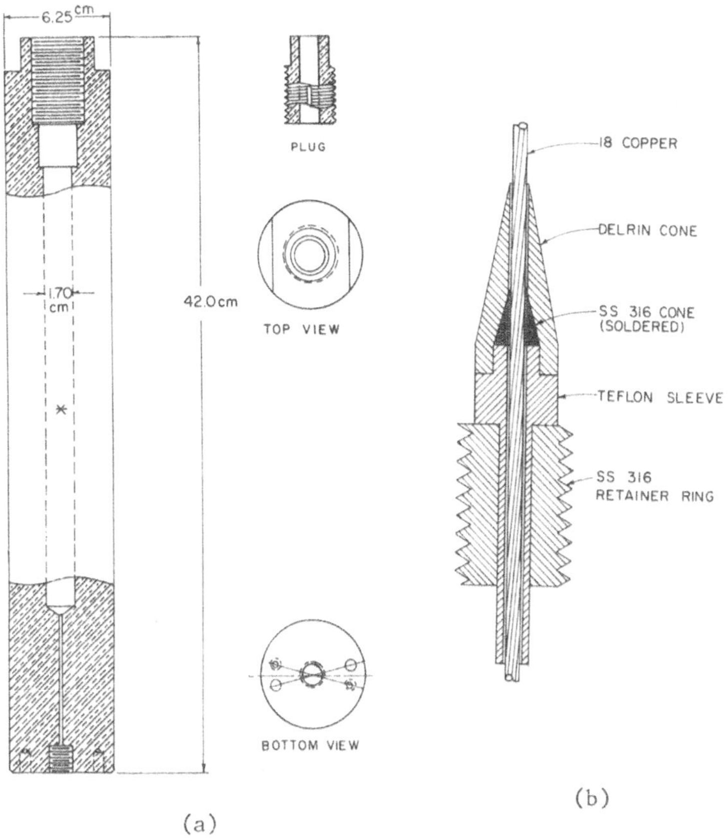

Fig. 7a. Titanium high pressure vessel.
b. RF lead terminal.

(Quartz Scientific, Inc., Palo Alto, Calif.) connects the actual glass sample chamber to the SS 316 tubing (o.d. 6 mm). Stainless steel bellows (Mechanized Science Seals, Inc., Los Angeles, Calif.) accommodate the volume change of the liquid due to compression. Hermetic closure of the sample cell was attained by a (60° angle) seat made of stainless steel. In order to ensure good mechanical strength of the plug and upper bellows adaptor, these parts were made from SS 316 which were brazed to the copper ring cones. We did not encounter difficulties with these hermetic closures. The spacer in the sample cell serves only to reduce the total volume of the sample (1.5 ml). Either SS 316 or high temperature polyimide plastic (Vespel, Du Pont de Nemours & Co., Wilmington, Del.) was used to make this spacer. In order to permit easy cleaning of the cell, the upper glass part of the cell was removable. The sample was degassed using the usual freeze-pump-thaw technique and

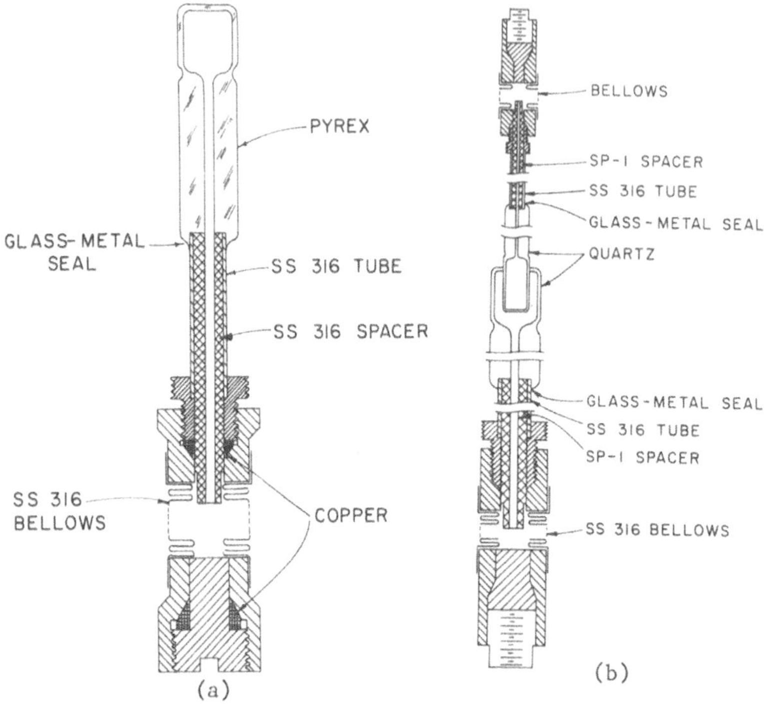

Fig. 8a. High pressure sample cell.
b. High pressure sample cell for use with external standard.

then sealed under vacuum in a glass ampoule. The ampoule was transferred to an inert gas dry box and the sample was filled into the sample cell. Repeated pressure cycles at high temperature had no detectable effect on the relaxation times of the liquid even when measuring T_1 longer than 30 sec. One detail about the design of the sample cell should be emphasized; the bulky metal parts with bellows should be at least 35 mm from the glass chamber. Otherwise, the homogeneity of the magnetic field over the sample volume is impaired.

Figure 8b shows a schematic drawing of the coaxial high pressure NMR sample cell which was used in our measurements with external reference[12].

The sample cell is simply inserted into the sample holder which is schematically depicted in Figure 9. The material used for its construction was either polytetrafluoroethylene (Teflon), nylon, or the high temperature polyimide plastic Vespel. We found

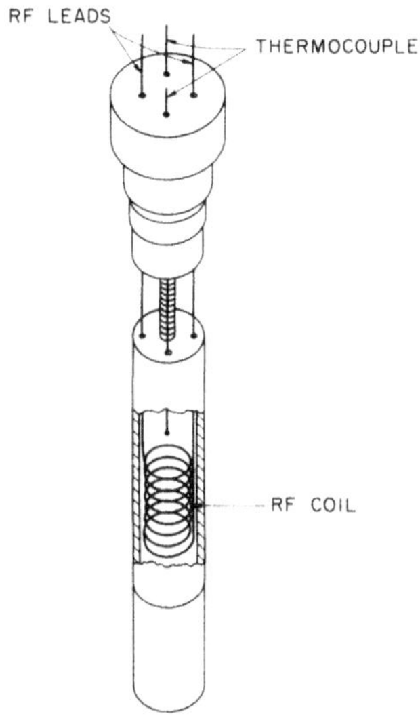

Fig. 9. Sample holder.

the mechanical stability and inertness of Vespel at high temperatures excellent for our purposes. The dimensions of the transmitter/receiver coil (No. 18 copper wire) were 20 mm length x 12 mm diameter, the number of turns depending on the operating frequency (3-20 turns). The actual sample volume was a cylinder of 12 mm length, 8 mm diameter so that a good homogeneity of the rotating magnetic H_1 field was achieved. For 10 kbar experiments, smaller sample and receiver coil sizes were used. We also found that the copper/constantan thermocouple located approximately 7 mm from the coil had no effect on the homogeneity of the magnetic field over the sample volume.

In our experiments on compressed gases[10,11] we used an NMR probe depicted schematically in Figure 10. The basic principle of the experimental set-up developed for NMR measurements on compressed supercritical water is illustrated in Figure 11.

In the case of NMR experiments on organic solids, one can use the above described high pressure system with either liquid or gas pressurizing medium. In our laboratory we used a sample container shown in Figure 12a. It is a separator with several stages where the carbon disulfide pressurizing fluid is excluded from the sample chamber using first a silicon grease seal of higher viscosity than the carbon disulfide. This is in turn separated from a sample seal by a tightly fitted Teflon plug. The Teflon plug is further spaced from the glass plug on the sample

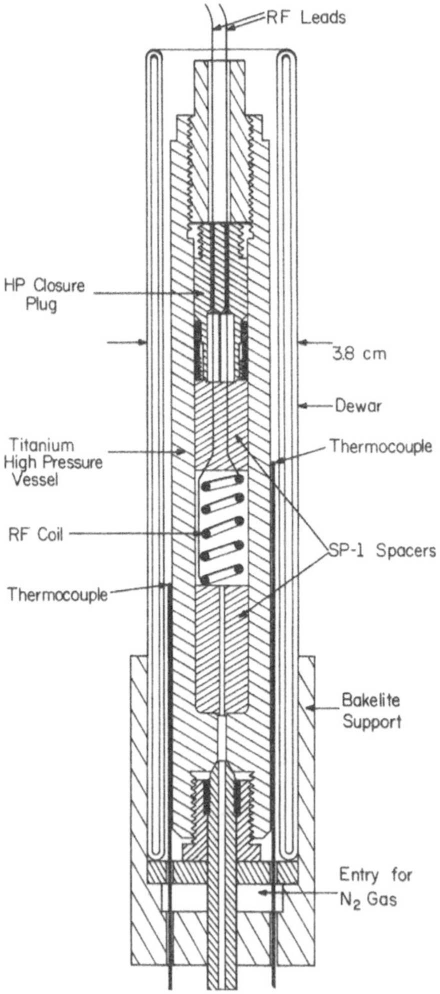

Fig. 10. NMR probe for work on compressed gases.

side so as to inhibit possible contamination of the sample. Precision ground, glass rods make up the major features of the separate chambers. The top of the sample chamber was left open so as to allow for the vacuum distillation of the sample and consequent sealing off.

In their extensive NMR studies of organic solids, Strange and collaborators[13] have employed sample cells depicted in Figures 12b and c.

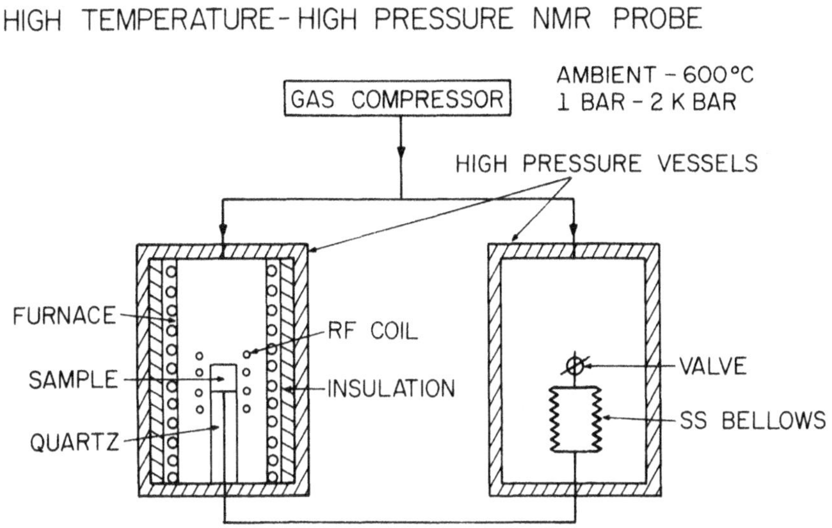

Fig. 11. General schematics of the NMR probe for measurements on compressed supercritical water.

The NMR experiments on solids at very high pressures up to 100 kbar require, of course, a different type of high pressure instrumentation. As an example, we present Figure 13 which shows the high pressure sample area within the Bridgmann anvil high pressure cell used by Vaughan et al.[14] in their wide line and pulsed NMR studies of solids.

iii) High resolution experiments

In the experimental setup discussed above, it has been possible to achieve a high homogeneity of the magnetic field over the sample volume even under extreme conditions of temperature and pressure[8]. Using the Fourier transform technique[15] enabled us to obtain high resolution spectra. A linewidth of 4-6 Hz at 60 mHz was achieved for 8 mm sample.

Yamada[16] described the use of HF etching techniques to raise the bursting pressure of glass tubing to the vicinity of 2 kbar. This approach has the advantage of using a standard, commercial NMR spectrometer for the NMR experiments at pressures up to 2 kbar. Another high pressure cell which may be used with standard equipment for high resolution NMR measurements under increased hydrostatic pressure up to 400 bar was reported by von Jouanne and Heidberg[17]. In their apparatus, a quartz capillary (1 mm i.d., 4 mm o.d., bursting pressures between 500 and 800 bar) was used

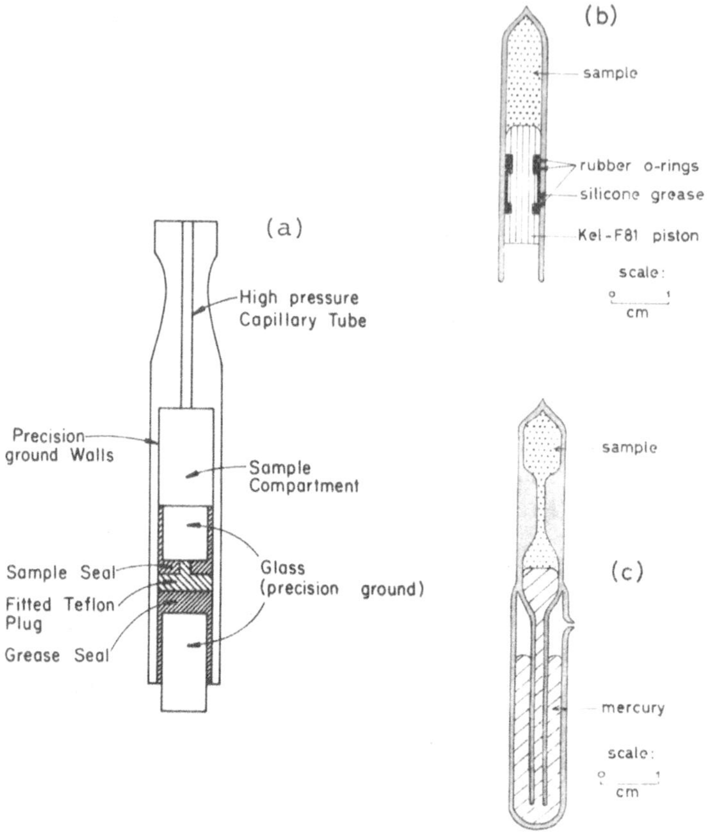

Fig. 12a. High pressure sample cell for work on organic plastic crystals.
b,c. Sample cells used by Folland, Ross and Strange; from ref. 13.

as a sample tube. This tube was fitted to a rotatable high pressure connection which was mounted above the magnet of the spectrometer in such a way that the capillary could be lowered precisely into the insert of the variable temperature probe. Hydrostatic pressures up to 350 bar were obtained with the aid of a pressure generator, the sample being separated from the pressure transmitting oil by a teflon membrane. The schematic drawing of the apparatus is shown in Figure 14.

Oldenziel and Trappeniers[18] have recently reported a high pressure vessel in which the sample can be rotated under high pressure. The resolution was 1 part in 10^8, the maximum pressure

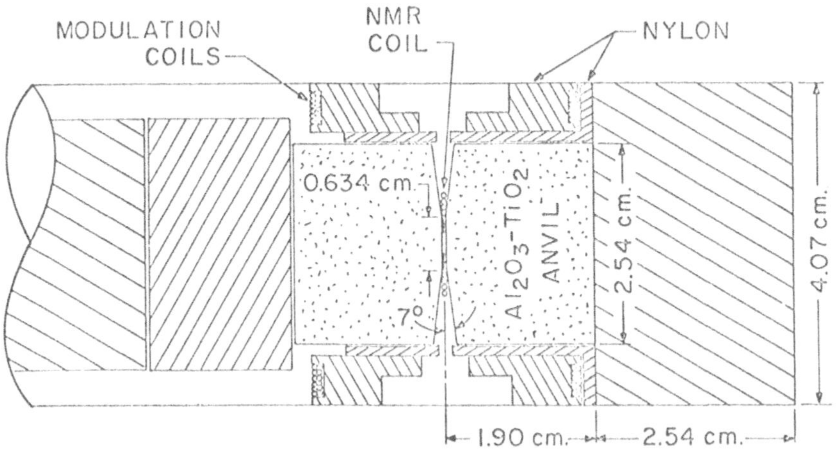

Fig. 13. High pressure sample area. From ref. 14.

2.5 kbar. A gas was used as a pressurizing medium and a small dc motor located inside the high pressure vessel was employed for spinning of the sample. The authors[18] mention that deterioration of the motor occurs rather rapidly under the influence of the pressurized gas and the motor was replaced frequently.

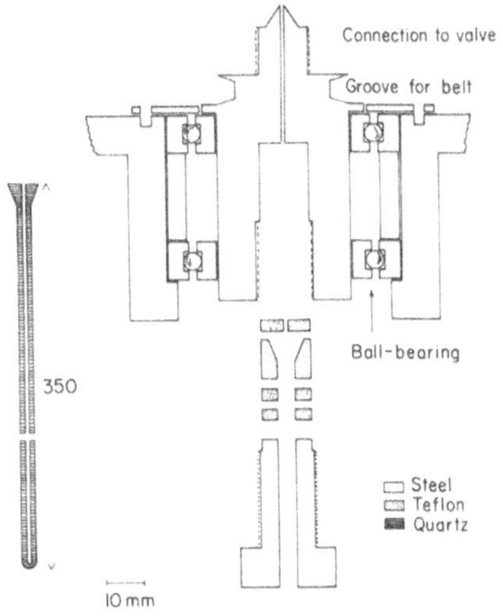

Fig. 14. Quartz capillary and rotatable high pressure connection. From ref. 17.

3. APPLICATIONS OF NMR AT HIGH PRESSURE

i) Molecular motions in liquids

Since 1970 we have performed NMR relaxation experiments on various liquids in a systematic way and the results are summarized in Table II. Three main classes of liquids were of interest. The

Table II. High pressure NMR studies of liquids.

Molecule	D	τ_J	τ_θ	η	ρ
Tetramethylsilane	x	-	-	x	x
Benzene	x	-	x	x	x
Neopentane	x	-	-	x	x
t-Butyl chloride	x	-	-	-	x
Pyridine	x	-	-	x	x
Monosubstituted benzenes	-	-	x	x	x
$F\emptyset D_5$	-	x	x	x	x
$CFCl_3$	x	x	-	x	x
SF_6	x	x	-	-	x
C_4F_8	x	x	-	x	x
CF_4	x	x	x	-	x
CF_4+inert gas mixtures	-	x	x	x	x
Methyl iodide-d_3	-	-	x	x	x
Methanol	x	-	x	x	x
t-Butyl alcohol	x	-	-	-	x
H_2O, D_2O	x	-	x	x	x
D_2O-electrolytes	x	-	-	x	x
Dioxane-H_2O (D_2O)	-	-	x	-	x

reorientational motions and self-diffusion were studied in several molecular liquids composed of molecules of spherical shape. The behavior of the angular momentum correlation time, τ_J, was of main interest in studies of liquids containing the ^{19}F nucleus, because the spin-rotation interactions very often represent the dominant relaxation mechanism for the fluorine nuclei. The last class of liquids investigated are the hydrogen bonded liquids including water and heavy water.

In this section, I shall discuss the results of several specific studies which will illustrate the type of information one may obtain from these NMR relaxation experiments at high pressure.

Our work on the effect of pressure on the overall and internal rotation in liquid benzyl cyanide[19] will be presented as the first example. The pressure dependence of the reorientation rates is valuable in determining the reorientational mechanism since the environment of the molecule and the forces on the molecule can be significantly altered without changing the kinetic energy of the molecules.

In our work[19] the deuterium spin-lattice relaxation times of the two selectively deuterated benzyl cyanides, benzyl-4-d_1 cyanide (I) and α,α-dideuteriobenzyl cyanide (II), were measured as a function of pressure. The relaxation time of I gives a measure of

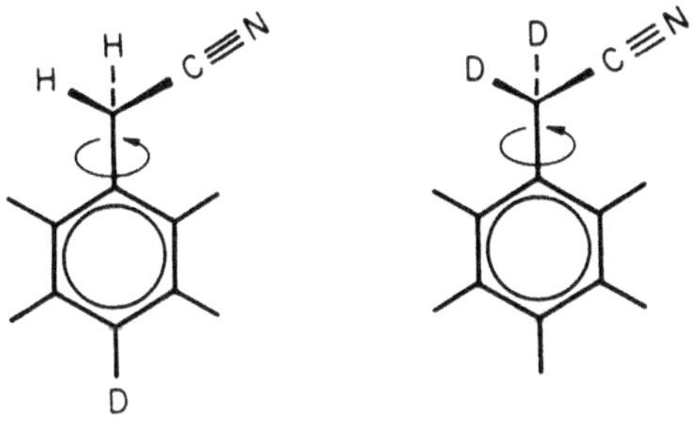

I II

the overall reorientation of the molecule, while the relaxation time of II reflects both the overall molecular reorientation and the internal rotation of the CD_2CN group. Using reported deuteron quadrupole coupling constants, the experimental spin-lattice relaxation times, T_1, are analyzed in terms of the rotational diffusion constants for overall molecular, D_M, and internal, D_I, rotation of the CD_2CN group. The calculated activation volumes for

D_M and D_I are given in Table III, which shows that the activation volumes ΔV^* for D_M and D_I are identical at 30° but at 150° the

Table III. Activation parameters.

Quantity	ΔE^* kcal/mol	ΔV^* (30°), cc/mol	ΔV^* (150°), cc/mol
D_M	3.31	14.4	11.2
D_I'	2.5	14.4	5.5
η/T	3.61	16.2	....

activation volume for the overall molecular diffusion is much larger than that for D_I'. This is analoguos to our earlier finding[6] in toluene-d_8 where at 100° the $\Delta V^*(D_M)$ was three times as large as $\Delta V^*(D_I)$. Again, these results reflect the difference in frictional torques connected with the overall and internal rotation and also indicate the presence of inertial effects influencing the rotation at higher temperatures.

Much recent evidence indicates that many physical properties of liquids may be determined to a large extent by the size and shape of the atoms or molecules composing them. In particular the molecular dynamics studies by Alder and collaborators[20] have elucidated many aspects of the static and dynamic behavior of hard sphere fluids and contain many of the basic physical ideas later developed by other authors. This picture of fluids is based on the idea that when the liquid density is high (approximately twice the critical density), the molecules are so closely packed that the forces between particles arise almost exclusively from the harsh short-ranged repulsive forces between nearest neighbors. Using this idea the static structures of several liquids have been successfully calculated by depicting them as hard sphere systems and neglecting the slowly varying attractive forces.

Based on a successful prediction of the static properties of liquids, we thought it worthwhile to find the extent to which their dynamic properties can be explained in terms of hard sphere systems. A major problem in this type of analysis, however, is to determine which hard sphere system (i.e., what hard sphere diameter) to associate with a particular liquid.

The purpose of our studies of self-diffusion in liquids was to determine the hard sphere diameter from the density dependence of the self-diffusion coefficients and to then compare the experimental self-diffusion coefficients with those predicted theoretically. In the determination of the effective hard sphere diameter, σ, and of its temperature dependence, we used the expression for the self-diffusion coefficient for a smooth hard sphere obtained by Dymond[21] by fitting the molecular dynamics results by Alder et al.[20] in their study of transport properties of the hard sphere fluid.

$$D_{SHS} = 0.1901\ (kT/m)^{\frac{1}{2}}\ \sigma[(\sqrt{2}/\rho\sigma^3) - 1.384]\ , \qquad (4)$$

where ρ is the number density and all the other symbols have their usual meaning. The experimental data was smoothed by a quadratic least-squares fit and subsequently a value for the slope, $(\partial \ln D/\partial \ln \rho)_T$ was computed at an average number density. It is easily shown that this slope is only a function of $\rho\sigma^3$. Since the density is known, σ can be extracted from a tabulated function of $\rho\sigma^3$.

As an illustrative example we give Table IV which lists the hard sphere diameter for several molecular liquids. The hard sphere diameter is found to decrease with increasing temperature.

Table IV. Hard sphere diameters obtained from self-diffusion data.

Tetramethyl silane		Benzene		$CFCl_3$	
T^oC	σ(Å)	T^oC	σ(Å)	T^oC	σ(Å)
25	5.68	30	5.13	68	5.03
50	5.67	75	5.10	106	4.96
75	5.65	120	5.08	187	4.90
100	5.63	160	5.05	-	-

Physically, this corresponds to the fact that the repulsive part of the potential has a finite slope. Assuming that at the turning point in a collision the translational kinetic energy is converted to potential energy, then the molecules with higher kinetic energy have higher potential energies corresponding to a smaller σ on the intermolecular potential curve.

In contrast to previous studies, Chandler[22] has considered the coupling between rotational and translational motions of molecules and has shown that the diffusion coefficient D of a real fluid is best approximated by the self-diffusion coefficient of a rough hard sphere fluid, D_{RHS},

$$D \approx D_{RHS} = A\, D_{SHS} = A \frac{kT}{m} \tau_E , \qquad (5)$$

where D_{SHS} is the diffusion coefficient for a smooth hard sphere fluid and A is a constant less than 1, whose magnitude reflects the effect of the translational-rotational coupling. The constant A equal to 1 would correspond to a perfectly smooth hard sphere fluid with no coupling. Our earlier experimental studies on the coupling between rotational and translational coupling in molecular liquids represent strong supporting evidence for the rough hard sphere model. In order to calculate the theoretical D_{RHS} we used the expression for D_{SHS} given in Eq. (4). Using the effective hard sphere diameter for benzene (Table IV) we calculated D_{SHS} and D_{RHS} and the agreement between theory and experiment[23] is shown in Figure 15. The constant A was found to be 0.77 ± 0.02.

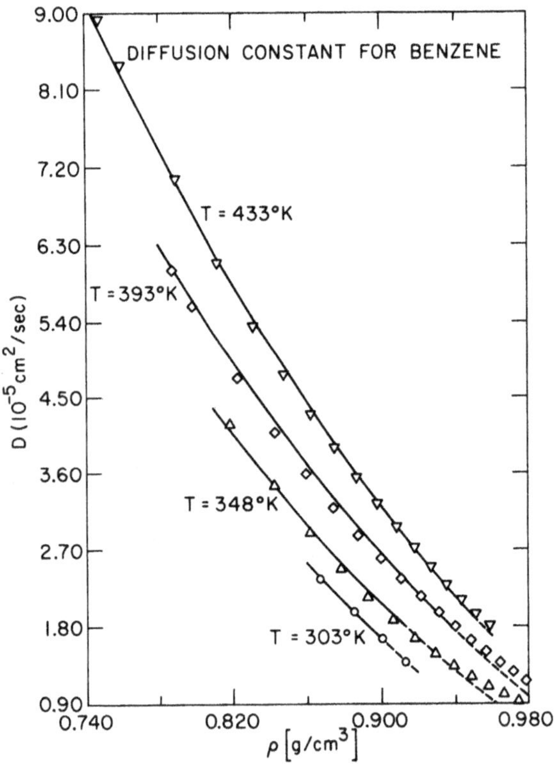

Fig. 15. Comparison of experimental with predicted self-diffusion coefficients for liquid benzene.

Figure 15 shows that the rough hard sphere model describes quite well the experimental self-diffusion data for benzene. Analogous results were found for all the molecular liquids as listed in Table II. However, a closer inspection of this figure reveals that at densities above $\rho\sigma^3 > 0.93$ there is a systematic deviation of the experimental and theoretical D's. At these densities the hard sphere fluid is metastable. In order to illustrate better this behavior of D at high packing fractions we include Figure 16, which gives Δ as a function of the packing fraction ξ.

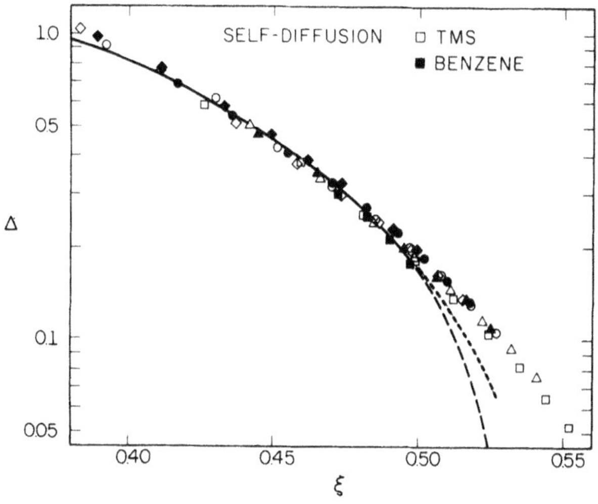

Fig. 16. Plot of function Δ vs. packing fraction ξ.

We call the function Δ a "reduced" self-diffusion coefficient defined as follows

$$\Delta = 10^{-3}\, D(M/T)^{\frac{1}{2}}/\sigma A\ , \qquad (6)$$

where all the symbols have previously been defined. It is clear that both liquids show the same behavior with density all the way to the highest packing fractions. The molecular dynamics results are shown by the full line and the extrapolation of the Dymond[21] is represented by the dashed line. For comparison we also include the extrapolation of the quadratic fit as reported by Chandler[22].

Let us discuss the possible reasons for the interesting behavior of the self-diffusion coefficients at high densities.

First, there is a distinct possibility that the extrapolated values from the molecular dynamics calculations may be in serious error at high densities. Secondly, the effective hard-core diameters can become a function of density at the high packing fractions. This possible dependence of σ upon density may have its origin in the following. One has to realize that the rough-hard sphere model is still a qualitative model, and that the real liquids we deal with in this study could perhaps be represented as the soft sphere liquid. A softness of the potential may have a significant effect on the self-diffusion coefficients because the relationship between the hard spheres and the soft spheres is not well understood for the transport processes. The density dependence of the effect hard core-diameter could also arise from orientational ordering at high packing fractions. It is interesting to note that Robb and Drickamer[24] have proposed in their study of compressed gases that the decrease of the effective hard-core diameter at higher densities results from the orientational effects due to non-spherical shape of the molecules.

The conclusion of our studies of self-diffusion can be summarized as follows. As far as our results are concerned, the predictions of the rough hard sphere model agree well with the experimental data. Our data confirm the molecular dynamics calculations for diffusion and shear viscosity of a hard sphere fluid and clearly show that many body correlations affect in a major way the self-diffusion coefficients and cause deviations from predictsion based on the simple Enskog theory. At intermediate densities the self-diffusion coefficient of a liquid is enhanced due to long persistence of velocity currents (vortex effect), whereas at high densities the self-diffusion coefficient is much lower than the Enskog self-diffusion coefficient. This effect can be ascribed to back-scattering caused by the high probability of reversal of velocity of a particle upon collision with its nearest neighbor in dense systems. The experimental data enabled us to calculate the effective hard sphere diameters for the various liquids studied and also obtain their temperature dependence. For all liquids we found that at high packing the predictions based on a hard sphere model deviate significantly from the observed values of self-diffusion coefficients and shear viscosities.

Since our experiments provided self-diffusion coefficients and shear viscosities of various liquids over a wide range of temperatures and densities it was possible to test the validity of the hydrodynamic Stokes-Einstein equation on a molecular scale[23]. This relation states that the product of the diffusion coefficient and the viscosity is a constant related to the diameter of the spherical particle moving through a fluid. The many body correlations which led to a decrease of diffusion below the Enskog theory are compensated for by correlations which enhance viscosity by nearly that much. Over a wide range in volume and temperature the Stokes-Einstein equation is remarkably precise and shows that the slipping boundary condition is satisfied well for most of the molecular liquids studied. It is interesting to note that this is true in spite of the fact that the velocity autocorrelation function is not exponential as required by the application of the Stokes theory. In Figure 17 the diffusion and viscosity data for benzene are presented as an illustrative example[23] of the validity of the Stokes-Einstein equation over a wide range of temperatures and pressure.

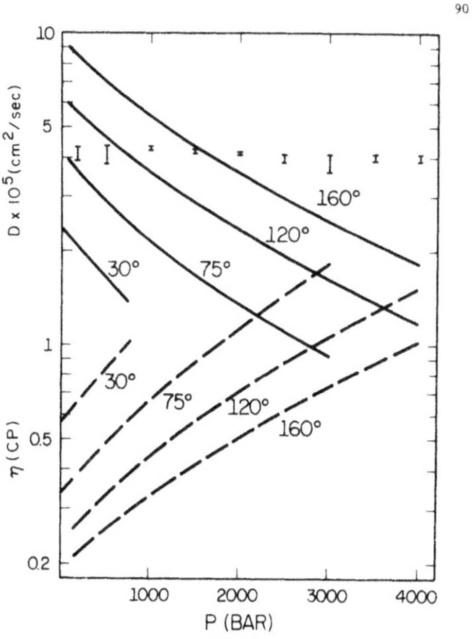

Fig. 17. Diffusion, viscosity and Stokes-Einstein relation for benzene. The solid lines show the experimental values for D and the dashed lines denote viscosity values. The vertical bars indicate the range of the calculated value of $kT/\pi D\eta a$ from 30°C to 160°C.

Angular momentum relaxation of small quasispherical molecules as determined from nuclear magnetic resonance relaxation times has contributed to our understanding of molecular dynamics both in the liquid and the gaseous states[4]. Let us repeat the definition of the angular momentum correlation time, τ_J,

$$\tau_J = \int_0^\infty dt < \vec{J}(0)\, \vec{J}(t) > / < J^2 > , \qquad (7)$$

where $\vec{J}(t)$ is the angular momentum of a molecule at time t, and $<\ldots>$ denotes an ensemble average. Physically, τ_J is the average time a molecule needs to lose memory of its initial angular momentum $\vec{J}(0)$. One would expect τ_J to be closely related to the time between collisions which randomize the initial angular momentum. In the NMR relaxation experiment τ_J can be directly determined from the spin-lattice relaxation time, T_1, of a nucleus which is relaxed by the spin-rotation interaction mechanism. For a spherical molecule, the spin-lattice relaxation rate arising from spin-rotation interactions can be written as

$$\left(\frac{1}{T_1}\right)_{SR} = \frac{8\pi^2 IkT}{\hbar^2} C_{eff}\tau_J , \tag{8}$$

where I is the moment of inertia, C_{eff} is the effective spin rotation interaction constant, and other symbols have their usual meaning.

If the events which modulate the magnetic field at the fluorine nuclei may be approximated as an uncorrelated sequence of binary collisions, Chandler[25] has shown that

$$\tau_J^{-1} = a(T)\ \tau_E^{-1} , \tag{9}$$

where τ_E^{-1} is the Enskog collision frequency for hard spheres and a(T) is a constant reflecting the efficiency with which angular momentum is transferred at a collision. This quantity, a(T), may be slightly temperature dependent but should be independent of density.

The Enskog collision frequency is given by the expression

$$\tau_E^{-1} = \frac{8}{3} (\pi kT/m)^{\frac{1}{2}} \sigma^2 \rho\chi , \tag{10}$$

where m is the molecular mass, σ is the hard sphere diameter, ρ is the number density, and χ is the Enskog correction factor related to the compressibility factor Z. In this analysis the Carnahan-Starling approximation for the equation of state of hard spheres was used. This allows χ to be expressed analytically in terms of the packing fraction ξ, where

$$\xi = \rho\pi\sigma^3/6 . \tag{11}$$

It should be noted that Eq. (10) gives the collision frequency for a system of hard spheres and may be expected to approximate the collision frequency of an experimental dense fluid only if molecules are so closely packed that nearest neighbor interactions arise exclusively from harsh repulsive forces. To obtain the best estimate of the hard sphere diameters we used either graphical or computer procedures.

From these hard sphere diameters, one calculates the Enskog relaxation times and by comparison with the experimental τ_J's one obtains the parameter a(T). For example, this parameter a(T) = 0.6 for SF_6.

The quantity a(T) is related to the "roughness" of the molecule and for a spherical molecule, such as SF_6, should be bounded by the perfectly smooth hard sphere and the perfectly rough hard sphere as described in ref. 25. Following Chandler[25], $0 < a(T) < 0.82$, where the zero corresponds to the perfectly smooth case and 0.82 to the perfectly rough hard sphere with the same moment of inertia as SF_6. Our value of 0.6 falls well within this range and indicates that SF_6 cannot be approximated as a smooth hard sphere system as far as angular relaxation is concerned. This value is significantly lower, however, than the 0.8-0.98 value previously reported[27] for the nonspherical CCl_3F molecule. Using the value of a = 0.6 for SF_6 we calculated theoretical spin-lattice relaxation times from the rough hard sphere theory and compared them with the experimental values in Figure 18.

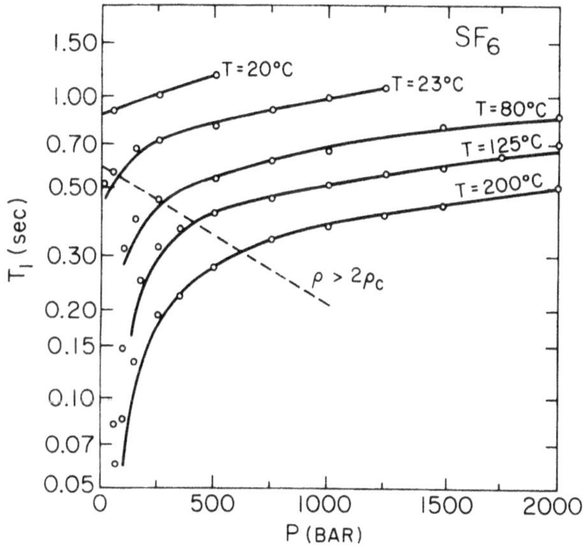

Fig. 18. Temperature and pressure dependence of the ^{19}F spin-lattice relaxation times in SF_6. The full lines represent theoretical values calculated from the rough hard sphere model. O denote experimental values.

Good agreement is obtained for the isothermal pressure dependence of all temperatures for the region $\rho > 2\rho_c$, which at most of the temperatures corresponds to a substantial density variation (1.5 to 1.90 g/cm^3). At temperatures above the critical temperature where lower fluid densities can be attained, the experimental relaxation times are larger than the predictions of the hard sphere theory. This indicates that the actual collision frequency is greater than that of a hard sphere fluid of the same density and is of course the behavior expected when intermolecular attractions can no longer be neglected and the hard sphere approximation breaks down.

For several additional molecules such as $CFCl_3$, C_4F_8, and CF_4, we found that the rough hard sphere theory predicts well the experimental behavior of the angular momentum correlation time in the dense fluid region. The important general result from these studies is that in contrast to the behavior of the self-diffusion coefficient and shear viscosity, no correlation effects were observed for the angular momentum correlation times.

However, the finding of deviation between the theoretical and experimental T_1 at densities lower than twice the critical density, led us to experiments[10] where we measured the T_1 in CF_4 over a very wide density range extending from dilute gas region to the dense fluid region at $\rho > 2\rho_c$. There were two main purposes of this study[10]: i) to test the applicability of the rough hard sphere (RHS) model at densities lower than $2\rho_c$; ii) to attempt to account for the effect of attractive forces which may become significant at lower densities. The NMR ^{19}F spin-lattice relaxation times, T_1, have been measured in CF_4 as a function of pressure and temperature over the wide density range $0.28 \leq \rho/\rho_c \leq 2.5$ and the temperature range $1.2 \leq T/T_c < 1.64$.

Using the hard sphere diameters obtained at densities above $2\rho_c$, we can calculate the ratio τ_E/τ_J over the complete density range. A plot of τ_E/τ_J vs. density along an isotherm is shown in Figure 19 and shows an interesting departure from the rough hard sphere model. Above twice the critical density, τ_E/τ_J is a constant within experimental error. However, at densities less than $2\rho_c$, there is a pronounced increase in this ratio. This breakdown of the RHS model at lower densities may be attributed to the increasing importance of intermolecular attractive forces. At high densities the long range attractive forces are negligible because of the "screening" of the slowly varying attractive potential by the short range and quickly varying repulsive forces. As the density is lowered, the screening of the attractive forces becomes less effective. The fact that τ_E/τ_J increases rather than decreases with decreasing density makes sense when one considers the effect of attractive forces on the radial distribution function at contact, $g(\sigma)$, which is equivalent to χ of Eq. (10). At high densities, since the attractive forces are negligible, $g(\sigma)$ can be well approximated by its value for a hard sphere fluid $g_{hs}(\sigma)$. At lower densities where attractive forces are more important, the true $g(\sigma)$ should be greater than that for a hard sphere fluid. This would tend to make τ_E/τ_J larger than that predicted by the RHS model.

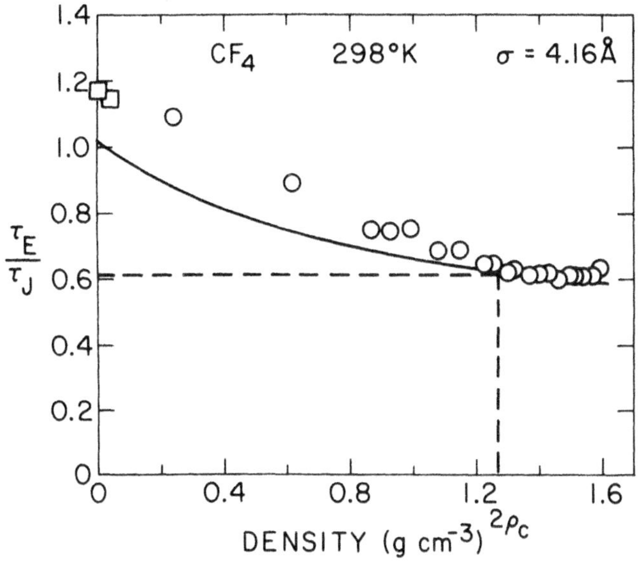

Fig. 19. Density dependence of τ_E/τ_J for CF_4. The dashed line represents average value of $a(T) = \tau_E/\tau_J$ for dense fluid region ($\rho > 2\rho_c$). The full line is predicted by using the rough hard sphere model modified to account for the attractive forces.

Another indication that this breakdown in the rough hard sphere model is caused by attractive forces is the variation of τ_E/τ_J with temperature at constant density. With increasing temperature, attractive forces will become less important and the deviations from the model should become smaller at higher temperatures. This seems to substantiate our conjecture that it is the attractive forces between the molecules causing the deviations from the RHS model. To account for the attractive forces we proposed a modified hard sphere model based on optimized cluster theory[28] with the parameter $b(\rho,T) = a(T)\exp(C_L(\sigma))$, where $a(T)$ is the limiting value of τ_E/τ_J for $\rho > 2\rho_c$, and $C_L(r)$ is the renormalized intermolecular potential based on Lennard-Jones potential for CF_4. By including the effect of attractive forces the modified rough hard sphere model accounts well for the experimental a(T) behavior over the entire density range. The theoretical prediction of τ_E/τ_J is also shown in Figure 19.

There are, of course, many important questions which should be answered in order to understand better the angular momentum relaxation in fluids. In the study[11] which I briefly discuss, we addressed ourselves to the following questions in particular: (1) Can an extension of the RHSM to angular momentum relaxation in binary mixtures successfully describe experimental results? Such an extension was shown to be valid for diffusion in binary mixtures of Lennard-Jones molecules in molecular dynamics calculations[29]. For molecular liquid mixtures, experimental Rayleigh light scattering measurements yielding mutual diffusion coefficients were successfully interpreted in terms of the RHSM for diffusion[30]. (2) How sensitive are the deviations from the RHSM for $\rho > 2\rho_c$ to the magnitude of the intermolecular attractive forces? (3) What are the factors which determine the magnitude of the parameter a(T) in the RHSM?

To answer these questions we have performed NMR experiments on mixtures of CF_4 with argon and with neon. The ^{19}F spin lattice relaxation times, T_1, were measured[11] for 0.1, 0.25 and 0.6 mole fractions of CF_4 in Ar and Ne from 273° to 373°K and pressures up to 2 kbar. The spin-rotation interactions provide the dominant relaxation mechanism for the fluorine nuclei. Typical experimental data are shown in Figure 20 which plots T_1 for CF_4 infinitely dilute in Ar at different temperatures.

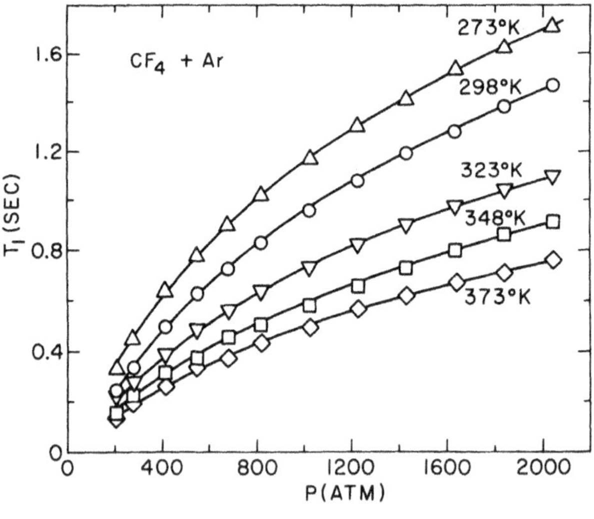

Fig. 20. Pressure dependence of T_1 for CF_4 infinitely dilute in Ar at different temperatures.

The experimental results were discussed from two perspectives: CF_4 infinitely dilute in the inert gases and T_1

s a function of mole fraction. The main conclusions of our theoretical interpretation of the experimental data were as follows: First, in the dense fluid region for $\rho > 2\rho_c$ the rough hard sphere model (RHSM) of liquids can be extended to binary mixtures. Second, one can modify RHSM to account for the effect of attractive forces at lower densities following the procedure employed in the case of pure CF_4. (For detailed discussion see the original study[11].) A typical plot of $b_{12} = (\tau_J \Gamma_{12})^{-1}$ vs. ρ/ρ_c, where b_{12} is the efficiency of angular momentum transfer, and Γ_{ij}, the collision frequency, is given in Figure 21 for CF_4 infinitely dilute for argon and neon. For comparison, the results for pure CF_4 are also included in this figure. These plots suggest that the efficiency of angular momentum transfer is constant at densities higher than $2\rho_c$ in the binary mixtures.

The trend of $b_{12} = (\tau_J \Gamma_{12})^{-1}$ with density for CF_4 infinitely dilute in argon is similar to that reported[10] for $b_{11} = (\tau_E/\tau_J)$ in pure CF_4, though not as pronounced. This trend of b_{12} increasing with decreasing density, is very small for the case of CF_4 infinitely dilute in neon. As shown in Figure 21, b_{12} varies only slightly. The changes are within experimental error. The point of interest here is that the magnitude of the deviations from the RHSM (where b_{11} and b_{12} are independent of density and equal to a_{11} and a_{12} respectively) parallels the magnitude of attractive forces between collision partners.

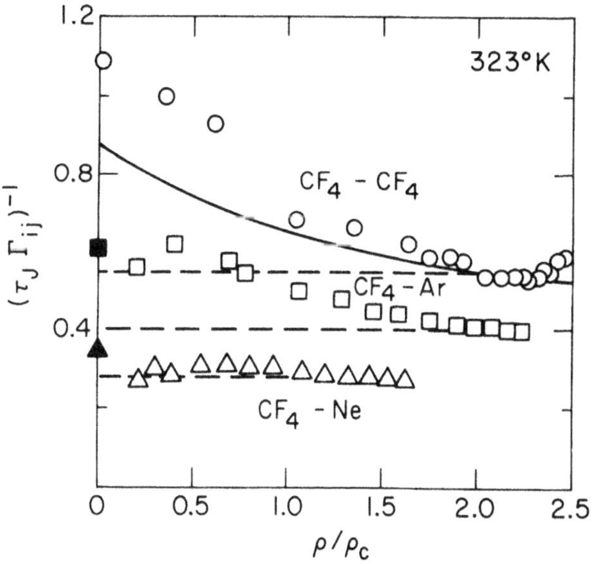

Fig. 21. Density dependence of $(\tau_J \Gamma_{ij})^{-1}$ for CF_4 infinitely dilute in Ar and Ne. Temperature T = 323°K.

Another interesting result following from Figure 21 are the relative magnitudes of the parameter a_{ij} for the different collision partners. It is readily seen that a_{ij} increases with increasing reduced mass of the collision partners. In addition, a_{ij} increases with increasing σ_{ij} of the collision partners. Of course, mass and size are interrelated making it difficult to attribute the trends in a_{ij} to any single parameter. However, we feel these trends are physically reasonable.

Before going into the discussion of specific experiments on water and heavy water it may prove useful to point out the main reasons why water behaves so differently from the so-called normal liquids. The current picture of normal, simple molecular liquids is based on the idea that when liquid density is high (approximately twice the critical density), the molecules are so closely packed that forces between particles arise almost exclusively from the harsh, short-ranged repulsive forces between nearest neighbors. One can neglect the slowly varying attractive forces. A very different situation exists in liquid water, for which the above simple picture is definitively not adequate. The presence of a well developed, random hydrogen-bond network is responsible for the anomalous properties. The most characteristic structural feature of liquid water and heavy water is the local tetrahedral environment of each water molecule beyond which there is a randomized, imperfect, space-filling network of hydrogen bonded molecules.

During systematic studies[4] of density effects on the dynamic structure of liquids, we observed, for water and heavy water, that the most interesting behavior of various transport and relaxation properties occurs at temperature between 10°C and 30°C. In several studies[7,31] we took advantage of the phase diagram of water and heavy water (see Figure 3) and measured various transport properties of these liquids down to temperatures of -15°C. Nuclear magnetic resonance spin-lattice relaxation times, T_1, self-diffusion coefficients, and shear viscosities were measured as a function of pressure in the temperature interval -15°C to 30°C. The pressure range was always adjusted according to the phase diagram in such a way that water always stays in the liquid phase. Before we discuss the results on water let us mention what happens with various motions in a normal liquid with compression. Simply, increased packing slows significantly all motions in the liquid. In contrast, spin-lattice relaxation time[32], T_1, self-diffusion coefficients[33], and fluidity, in water and/or heavy water go through a maximum with initial compression and only a further increase of pressure begins to restrict motional freedom and consequently results in shorter T_1 and lower values of diffusion coefficients and fluidity.

The anomalous motional behavior of water molecules with initial compression can be qualitatively interpreted in terms of a simple physical picture based on changes in the random hydrogen bond network. The most characteristic structural feature of liquid water is the local tetrahedral environment of each molecule beyond which there is a randomized imperfect space filling network of hydrogen bonded molecules. As it is clear from the phase diagram of water the boundaries of the measurements were, at low pressure, ice I and, at high pressures, ice V and also ice VI. The most important structural differences between ice I and the high pressure ices are the distorted hydrogen bond angles and the close approach between the nonhydrogen-bonded neighbors. By compressing liquid water in the selected temperature range, we affect the hydrogen bond network and gradually go from optimal tetrahedral order towards a more compact packing arrangement. There is a competition in water between the tendency of strongly directional forces to build an open hydrogen bond network and the tendency for external pressure to pack the molecules more efficiently together. Since the process for self-diffusion, shear viscosity and reorientation of the water molecules necessitates breaking and reforming of hydrogen bonds, one can expect that it is easier to break an already bent hydrogen bond than an undistorted one. Initial compression, therefore, increases motional freedom of molecules and causes diffusion, fluidity, and T_1 to go through a maximum with increasing pressure under isothermal conditions (Figure 22). Further compression slows down

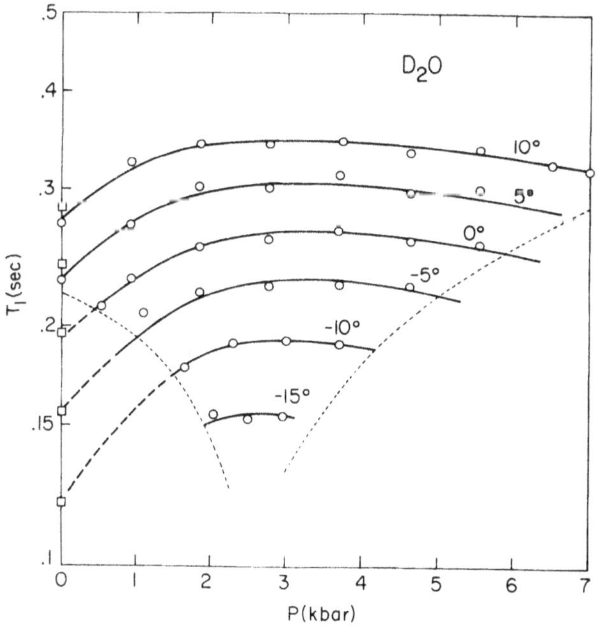

Fig. 22. Pressure dependence of the deuteron T_1 in liquid D_2O. Dotted lines denote phase boundaries.

all the dynamic processes because of a more compact packing of the water molecules. Under high compression the repulsive hard-core interactions begin to compete strongly with the directional forces that are responsible for the open structure at low pressures and low temperatures (< 30-40°C).

In order to illustrate the general effect of pressure and temperature on the dynamic properties of water, we present Figure 23, which shows the effects of pressure on self-diffusion

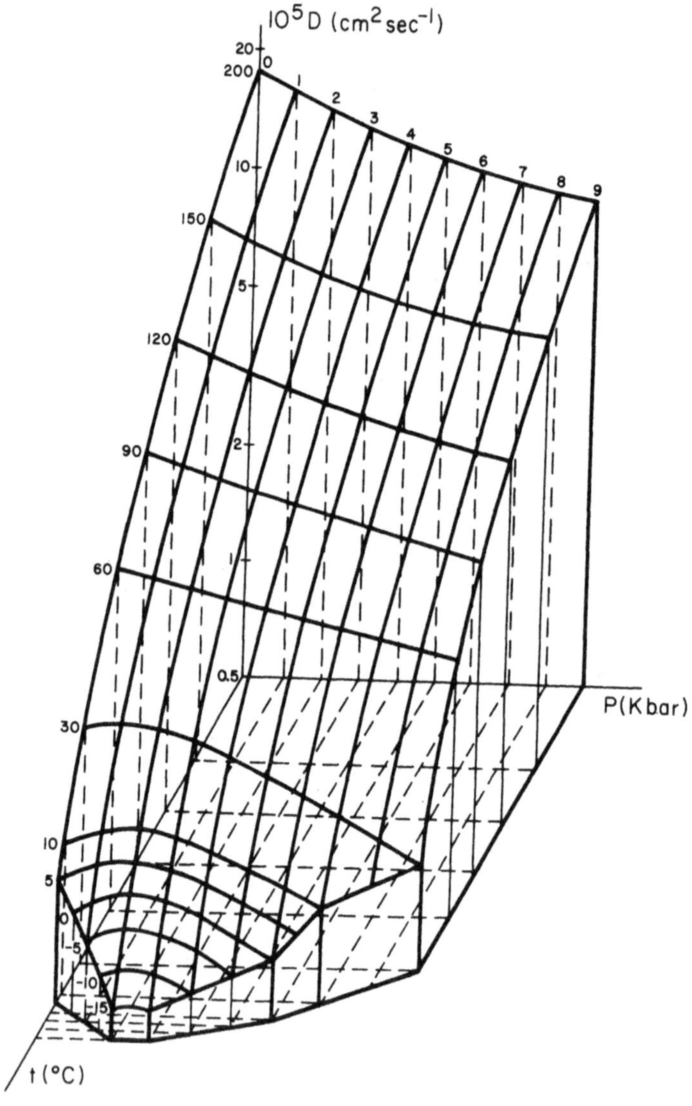

Fig. 23. Self-diffusion relief map as a function of temperature and pressure in liquid D_2O.

in liquid heavy water[7] in the temperature range -15°C to 200°C. This behavior of self-diffusion coefficient, D, vs. pressure, P, at different temperature is characteristic for D_2O. We emphasize that analogous plots can be drawn for any dynamic property of D_2O and/or H_2O. The important finding of general validity that is clear from Figure 23 is that both temperature and pressure have a parallel effect on the dynamic structure of water -- increases in P and T promote motional freedome at low temperatures and pressures. On the other hand, above ∿2 kbar and at temperatures above ∿40°C, pressure and temperature increases have opposite effects on motional freedom, as is the case for normal liquids. It is understandable that compression of a normal liquid slows down the motional processes owing to increased packing and stronger repulsive interactions. One may conclude that at high temperatures and under high compression the behavior of water resembles that of normal molecular liquids.

Another result of these studies is related to the applicability of the hydrodynamic equations at the molecular level. Several studies of dielectric and NMR relaxation in liquid H_2O and D_2O indicated that the Debye equation describes well the temperature dependence of the reorientation of water molecules at atmospheric pressure. However, the results of our high pressure studies show convincingly that the Debye equation fails to account for the density effects on reorientation of water molecules. One has to conclude that the success of the Debye equation for reorientation of water at 1 bar is accidental. Nevertheless, it is remarkable to have such a coincidence and it remains an experimental fact that both the proton and deuteron relaxation rates in water are linearly dependent on η/T, where η is the viscosity and T temperature (Figure 24).

Another important hydrodynamic equation is the Stokes-Einstein equation which relates the self-diffusion coefficient to shear viscosity. Both molecular dynamics calculations and experiments have shown that this equation is well obeyed in the slipping boundary condition over a wide range of densities and temperatures in simple liquids approaching the model of hard sphere fluids. In view of the strongly directional forces in water it is not surprising to find that the Stokes-Einstein equation is not valid for liquid water (Figure 25). One may summarize the main conclusions of the experimental studies of liquid water and heavy water at high pressures as follows: (1) All dynamic properties investigated behave anomalously with initial compression in the temperature range -15°C to ∿40°C. (2) Only under higher compression does temperature and pressure have opposite effects on the transport and relaxation properties of H_2O and D_2O. (3) The Debye, and the Stokes and Einstein hydrodynamic equations are not valid for liquid H_2O and D_2O. (4) Plots of relaxation rate ($1/T_1$) vs. η/T are linear under isochoric and isobaric conditions. (5) Compression reduces the coupling between the rotational and translational motions.

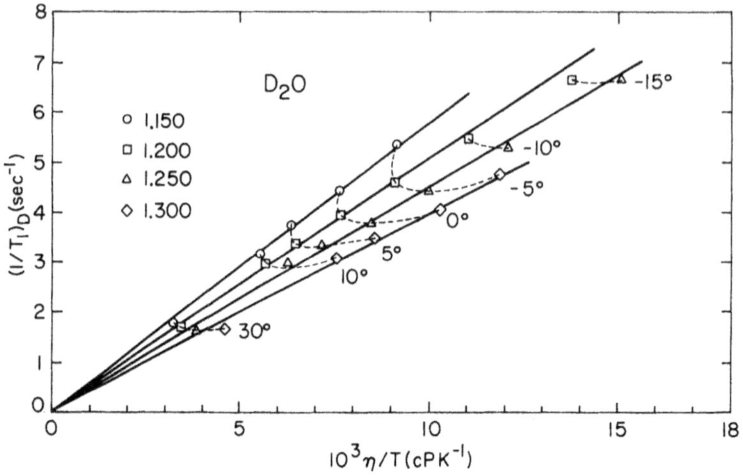

Fig. 24. Dependence of the observed deuteron relaxation rate $(1/T_1)_D$ upon η/T in liquid D_2O at several constant densities.

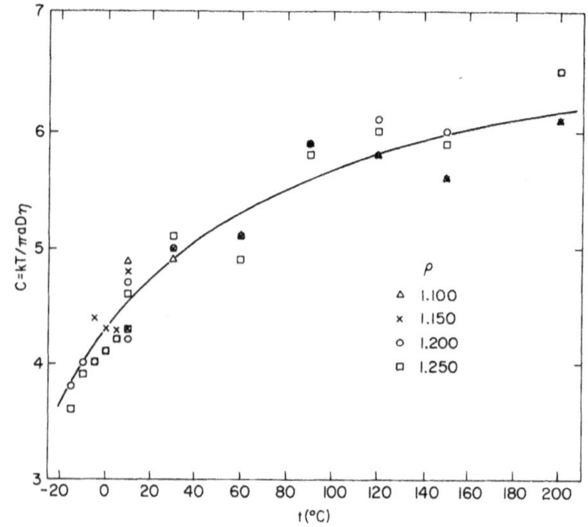

Fig. 25. Temperature dependence of the constant $c = kT/\pi D\eta a$ in the Stokes-Einstein equation in liquid heavy water at several constant densities.

ii) High resolution spectroscopy

Few investigations of the high resolution spectra of liquids at high pressure have so far appeared in the literature. Main applications of the high resolution NMR spectroscopy in liquids at high pressure include studies of the effects of pressure (density) on: a) chemical shifts; b) spin-spin coupling constants; c) relaxation of chemically shifted nuclei; and d) chemical exchange processes.

The purpose of our study[34] of chemical shifts in benzotrifluorides using pressure as a variable was twofold. First, to provide additional experimental evidence which would allow us to explain the origin of the large temperature dpeendence of the ^{19}F chemical shift in benzotrifluoride and also to determine the relative role of the intermolecular and intramolecular effects for the fluorine shift in ortho-chlorobenzotrifluoride. The second, more general, aspect of this work was to illustrate the usefulness of the Fourier transform NMR technique in liquids at high pressure in studying intermolecular and intramolecular interactions as we separate the effect of temperature and density on chemical shifts.

Using the Fourier transform technique the ^{19}F chemical shifts of benzotrifluoride (BTF), ortho-chlorobenzotrifluoride (o-CℓBTF) and p-chlorobenzotrifluoride (p-CℓBTF) were measured as a function of pressure up to 4 kbar at 30° and 150°C relative to the internal standard of $CFCl_3$. The ^{19}F chemical shifts of the o-CℓBTF and p-CℓBTF relative to BTF were also measured in binary mixtures of these compounds. Since literature data for densities of the previous mixtures were not available, densities were also measured as a function of pressure and temperature. Using $CFCl_3$ as an internal reference the ^{19}F chemical shifts of BTF, o-CℓBTF, and p-CℓBTF were found to shift to lower fields with increasing pressure, i.e., the shielding was decreased with increasing density (see Table V). The change in shielding with pressure is

Table V. Pressure and temperature dependence of σ.

		BTF	o-CℓBTF	p-CℓBTF
$(\frac{\partial\sigma}{\partial T})_p^a$	1 bar	0.28	0.033	0.30
	2 kbar	0.29	0.07	0.32
$(\frac{\partial\sigma}{\partial p})_T^b$	30°C	-0.026	-0.022	-0.027
	150°C	-0.026	-0.020	-0.025
$(\frac{\partial\sigma}{\partial T})_\rho^a$	1.20g/cc	+0.041	-0.17	--
	1.30g/cc	-0.025	-0.18	--

[a] Hz/degree.

[b] Hz/bar.

slightly smaller for o-CℓBTF than that for BTF and p-CℓBTF, while the changes for the latter two compound were the same within experimental error. The reason for the smaller pressure dependence of the ^{19}F shift in o-CℓBTF can be attributed to the steric protection of the CF_3 group by the relatively bulky o-chlorine atom. It seems that the medium effect depends strongly on the molecular geometry of the fluorine compound and the ortho substitution shields to some extent the CF_3 group from surrounding molecules. An increase in temperature increased the shielding for all three compounds but again the o-CℓBTF showed less change than the other two compounds. The pressure dependence of the ^{19}F chemical shift was interpreted in terms of the intermolecular effect originating predominantly from the dispersion interaction (σ_W) between the CF_3 group and surrounding molecules. When density is increased the average intermolecular separation between molecules is reduced, resulting in a larger σ_W term which shifts the resonance to lower fields. By separating the effect of density and temperature on the ^{19}F chemical shift, we were able to estimate the relative roles of the intramolecular effect due to the ortho substituent and the intermolecular dispersive interactions. The intermolecular dispersive interactions (σ_W) are mainly responsible for the temperature dependence of the ^{19}F chemical shift in BTF and p-CℓBTF whereas both σ_W and the intramolecular repulsive interactions between the chlorine and the CF_3 group contribute to the temperature dependence of the ^{19}F shift in o-CℓBTF. This work illustrates the capabilities of using the FT NMR spectroscopy in liquids at high pressure to study the nature of the solvent effects.

In view of the fact that the anomalous behavior of various transport and relaxation properties of water due to initial compression was discussed in some detail, it seems appropriate to mention the pressure dependence of the proton NMR chemical shift in liquid water[12].

For example, we observed that initial compression at temperatures of 0 to approximately 30°C produces faster reorientation of water molecules, higher self-diffusion coefficients, and lower shear viscosity, i.e., the pressure effect parallels the temperature effect. However, at temperatures above ∿40°C and/or at pressures in excess of 2-3 kbar the pressure effect is just opposite to the temperature effect, i.e., water begins to behave more like a normal liquid. Analogous behavior is observed[12] for the proton chemical shift in water as evidenced by Figure 26. At 0 and 20°C the initial compression produces an upfield shift parallel to the shift produced by increasing temperature. At higher temperatures and pressures these two effects are in the opposite direction. From Figure 26 we also notice that the higher the temperature the larger the relative change of the proton chemical shift with density.

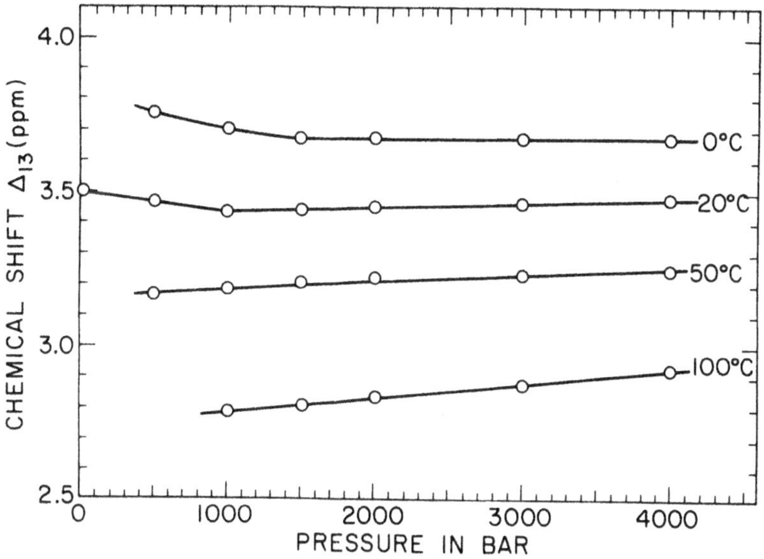

Fig. 26. Pressure dependence of the proton chemical shift Δ_{13} in liquid H_2O at several temperatures. (Diethyl ether was used as an external standard.)

The density dependence of the ^{13}C-H spin-spin coupling constant in $^{13}CH_4$ has been measured by Trappeniers and Oldenziel[35] in a methane sample over a density range 0 to 575 amagat. A linear increase of J(^{13}C-H) with increasing density was found. This experiment requires extremely high resolution because of the small density effect on the spin-spin coupling constants. Over the density range covered the total variation of J(^{13}C-H) was 0.24 Hz (0.2%).

A study[6] of spin-lattice relaxation in toluene-d_8 is used as an example of high resolution experiments which measure T_1 behavior of chemically shifted nuclei. The spin-lattice relaxation time (T_1) of both ring and methyl deuterons have measured in toluene-d_8 by the Fourier transform inversion recovery method. Measurements were made over the temperature range -35 to 200°C, and over the pressure range 1 bar to 4 kbar. Viscosities and densities were also measured over most of this range. The T_1 data are analyzed in terms of correlation times for isotropic overall molecular reorientation and internal rotation of the methyl group. Using pressure as a variable enables us to separate effects of density and temperature on overall and internal motion of the toluene molecule. Activation energies and activation volumes were reported. For the case of internal motion of the methyl groups, large angle reorientation appeared to be important in the overall motion of the toluene molecule at temperatures above 50°C, whereas the motion of the methyl group is characterized by large angle steps even at lower temperatures. At high temperatures, the motion of the methyl group approached that of a free rotor.

Von Jouanne and Heidberg[17] have used their apparatus, described in section 2, to investigate the effect of pressure on the keto-enol equilibrium of acetyl acetone. The difference ΔV' of the partial molal volumes of the keto and enol species of acetyl acetone was found to be -4.7 ± 0.7 ml/mole in the temperature range $290^{o}K \leq T \leq 388.9^{o}K$. No significant temperature dependence of ΔV' was observed.

iii) Dynamic structure of solids

In this section, I shall discuss briefly some recent work on dynamics of so-called "plastic" crystals, i.e., disordered organic solids whose molecules in the high temperature rotor phase show a great amount of motional freedom and resemble in many ways typical dense liquids. In view of of high degree of symmetry, sometimes one uses the name globular solids for plastic crystals. Their close proximity to the liquid is exemplified by a low entropy of fusion. At temperatures below the melting point these crystals undergo at least one, if not several, solid state phase transitions marked for the most part by larger changes in the entropy and dielectric properties than found in the melting of the plastic crystals. The molecules undergoing such a transition pass from a disordered orirentation within the rotor phase to a more ordered arrangement in the rigid phase.

An excellent representant of such a class of plastic crystals is adamantane, a saturated hydrocarbon $C_{10}H_{16}$, which has a highly spherical shape. The behavior of the proton spin lattice relaxation time T_1 with pressure at different temperatures in adamantane[36] is depicted in Figure 27. Adamantane has a fcc crystal structure in its plastic phase (α-phase) and undergoes a solid-solid phase transition to a tetragonal structure (β-phase) of lower symmetry and of higher order in which the rotational barrier is higher. The most interesting feature in this plot is the rapid change in T_1 accompanying the order-disorder transition and the marked hysteresis accompanying this transition. We have done studies of this hysteresis on this and other systems and feel that with more detailed investigation we may learn more about the dynamics of the phase transitions.

The second example is the NMR study.of neopentane[37] - a highly symmetric molecule which exhibits a plastic phase. Gutowsky and coworkers[38,39] have studied second moments in neopentane and concluded that at temperatures near the melting point, both molecular reorientation and self-diffusion contribute to the observed

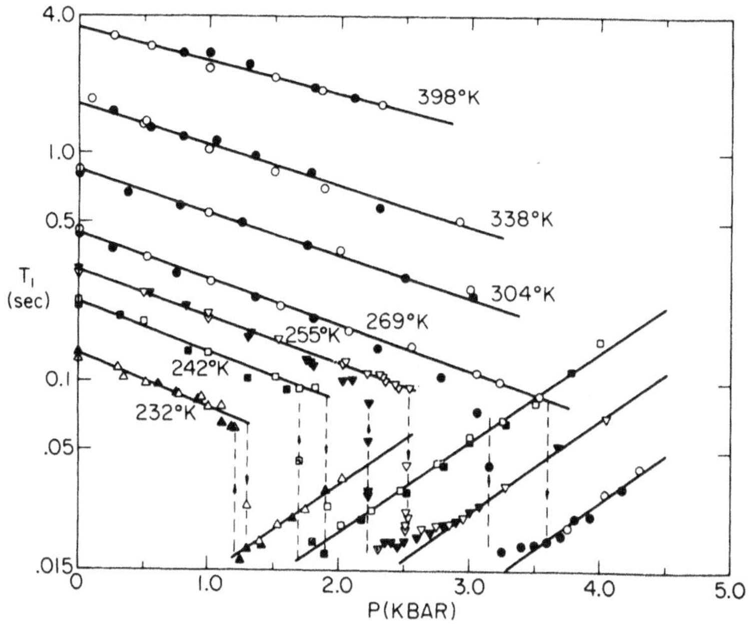

Fig. 27. Pressure dependence of the proton T_1 of solid adamantane in its two phases at different temperatures. The open symbols represent values obtained in the forward (compression) run whereas the full symbols give values for the reverse (decompression) run.

proton spin-lattice relaxation rate. Since the time scales of these processes are very different, this system offered an opportunity to use pressure as a means of separating these contributions. Simply, pressure has opposite effects on the contribution to $1/T_1$ due to self-diffusion than on the contribution due to overall molecular orientation.

There are two main contributions to the spin-lattice relaxation in neopentane near its melting point. One may write

$$T^{-1}_{1,OBS} = T^{-1}_{1,SD} + T^{-1}_{1,RO} , \qquad (12)$$

where $T_{1,OBS}$ is the observed relaxation time, and $T_{1,SD}$ and $T_{1,RO}$ are the spin-lattice relaxation times due to self-diffusion and over-all molecular reorientation, respectively. An increase in pressure slows down the self-diffusion and consequently its contribution to overall relaxation becomes smaller. On the other hand, the contribution due to over-all molecular reorientation increases with increasing pressure. Therefore, the pressure dependence of T_1 shows the maximum as shown in Figure 28. The opposite effect of pressure on $T_{1,SD}$ and $T_{1,RO}$ is readily understood when one considers the time scale of the respective molecular motions. One can assume the Arrhenius-type behavior for the correlation times involved and then

$$T_{1,RO} \propto \exp(-\Delta H^{\dagger}_{RO}/RT) \quad \omega_0\tau \ll 1 \tag{13}$$

$$T_{1,SD} \propto \exp(\Delta H^{\dagger}_{SD}/RT) \quad \omega_0\tau \gg 1 , \tag{14}$$

where τ denotes the correlation times, ω_0 is the resonant frequency, and $\Delta H^{\ddagger}_{RO}$ and $\Delta H^{\ddagger}_{SD}$ are the activation enthalpies for molecular reorientation and self-diffusion, respectively. The data also enabled us to calculate the activation volumes $\Delta V^{\ddagger}_{RO}$ and $\Delta V^{\ddagger}_{SD}$ for the molecular reorientation and self-diffusion given in Table VI.

The importance of high pressure NMR experiments in elucidating the mechanism of self-diffusion in organic plastic crystals is well documented in several studies by Strange and collaborators[13,40]. Folland and Strange[40] and also Folland, Ross and

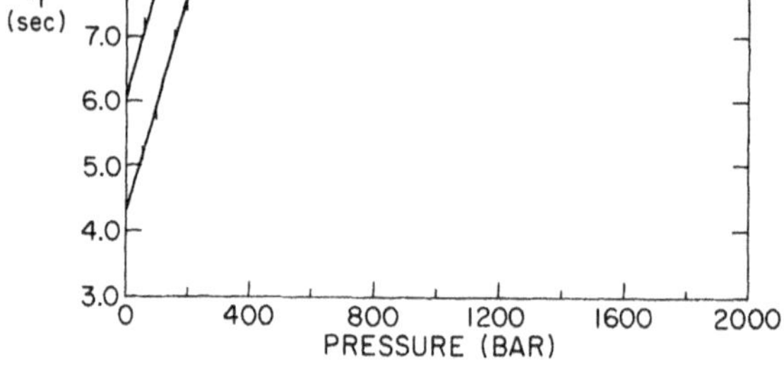

Fig. 28. Pressure dependence of the proton T_1 in the plastic phase of neopentane at temperatures 245.3°K and 253.5°K.

Strange[13] have reported a detailed investigation of the pressure dependence of molecular self-diffusion and reorientation in a number of organic plastic crystals.

Table VI. Activation parameters for molecular reorientation and self-diffusion in the plastic phase of neopentane.

$\Delta V^{\dagger}_{SD}$ cm^3/mole	$\Delta V^{\dagger}_{RO}$ cm^3/mole	$\Delta H^{\dagger}_{SD}$ kcal/mole	$\Delta H^{\dagger}_{RO}$ kcal/mole
110 $\pm$ 20[a]	5.0 $\pm$ 1.5	8.11 $\pm$ 1.5	0.9 $\pm$ 0.15

[a]Molecular volume is 102 cm^3/mole.

The proton relaxation times T_1 and $T_{1\rho}$ have been measured as a function of pressure at several temperatures in various organic plastic crystals. The T_1 measurements were analyzed in terms of reorientational motion, whereas $T_{1\rho}$ gave information about self-diffusion of the molecules. The nature of the pressure dependence of self-diffusion exhibited a correlation with the entropy of fusion of the plastic crystals. The pressure dependences of the mean jump times for self diffusion $\tau(P)$, normalized with respect to their value at atmospheric pressure $\tau(0)$, are given in Figure 29 for hexamethylethane and cyclohexane. The entropy of fusion

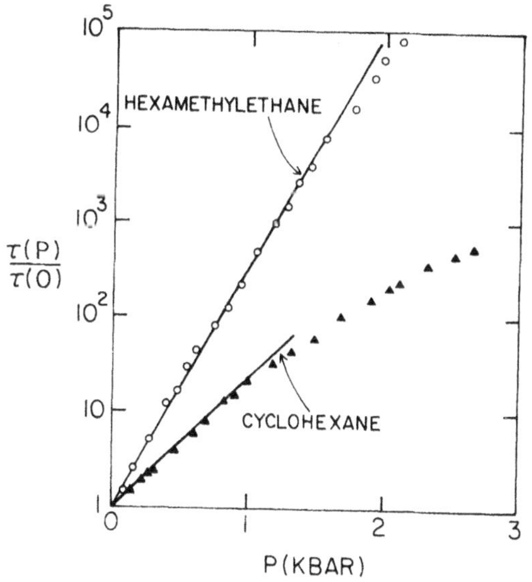

Fig. 29. The pressure dependence of the derived mean jump times for self-diffusion, $\tau(P)$, normalized with respect to their values at atmospheric pressure for hexamethylethane and cyclohexane. From ref. 13.

ΔS_f = 2.42 for hexamethylmethane and only 1.1 R for cyclohexane. The $\tau(P)/\tau(0)$ show very different behavior for these two plastic crystals. The activation volume for hexamethylethane (high ΔS_f) is nearly pressure independent and considerably larger than the activation volume for cyclohexane (low ΔS_f) which in addition shows a strong pressure dependence. These results suggest that the lower ΔS_f materials exhibit greater complexity of vacancies than the high ΔS_f materials. One additional comment about the data[13] in Figure 29 is worthwhile. The $\tau(P)$ shows a great sensitivity to pressure as an increase of pressure to 2.1 kbar changes $\tau(P)$ by a factor of 10^5 in the case of hexamethylethane.

4. PURE NUCLEAR QUADRUPOLE RESONANCE AND ELECTRON SPIN RESONANCE SPECTROSCOPY AT HIGH PRESSURE

i) Experimental techniques

High pressure experiments in pure nuclear quadrupole resonance spectroscopy[2] do not require any different instrumental considerations than nuclear magnetic resonance experiments. As a matter of fact, one can use steel for the construction of the high pressure vessel because no external magnetic field is necessary.

In contrast, electron spin resonance experiments at high pressure are relatively difficult from experimental point of view. This is due to the fact that in ESR one operates in the microwave region; an X band ESR spectrometer operates at ∿9,500 MHz at a field of 3,500 G.

Several authors[2,41,42] have used high pressure ESR vessel design which used the cylindrical microwave cavity. However, one may also use a microwave helix; i.e., so called helical slow wave device. For their ease of construction, microwave helices are well suited for high pressure X-band ESR studies[43]. The cross sectional view of the ESR high pressure vessel and the detailed diagram of the microwave helix are given in Figures 30a and 30b. These figures are reproduced from the paper by Plachy and Schaafsma[43]. This experimental study was used for ESR studies of liquids at high pressure. Either the solution under study was used directly as a pressurizing medium or the liquid sample was sealed in a teflon sample holder which also contained slow-wave helix.

iii) Applications of PNQR and ESR

One of the very first studies of the pressure dependence of PNQR frequency by Kushida, Benedek and Bloembergen[44] serves as

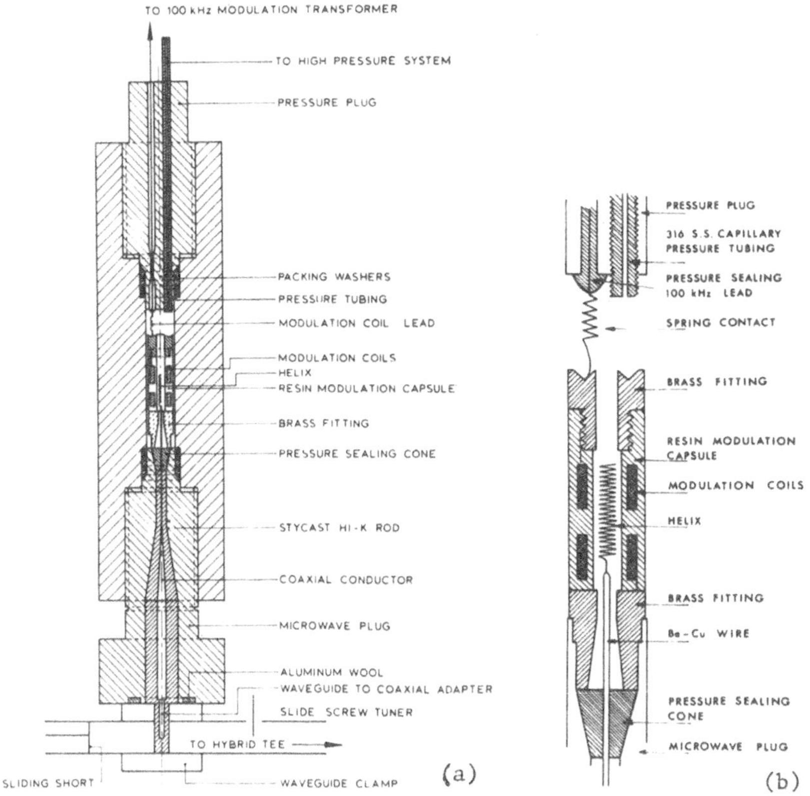

Fig. 30a. Schematic diagram of the ESR high pressure vessel. From ref. 43.
b. Schematic diagram of the microwave helix and modulation coils. From ref. 43.

an excellent example of the importance of using pressure as a variable in PNQR experiments. Since the very beginning of PNQR, the temperature dependence of the nuclear quadrupole resonance frequency was studied. The theory for the temperature dependence of the NQR frequency was proposed by Bayer[45] but discrepancy existed between theory and experiment. According to this theory, the average internal field gradient at the nucleus decreases with increasing temperature because of the increase in the amplitude of thermal vibrations. Kushida et al.[44] have observed the pressure dependence of the resonance frequency at various temperatures and pointed out that the temperature dependence at constant pressure includes not only the explicit temperature dependence of a particular mode of lattice vibration $\xi_1{}^0$, but also an implicit temperature dependence through the volume dependence of the static field gradient q as well as $\xi_1{}^0$. Kushida et al.[44] have obtained the value of q and $\xi_1{}^0$ as a function of volume and concluded that the observed temperature dependence of the resonance frequency is correctly accounted for by the Bayer type theory, if the effects of the thermal expansion on q and $\xi_1{}^0$ were accounted for.

Several PNQR studies on polycrystalline solids at high pressure have been reviewed by Smith[46]. The effect of pressure on molecular reorientation and phase transitions in several molecular organic solids have been investigated by Zussman and Rapoport[47,48].

In his monograph, Benedek[2] has discussed in considerable detail the results of several ESR studies of solids at high pressure. The ESR spectra of transition metal ions are very sensitive to their electrical environment in the crystal. Therefore, pressure changes in the ESR spectrum provide valuable and detailed information. For example Walsh[49] has studied the pressure dependence of the crystalline field splitting of Ni^{2+} and Ci^{3+} ions substituted into a cubic MgO crystal. Using the experimental arrangement described in the preceding section, Hwang et al.[50] have measured the ESR linewidths of vanadyl acetylacetonate in a number of non-hydrogen bonded solvents as a function of temperature and pressure. The experimental reorientational correlation times related to shear viscosity were discussed via the modified Debye equation. In spite of the fact that there are very few chemical applications of ESR at high pressure, the future of this specific high pressure field appears very promising.

ACKNOWLEDGMENTS

The work from our laboratory discussed in these lectures was supported under various grants by the National Science Foundation, the Energy Research and Development Administration and the Air Force Office of Scientific Research. The grant numbers are given in the original papers. I wish to express my thanks to several colleagues and the journals of Review of Scientific Instruments, Journal of Magnetic Resonance and Molecular Physics for permission to reproduce Figures 12-14, 29 and 30.

REFERENCES

1. G. B. Benedek and E. M. Purcell, J. Chem. Phys. 22, 2003 (1954).
2. G. B. Benedek, "Magnetic Resonance at High Pressure", Wiley (Interscience), New York, 1963.
3. J. Jonas, "NMR Studies in Liquids at High Pressure", in Adv. Magn. Resonance 6, 73 (1973).

4. J. Jonas, Ann. Revs. Phys. Chem. 26, 167 (1975).
5. J. H. Campbell, J. F. Fisher and J. Jonas, J. Chem. Phys. 61, 346 (1974).
6. D. J. Wilbur and J. Jonas, J. Chem. Phys. 62, 2800 (1975).
7. T. DeFries and J. Jonas, J. Chem. Phys. 66, 5393 (1977).
8. J. Jonas, Rev. Sci. Instr. 43, 643 (1972).
9. D. M. Cantor and J. Jonas, Anal. Chem. 48, 1904 (1976).
10. R. J. Finney, M. Wolfe and J. Jonas, J. Chem. Phys. 67, November issue (1977).
11. M. Wolfe, E. Arndt and J. Jonas, J. Chem. Phys. 67, November issue (1977).
12. J. W. Linowski, Nan-I Liu and J. Jonas, J. Chem. Phys. 65, 3383 (1976).
13. R. Folland, S. M. Ross and J. M. Strange, Mol. Phys. 26, 27 (1973).
14. R. W. Vaughan, C. F. Lai and D. D. Elleman, Rev. Sci. Instr. 42, 626 (1971).
15. D. J. Wilbur and J. Jonas, J. Chem. Phys. 55, 5840 (1971).
16. H. Yamada, Rev. Sci. Instr. 45, 640 (1974).
17. J. von Jouanne and J. Heidberg, J. Magn. Resonance 7, 1 (1972).
18. J. G. Oldenziel and N. J. Trappeniers, Physica 82A, 565 (1976).
19. J. DeZwaan and J. Jonas, J. Phys. Chem. 77, 1768 (1973).
20. B. J. Alder, D. M. Gass and T. E. Wainwright, J. Chem. Phys. 53, 3813 (1970).
21. J. M. Dymond, J. Chem. Phys. 60, 969 (1974).
22. D. Chandler, J. Chem. Phys. 62, 1358 (1975).
23. H. J. Parkhurst, Jr. and J. Jonas, J. Chem. Phys. 63, 2698, 2705 (1975).
24. W. L. Robb and H. G. Drickamer, J. Chem. Phys. 12, 1504 (1951).
25. D. Chandler, J. Chem. Phys. 60, 3500, 3508 (1974).
26. J. Dezwaan and J. Jonas, J. Chem. Phys. 63, 4606 (1975).
27. J. DeZwaan and J. Jonas, J. Chem. Phys. 62, 4036 (1975).
28. H. C. Andersen, D. Chandler and J. D. Weeks, Adv. Chem. Phys. 34, 105 (1976).
29. G. Jacucci and I. R. McDonald, Physica 80A, 607 (1975).
30. S. J. Bertucci and W. H. Flygare, J. Chem. Phys. 63, 1 (1975).
31. T. DeFries and J. Jonas, J. Chem. Phys. 66, 896 (1977).
32. J. Jonas, T. DeFries and D. J. Wilbur, J. Chem. Phys. 65, 582 (1976).
33. D. J. Wilbur, T. DeFries and J. Jonas, J. Chem. Phys. 65, 1783 (1976).
34. D. J. Wilbur and J. Jonas, J. Magn. Resonance 10, 279 (1973).
35. J. G. Oldenziel and N. J. Trappeniers, Physica 83A, 161 (1976).
36. Nan-I Liu and J. Jonas, Chem. Phys. Letters 14, 555 (1972).
37. W. C. Allen, Nan-I Liu and J. Jonas, J. Chem. Phys. 63, 3317 (1975).
38. J. G. Powles and H. S. Gutowsky, J. Chem. Phys. 21, 169 (1953).
39. E. O. Stejskal, D. E. Woessner, T. C. Farrar and H. S. Gutowsky, J. Chem. Phys. 31, 55 (1959).
40. R. Folland and J. H. Strange, J. Phys. Chem. 5, L50-55 (1972).
41. W. M. Walsh and N. Bloembergen, Phys. Rev. 107, 904 (1957).

42. C. P. Poole, Jr., Electron Spin Resonance, A Comprehensive Treatise on Experimental Techniques, Wiley (Interscience), New York, 1967, pp. 653-659.
43. W. Z. Plachy and T. J. Schaafsma, Rev. Sci. Instr. 40, 1590 (1969).
44. T. Kushida, G. B. Benedek and N. Bloembergen, Phys. Rev. 104, 1364 (1956).
45. H. Bayer, Z. Phys. 130, 227 (1951).
46. J. A. S. Smith, in "High Pressure Physics and Chemistry" (R. S. Bradley, Ed.), Vol. 2, p. 293, Academic Press, New York, New York.
47. A. Zussman and E. Rapoport, J. Chem. Phys. 61, 5098 (1974).
48. A. Zussman and E. Rapoport, J. Chem. Phys. 66, 5330 (1977).
49. W. M. Walsh, Phys. Rev. 122, 762 (1961).
50. J. Hwang, D. Kivelson and W. Plachy, J. Chem. Phys. 58, 1753 (1973).

NEUTRON SCATTERING AT HIGH PRESSURE

D. Bloch[+] and J. Paureau[++]

[+]Laboratoire Louis Néel, [++]Service National des Champs Intenses, C.N.R.S., 166X, 38042 Grenoble-Cedex, France.

ABSTRACT. A review is presented of neutron techniques for high pressure experiments. Special emphasis is given to materials, to pressure transmitting media and to the accessible temperature range. Applications to the physics of crystallographic and magnetic phase changes, lattice dynamics and magnetic excitations are indicated.

1. NEUTRON TECHNIQUES

The purpose of this paragraph is to introduce briefly the features which are needed for the understanding of most up to date high pressure neutron experiments |1|.

1.1 Neutron sources

Neutrons ($m = 1.674 \times 10^{-27}$ kg), in the 10^6 eV range* can be produced in reactors through the fission of $^{235}_{92}$U**. Since neutrons have no electrical charge, they are able to penetrate deeply in condensed matter. Thereby they will be peculiarly suited for high pressure studies on samples enclosed in a thick wall cylinder. The high energy neutrons are slowed down in a diffusing medium (light or heavy water, graphite...) where they take the speed, and associated wave length ($\lambda = h/mv$) distribution at statistical equilibrium of the temperature T of the medium

* 1 eV = 1.602×10^{-19} J

** $^{235}_{92}U + ^{1}_{0}n \rightarrow ^{236}_{92}U^* \rightarrow ^{94}_{38}Sr + ^{140}_{54}Xe + 2^{1}_{0}n + \gamma$

Kelm, H. (Ed.) High Pressure Chemistry. 111-126. *Copyright © 1978. D. Reidel Publishing Company. All Rights Reserved.*

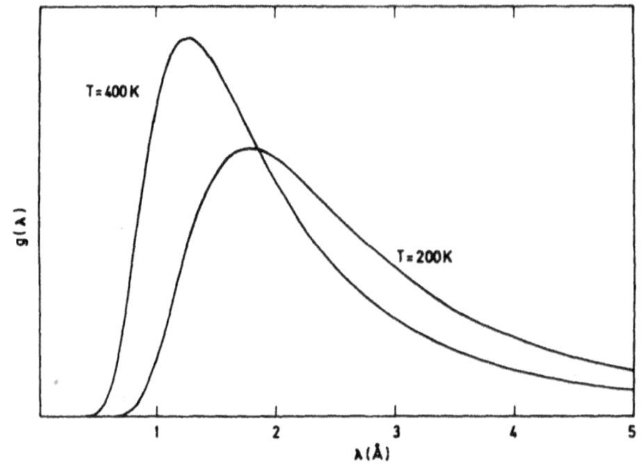

Fig. 1. Thermal dependence of wave length distribution

(fig. 1). Their energy $W = 1/2\ mv^2$ is close to 3/2 kT, that is for usual condition in the 0.01 eV range (Table I). The wave lengths as shown in Table I are of the order of magnitude of interatomic distances in condensed matter. A monochromatic neutron beam can be selected using for instance a single crystal which scatters only those wave lengths which satisfy Bragg – Laue equation

$$2d \sin \Theta_m = \lambda \tag{1}$$

T(K)	20	100	300	1000
v (km/s)	0.705	1.57	2.73	4.98
W (eV)	2.58×10^{-3}	1.29×10^{-2}	3.87×20^{-2}	1.29×10^{-1}
λ (Å)	5.62	2.52	1.45	0.793

Table I.

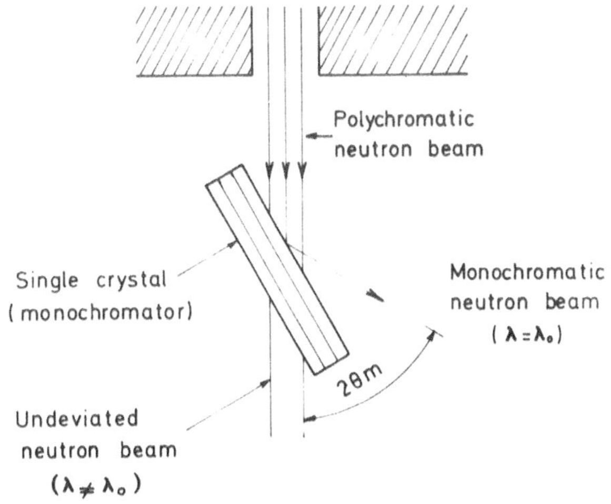

Fig. 2. Monochromator

where d is the interplanar spacing of planes parallel to the crystal surface (fig. 2). Wave length can equally be separated in a burst of polychromatic neutrons making use of their different time of flight between fixed points, due to their different speed (Table I).

The neutron sources are, as far as the number of particles are concerned, much less efficient than X-ray sources, since high flux reactor have a flux of 10^{15} neutrons/cm^2.s to be compared with 10^{20} photons/cm^2.s for a standard X-ray anticathode and about 10^7 neutrons/cm^2.s at the level of the monochromatic neutron beam.

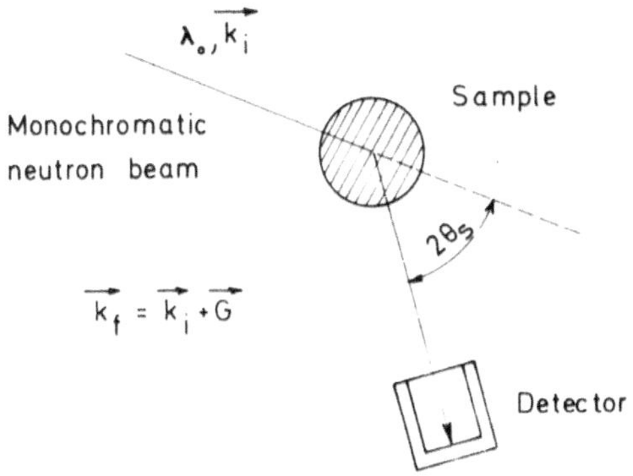

Fig. 3. Elastic scattering.

1.2 Nuclear and magnetic elastic scattering

Due to the short range nature of the interaction of the neutron with the nucleus, the scattering of incident neutrons from nuclei is isotropic. The scattering cross section, defined as the number of neutrons scattered per second to the incident neutron flux, $\sigma = 4\Pi |b|^2$, where b is the Fermi scattering length, varies with the considered nuclei, with $|b| \sim 10^{-12}$ cm. For a crystalline solid, coherent scattering and diffraction patterns characteristic of the crystal can be observed in special directions $\vec{k}_f$ (with $k_f = 2\Pi/\lambda$), which satisfy to the Bragg - Laue condition

$$\vec{k}_f = \vec{k}_i + \vec{G} \qquad (1') \text{ similar to } (1)$$

where G is a reciprocal lattice vector (fig. 3). In fact, due to the presence of various isotopes, and nuclear spin orientations, part of the scattering will be coherent and part will be incoherent, and thus, the total cross section σ is partly associated with a coherent cross section and partly associated with an incoherent one. For hydrogen (but not for deuterium) the total cross section is to a few percent mostly an incoherent cross section, whereas for iron for instance, it is mostly coherent. Neutron and X-ray scattering are in fact complementary in order to determine crystallographic structures. For instance neutrons permit atoms of close atomic numbers to be distinguished : Fe and Co have different neutron cross sections but a similar X-ray behaviour. Neutron techniques are useful too, to determine the atomic positions for those materials which contains atoms with largely different atomic numbers. Since X-ray scattering factor is a rapidly increasing function of atomic number, the contribution to the scattered radiation associated with the lighter atoms is too weak compared to that associated with the heavy nuclei. Such a drastic dependence of atomic number does not exist for neutron scattering and therefore structures containing both heavy and light atoms can be determined more accurately.

Neutron have a magnetic moment, and therefore interact with magnetic atoms. This in turn, leads in magnetic solids to diffraction patterns which depend on the amplitude of the atomic magnetic moments, as well as on their distribution, direction and order. Neutron techniques are of major importance for the determination of the magnetic moments in complex magnetic structures such as antiferromagnets or helimagnets, where no overall magnetic moment exist, and therefore magnetization measurements yield little information about the local magnetic properties. The magnetic cross section, associated with the interaction of neutrons with magnetic shells (3d, 4f, 5f), and the nuclear cross section, associated with the interaction of neutrons with the nuclei are often of comparable order of magnitude and can be separated using various specific properties (different periodicity, magnetic field effect, stress effects...).

1.3 Inelastic scattering

A large variety of excitations such as collective excitations in solids, phonons (collective atomic motion) or magnons (collective magnetic moment motion) have energies in the 10 meV range. This is the usual neutron energy range (Table I). Neutron can exchange a large part of their energy through inelastic collisions with nuclei (or atomic moments) systems. This leads to a considerable change of their wave length. It should be noticed that a different situation occurs with X-ray, since their energies, for similar wave lengths, is 10^6 times larger. For instance the energy equation will be for lattice vibrations at frequency $\frac{\omega}{2\Pi}$, wave vector $\vec{q}$ and energy h

$$\frac{h^2k_i^2}{2m} = \frac{h^2k_f^2}{2m} + h\omega(q) \qquad (2)$$

and for the momentum equation :

$$\vec{k}_i = \vec{k}_f + \vec{G} + \vec{q} \qquad (3)$$

where q = 0 corresponds to the elastic Bragg-Laue scattering. Similar relations hold for magnetic moment excitations.

A single crystal analyser permits the different wave lengths to be separated, and a detector compares the number of neutrons for each wave length. For a fixed q, or fixed $h\omega(q)$, for instance, the corresponding $h\omega(q)$ or q can be determined. The relations between these quantities are known as dispersion relations (fig. 4). The analysis of the detailed shape of these dispersion relations allow for the determination of interatomic forces (phonons) or magnetic

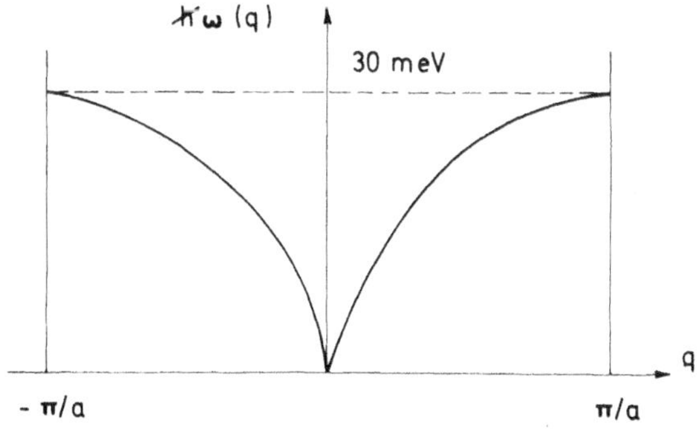

Fig. 4. Phonon dispersion relation.

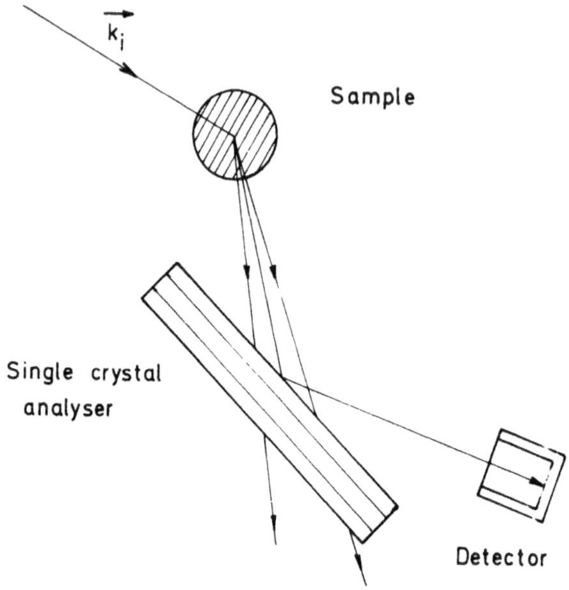

Fig. 5. Energy analysis.

coupling forces (magnons). Crystal field effect can similarly be studied through inelastic neutron scattering.

2. EXPERIMENTAL TECHNIQUES

The cell design for high pressure neutron scattering depends on various parameters, such as sample dimensions, informations and sample of interest, temperature range.

2.1 Materials for high pressure cells : mechanical properties and transparency

These materials should have sufficiently good transparency for neutrons and sufficiently high mechanical properties for pressure. They are mainly aluminium alloys (Al-Zn-Mg-Cu), copper alloys (Cu-Be), maraging steels (Fe-Ni-Co-Mo) or alumina (Al_2O_3). The mechanical properties of these materials, as well as their associated maximum working pressures, decrease rapidly above a certain temperature equal to about 100°C for Al-alloys, 200°C for Be-Cu alloys, 300°C for maraging steels and 700°C for Al_2O_3.

For example, with an internal diameter $\emptyset_i$ = 5 mm and pressure P = 8 kbar, with h_1 = 5 mm or h_2 = 20 mm, the external diameter $\emptyset_e$

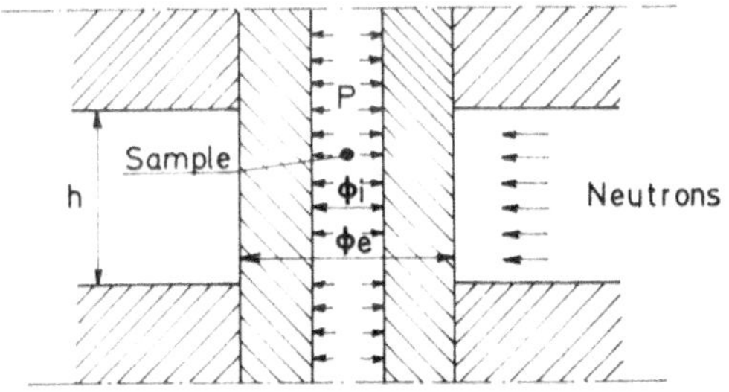

Fig. 6. Neutron window

for aluminium alloys at room temperature will be respectively about 11 mm and 25 mm (fig. 6). The complete numbers are given table II, line 1, whereas the neutron absorption is given in line 2. At low pressure, aluminium is better than copper alloys and maraging steel, as far as transparency is concerned. Another problem is to obtain high pressures. Table III gives approximate values of the bursting pressures and absorption for these alloys in similar conditions, that is $\emptyset_i$ = 5 mm, $\emptyset_e$ = 20 mm with h = 5 or 20 mm.

2.2 Pressure transmitting media and pressure measurements

The pressure and temperature range and the specific neutron properties are of prime importance to determine the pressure transmitting media : for instance hydrogen compounds are not used, due to their large incoherent scattering. The most common pressure

		Al	Cu	Maraging Steel	Al_2O_3
h_1= 5 mm	$\emptyset_e$ (mm)	11	7	6.5	9
h_2=20 mm		25	9.5	7.5	17.5
h_1= 5 mm	absorption (%)	6	14	16	16
h_2=20 mm		19	28	25	43

Table II. P = 8 kbar. Outside diameter and absorption.

	Al	Cu	Maraging Steel	Al_2O_3
h_1= 5 mm P_{max}	27	64	85	43
h_2=20 mm (kbar)	10	24	30	9
Absorption (%)	16	67	82	48

Table III. Bursting pressures and absorption.

media are helium with freezing pressure estimated to 120 kbar at room temperature, C_6F_{12}, "fluorinert" with a freezing pressure 20 kbar at room temperature as well as deuterated ethanol-methanol mixture, which is believed to solidify at about 100 kbar. In fact experiments performed on a single crystal at the highest pressure necessitate, in order to keep the orientation of the crystal and to leave the mosaïc spreads at acceptable level, a transmitting media which is fluid or sufficiently soft, as solid helium.
When helium or other fluids are used, pressure on the sample can be determined through a standard manometer connected to the high pressure cell. However, neutrons themselves can be used for the determination of internal pressure by means of an internal probe (NaCl, CsCl) for which the P, V, T relationships are well known, from a measurement of the temperature T of the cell and of volume V of the probe through an elastic neutron scattering experiment. Furthermore the analysis of the pressure dependence of the diffraction pattern (linewidth) allows for a direct evaluation of the pressure homogeneity within the neutron cell |2|.

2.3 Low temperature experiments

Most of the actual set ups have been designed in order to allow for low temperature experiments. This necessitates cells of moderate dimensions and weight as well as sufficient thermal insulation. Standard cryostats are made of aluminium, which is transparent to neutrons. Hydrostatic cells can be connected to an high pressure generator by means of an high pressure tubing, with suitable heaters in order to enable pressure changes. Clamped systems, with sufficiently small dimensions, allow conventional neutron cryostats to be used, as available in most reactors, without major consideration for low temperature problems |2|. Another technique is to transmitt the applied load from hydraulic

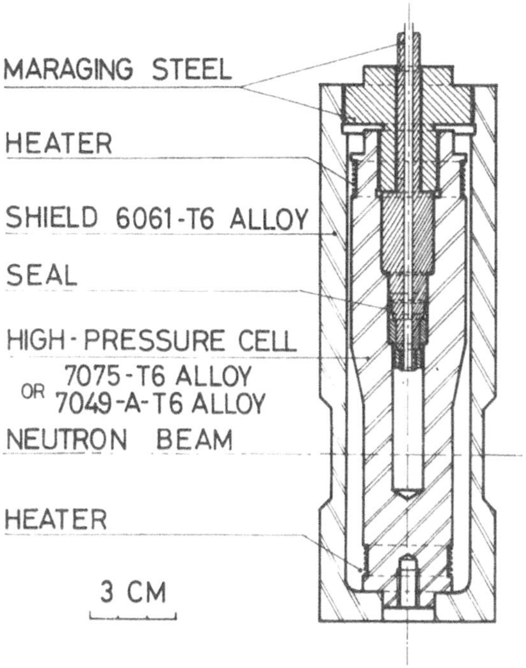

Fig. 7. Aluminium cell.

ram at room temperature |3| to the pressure cell at low temperature via insulating systems. This has the advantage on clamped system to allow for in-situ pressure modifications.

2.4 Some typical realisation

Aluminium cells. The cell given figure 7 |4| has been used from 4 to 300 K and working pressures up to 4 kbar. The sample has a volume up to 5 cm^3. A shielding allows for safety in a reactor context. These cells have been used for 2-axes elastic scattering, 3-axes inelastic scattering as well as 4-circles elastic scattering experiments. Maraging belts allow pressures in the 20 kbar range to be obtained as seen from Table II.

Clamped alumina cells. This portable (4.5 kg) clamped cell |2| (fig. 8) has a small outside diameter (68 mm). The maximum working pressure is 45 kbar, for an internal diameter $\emptyset_i$ = 5 mm and a window h = 6 mm. The alumina cell is similar to that introduced by McWhan et al |5| but has a clamped external support as well as a clamped internal pressure. This type of cell has been used down to 4 K for compressibility measurements as well as magnetic measurements

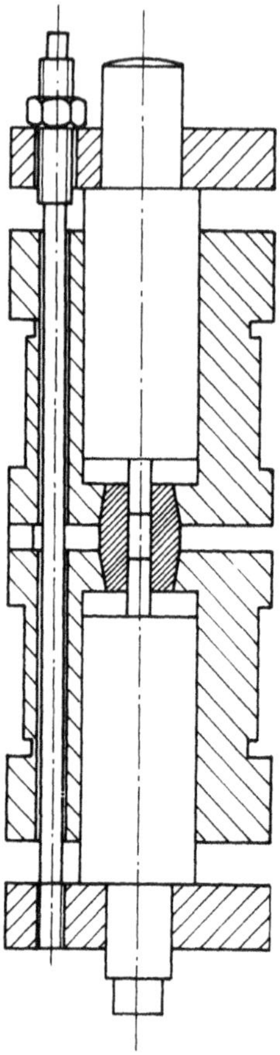

Fig. 8. Clamped alumina cell.

Low temperature press. The cell |3| (A) described figure 9 is similar to that described figure 8. However a close cycle helium refrigerator (B) permits temperature down to 20 K to be obtained. The applied load from the hydraulic press (C) is transmitted through fiber glass epoxy columns (D), whereas the vaccum jacket for the cryostat (E) also serves as the frame of the hydraulic press. This ensemble has been used for inelastic neutron scattering up to 20 kbar at 20 K.

3. APPLICATIONS

3.1 Crystallographic phase change

Continuous phase change. The low pressure tetragonal ($a = b \neq c$; $\alpha = \beta = \gamma = \Pi/2$) form of TeO_2 |6-8| and NiF_2 |9| changes at critical pressure P_c = 9 kbar (TeO_2) and 18 kbar (NiF_2) to an

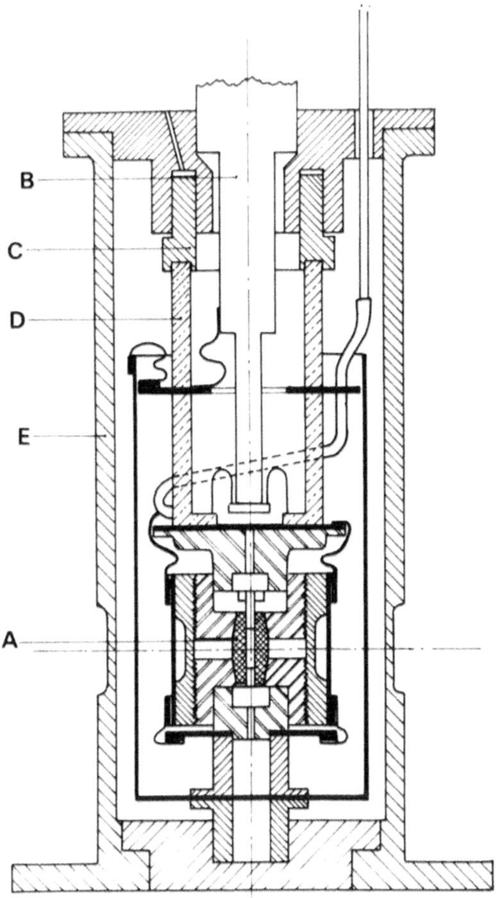

Fig. 9. Alumina cell.

orthorhombic structure ($a \neq b \neq c$; $\alpha = \beta = \gamma = \Pi/2$). The neutron scattering experiments are particularly well adapted to determine the atomic position of O and F in the presence of Te or Ni atoms. This pressure induced second order (or continuous) transition is characterized with the strain (b - a) as ordering parameter |10| with $(b - a)^2$ linear as a function of applied pressure (fig. 10).

Discontinuous phase change. Neutron are used to define the high pressure crystallographic structure when they are more efficient than X-ray techniques. In this way, it has been possible to determine crystallographic structure and atomic positions in the high pressure phases of KCN |11| or BiII |12|. In the case of

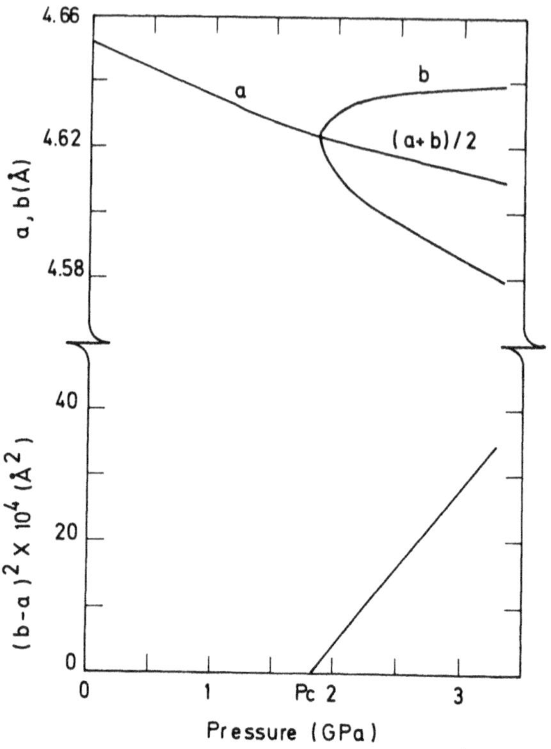

Fig. 10. Lattice parameters of NiF_2 |9|.

Bi for instance, the best X-ray patterns have at most 8 lines, whereas high pressure neutron time of flight techniques have ∿40, with an excellent accuracy. Another example is given from the lamellar compound $FeCl_2$, which presents, at 5.8 kbar |13| the hexagonal structure of $FeBr_2$. Thus the properties of isomorphous compounds $FeBr_2$ (P = 0) and $FeCl_2$ (P > 5.8 kbar) can be compared to study the effect of Br-Cl substitution |14|.

3.2 Magnetic phase change

Neutrons constitue a unique method to study the high pressure behaviour of a complex magnetic structure. We take as an example an helimagnetic structure (fig. 11a) which exists for instance in $MnAu_2$ where the moment direction turns from 51° between successive ferromagnetic plane. This turn angle is deduced from the 002^- to 002^+ satellite angle (fig. 11b). Pressure of 8.8 kbar |15| decreases the interlayer angle to 42°. Further pressure increase would eventually transform the helimagnetic structure to a ferromagnetic one.

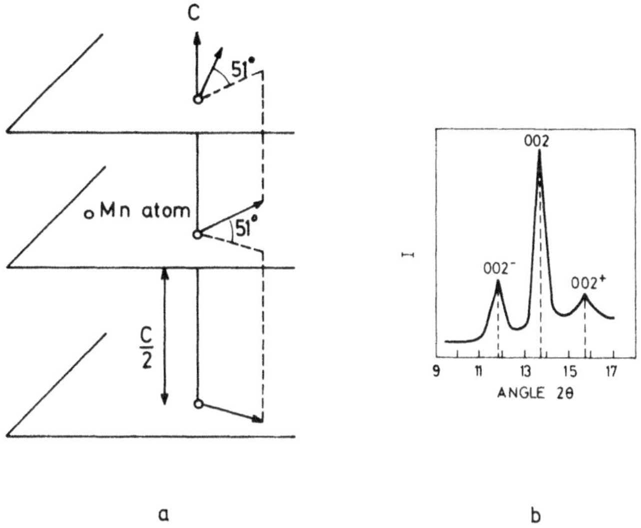

Fig. 11. $MnAu_2$ |15|.

Similarly, high pressure neutron scattering is an excellent tool to study the high pressure occurence of a magnetic moment. This has been proved for instance in PrSb, where a magnetic moment, with antiferromagnetic structure, appears on Pr ions at a pressure of about 30 kbar |16|.

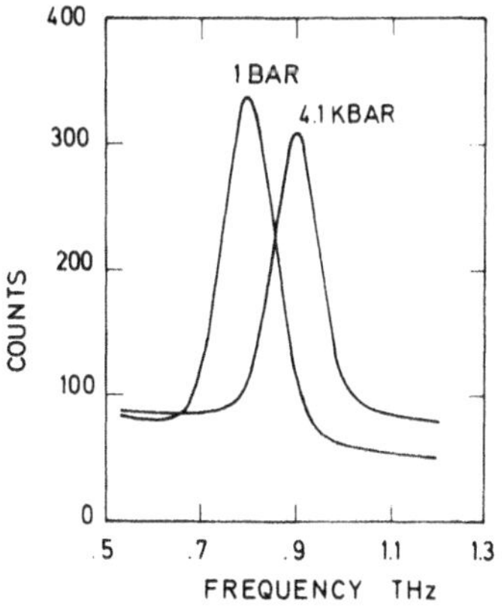

Fig. 12. Counts versus frequency for K at various pressures |22|.

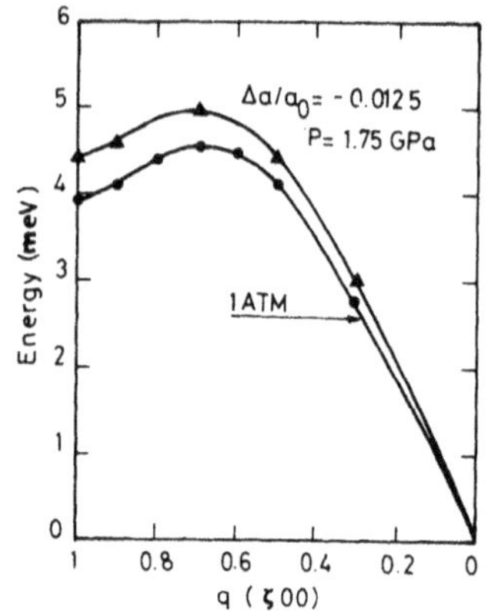

Fig. 13. Phonons in lead at various pressures |23|.

3.3 Phonons and magnons

Rare gases. High pressures and low temperatures are used simultaneously in order to prepare large single crystals of rare gases. Inelastic in-situ neutron scattering allows the phonon dispersion curves to be determined. This in turn leads to useful informations on elastic coefficients, interatomic forces (Lennard-Jones potential, effect of next nearest neighbours...) and their associated harmonic and anharmonic effect obtained for instance from a comparison of thermal effects associated with constant volume and constant pressure experiments. Such experiments have been performed on Kr |17|, He |18|, Ne |19|.

Alkali metals. Alkali metals form an interesting class of metallic materials for high pressure studies, due to their large compressibility and rather simple electronic structure. Experiments directed on the lattice dynamics properties and interatomic forces have been performed on Na |20|, Rb |21| and K |22| (fig. 12).

Lead. Lead has been the subject of considerable attention due to its large compressibility, and its high superconducting temperature (7.2 K). In fact lead probes are often used at low temperatures as pressure manometer, using the pressure dependence of its superconducting temperature. The rapid hardening of the lattice is clear from fig. 13.

Soft modes. In a large variety of crystalline structure a softening of the lattice ($\omega \downarrow 0$) is observed when approaching a crystallographic phase transition. Such a softening has been observed using neutron scattering, for TeO_2 |8| and KBr |22|. It should be noticed that this type of crystallographic phase transition is among the few ones which are really well understood.

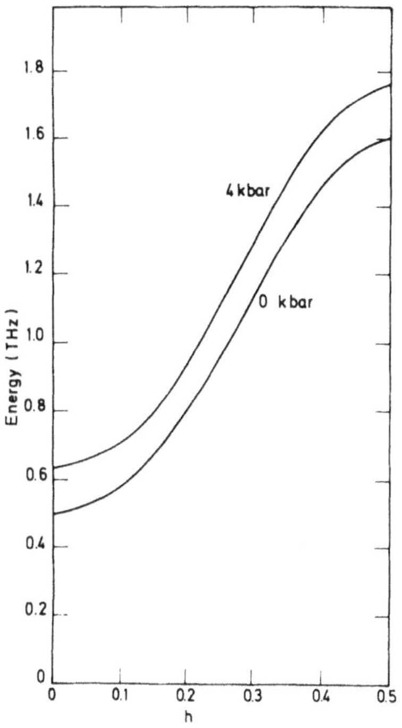

Fig. 14. Dispersion curves for magnons in $FeCl_2$ |24|.

Magnons. Dependence of the exchange interactions on inter-atomic distances can be determined directly using magnons dispersion curves determined at various pressures. Few experiments of this kind have been done at the present time, although indication of their usefulness has been evidenced for $FeCl_2$ |24| (fig. 14).

ACKNOWLEDGEMENTS

The authors acknowledge D.B. McWhan, G. Parisot and C. Vettier for fruitful discussions.

REFERENCES

1. for reviews on neutron scattering see for instance :
   a - G.E. Bacon, Neutron diffraction, Clarendon Press, 1967.
   b - Thermal neutron scattering, P.A. Egelstaff Edit., Academic Press, 1965.
   c - W. Marshall and S.W. Lovesey, Theory of thermal neutron scattering, Clarendon Press, 1971.

d - I.I. Gurevich and L.V. Tarasov, Low energy neutron physics, North Holland, 1968.
2. D. Bloch, J. Paureau, J. Voiron and G. Parisot, Rev. Sc. Instrum., 47, 296, 1976.
3. D.B. McWhan and C. Vettier, to be published.
4. J. Paureau and C. Vettier, Rev. Sc. Instrum., 46, 1484, 1975.
5. D.B. McWhan, D. Bloch and G. Parisot, Rev. Sc. Instrum., 45, 643, 1974.
6. P.S. Peercy and I.J. Fritz, Phys. Rev. Letters, 32, 466, 1974.
7. T.G. Worlton and R.A. Beyerlein, Phys. Rev., B12, 1899, 1975.
8. D.B. McWhan, R.J. Birgeneau, W.A. Bonner, H. Taub and J.D. Axe, J. Phys. C : Sol. Stat. Phys., 8, L81, 1975.
9. J.D. Jorgensen, T.G. Worlton and J.C. Jamieson in Intern. Conf. on High Pressures,, Boulder, 1977, K. Timmerhaus Edit., University of Colorado, to be published.
10. J.F. Nye, Physical properties of crystals, Pergamon Press, 1969.
11. D.L. Decker, R.A. Beyerlein, G. Roult and T.G. Worlton, Phys. Rev., B10, 3584, 1974.
12. R.M. Brugger, R.D. Bennion and T.G. Worlton, Phys. Letters, 24A, 714, 1967.
13. C. Vettier and W.B. Yelon, J. Phys. Chem. Solids, 36, 401, 1975.
14. C. Vettier, H.L. Alberts, D. Bloch, Phys. Rev. Letters, 31, 1414, 1973.
15. F.A. Smith, C.C. Bradley and G.E. Bacon, J. Phys. Chem. Solids, 27, 925, 1966.
16. C. Vettier and D.B. McWhan, to be published.
17. W.B. Daniels, G. Shirane, B.C. Frazer, H. Umebayashi and J.A. Leake, Phys. Rev. Letters, 18, 548, 1967.
18. J.G. Traylor, C. Stassis, R.A. Reese and S.K. Sinha, Symp. on Inelastic scattering of neutrons in solids and liquids, Vienne, IAEA, 129, 1972.
19. J. Skalyo, V.J. Minkiewicz, G. Shirane and W.B. Daniels, Phys. Rev., B6, 4766, 1972.
20. G. Ernst, Acta Phys. Austria, 33, 27, 1971.
21. J.R.D. Copley, C.A. Rotter, H.G. Smith and W.A. Kamitakahara, Phys. Rev. Letters, 33, 365, 1974.
22. J. Meyer, G. Dolling, J. Kalus, C. Vettier and J. Paureau, Conf. on Neutron scattering, Gatlinburg, 1976, to be published.
23. D.B. McWhan, R.C. Dynes and S.M. Shapiro, to be published.
24. C. Vettier and W.B. Yelon, Phys. Rev., B11, 4700, 1975.

HIGH-PRESSURE INFRARED AND RAMAN SPECTROSCOPY

E. Whalley

Division of Chemistry, National Research Council,
Ottawa K1A 0R9, Canada

ABSTRACT

The experimental techniques of high-pressure infrared and Raman spectroscopy, and selected applications to the study of the physical-chemical behavior of matter under pressure are reviewed.

1. INTRODUCTION

Atoms and molecules are continually moving, and the macroscopic properties of matter are continually fluctuating. Molecular science is concerned largely with measuring the properties of these fluctuations. There are two principal ways to measure them: to study the spontaneous emission of particles or radiation, or to perturb them with a perturbing agent and study the response of the sample or of the perturbing agent.

In spectroscopy using electromagnetic radiation, photons are hurled at the sample, and their fate investigated. In absorption spectroscopy, the fraction of photons that survive transmission through the sample is measured. In scattering spectroscopy it is the number and energy of photons that are scattered, either elastically or inelastically, and in fluorescence and phosphorescence spectroscopy it is the number and energy of photons that are emitted after a photon has been absorbed that are measured.

This paper is concerned with the absorption and

Kelm, H. (Ed.) High Pressure Chemistry. 127–158. *Copyright © 1978. D. Reidel Publishing Company. All Rights Reserved.*

scattering of light by nuclear motions. An isolated molecule having N atoms has 3N degrees of freedom of nuclear motion. Three of them are pure translations of the molecule and three (two if the molecule is linear) are pure rotations. There are therefore 3N-6 (3N-5 if it is linear) internal motions. Frequently, the potential energy V of the nuclei relative to the minimum potential energy can be expanded in a power series in the displacements from the minimum

$$V = \tfrac{1}{2}\sum_{ij} k_{ij}x_i x_j + \frac{1}{3!}\sum_{ijk} k_{ijk}x_i x_j x_k + \text{-----},$$

where x_i, i = 1-3N, are the cartesian displacements. If the displacements are small enough, only the quadratic term is important. Even so, the atomic motions are very complicated, and may have no simple period. They can, however, be described in terms of normal coordinates. These are coordinates which, when excited, will oscillate indefinitely with a fixed amplitude and frequency. They can be calculated from the geometry, masses, and force constants of the system, and are usually written in terms of the mass-reduced cartesian displacements $m_i^{\frac{1}{2}}x_i$ where x_i is the displacement of the ith cartesian coordinate and m_i the mass corresponding to the ith coordinate. The normal coordinate for the kth vibration is then

$$Q_k = \sum_i a_{ki}\, m_i^{\frac{1}{2}} x_i,$$

where N is the number of atoms and a_{ki} are coefficients that depend on the geometry, masses, and force constants. The normal coordinates form an orthonormal set. Any motion of the vibrating system can be described as a superposition of motions of the normal coordinates with appropriate phases and amplitudes. If the molecule is harmonic, the potential energy in terms of the normal coordinates is

$$V = \tfrac{1}{2}\sum_k Q_k^{\,2}\omega_k^{\,2},$$

where ω_k is the circular frequency of oscillation of normal coordinate k. While the harmonic approximation is adequate for many purposes, the anharmonicity must often be taken account of. Then terms in $Q_k Q_{k'} Q_{k''}$ and higher-order terms occur in the potential energy.

The absorption of electromagnetic radiation can be described in classical mechanics as follows. A normal vibration of a molecule having normal coordinate Q is represented as a damped harmonic oscillator having circular frequency ω_o and damping constant γ. The electric field of the light is oscillating and can be written $E_o e^{i\omega t}$ where E_o is the amplitude and ω the frequency. It can drive the molecular oscillation by interacting with the oscillating component of the molecular dipole moment, and the interaction force is $(\partial\mu/\partial Q)E\cos\theta$ where θ is the angle between the moment and the field. The equation of motion of the oscillator is

$$\ddot{Q} + \gamma\dot{Q} + \omega_o^2 Q = (\partial\mu/\partial Q)E\cos\theta. \tag{1.1}$$

At steady state, Q must oscillate at the driving frequency ω

$$Q = Q_o e^{i\omega t}, \tag{1.2}$$

where Q_o is in general complex and the real part represents the amplitude and the imaginary part represents the phase. By inserting equation (1.2) into (1.1) and solving for Q_o we find

$$Q_o = \frac{(\partial\mu/\partial Q)E_o\cos\theta}{\omega_o^2-\omega^2 + i\gamma\omega} \tag{1.3}$$

This corresponds to an amplitude Q_o', which is the real part of Q_o,

$$Q_o'^2 = \frac{(\partial\mu/\partial Q)^2 E_o^2 \cos^2\theta}{(\omega_o^2-\omega^2)^2 + \gamma^2\omega^2}, \tag{1.4}$$

and a phase shift θ, which is related to the complex part,

$$\theta = \arctan \gamma\omega(\omega_o^2-\omega^2).$$

If the damping is small so that $\gamma << \omega_o$, the amplitude Q_o' is a maximum when the light has the same frequency as the molecule. The light is absorbed (or emitted if the molecule has enough energy to drive the electric field), and in the steady state its energy is dissipated by the damping.

The classical description is readily translated into quantum mechanics. The results are the same as given by the classical treatment, as might be expected as Planck's constant does not occur, except that the damping constant is reinterpreted as the reciprocal lifetime of the excited state.

Molecular vibrations have the same frequencies as infrared

light, and they can be investigated by measuring the absorption of the light as it traverses the sample.

When the frequencies of the light and the molecules are far apart, nothing much happens most of the time. But sometimes energy is exchanged between the molecule and the light, and Raman spectroscopy sets out to measure the results of this exchange. The way it occurs is as follows.

A molecule or a collection of molecules is irradiated by light of frequency ω. The electric field of the light polarizes the molecule by causing an electric dipole moment in addition to the permanent moment μ_o. The moment is

$$\mu = \mu_o + \alpha E,$$

where α is the polarizability. The induced moment oscillates with the frequency of the incoming light

$$\mu = \mu_o + \alpha E_o \cos\omega t. \tag{1.5}$$

An oscillating dipole radiates light in all directions except along the dipolar axis and so the light is scattered in different directions without change of frequency. This is Rayleigh scattering.

But molecules also oscillate, and during these oscillations the polarizability may change

$$\alpha = \alpha_o + \alpha_1 \cos\omega_o t. \tag{1.6}$$

From equations (1.5) and (1.6), the induced dipole moment is given by

$$\mu = \mu_o + \alpha_o E_o \cos\omega t + \alpha_1 E_o \cos\omega t \cos\omega_o t. \tag{1.7}$$

The product of cosines in equation (1.7) can be expanded to give

$$\mu = \mu_o + \alpha_o E_o \cos\omega t + \tfrac{1}{2}\alpha_1 E_o [\cos(\omega-\omega_o)t + \cos(\omega+\omega_o)t].$$

Light is now radiated at several frequencies. The Rayleigh line has unchanged frequency. The Stokes line is at a lower frequency than the original light - the light has lost energy, which has been given to the molecule, and the molecule is now vibrating more vigorously than before. The anti-Stokes line is at higher frequency, and the light has gained energy which has been taken from the molecule, and the molecule is now vibrating less vigorously than before. Usually $\alpha_1 \ll \alpha_o$

and the Rayleigh line is much stronger than the Raman lines.

This theory is a classical theory, and only quantum mechanics can properly describe the absorption and scattering. The main difference between classical and quantum theory is that quantum theory requires the vibrations of molecules to be quantized. This causes the Stokes line to be stronger than the anti-Stokes by the Boltzmann factor, instead of having the same intensity. But the classical description is accurate enough to give a reasonable picture of the mechanism of the scattering.

In a Raman experiment, monochromatic light is shone on a sample and the scattered light is examined for shifted frequencies. These frequencies tell about the motion of the atoms, and high-pressure Raman spectroscopy is concerned with the effect of pressure on these motions.

There are therefore three parameters that can be determined from the effect of pressure on infrared and Raman spectroscopy, the effect of pressure on the frequency ω, the intensity $\partial\mu/\partial Q$ or $\partial\alpha/\partial Q$, and the damping constant or bandwidth, or more generally the bandshape. All have something useful to tell us.

2. EXPERIMENTAL TECHNIQUES

The principal problem of high-pressure optical spectroscopy is to lead light into and out of a high-pressure cell. In an absorption experiment, a light beam usually passes straight through the cell, and the intensity is measured as a function of frequency before and after. In a scattering experiment, the incident light is monochromatic, and the intensity of the scattered light is measured as a function of frequency. Usually, the scattering angle is 90°, but either forward or backward scattering are occasionally measured.

The amount of light absorbed depends on the length of material traversed. In the infrared, strong vibrations absorb nearly completely in 1 μm, and path lengths of 20-100 μm are common. Weak absorbers or dilute solutions may require much longer path lengths. Different cells may, therefore, be required for different materials, and strong and weak bands of the same material may require different path lengths. In Raman spectroscopy, only about 10^{-6}-10^{-9} of the exciting light is scattered. Hence, intense light sources and efficient collecting optics are required. Lasers provide the light sources, and it is only since the invention of lasers that high-pressure Raman spectroscopy has been really practicable, except as a tour de force. As the Raman intensity does not depend greatly on the path length, strong and weak bands require only a different sensitivity in the detecting system.

Windows that are transparent to the radiation being used are needed. There are three principal kinds of cell, namely transparent pressure vessels, pressure vessels fitted with relatively small windows, and anvil apparatuses. Sometimes, it is the spectrum of a high-pressure phase that is wanted, not necessarily the effect of pressure on the spectrum. Sometimes the high-pressure phase can be recovered in a metastable state at room temperature, and it can then be examined by the usual techniques. Sometimes, however, it is unstable at room temperature. Then, it can sometimes be recovered in a metastable state at low temperatures. These techniques can be classified and reviewed as follows.

2.1 Transparent pressure vessels.

Glass tubes have long been used as transparent pressure vessels up to a few hundred bars. This pressure range has recently been increased to 2 kbar by Yamada (1974) by treating glass tubes internally and externally with hydrogen fluoride. His cell was intended for nmr spectroscopy, but could easily be used for infrared and Raman spectroscopy. Stryland and May (1960) made use of the finding (Bridgman 1918) that glass tubes can stand much higher pressures on the outside than they can on the inside because they can withstand much higher compressive stresses than tensile stresses. They made the only successful cell for Raman spectroscopy in pre-laser days by fitting a pressure vessel with a re-entrant glass tube, so that pressure was on the outside of the tube and the glass was under compression, as was first done by Madelung and Fuchs (1921). A mercury lamp was placed inside the re-entrant tube and the pressurized fluid was placed on the outside. The Raman radiation was removed from the pressure vessel through a separate window.

A diamond pressure vessel of the kind sometimes used for x-ray diffraction has been adapted to Raman spectroscopy (Wong and Whalley 1974). A hole 400 μm in diameter was drilled through the diamond, and four plane faces were ground parallel to the bore for the passage of light. The sample was placed in the hole and pressurized by a steel piston in a small hydraulic press. It can be used to about 18 kbar. Frequently, the effect of density on the spectrum is more important than the effect of pressure, and, because the compression of the sample can be measured by the relative displacement of the pistons, the effect of density on the spectrum can be measured directly. As an example, the frequencies of the rotational vibrations of anthracene are plotted against the density in Fig. 1.

2.2 Windowed hydrostatic vessels.

Windowed pressure vessels have been used for over a hundred years (Amagat 1877) and were first used for spectroscopy fifty-five years ago (Wick 1923). A typical window mounted on its plug is illustrated in Fig. 2. It is clear that the stronger the window or the smaller the aperture, the higher the pressure the window can withstand. Unfortunately, no detailed theoretical treatment of windows on an aperture has been published, although there are some undocumented formulas for thin windows. The highest shear stress in the window is usually at the edge of the aperture, because the window is well supported where it rests on the plug and is unsupported otherwise. The sharper the edge the bigger the shear stress, and so the edge of the aperture should be rounded. The shear stresses can also be reduced by shaping the sealing surface so that it dips towards the center. By doing this empirically, Whalley and Lavergne (unpublished) were able to test glass windows on an aperture of 0.5 mm to 45 kbar. For the highest pressures it is probably best to make the surfaces of the plug and window flat enough that they will seal without gaskets (Poulter 1932a). Sometimes, a thin gasket or a film of Canada balsam or the like are used to help sealing. Sometimes conical windows fitting into a conical recess in the plug, as was originally used by Amagat (1877), are used; a recent design for 10 kbar is described by Stromberg and Schock (1970).

For Raman spectroscopy, the best windows are probably glass. So-called single-crystal sapphire is often used; it is stronger than glass but often depolarizes the light and so cannot be used for polarization experiments. Diamond windows were used originally fifty-five years ago to 21 kbar on an aperture of 1.5 mm (Poulter 1932b), and could undoubtedly be used for much higher hydrostatic pressures, as they have in anvil apparatuses (see Section 2.4 and 2.5). However, many have enough internal strain to depolarize light.

For infrared spectroscopy, the strongest windows are undoubtedly diamond. Silicon and germanium are strong and usable in the far infrared. Sapphire is transparent above ~2.5 μm, and brucite (MgO) and the Irtrans (compacted ZnS etc. powders), sodium chloride, calcium fluoride etc. can be used.

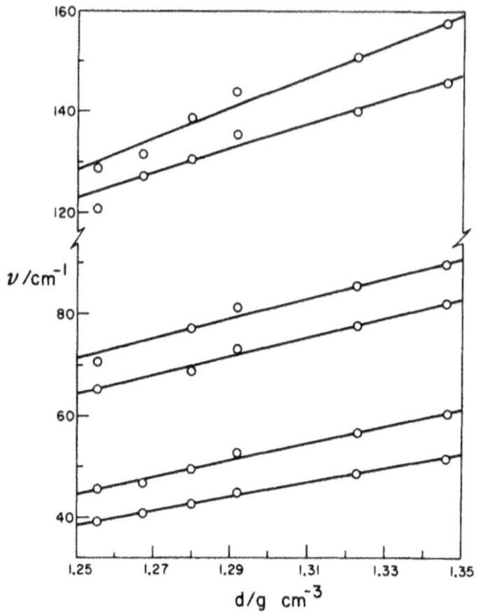

Fig. 1 Effect of density on the frequencies of the rotational vibrations of anthracene (Wong and Whalley 1974).

Salts like sodium chloride are opaque in the far infrared at room temperature, but become transparent near liquid helium temperature. A number of far infrared windows and the pressures they have withstood have been reviewed by Paul et al. (1968). Glass windows are frequently used for Raman spectroscopy up to about 10 kbar and several cells have been described (see, for example, Brafman et al. 1969, Stromberg and Schock 1970, Walrafen 1973), and one with sapphire windows has been used to 400°C and 5 kbar (Lindner 1972). A vessel used in the author's laboratory to 25 kbar is illustrated in Fig. 3 (Whalley et al. 1976). The aperture was 0.8 mm but has recently been increased to 1 mm, and the windows are glass.

2.3 Windowed non-hydrostatic vessels.

Either the aperture of a window or the pressure it can withstand can be increased greatly by protecting it from the full force of the pressure. This can be done by placing it at the end of a narrow bore filled with a transparent moderately plastic material (Fishman and Drickamer 1956, Fitch et al. 1957, Balchan and Drickamer 1960) as is illustrated in Fig. 4. The window protector, which is usually sodium chloride, carries a big enough pressure drop to allow over 50 kbar to be applied to the high-pressure end without cracking the window. The cell has been modified for low

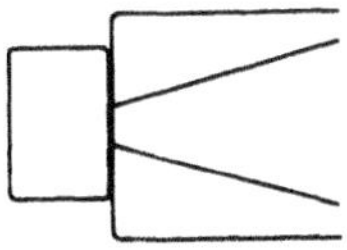

Fig. 2 Window mounted on a plug.

temperature (Sherman 1966, Offen et al. 1967, Lewis and Sherman 1974) and for Raman spectroscopy (Nicol 1965, Nicol et al. 1972).

2.4 Circular-anvil cells.

The highest static pressures are obtained by pushing together circular anvils of a hard material (Bridgman 1941) and many experiments have been done using anvils of steel or tungsten carbide. Spectroscopy under pressure can be done with transparent anvils, as was first shown by Runcorn (1956), who used anvils of fused quartz to 15 kbar and of sapphire to 30 kbar. Diamond anvils are the most common, and have been used for infrared spectroscopy for over twenty years (Weir et al. 1959). They were tried for Raman spectroscopy about ten years ago (Brasche et al. 1968, Postmus et al. 1968), but subsequently have been made really practicable (Adams and Payne 1973). The principles are illustrated in Fig. 5 for Raman spectroscopy. For infrared spectroscopy, the beam is condensed on the anvils. Recently Hawke et al. (1974) have used sapphire anvils for Raman spectroscopy to 80 kbar. Many of the measurements are done using as the pressure transmitter the (solid) sample itself, and sometimes the sample is dispersed in solid sodium chloride or other plastic material. This must be done if the pressure is over about 100 kbar at room temperature, but a 4:1 mixture of methanol and ethanol can be used for hydrostatic pressure up to 100 kbar. The pressure can (Barnett et al. 1973) be measured directly by inserting a small piece of ruby and measuring the frequency of the fluorescence R line, which shifts at the rate $-0.753\ cm^{-1}\ kbar^{-1}$ (Piermarini et al. 1975).

2.5 Multianvil cells.

Larger volumes of sample can be accommodated if, instead of pushing on a thin slice in a circular-anvil cell, a tetrahedron is pushed by four anvils with triangular faces, or a cube is pushed by six anvils with square faces. The tetrahedral-anvil cell is usually quite inconvenient for most spectroscopic work, but has been used for Raman spectroscopy (Clark and Wedlake 1973, Parsons 1977). The

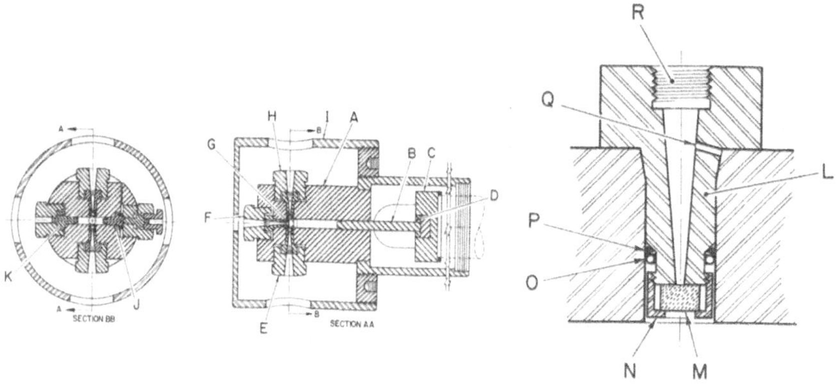

Fig. 3 Raman cell for hydrostatic pressures of 25 kbar (Whalley et al. 1976). The window assembly is shown enlarged. The vessel for the large piston of the intensifier is not shown.

cubic-anvil apparatus is more convenient for absorption measurements because light can pass straight through. Such a cell has been described by Owen (1966) and used with diamond windows on a 1-mm aperture to 55 kbar.

2.6 Spectroscopy of high-pressure phases recovered at low temperatures.

Frequently, a high-pressure phase can be recovered at zero pressure at a suitable temperature. Then, if the spectrum of the phase is required rather than the effect of pressure on the spectrum, it is often better to measure the spectrum of the recovered phase. The quality is usually higher because it is not determined by the constraints of a pressure vessel, and it is more suitable for comparing with the spectrum of the stable phase at zero pressure. In addition, accurate crystal structures are often measured for the recovered phases.

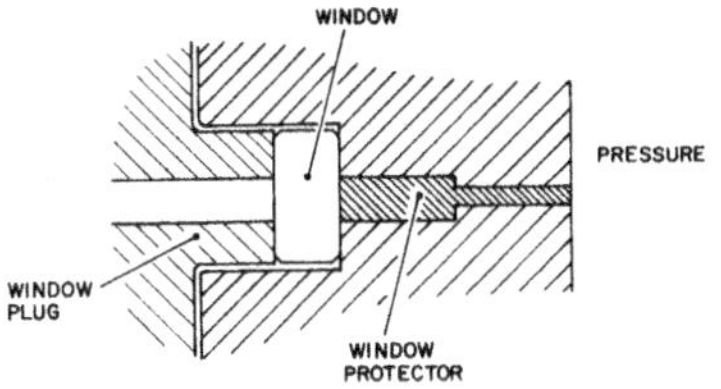

Fig. 4 Window protected from the full force of the pressure by a plastic material.

Frequently, the high-pressure phase is stable at room temperature, as is true of diamond, and then no special techniques need be used. Frequently, however, the samples must be recovered at low temperatures. The usual technique of making mulls for infrared spectroscopy has been adapted for temperatures down to 77 K (Bertie and Whalley 1964) using materials such as propane or propylene as mulling agents. The manipulations required to make the mulls are carried out in a refrigerator consisting of an uninsulated steel can with liquid nitrogen boiling rapidly in the bottom.

To make a sample of a recovered phase suitable for Raman spectroscopy (Wong and Whalley 1972), a small length of copper channel is immersed in the sample and the sample transformed to the high pressure phase, which is then cooled and removed from the pressure vessel. The channel is recovered filled with the high-pressure phase, the faces of the sample are polished with polishing paper and it is mounted on the cold finger of a low-temperature Raman cell.

3. SELECTED APPLICATIONS OF HIGH-PRESSURE INFRARED AND RAMAN SPECTROSCOPY

High-pressure infrared and Raman spectroscopy can be used to study many fields of chemistry, physics, and biology. Only a few of these can be reviewed in the space available, and I have selected a few examples to illustrate the kind of information that can be obtained. If an undue number of the examples are from our own work, it is only because I know them well.

3.1 Effect of pressure on vibration frequencies

3.1.1 Diamond. The potential energy of an atomic system can be expanded in the displacements from the equilibrium positions of the atoms when no force is applied to the system. The principles can be illustrated by the effect of pressure on the Raman spectrum of diamond. Diamond has one optically

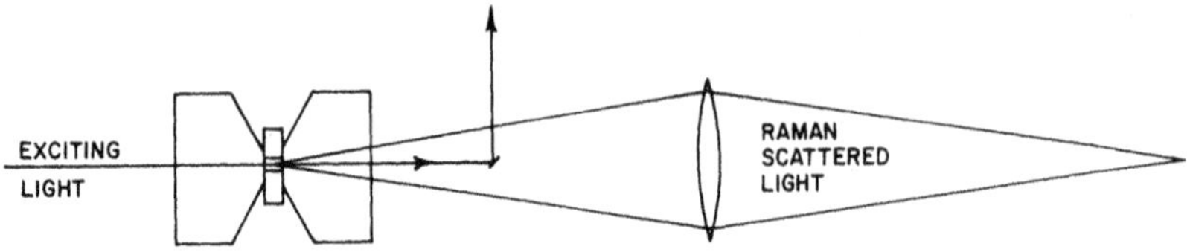

Fig. 5 Principle of anvil cells.

active fundamental vibration. It is triply degenerate, and each component can be considered as alternate planes perpendicular to the cubic axes vibrating against one another. Its frequency at zero pressure is 1334.0 cm^{-1} and it shifts with pressure at the rate 0.296 ± 0.011 cm^{-1} $kbar^{-1}$, as is illustrated in Fig. 6 (Whalley et al. 1976). The most elaborate potential that can be analyzed by the data is an interplanar potential

$$V = \frac{1}{2} k_2(r-r_e)^2 + \frac{1}{3!} k_3(r-r_e)^3 + \frac{1}{4!} k_4(r-r_e)^4 + \ldots, \qquad (3.1)$$

where r is the instantaneous distance between nearest neighbor planes and r_e the equilibrium distance. To terms in $(k_3/k_2)^2$, the force constant k_2 per atom is $4\pi^2c^2\nu^2m/2$ where c is the speed of light, ν the frequency in wave numbers, and m the mass of a carbon atom.

The effect of hydrostatic pressure is to shift the equilibrium internuclear distance from r_e to $r_e' = r_e - \delta$. Then the potential energy must be expanded about the new equilibrium position r_e'. When $r_e = r_e' + \delta$ is substituted into equation (3.1) we find

$$V = \frac{1}{2}(k_2-k_3\delta-k_4\delta^2-\ldots)(r-r_e')^2+\frac{1}{3!}(k_3-k_4\delta-\ldots)(r-r_e')^3+\ldots(3.2)$$

It follows from this equation that if ρ is the density

$$\left(\frac{\partial \ln \nu}{\partial \ln \rho}\right)_{T,p=o} = \frac{1}{6} r_e \frac{k_3}{k_2}$$

If the potential function is purely harmonic, $k_3 = 0$ and the frequency is independent of the density. This is a general conclusion, and so it follows that the effect of pressure on vibration frequencies is a direct measure of the anharmonicity of the vibrations, and is a useful method of determining it.

From the experimental results for diamond,

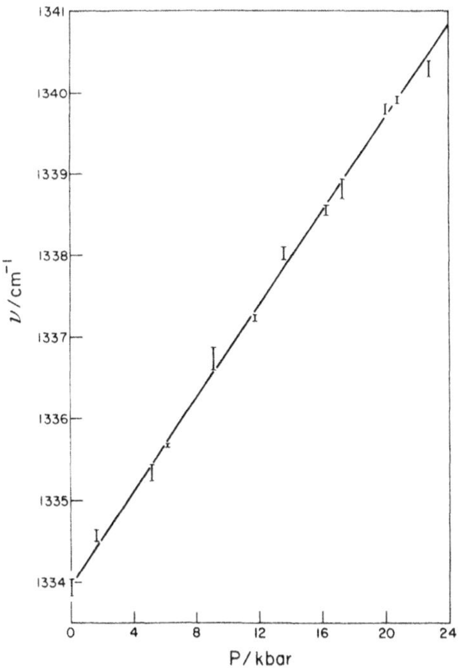

Fig. 6 Effect of pressure on the frequency of the fundamental Raman band of diamond (Whalley et al. 1976).

$$k_2 = 6.28 \text{ mdyn } \text{Å}^{-1},$$

$$k_3 = -42 \text{ mdyn } \text{Å}^{-2}.$$

If the change of potential energy caused by compressing a set of planes is independent of whether or not the planes perpendicular to this set are compressed, these force constants lead to the compressibility

$$\kappa/\text{Mbar}^{-1} = 0.227 - 0.359\ p/\text{Mbar}^{-1}.$$

The compressibility at zero pressure agrees well with the best value of 0.226 Mbar^{-1} (Grimsditch and Ramdas 1975).

3.1.2 Water. The Raman spectrum of water up to 10 kbar tells about how water responds to pressure. According to Walrafen (1973) the O-D stretching frequency ν_{OD} of a dilute solution of HDO in H_2O, so that inter- and intra-molecular coupling is eliminated, decreases with increasing pressure at the rate of about 2.3 cm^{-1} kbar^{-1}. From the spectra of various phases of ice and of ice VIII under pressure it appears that the O-D frequency is a function principally of the O--O distance, and much less of the O-H--O angle and other angles, and that it increases with increasing O--O distance at the approximate rate 770 cm^{-1} $Å^{-1}$ (Kamb et al. 1971). It follows from the effect of pressure and distance on the frequency that the O--O distance changes with pressure approximately as (Whalley 1976)

$$\partial r_{O--O}/\partial p = -0.0030 \text{ Å kbar}^{-1}.$$

The corresponding quantity for ice I, which can be calculated from the compressibility of 12 $Mbar^{-1}$ near the melting point, is -0.010 Å $kbar^{-1}$. The hydrogen bond in water, therefore, contracts at about 0.3 of the rate that the hydrogen bond in ice contracts. The hydrogen bond in water is weaker than in ice, but there are more of them per unit volume. Consequently, a simple model would predict that the hydrogen bond in water would contract at about the rate it does in ice, 0.010 Å $kbar^{-1}$, which is 3.3 times more than is observed.

The compressibility of water near the melting point is about four times that of ice. The compressibilities would be similar if pressure did not cause the O--O--O angles in water to bend, and so angle bending must contribute a large fraction of the compressibility of water. In ice I, the O--O--O angles are nearly tetrahedral, and the bond distances about 2.76 Å. In the high-pressure phases II, IX, V, and VI, the angles deviate greatly from tetrahedral, and the O--O bond lengths are longer than in ice I, although the phases are over 20% denser. It seems likely therefore, that bending O--O--O angles tends to make O--O distances increase.

The unexpectedly small decrease in O--O distance of 0.030 Å $kbar^{-1}$ can then be explained (Whalley 1976) as due to a compression of about 0.010 Å $kbar^{-1}$ caused by the direct effect of pressure on the bond, and an expansion of about 0.07 Å $kbar^{-1}$ due to an indirect effect of the pressure caused by bending the O--O--O angles.

3.1.3 Effective charge. The effective charge e^* of a pair of atoms measures the change of dipole moment μ when the internuclear distance r changes

$$e^* = \partial\mu/\partial r.$$

For an ideal sodium chloride crystal, say, the effective charge of the sodium and chloride ions would be one electronic charge. In fact, it is a little less because the ions polarize one another (Boer and Huang 1954). The first estimates of the effect of pressure on the effective charge were made by Jones (1967) and Cundill and Sherman (1968). The most comprehensive investigation was made by Mitra and Namjoshi (1971). According to Szigetti, the effective charge is related to the transverse optic frequency ω_{TO} in radians per unit time and to the limiting low- and high-frequency permittivities ε_o and ε_∞ by the relation

$$e^{*2} = (9V\mu/4\pi)\ [(\varepsilon_o-\varepsilon_\infty)/(\varepsilon_\infty+2)^2]\omega_{TO}^{\ 2},$$

where V is the volume of a unit cell and μ the reduced mass of the ion pair. The effect of pressure on e* can therefore be obtained from the effects of pressure on ε_O, ε_∞, and ω_{TO}, all of which have been measured. For potassium bromide the rate

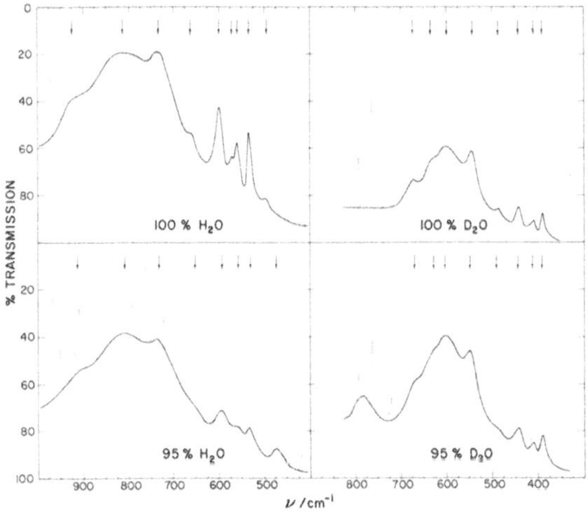

Fig. 7 Infrared spectrum of the rotational vibrations of ice IX (Bertie and Whalley 1964b). The sample was recovered and examined at ∿100 K and zero pressure.

of change at zero pressure is -1.89 Mbar^{-1}, and the largest effect occurred for cesium bromide, -8.20 Mbar^{-1}. If the effective charge were linear in the pressure, cesium bromide would become a covalent crystal at ∿100 kbar. It is probably a better approximation, however, to suppose that it is linear in the volume. Then cesium bromide would become covalent at about a megabar, if nothing else happened first.

3.1.4 Recovered phases. High-pressure phases are often better studied spectroscopically if they can be recovered at low temperature. Then better spectra can often be obtained than at high pressure. The spectra of several phases of ice have been studied in this way (e.g. Bertie and Whalley 1964b, Bertie and Bates 1977). One example is ice IX, whose infrared spectrum in the region of the rotational vibrations (Bertie and Whalley 1964b) is shown in Fig. 7. Although much of the band is broad and relatively featureless, it has several sharp components that broaden greatly when 5% of D_2O is added to the ice. This can only occur if the phase is not orientationally disordered, as in ice I, but has ordered orientations. This was the first identification of an ordered phase of ice.

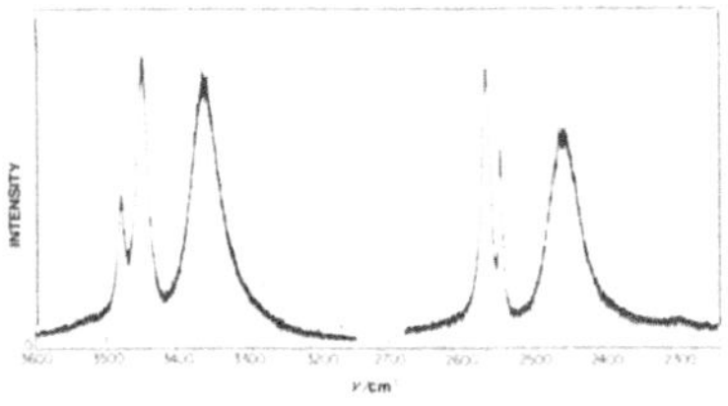

Fig. 8 Raman spectrum of the O-H and O-D stretching vibration of ice VIII (Wong and Whalley 1976).

Another example is the Raman spectrum of ice VIII, whose spectrum in the region of 3 μm is shown in Fig. 8 (Wong and Whalley 1976). The normal coordinates of the vibrations are determined by symmetry, apart from the small coupling with the bending and lower-frequency vibrations, and when the infrared spectrum becomes known, much can be learned about the forces between the water molecules.

3.2 Effect of pressure on intensities

The intensity of a band depends on the concentration of the molecule that carries it, and if the concentration is a function of pressure it can be determined from the intensity.

But intensities also change even if the concentration does not change. The dipole moment or the polarizability of a molecule or crystal can be expanded in the coordinates in much the same way as the potential energy, equation (3.3). For a single normal coordinate the expansion of the dipole moment μ is

$$\mu = \mu_0 + \frac{\partial \mu}{\partial Q} Q + \frac{1}{2} \frac{\partial^2 \mu}{\partial Q^2} Q^2 + \ldots,$$

and there are corresponding equations for the components of the polarizability. If pressure shifts the equilibrium value of the normal coordinate Q from zero to q, the intensity under pressure is determined by the coefficients of the expansion about the new equilibrium value. This expansion is

$$\mu = \mu_0 + \frac{\partial \mu}{\partial Q} q + \ldots + \left(\frac{\partial \mu}{\partial Q} + \frac{\partial^2 \mu}{\partial Q^2} q + \ldots\right)(Q-q) + \ldots$$

The new dipole-moment derivative is then the value at zero pressure plus the term $(\partial^2\mu/\partial Q^2)q$ and higher-order terms. The absorption (or scattering) intensity depends therefore on the pressure, and can be used to learn about the effects of pressure.

3.2.1 Chemical analysis. As the absorption or scattering intensity of a substance in dilute solution is proportional to its concentration, they can be used for chemical analysis. Of course, the intensity at fixed concentration depends on both the temperature and the pressure, on the concentration if the solution is not dilute, and on the nature of the solid if the material is in a solid phase. Sometimes, allowance can be made for these effects, but sometimes it cannot, and then the accuracy of the final conclusions must be reduced to allow for this uncertainty. The intensities of Raman bands often depend less on the environment than the intensities of infrared bands, and so are often more useful for measuring concentrations.

An example is the measurement of the dissociation of the bisulphate ion in water

$$HSO_4^- + H_2O \rightleftharpoons SO_4^{2-} + H_3O^+,$$

an important equilibrium in sea water. The ratio of the integrated intensities of the SO_4^{2-} band at 982 cm^{-1} and the HSO_4^- band at 1052 cm^{-1} increases from 0.36 to 0.54 between 0 and 1 kbar at 300 K (Davis et al. 1974). The bisulphate ion clearly dissociates under pressure, but the extent is uncertain because the change in the relative scattering cross-sections of the two ions is unknown.

Crystalline phosphorus pentachloride and pentabromide have quite different chemical constitutions, the chloride being $PCl_4^+PCl_6^-$ and the bromide $PBr_4^+Br^-$ (see Wyckoff 1964, Vol. 2, pp. 170, 171) and the reason for the difference has been widely discussed. The two structures can be considered to be related by the equilibrium

$$PCl_6^- \rightleftharpoons PCl_4^+ + 2Cl^-. \tag{3.2}$$

Pressure will of course favor the side with the lower volume. A rough calculation based on the known and estimated interatomic distances suggested that $PCl_4^+Cl^-$ having the $PBr_4^+Br^-$ structure would be denser than the ordinary phase $PCl_4^+PCl_6^-$. Pressure should then tend to favor $PCl_4^+Cl^-$ over $PCl_4^+PCl_6^-$, and so to induce the ionization described by reaction (3.2). Phosphorus pentachloride forms several phases under pressure (Cahay and Whalley, unpublished). In the spectrum of one of them (Cahay et al., unpublished) the intensities of the PCl_6^- bands relative to the PCl_4^+ bands are much smaller than in the ordinary phase, showing clearly that the ionization (3.2) has occurred, but not completely. It appears to be a unique example of an ionization occurring in a solid phase.

Kluyev and Vorob'ev (1964) have reported that when potassium ferricyanide was compressed to $\sim$24 kbar, it was reduced to ferrocyanide, as detected by the infrared spectrum, and in many papers Drickamer and coworkers (Drickamer and Frank 1973) have shown by Mössbauer spectroscopy that many FeIII salts are reduced to FeII at high pressures. In all these experiments, the pressure-transmitting medium was a solid, and so there must have been significant shear. Chemical changes can be induced by shear (Bridgman 1936, 1937) and in particular, potassium ferricyanide is readily reduced by shearing at pressures between 3 and 40 kbar (Larsen and Drickamer 1957). Gardiner et al. (1970) showed by infrared spectroscopy that cycling the pressure increased the extent of reduction, and their experiment gave no evidence for a purely pressure-induced reduction.

Lewis and Whalley (unpublished) have recorded the Raman spectrum of single-crystal potassium ferricyanide at purely hydrostatic pressures up to 18 kbar and found none of the spectral changes expected if ferrocyanide were formed, but only a completely reversible shift of frequencies of the kind expected if no chemical changes occurred. It seems certain that no pressure-induced reduction of potassium ferricyanide occurs up to 18 kbar, and suggests that much of the reported reduction should be considered sub-judice until verified under hydrostatic pressures.

Raman spectroscopy has recently been used to measure the effect of pressure on the equilibrium between conformational isomers (Taniguchi et al., unpublished). Others have used infrared spectroscopy for this purpose (Christian et al. 1975, 1976). If the C-Cl and C-Br stretching vibrations of dichloroethane and dibromoethane are not significantly coupled to the rest of the molecule, it can be shown that the sum of the intensities of the symmetric (i.e. in-phase) and asymmetric (i.e. out-of-phase) stretching vibrations is independent of the rotation angle. Hence, the ratio of the integrated intensities of the C-halogen stretching vibrations of the gauche and trans conformers is a measure of the equilibrium constant K, and the effect of pressure on the equilibrium constant is determined by the volume difference ΔV between the conformers,

$$\frac{\partial \ln K}{\partial p} = -\frac{\Delta V}{RT},$$

where R is the gas constant and T the temperature. Preliminary analysis of the ratio of intensities yields the volume change between the gauche and trans isomers of ~ -3.8 and ~ -4.5 cm^3 $mole^{-1}$ for the chloride and bromide respectively. Part of this contraction can be accounted for by the compression of the halogen atom in the gauche conformation, and part from the effects of the dipole moment and the anisotropic invariant of the quadrupole moment on the volume of solvation of the conformers. The effect of the electrostatic moments is relatively small as, although the trans conformer has no dipole moment, it has a smaller anisotropic invariant

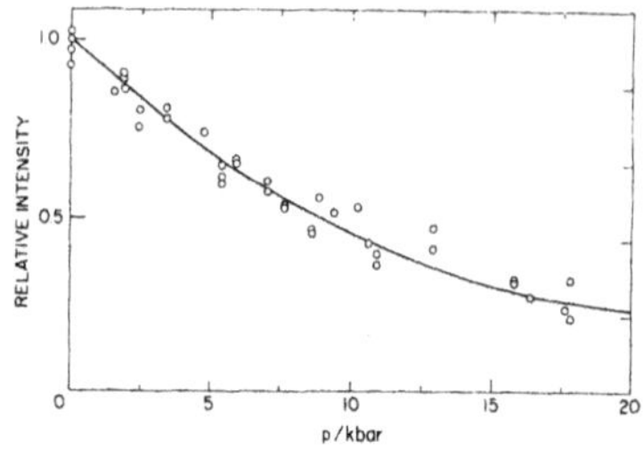

Fig. 9 Effect of pressure on the integrated Raman intensity of the ring-flattening vibration of trithiane relative to the value at zero pressure (Lewis and Whalley 1978).

of the quadrupole moment than the gauche conformer has.

3.2.2 Conformational changes. If internal rotation angles are not determined by symmetry, as they are for example in ethane, they may be changed by pressure. Even in ethane, high enough pressure will probably change the conformation from staggered to eclipsed. One example that has been studied recently is the flattening of trithiane under pressure. The first evidence that trithiane might flatten was reported by Brasch et al. (1970) and Hamann et al. (1973). Several bands in the infrared spectra of trioxane and trithiane were greatly reduced in intensity under pressures of the magnitude of 50 kbar, and it was suggested purely empirically that the molecules became flat. The observed effects and the assumed vibrational assignments were not, in fact, consistent with flattening, but the assignment for trithiane could be changed in a plausible way to make it so (Lewis & Whalley 1978).

If trithiane were to flatten, the principal effect on the Raman spectrum would be the disappearance of the A_1, based on the symmetry of the isolated molecule, internal rotation at 309 cm^{-1}. Experimentally (Lewis and Whalley 1978) its intensity relative to its neighbors at 18 kbar is about one quarter of its value of zero pressure, as shown in Fig. 9. For small degrees of puckering, symmetry requires the intensity of this vibration to be proportional to the square of the degree of puckering. Hence the degree of puckering at 18 kbar is about one half of its value at zero pressure, and it is predicted that the molecule should become flat at $\sim$38 kbar.

3.2.3 Fermi resonance. When a combination band is near a fundamental having the same symmetry, they can interact through anharmonic terms in the potential energy. This causes them to mix and repel one another in frequency space, and to transfer intensity from the fundamental to the overtone. If two bands in Fermi resonance have different pressure coefficients, Fermi resonance can be modified by changing the pressure. This is illustrated in Fig. 10. If the fundamental and combination had different symmetries, they would cross one another as indicated by the dashed lines. If they have the same symmetry, they cannot cross, and they behave as indicated by the solid lines. Then the fundamental gradually loses intensity as the pressure is increased and becomes the combination, and the combination gains the lost intensity and becomes the fundamental. The effect of pressure was first reported by Sherman and Smulovitch (1970) for the ν_3 and $\nu_1 + \nu_T$ bands of the ammonium ion dissolved in cesium bromide, as shown in Fig. 11. The gradual change in the relative intensity of the two bands is very striking.

3.2.4 Amorphous silicon and germanium. Work at high pressures often points the way to progress in quite a different field. One example is the spectra of highly disordered materials, which we have been concerned with for some years. One of the problems in this field is that amorphous silicon and germanium absorb infrared radiation throughout the whole region of their lattice vibrations, whereas crystalline silicon and germanium do not, but the origin of the absorption in the amorphous phases was unknown. The treatment of the spectra of solids under pressure, in which the potential function for displacements from equilibrium under pressure can be related to the function at zero pressure, suggested that a similar treatment can be applied to the displacements required to convert a crystal to an amorphous phase (Klug and Whalley 1977). By symmetry, the first term in the expansion of the dipole moment $\underset{\sim}{\mu}$ of a crystal of silicon or germanium in the cartesian displacements r_i is the quadratic

$$\underset{\sim}{\mu} = \frac{1}{2} \sum_{ij} \underset{\sim}{e}_{ij} \, r_i \, r_j + ---,$$

where

$$\underset{\sim}{e}_{ij} = \partial^2 \underset{\sim}{\mu} / \partial r_i \partial r_j .$$

If an amorphous phase can be made by permanent cartesian displacements ρ_i of the coordinates r_i, its dipole moment has first-order derivatives

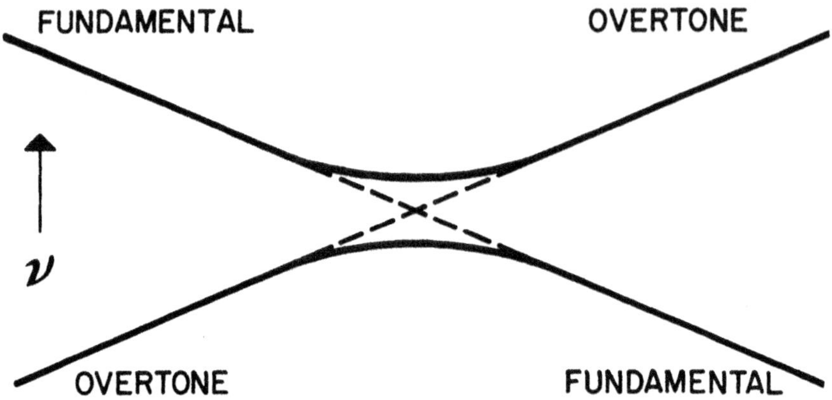

Fig. 10 Variation of the frequency and type of motion of two bands in Fermi resonance as the resonance is perturbed by pressure, for example.

$$\partial \underset{\sim}{\mu}/\partial(r_i - \rho_i) = \sum_{ij} \underset{\sim}{e}_{ij}\, \rho_j,$$

by analogy with equation (3.1). The second-order derivatives in the crystal are closely related to the intensities of binary combinations, and it can be shown that the ratio of the zeroth moment of the fundamental absorptivity of the amorphous phase to the first moment of the binary-combination absorptivity of the crystal is

$$\frac{N_c}{N_a}\, \frac{\pi m c}{\hbar \alpha} \langle \rho_i^{\,2} \rangle_i,$$

where N_c and N_a are the number densities of atoms of the crystalline and amorphous phases, m the mass of the atoms, c the speed of light, $\hbar$ Planck's constant divided by 2π, α a quantity that can be readily obtained from the spectrum, and $\langle \rho_i^{\,2} \rangle_i$ the mean square displacement. From the measured spectra, $\langle \rho_i^{\,2} \rangle_i^{\frac{1}{2}}$ is about 0.6 Å for silicon and 0.3 Å for germanium.

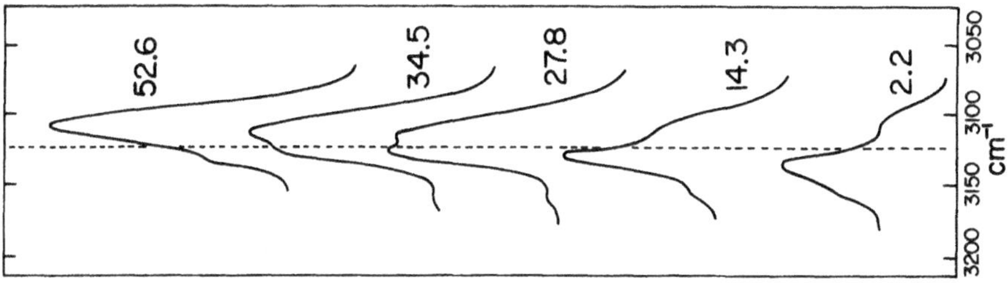

Fig. 11 Fermi resonance of NH_4^+ dissolved in cesium bromide (after Sherman and Smulovitch 1970). The numbers attached to the spectra are the pressures in kbars, and the dashed line indicates the mean frequency of the resonance doublet.

3.2.5 Structure of solids. The amount of information about crystal structure that can be obtained from infrared and Raman spectroscopy is quite limited because vibrational frequencies and intensities are for the most part only remotely related to the crystal structure. Examples in which useful and surprising information has been obtained are mercurous chloride and bromide. According to x-ray diffraction, both crystallize at room temperature in a body-centered tetragonal cell in space group I4/mmm. There are translationally identical molecules at the corners and the center of the cell having their axes parallel to [001]. Both have Raman-active bands at about 40 cm^{-1} due to an E_g rotational vibration about the axes perpendicular to the molecular axis. No other Raman band is expected in this region.

The Raman spectra of both crystals at pressures up to about 15 kbar (Richter et al. 1977) is shown in Fig. 12. At high pressures, each band acquires a low-frequency component whose intensity relative to the principal band increases smoothly to ∿0.25 and ∿0.20 for the chloride and bromide respectively. Knowing that it occurs under pressure allowed us to resolve a weak component of relative intensity ∿0.05 at zero pressure.

These new bands cannot be explained on the basis of space group I4/mmm, and so there is either a distortion of the unit

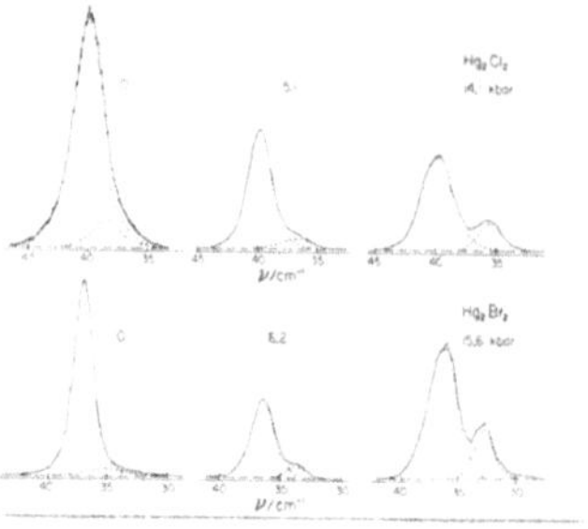

Fig. 12 Raman spectrum of the rotational bands of mercurous chloride and bromide (Richter, Wong, and Whalley 1977).

cell to remove the degeneracy, or the formation of a super cell which would allow zone-boundary vibrations of the tetragonal cell to become active. If the degeneracy were removed, the band would split into two of approximately equal intensities. If the unit cell were (say) doubled by a distortion that increased with increasing pressure, it would cause a weak side band that would increase in intensity with increasing pressure. It seems reasonably certain that the structure of mercurous chloride and bromide is distorted from the ideal body-centered tetragonal structure even at zero pressure, and the observed tetragonal diffraction pattern must be due to disorder in the distortion. The chloride and bromide transform at 185 and 143 K respectively and zero pressure to a phase of lower symmetry (Barta et al. 1976). Perhaps the disordered structure at room temperature is related to the (at present unknown) structure of the low-temperature phases.

3.2.6 Brillouin spectroscopy. The Brillouin spectrum is the Raman spectrum of the sound waves. The shift of frequency and wave vector of the exciting light is the frequency and wave vector of the sound wave that interacts with it. In solids, both compression and shear waves give a Brillouin spectrum, and from it the elastic constants at the frequency of the sound waves, which is in the region of 10 GHz, can be measured. In liquids, the shear waves give a spectrum only if the shear relaxation time is greater than about 10 ps. Toluene has no such components at zero pressure, but develops them as the pressure is increased, as shown in Fig. 13 (Buccaro et al. 1974).

3.3 Effect of pressure on the band shapes

The breadths of vibrational lines in liquids depends principally on the life-times of vibrational states caused by the diffusion and reorientation of the molecules and on intermolecular vibrational coupling. In some simple liquids,

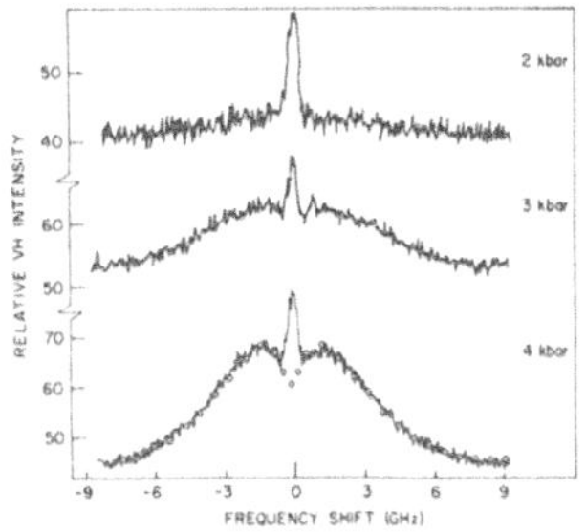

Fig. 13 Effect of pressure up to 4 kbar on the Brillouin spectrum of toluene (Buccaro et al. 1974).

the effect of intermolecular coupling and diffusion is small, and then the breadth tells about the rate of reorientation. The normalized Fourier transform of the band of a polarized vibration is related to the orientational correlation function

$$P_2[\cos\theta(t)] = \tfrac{1}{2}[3\cos^2\theta(t) - 1]$$

where P_2 is the second Legendre polynomial and $\theta(t)$ the angle between the directions of a molecule at times 0 and t. The ensemble average value of the function is

$$\langle P_2[\cos\theta(t)]\rangle = \int I(\omega)\cos\omega t d\omega / \int I_p(\omega)d\omega,$$

where $I(\omega)$ is the intensity of the scattering by a polarized vibration at frequency (in units of radians per unit time) ω. Other effects on the band shape than orientational correlation can be eliminated by dividing the Fourier transform by the Fourier transform of a depolarized band.

The orientational autocorrelation function for chloroform at pressures up to 2 kbar (Campbell and Jonas 1973), based on the A_1 C-H stretching band at 3019 cm^{-1}, is shown in Fig. 14. In times up to ∿0.3 ps, the molecule behaves like a free rotor at all pressures, but at times greater than ∿0.6 ps, the autocorrelation function decays by a first-order law, and the rate is greatly decreased by increasing the pressure, as might be expected from the increased viscosity.

3.4 Effect of non-hydrostatic stresses

The effect of hydrostatic pressure is, of course, only a small part of the more general field, the effect of stress. A general stress field has three independent components, which might be taken as the three principal normal stresses, and consequently there are three variables, in place of the single

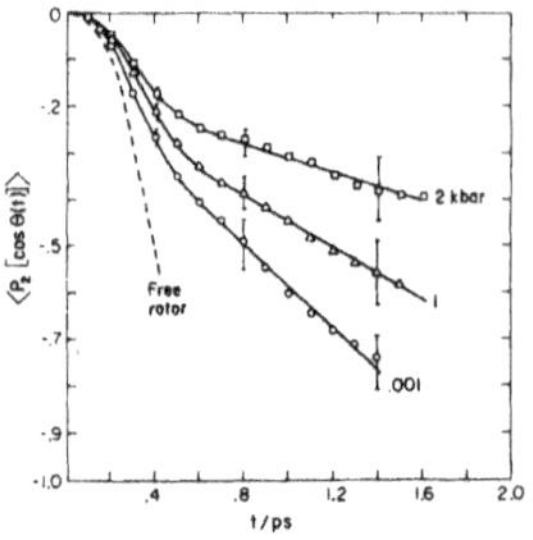

Fig. 14 Orientational correlation function $P_2[\cos\theta(t)]$ of chloroform at °C (after Campbell and Jonas 1973).

variable when hydrostatic pressure is used. There is therefore more information to be obtained from varying the stress than from varying the hydrostatic pressure. In infrared and Raman spectroscopy the effect of stress on the frequencies, intensities, and half widths gives information on the effect of stress on the force constants, dipole-moment and polarizability derivatives, and lifetimes of the normal oscillations. In turn these can tell about the anharmonic terms in the force constants and the non-linearities in the dipole-moment and polarizability derivatives. The first experiments were done long before any high-pressure Raman spectra were obtained (Mariée and Mathieu 1946).

Cubic crystals having the diamond structure have two atoms per unit cell and one triply degenerate F_{2g} Raman-active vibration. If a uniaxial stress is applied along a cube edge, the symmetry is reduced to tetragonal, and the triply degenerate vibration splits into a B_{1g} vibration having displacements parallel to the applied stress, and an E_g vibration having displacements perpendicular to the applied stress. A stress of 10 kbar along the [100] direction causes the B_{1g} and E_g bands of germanium to shift upwards by 0.6 cm^{-1} and 1.4 cm^{-1} respectively (Cerdiera et al. 1972), as shown in Fig. 15. A similar, but larger splitting occurs when a stress is applied along the cube diagonal, as is also shown in Fig. 15.

Non-hydrostatic stresses can also induce phase transitions. Strontium titanate undergoes a transition from a tetragonal to a trigonal structure under stress along the [111] direction (Burke and Pressley 1969). The effect of pressure on the transition can be investigated by Raman spectroscopy, as Fig. 16 shows. The transition occurs at 110 K at zero pressure, and moves to lower temperatures at higher pressures (Burke et al. 1971).

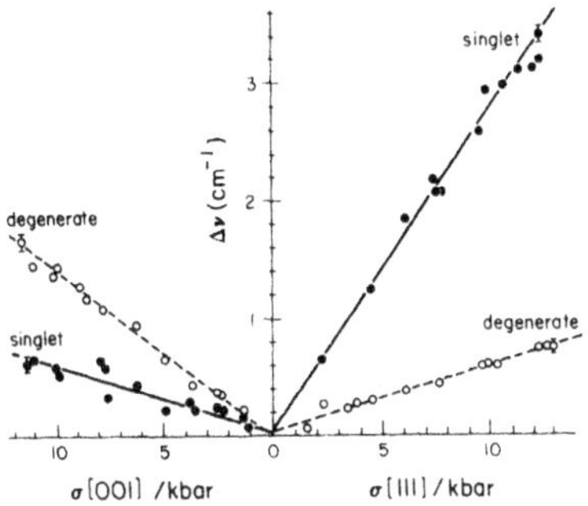

Fig. 15 Effect of uniaxial stress along the [001] and [111] directions on the frequencies of the fundamental Raman spectrum of germanium (after Cerdiera et al. 1972)

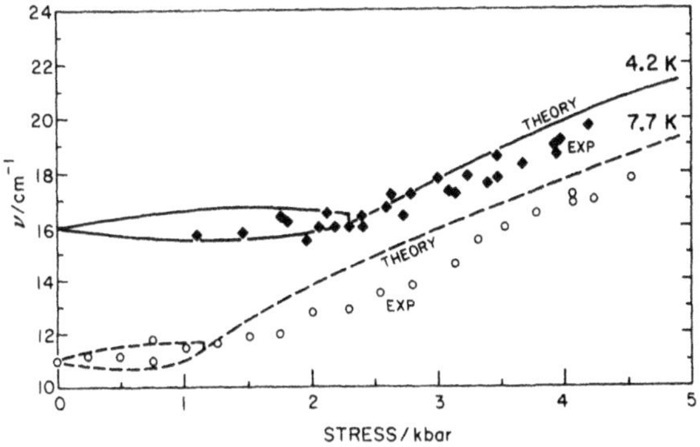

Fig. 16 Effect of uniaxial stress on the phase transformation in strontium titanate, as detected by the Raman spectrum (after Burke et al. 1971).

4. CONCLUSION

I have attempted to review some of the things that can be learned from infrared and Raman spectroscopy under pressure, particularly in the field of physical chemisdry. A large number of interesting fields have been omitted because time and space is limited, and so I have concentrated more on fields that have interested me. An even larger number of fields have not yet been touched. The subject has slowly increased in vigor since the first experiments were done in the 1950s, and since lasers were first used for Raman spectroscopy in the late 1960s. It will no doubt continue to increase, particularly perhaps in the infrared, now that very efficient Michelson interferometers are becoming common, and particularly if a practical tunable wide-band infrared laser is produced. We can look forward to learning much more about the effects of pressure in many fields.

REFERENCES

D.M. Adams and S.J. Payne 1973. The fluorescence of diamond and Raman spectroscopy at high pressures using a new design of anvil cell. Appl. Spect. 27, 377-381.

E.H. Amagat 1893. Mémoires sur l'elasticité et la dilatabilité des fluides jusqu'aux très hautes pressions. Ann. Chim. Phys. 29, 68-136.

A.S. Balchan and H.G. Drickamer 1960. High pressure high temperature optical device. Rev. Sci. Instrum. 31, 511-513.

J.D. Barnett, S. Block, and G.J. Piermarini 1973. An optical fluorescence system for quantitative pressure measurement in the diamond-anvil cell. Rev. Sci. Instr. 44, 1-9.

C. Barta, A.A. Kaplyanskii, V.V. Kulakov, B.Z. Malkin, and Y.F. Markov 1976. Raman scattering spectra and structural phase transitions in the improper ferroelastics Hg_2Cl_2 and Hg_2Br_2. Zh. Eksp. Teor. Fiz. 70, 1429-1444. English translation, Soviet Phys. JETP 43, 744-752.

J.E. Bertie and F.E. Bates 1977. Mid-infrared spectra of deuterated ices at 10°K and interpretation of the OD stretching bands of ices II and IX. J. Chem. Phys. 67, 1511-1518.

J.E. Bertie and E. Whalley 1964a. Infrared spectra by mulling techniques at liquid-nitrogen temperature. Spectrochim. Acta 20, 1349-1356.

J.E. Bertie and E. Whalley 1964b. Infrared spectra of ices II, III, and V in the range 4000-350 cm^{-1}. J. Chem. Phys. 40, 1646-1649.

M. Born and K. Huang 1954. Dynamical theory of crystal lattices. Oxford University Press, London.

O. Brafman, S.S. Mitra, R.K. Crawford, W.B. Daniels, C. Postmus, and J.R. Ferraro 1969. Pressure dependence of the Raman spectra of solids. Phase transition in TℓI. Solid State communications 7, 449-452 (1969).

J.W. Brasch, A.J. Melveger, and E.R. Lippincott 1968. Laser excited Raman spectra of samples under very high pressure. Chem. Phys. Lett. 2, 99-100.

J.W. Brasch, A.J. Melveger, E.R. Lippincott, and S.D. Hamann 1970. Spectroscopic monitoring of high pressure effects in trioxane. Appl. Spect. 24, 184-186.

P.W. Bridgman 1918. The failure of cavities in crystals and rocks under pressure. Am. J. Sci. 45, 243-268.

P.W. Bridgman 1936. Shearing phenomena at high pressure of possible importance for geology. J. Geol. 44, 653-669.

P.W. Bridgman 1937. Shearing phenomena at high pressures, particularly in inorganic compounds, Proc. Am. Acad. Arts Sci. 71, 387-460.

P.W. Bridgman 1941. Explorations towards the limit of utilizable pressures. J. Appl. Phys. 12, 461-469.

J.A. Buccaro, H.D. Hardy, and R.F. Horton 1974. Depolarized doublet spectra of toluene at high pressures. J. Chem. Phys. 60, 1671-1672.

W.J. Burke and R.J. Pressley 1971. Stress-induced ferroelasticity in strontium titanate. Solid State Commun. 9, 191-195.

W.J. Burke, R.J. Pressley, and J.C. Slonczewski 1971. Raman scattering and phase transition in stressed $SrTiO_3$. Solid State Commun. 9, 121-124.

R. Cahay and E. Whalley, unpublished. Ionization of PCl_6^- in the solid state under pressure. I. Phase diagram of phosphorous pentachloride.

R. Cahay, P.T.T. Wong, and E. Whalley, unpublished. Ionization of PCl_6^- in the solid state under pressure. II. Raman spectra of phosphorous pentachloride II and III.

J.H. Campbell and J. Jonas 1973. Raman spectroscopy in liquids at high pressure. Chem. Phys. Lett. 18, 441-445.

F. Cerdiera, C.J. Buchenauer, F.H. Pollak, and M. Cardona 1972. Stress-induced shifts of first-order Raman frequencies of diamond and zinc-blend type semiconductors. Phys. Rev. B 5, 580-593.

C.D. Clark and R.J. Wedlake 1973. A tetrahedral anvil apparatus for optical studies under high hydrostatic pressure. J. Phys. E. 6, 43-47.

S.D. Christian, J. Grundnes, and P. Klaboe 1975. Pressure effects on conformational equilibrium in solution. J. Am. Chem. Soc. 97, 3864-3865.

S.D. Christian, J. Grundnes, and P. Klaboe 1976. Pressure effects on conformational equilibria. Difference in volume between conformers of 1,1,2-trichloroethane. J. Chem. Phys. 65, 496-498.

M.A. Cundill and W.F. Sherman 1968. Lattice sidebands of vibrational spectra and their pressure dependence. Phys. Rev. 168, 1007-1025.

A.R. Davis, W.A. Adams, and M.J. McGuire 1974. High pressure laser Raman study of the dissociation of aqueous bisulphate ion. J. Chem. Phys. 60, 1751-1753.

H.G. Drickamer and C.W. Frank 1973. Electronic transitions and the high pressure chemistry and physics of solids. Chapman and Hall, London.

E. Fishman and H.G. Drickamer 1956. Equipment for high pressure optical and spectroscopic studies. Anal. Chem. 28, 804-806 (1956).

R.A. Fitch, T.E. Slykhouse, and H.G. Drickamer 1957. Apparatus for optical studies to very high pressures. J. Opt. Soc. Am. 47, 1015-1017.

R.G. Gardiner, S.D. Hamann, and M. Linton 1970. Oxidation state of ferric salts at high pressures. Austral. J. Chem. 23, 2369-2373.

M.H. Grimsditch and A.K. Ramdas 1975. Brillouin scattering in diamond. Phys. Rev. B 11, 3139-3148.

S.D. Hamann, M. Linton, and C.W.F.T. Pistorius 1973. Infrared and volumetric studies of the high-pressure phase transitions of ammonium acetate and formate, trioxane, and trithiane. High Temp. High Pres. 5, 575-580.

S. Hawke, K. Syassen, and W.B. Holzapfel 1974. An apparatus for high pressure Raman spectroscopy. Rev. Sci. Instrum. 45, 1593-1601.

B.W. Jones 1967. The dielectric constant of ionic solids and its change with hydrostatic pressure. Phil. Mag. 16, 1085-1096.

B. Kamb, W.C. Hamilton, S.J. La Placa, and A. Prakash 1971. Ordered proton configuration in ice II from single-crystal neutron diffraction. J. Chem. Phys. 55, 1934-1945.

D.D. Klug and E. Whalley 1977. Origin of the effective charges of amorphous silicon and germanium. Phys. Rev. B 15, 2091-2098.

Yu.A. Kluyev and A.M. Vorob'ev 1964. Study of the 4197-cm^{-1} absorption band of potassium ferricyanide at high pressure. Doklady Akad. Nauk. S.S.S.R. 158, 1396-1398. English Transl. Soviet Phys. Doklady 158, 975-976.

H.A. Larsen and H.G. Drickamer 1957. Chemical effects of plastic deformation. J. Phys. Chem. 61, 1249-1252.

G.J. Lewis and W.F. Sherman 1974. Liquid helium cooled high pressure optical cell. J. Phys. E. 7, 300-304.

G.J. Lewis and E. Whalley 1978. Effect of pressure on molecular conformations. I. Internal rotation angle in trithiane. J. Chem. Phys. 68, in press.

H. Lindner 1972. Quoted by K. Tödheide. Water at high temperatures and pressures. In F. Franks, Editor, Water a comprehensive treatise. Plenum Press, New York, p.463-514.

E. Madelung and R. Fuchs 1921. Compressibilitäts-messungen an festen Körpern. Ann. Physik. 65, 289-309.

M. Mariée and J.P. Mathieu 1946. Modifications apportées au spectra de Raman d'un crystal de quartz par une compression dirigée suivant un axe binain. Compt. Rend. 223, 147.

S.S. Mitra and K.V. Namjoshi 1971. Pressure dependence of the effective ionic charge of solids. J. Chem. Phys. 55, 1817-1820.

M. Nicol, Y. Ebisuzaki, W.D. Ellenson, and A. Karim 1972. A versatile high-pressure apparatus for optical spectroscopy between 77 and 500 K. Rev. Sci. Instrum. 43, 1368-1371.

M. Nicol 1965. Fluorescence of anthracene in lucite at very high pressures. J. Opt. Soc. Am. 55, 1176-1178.

H.W. Offen, R.C. Tanquary, and K.F. Sterrett 1967. High-pressure apparatus for optical studies at 77 K. J. Applied Phys. 38, 5245-5248.

N.B. Owen 1966. A high-pressure apparatus for optical measurements. J. Sci. Instrum. 43, 765-766.

B.J. Parsons 1977. Spectroscopic mode Grüneisen parameters for diamond. Proc. Roy. Soc. A 352, 397-417.

W. Paul 1966. Influence of optical and anisotropic stress on the optical properties of solids. Proc. Int. School Phys. "Enrico Fermi", Course XXXIV, Academic Press, New York, 1966, pp.257-309.

W. Paul, W.M. De Mies, and J.M. Besson 1968. Windows for optical measurements at high pressures and long infrared wavelengths. Rev. Sci. Instrum. 39, 928-930.

G.J. Piermarini, S. Block, J.D. Barnett, and R.A. Forman 1975. Calibration of the pressure dependence of the R_1 ruby fluorescence line to 195 kbar. Appl. Phys. 46, 2774-2780 (1975).

C. Postmus, V.A. Maroni, J.R. Ferraro, and S.S. Mitra 1968. High pressure-laser Raman spectra. Inorg. Nucl. Chem. Lett. 4, 269-274.

T.C. Poulter 1930. A glass window mount for withstanding pressures of 30,000 atmospheres. Phys. Rev. 35, 297.

T.C. Poulter 1932a. Apparatus for optical studies at high pressure. Phys. Rev. 40, 860-869.

T.C. Poulter and F. Buckley 1932b. Diamond windows for withstanding very high pressures. Phys. Rev. 41, 364-365 (1932).

P.W. Richter, P.T.T. Wong, and E. Whalley 1976. The effect of pressure on the Raman spectra of mercurous chloride and bromide. J. Chem. Phys. 67, 2348-2354.

S.K. Runcorn 1956. Experiments on the displacement of the ultraviolet absorption edge of olivine at high pressures. J. Appl. Phys. 27, 598-602.

W.F. Sherman 1966. A high-pressure absorption cell for obtaining well resolved infrared spectra of samples under 50 kb pressures at 90 K. J. Sci. Instrum. 43, 462-465.

W.F. Sherman and P.P. Smulovitch 1970. Pressure-scanned Fermi resonance in the spectrum of NH_4^+ isolated in CsBr. J. Chem. Phys. 52, 5187-5193.

H.D. Stromberg and R.N. Schock 1970. A window configuration for high pressure optical cells. Rev. Sci. Instrum. 41, 1880-1881.

J.C. Stryland and A.D. May 1960. Optical cell for the observation of Raman scattering in gases at high pressures. Rev. Sci. Instrum. 31, 414-415 (1960).

Y. Taniguchi, H. Takaya, P.T.T. Wong, and E. Whalley, unpublished. Effect of pressure on molecular conformations. Trans-gauche equilibrium of 1,2-dichloroethane and 1,2-dibromoethane.

G.E. Walrafen 1973. Raman spectra from partially deuterated water and ice VI to 10.1 kbar at 28°C. J. Solution Chem. 2, 159-171.

C.E. Weir, E.R. Lippincott, A. Van Valkenburg, and E.N. Bunting 1959. Infrared studies in the 1- to 15-micron region to 30,000 atmospheres. J. Res. Natl. Bur. Standards A 63, 55-62.

E. Whalley 1976. Effect of pressure on the near-neighbor distance in water. J. Chem. Phys. 63, 5205-5207 (1975).

E. Whalley, A. Lavergne, and P.T.T. Wong 1976. Hydrostatic optical cell with glass windows for 25 kbar. Rev. Sci. Instrum. 47, 845-848. (1976).

F.G. Wick 1923. The effect of pressure upon optical absorption. Proc. Am. Acad. Arts Sci. 58, 557-573.

P.T.T. Wong and E. Whalley 1972. A technique for the Raman spectroscopy of high-pressure phases recovered at low temperatures. The low-frequency Raman spectrum of ammonium fluoride II. Rev. Sci. Instrum. 43, 935-937.

P.T.T. Wong and E. Whalley 1974. Raman Spectroscopy under pressure using a diamond pressure vessel. Rev. Sci. Instrum. 45, 904-906.

P.T.T. Wong and E. Whalley 1976. Raman spectrum of ice VIII. J. Chem. Phys. 64, 2359-2366.

R.G.W. Wyckoff 1964. Crystal Structures, Vol. 2, 2nd Ed. Interscience, New York.

H. Yamada 1974. Pressure-resisting glass cell for high pressure, high resolution nmr measurement. Rev. Sci. Instrum. 45, 640-642 (1974).

MÖSSBAUER SPECTROSCOPY ON SOLIDS UNDER PRESSURE+)

Wilfried B. Holzapfel

Max-Planck-Institut für Festkörperforschung
Stuttgart, Federal Republic of Germany

1. Introduction
2. Theoretical Background
   - The Mössbauer Effect
   - Lattice Dynamical Aspects
   - Hyperfine Interactions
3. High-pressure Techniques
4. High-pressure Mössbauer Studies
   - Europium compounds
   - Magnetic splitting and 3d wave functions of iron
   - Electronic transitions in iron compounds
   - Mixed valence compounds
5. Perspectives

+) This work has been supported in part by the Deutsche Forschungsgemeinschaft

Kelm, H. (Ed.) High Pressure Chemistry. 159–175. Copyright © 1978. D. Reidel Publishing Company. All Rights Reserved.

1. INTRODUCTION

Many excellent textbooks describe the general ideas of Mössbauer spectroscopy / 1-9 /. The present introduction into high pressure Mössbauer spectroscopy therefore includes only a very short recapitulation of this general theoretical background and the common experimental details. Various reviews on high pressure Mössbauer spectroscopy have also been published rather recently / 10-18 /. It is therefore not intended to give a complete review on all the experimental data here, but rather to present some typical aspects of the special high pressure techniques and more details on just a few typical applications.

2. THEORETICAL BACKGROUND

Mössbauer spectroscopy is often called "Recoilless Nuclear Gamma Resonance Absorption" / 19 /. This name points to the fact that Mössbauer spectroscopy is based on the emission and reabsorption of gamma quanta, whereby the gamma quanta are emitted from excited nuclei in radioactive decay processes. In the "recoilless" resonance absorption, similar nuclei are excited from their ground state into the same excited state without any significant transfer of recoil energy to the nucleus or to the lattice to which the nucleus is coupled by the surrounding electrons or by the equivalent chemical bonding. The relative probability f for gamma ray emission (or absorption) without phonon emission (or absorption) depends primarily on the energy of the gamma ray E_γ, on the phonon spectrum of the solid and on temperature. In the Debye approximation for a harmonic lattice, this f factor is equal to the Debye-Waller-factor which is well known from x-ray diffraction. This f factor is responsible for the strong decrease in intensity of Mössbauer spectra recorded at elevated temperatures. This temperature dependence of the intensity is very pronounced already below the Debye-temperature of the solid, if the gamma ray energy is $E_\gamma \gtrsim 20$ keV and it can be used in favorable cases to study the local lattice dynamics around the Mössbauer nuclei.

The line width of the "recoilless" nuclear excitations can be very small. In fact, it can be even smaller than the shifts and splittings of the nuclear energy levels due to interactions of the nuclear charge and spin with the surrounding electrons. Just in these favorable cases, the Mössbauer spectroscopy gives very valuable information on the local electronic and magnetic fields around these special "Mössbauer nuclei" in a given solid.

Let us consider the case, where the source emits gamma rays of an average energy E_γ with just the "natural" line width $\Gamma_{nat} = \hbar/\tau$. $\hbar$ is Planck's constant and τ the life time of the excited nuclear state. Energy shifts and splittings in the absorber can be observed by the Mössbauer spectroscopy if these shifts or splittings are larger, or at least not too much smaller than this natural line width. In favorable cases, for instance for the 6,2 keV transition in ^{181}Ta, Γ_{nat} is smaller than 10^{-10} eV and the relative width $\Gamma_{nat}/E_\gamma \approx 10^{-14}$. In these cases, the energy E_γ of the emitted gamma rays can be tuned through the energy levels of the absorbing nuclei by the use of the Doppler effect. Typically, relative velocities between the source and the absorber of a few mm/sec are sufficient in these cases to tune through the resonance absorption. From this experimental point of view, it is convenient to present Mössbauer spectra always on a velocity scale v, which is simply related to the energy scale E by the formula for the Doppler shift $E = (v/c)\cdot E_\gamma$.

The effects of the different hyperfine interactions on the nuclear energy levels and on the corresponding Mössbauer spectra are illustrated in Fig. 1 / 19 /. The isomer shift, IS, is noticed as a shift of the center of gravity in the first spectrum of Fig. 1 with respect to 0 on the velocity scale. The isomer shift measures the difference in electron density at the sites of the Mössbauer nuclei in the absorber with respect to the source. If the source emits only one sharp line, the splittings of the spectra are simply related to the splittings of the energy levels in the absorber and vice versa. The isomer shift (with respect to a standard), the magnetic hyperfine splitting and the electric quadrupole splitting give detailed information on the electronic configuration of the Mössbauer isotopes in the given solid. Examples will be presented in Section 4.

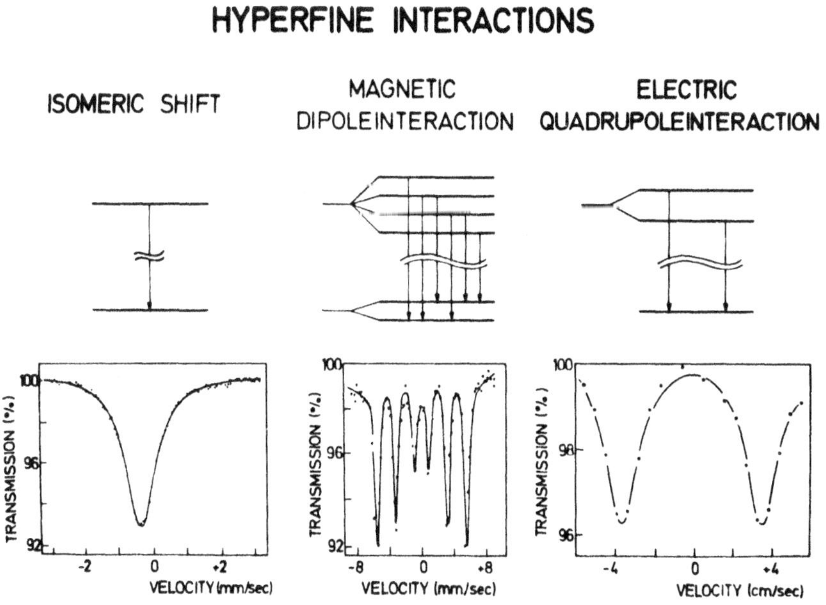

Fig.1. Effects of hyperfine interactions on Mössbauer spectra /19/.

3. HIGH PRESSURE TECHNIQUES

A wide variety of high pressure techniques has been developed for Mössbauer measurements on either sources or absorbers under pressure / 20-34 /. In some cases, hydrostatic pressures were necessary / 32,33 / and the pressure range up to 2,5 kbar was quite sufficient. A pressure cell for this kind of experiments is shown in Fig. 2A / 16 /. Truly hydrostatic pressures up to 10 kbar were generated in the cell shown in Fig. 2B / 26 /. With some improvements / 16 /, the pressure range of this cell can be expended to 20 kbar and above.

In many cases, higher pressures are necessary to notice changes in the spectra or to reach a special phase transition. The techniques for these applications use either one or the other way to compress the sample between high pressure anvils as shown in Fig. 2C-F. The different geometries resulted mostly from special adaptations to an optimized performance with one or the other Mössbauer isotope. In fact, the last example of Fig. 2, a gasketed diamond anvil cell with liquid pressure transmitting medium, has not yet been used in high pressure Mössbauer studies, however, it would be very interesting for applications where either non-hydrostatic stresses should be avoided up to 100 kbar or where pressures far above 100 kbar should be obtained.

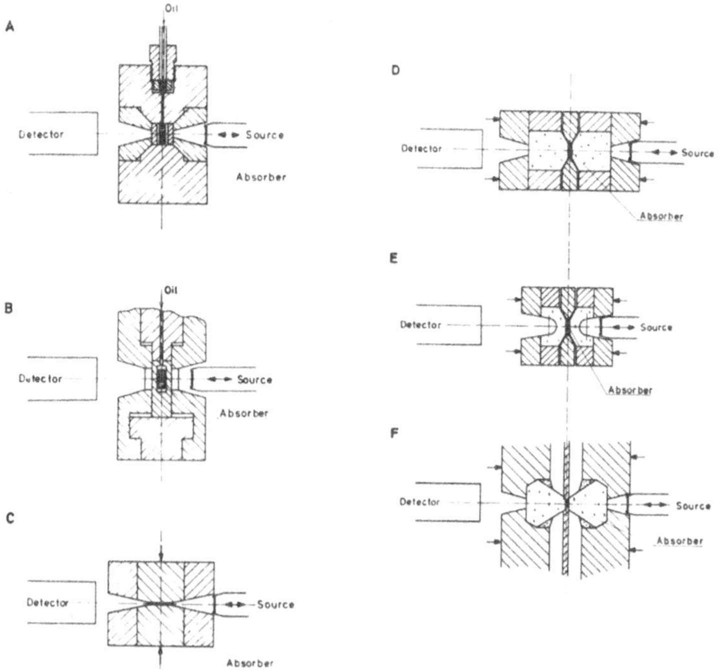

Fig. 2. High pressure techniques for Mössbauer spectroscopy using either hydrolic systems A /20,16/ and B /26/ or supported anvils C /23/, D /28/, E /34/ and F /16/.

All the anvil devices, Fig. 2C-F, require a mechanism to apply a force onto these anvils. Hydraulic presses /21,23-25, 28,31/ as well as purely mechanical devices /22,27,29,34/ have been developed for this purpose. High pressure Mössbauer studies at low temperatures are preformed most economically with so-called "clamp" devices. These devices include springs which are compressed and "clamped" in conventional presses at room temperature. The applied force and pressure is preserved largely when these devices are inserted into cryostats and used at low temperatures. One of these devices / 34 / is shown in Fig. 3. Various different internal pressure sensors can be used in these devices / 27,29,34,35 /.

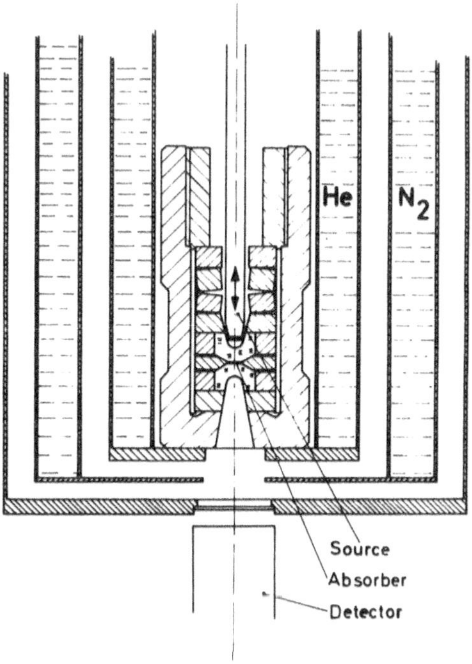

Fig. 3. Cryostat with high pressure clamp device for Mössbauer studies /34/.

4. HIGH PRESSURE MÖSSBAUER STUDIES

A very simple behaviour has been observed with europium compounds under pressure / 36,37,16,17 /. The linear changes in the isomer shifts, ΔIS, in Fig. 4 indicate a linear increase of the electron density at the Eu nuclei in these compounds under pressure. Obviously, the Eu^{2+} compounds show a stronger pressure dependence than the Eu^{3+} compounds. If one takes the different compressibilities into account and plots these changes in ΔIS versus the relative volume V/V_O, as shown in Fig. 5 / 37 /, one observes only minor differences between Eu^{2+} and Eu^{3+} compounds. These data support therefore the assumption that the 4f-electrons of Eu in these compounds are not affected noticably up to 100 kbar. This conclusion is in good agreement with x-ray measurements on europium-monochalcogenides under pressures up to 400 kbar / 38 /, where a normal behaviour was observed for EuO up to 280 kbar followed at 300 kbar by a rather sudden isostructural transition due to the delocalization of one 4f electron per Eu ion. The other EuX compounds, EuS, EuSe and EuTe, transformed without any indication of an electronic instability from the NaCl to the CsCl structure at transition pressures of 210, 150 and 100 kbar, respectively. If, however, one plots the absolute values of IS versus the lattice parameters of these compounds, and extrapolates the experimental results up to the transition pressures,

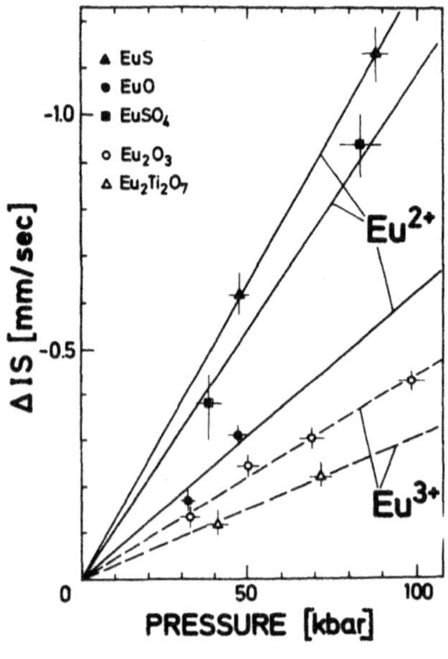

Fig. 4. Effect of pressure on isomer shifts of europium compounds / 36,37 /.

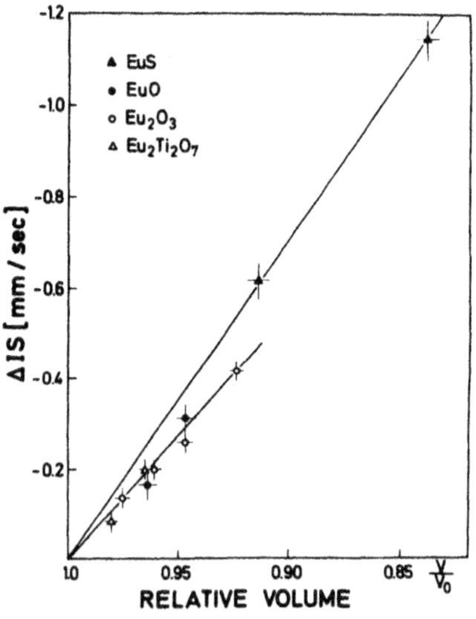

Fig. 5. Effect of volume compression on isomer shifts for europium compounds / 37 /.

Fig. 6, one finds that the electronic transition in EuO as well as the structural transitions in the other EuX compounds occur at about the same value of IS. This observation supports the assumption that the structural transitions in EuS, EuSe and EuTe are initiated by the same electronic instability which is responsible for the isostructural transition in EuO. Due to the denser packing in the high pressure CsCl-phase, however, EuS, EuSe and EuTe remain still semiconducting divalent compounds and can only speculate that the delocalization of 4f electron will occur in these compounds at even higher pressures.

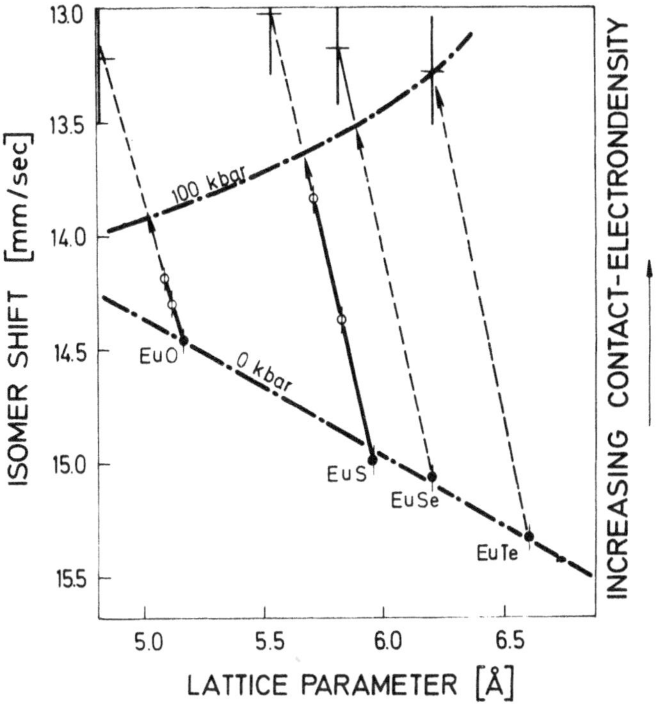

Fig. 6. Extrapolation of isomer shifts to the values at the phase transitions for the europium-monochalcogenides / 37 /.

Changes in the magnetic hyperfine splitting have been studied in detail by Mössbauer measurements on Fe^{2+} in CoO under pressure in a wide range of temperatures / 39 /. Experimental results of this investigation are shown in Fig. 7 /39/. Smooth curves connect the experimental points at constant pressure. The dashed lines represent theoretical curves, which follow from a molecular field model with three adjustable parameters / 39 /. The characteristic features of the experimental results are reasonably described by these theoretical curves, and the systematic differences are well understood. The unusual minimum in the hyperfine field at low temperatures gives valuable information on the electronic structure of Fe^{2+} in CoO.

Let us consider first the electronic structure of a free Fe^{2+} ion. The electron configuration is [Ar] $3d^6$. Due to Hund's rule, one half 3d shell is filled by 5 3d electrons with parallel spins and only the last 3d electron occupies one orbital of the second half 3d shell with opposite spin. In a free iron ion, this state is fivefold degenerate. In a crystal field of octahedral symmetry, this degeneracy is partly lifted. One can distinguish two orbitals with e_g symmetry from three orbitals with t_{2g} symmetry. The different overlap of the σ-antibonding $e_g(\sigma^+)$ and the π-antibonding $t_{2g}(\pi^+)$ molecular orbitals of Fe^{2+} in CoO is illustrated in Fig. 8 / 16 /.

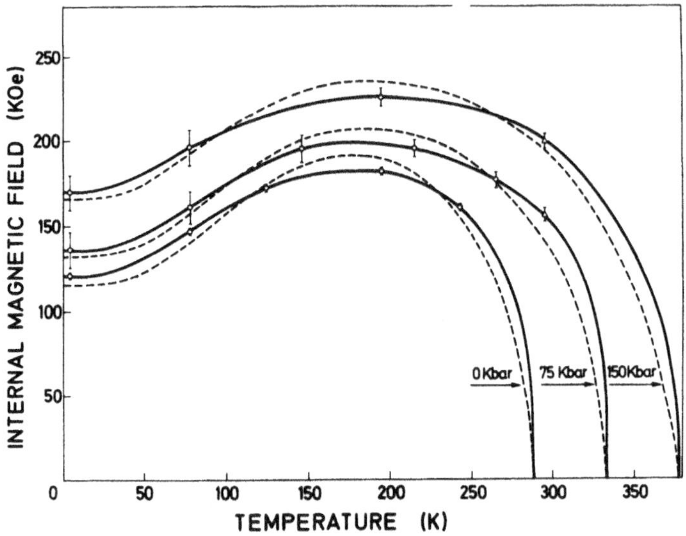

Fig. 7. Effect of pressure and temperature on the magnetic hyperfine fields for Fe^{2+} in CoO / 39 /.

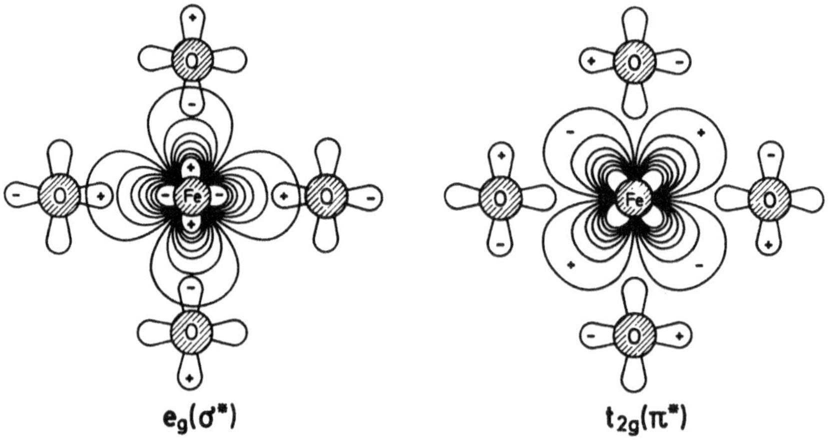

Fig. 8. Schematic representation of $e_g(\sigma^+)$ and $t_{2g}(\pi^+)$ molecular orbitals of Fe^{2+} in CoO / 16 /.

The strong σ-antibonding overlap lifts the $e_g(\sigma^+)$ levels with respect to the $t_{2g}(\pi^+)$ levels. The single electron in the second half 3d shell occupies, therefore, a t_{2g} orbital with spin up as shown in Fig. 9. The effective hyperfine field H_{eff} for Fe^{2+} in CoO can be divided into different contributions as indicated by the first formula in Fig. 9. For Fe^{2+} in CoO, the external field H_{ext} is 0 and the dipolar contribution H_D is negligible. The remaining components H_S and H_L have different signs and the thermal averages for the spin $\langle S \rangle$ and for the orbital moment $\langle L \rangle$ show different temperature dependences / 40 /. The detailed model for the crystal field splitting shows / 40 / that the orbital momentum $\langle L \rangle$ is quensched rapidly with increasing temperature at about 150 K. Thus, the minimum in H_{eff} results from a negative contribution of H_L which vanishes rapidly when the temperature exceeds 150 K. The special shape of the curves in Fig. 7 determines, therefore, the values of both coupling parameters H_c and $\langle r^{-3} \rangle_L$ at each pressure. The Fermi contact parameter H_c depends strongly on the average density of unpaired electrons, $\langle r^{-3} \rangle_{eg} + \langle r^{-3} \rangle_{t2g}$, whereas $\langle r^{-3} \rangle_L$ includes only contributions from the single t_{2g} electron in the second half 3d shell. Therefore, the effects of pressure or volume on

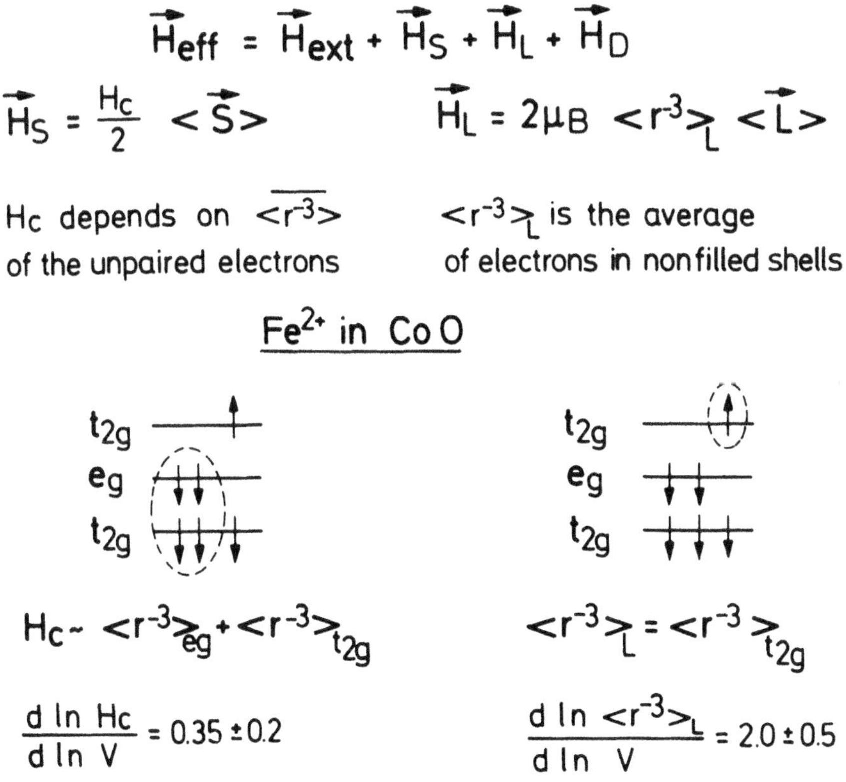

Fig. 9. Contributions to the effective magnetic hyperfine field, H_{eff}, and the results for Fe^{2+} in CoO.

H_C and $\langle r^{-3} \rangle_L$ are directly related to the different variations of $\langle r^{-3} \rangle_{eg}$ and $\langle r^{-3} \rangle_{t2g}$.

The rather large value of dln $\langle r^{-3} \rangle_L$/dlnV = 2.0±0.5 /39/ indicates a strong expansion of the t_{2g} electrons of Fe^{2+} under pressure and the small value of dlnH_C/dlnV = 0.35±0.2 reflects the very plausible fact that this expansion of the t_{2g} electrons results just from an equivalent compression of the e_g electrons, and it is clear that the difference in these average radii $\langle r^{-3} \rangle_{eg} - \langle r^{-3} \rangle_{t2g}$ should be proportional to the same crystal field strength which causes also the more familiar crystal field splitting between the e_g and t_{2g} levels.

Let us consider the electronic structure of octahedral iron complexes in even more detail: if one includes in the usual molecular orbital model / 41 / the spin pairing interaction schematically as a splitting ΔSP between the single electron levels of opposite spin, one gets the picture which is shown in Fig. 10 / 17 /. The crystal field splitting between e_g and t_{2g} levels with equal spin is denoted as usually by Δ. Filled and open half circles represent occupied and empty single electron states. Pressure-induced electronic transitions can easily be generated in this picture by changes in the relative strengths of the different splittings. In fact, one expects under pressure first of

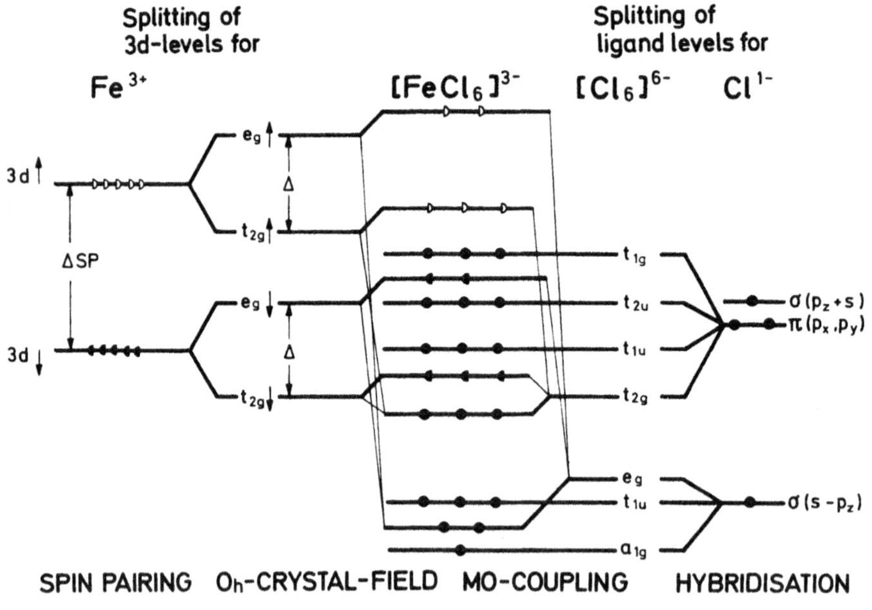

Fig. 10. Schematic representation of the molecular orbital levels for Fe^{3+} in octahedral symmetry.

all a strong increase in the crystal field splitting, Δ, a medium increase of the splitting between the different $\pi(p_x, p_y)$ ligand levels and only a minor change, most likely a decrease, in the spin-pairing splitting ΔSP. Let us consider first the case represented in Fig. 10, where the splitting of the $\pi(p_x, p_y)$ ligand levels is rather large and the occupied t_{1g} ligand level lies above the occupied $e_g\downarrow$ level which has primarily iron 3d character. When crystal field splittings Δ as well as the splitting of the ligand $\pi(p_x, p_y)$ levels increase under pressure, the empty $t_{2g}\uparrow$ level can be shifted below the filled t_{1g} level and an electron from this level can be back-donated to the t_{2g} level. The electronic structure of iron in this complex would resemble a typical high spin Fe^{2+} configuration.

On the other hand, if the $\pi(p_x, p_y)$ electrons are more strongly bonded than the 3d↓ electrons and the $\pi(p_x, p_y)$ level splitting is not too large, the t_{1g} level will stay below the occupied $e_g\downarrow$ level, and this level will crossover with the empty $t_{2g}\uparrow$ level first. In this case, the electronic transition involves only levels which have predominantly iron 3d character and the transition can be characterized as a Fe^{3+} high spin → Fe^{3+} low spin transition.

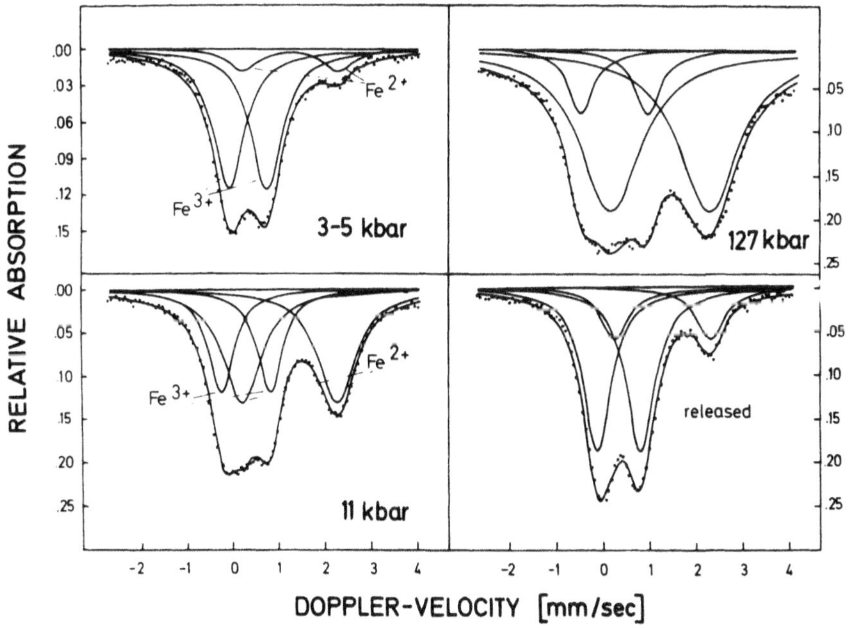

Fig. 11. Effect of pressure on $Fe^{3+} \rightarrow Fe^{2+}$ conversion / 11 /.

This model is definitely very crude. A decrease of the symmetry due to Jahn-Teller-type electron lattice interactions and distortions is neglected as well as local stresses and local polarisations. Nevertheless, this model points toward the possibility of observing various different kinds of electronic transitions in transition metal complexes under pressure / 15 /. Drickamer and coworkers / 11,14,15 / studied a wide variety of iron compounds under pressure by Mössbauer spectroscopy and observed in many cases this kind of electronic transitions. Just one example from these studies is shown in Fig. 11 / 11 /, which represents Mössbauer spectra of $FeCl_3$ at different pressures. The reversible increase of the Fe^{2+} contribution due to the $Fe^{3+} \rightarrow Fe^{2+}$ transition can be noticed clearly in these spectra.

At this point it may be interesting to compare the electronic transition in iron complexes with changes in the electronic properties of stochiometric mixed valence compounds. As an example, let us consider $Cs_2Au^{I}Au^{III}Cl_6$. Resistivity measurements up to 100 kbar at different temperatures / 41 / indicate that this material transforms continuously from an insulating state at 1 bar to semiconducting behavior at intermediate pressures and to a metallic state at 60 kbar and above. The structure of $Cs_2Au_2Cl_6$ at 1 bar is well known / 43 / and can be described as a distorted perovskite structure with a regular arrangement of uniaxially contracted $[Au^{I}Cl_6]$ and elongated $[Au^{III}Cl_6]$ octahedra, Fig. 12. Not only the structure but also

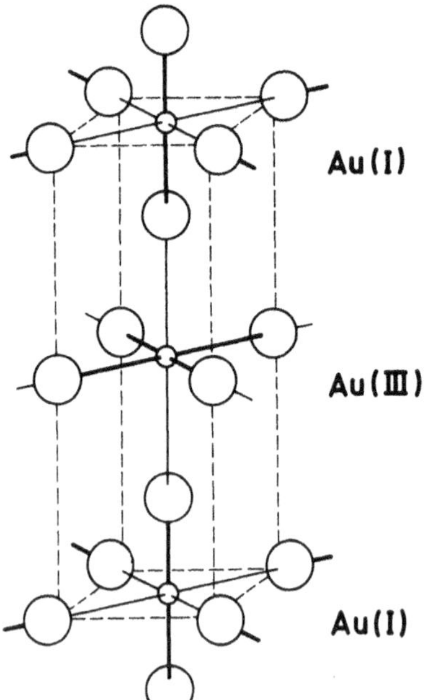

Fig. 12. Distorted $[AuCl_6]$ octahedra in $Cs_2Au_2Cl_6$ /43/.

Mössbauer spectra point clearly to the presence of two different Au species / 44,45 /, one species Au^{I} with a negative isomer shift and a large (negative) quadrupole splitting and the other Au^{III} with a small positive isomer shift and a small (positive) quadrupole splitting. With these tentative signs of the quadrupole splittings, both species fit perfectly into the well established correlation / 46 / of isomer shifts and quadrupole splittings for Au^{I} and Au^{III} compounds. Figure 13 / 45,16,17 / illustrates this correlation and also represents the changes that have been observed previously on "regular" Au^{I} and Au^{III} compounds under pressure / 35,47 /. In all the "regular" compounds, the isomer shifts increase under pressure whereas both the Au^{I} as well as the Au^{III} sites in $Cs_2Au_2Cl_6$ display an unusual decrease in the isomer shift. The decrease in the quadrupole splitting for the Au^{III} site is also quite unusual / 36 /. These results find a simple explanation, if one assumes that the overall decrease in the lattice constants under pressure leads at least up to 40 kbar to an increase in the nearest neighbor distances for both the two nearest neighbors of Au^{I} as well as for the four nearest neighbors of Au^{III} (see Fig. 12). In fact, recent complete x-ray structure determinations on $Cs_2Au_2Cl_6$ under pressures up to 50 kbar / 48 / support this interpretation of the Mössbauer data and allow for an even more detailed analysis of these results.

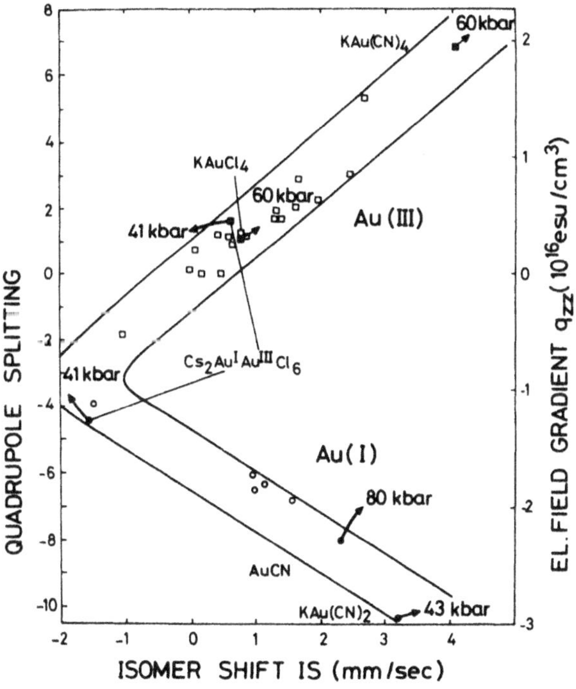

Fig. 13. Correlation between isomer shift and quadrupole splitting for gold compounds and the effect of pressure /45,16/.

5. PERSPECTIVES

If one considers the capabilities of the present techniques for high pressure Mössbauer spectroscopy, one may conclude that these techniques have just been used in only a very limited number of examples. However, these few above-mentioned studies demonstrate clearly that this method has various unique advantages and that there is still a wide field of very interesting applications of this technique.

ACKNOWLEDGMENTS

I am very grateful to H.G. Drickamer for introducing me to this field some years ago. I enjoyed very stimulating cooperation and discussions on this subject with H.D. Bartunik, C.W. Christoe, U.F.G. Klein, R.L. Mössbauer, H. Prosser, K. Syassen and G. Wortmann.

REFERENCES

1. H.Frauenfelder, The Mössbauer Effect, Benjamin, New York,1963.
2. G.K. Wertheim, Mössbauer Effect: Principles and Applications, Academic Press, New York, 1964.
3. H. Wegener, Der Mössbauer-Effekt, Bibliographisches Institut, Mannheim, 1965.
4. V.I. Goldanskij and R.H. Herber, Eds., Chemical Applications of Mössbauer Spectroscopy, Academic Press, New York, 1968.
5. N.N. Greenwood and T.C. Gibb, Mössbauer Spectroscopy, Chapman and Hall, London, 1971.
6. J.G. Stevens, "Mössbauer Spectroscopy", Magnet.Resonance Rev. 2, 97, 1973.
7. I.J. Gruverman and C.W. Seidell, Eds., Mössbauer Effect Methodology, Vols. 1-8, Plenum Press, New York, 1965-1972.
8. J.G. Stevens and V.E. Stevens, Eds., Mössbauer Effect Data Index, Plenum Press, New York, 1969-1972.
9. G.K. Shenoy and F.E. Wagner, Eds., Mössbauer Isomer Shifts, North Holland Publ., Amsterdam, 1977.
10. H.G. Drickamer, R.L. Ingalls and C.J. Coston, "The Mössbauer Effect at High Pressure," in Physics of Solids at High Pressure, Tomizuka and R. Emrick, Eds., Academic Press, New York, p.313, 1965.
11. H.G. Drickamer, S.C. Fung and G.K. Lewis, Jr., "High Pressure Mössbauer Studies", p. 1, in Advances in High Pressure Research, Vol. 3, Bradley, R.S., Ed., Academic Press, New York, 1969.
12. H.G. Drickamer, V.C. Bastron, D.C. Fisher and D.C. Grenoble, "The High Pressure Chemistry or Iron", J. Solid State Chem., 2, 94, 1970.

13. W.B. Holzapfel, "Mössbauer Studies on Solids Under High Pressures" (a review), High Temp. High Pressures 2, 241, 1970.
14. H.G. Drickamer, and C.W. Frank, "Electronic Structure, Electronic Transitions, and the High Pressure Chemistry and Physics of Solids", Ann. Rev. Phys. Chem. 23, 39, 1972.
15. H.G. Drickamer and C.W. Frank, Electronic Transitions and the High Pressure Chemistry and Physics of Solids, Chapman and Hall, London, 1973.
16. W.B. Holzapfel, "Mössbauer Studies at High Pressure", CRC Crit. Rev. Solid State Sci. 5, 89, 1975.
17. W.B. Holzapfel, Mössbauer-Spektroskopie an Festkörpern unter hohen Drucken, Habilitationsschrift, Universität Stuttgart, 1976.
18. D.L. Williamson, "Influence of Pressure on Isomer Shifts," in Ref. 9, 1977.
19. R.L. Mössbauer, "Resonance Spectroscopy of Gamma-Radiation", p. 77 in Proc. R. A. Welch Found. Conf. on Chem. Res. XI, Radiation and the Structure of Matter, Houston, 1967.
20. G.B. Benedek, "The Isomer Shift" in Magnetic Resonance at High Pressure, Wiley & Sons, New York, 44, 1963.
21. M. Nicol and G. Jura, "Mössbauer Spectrum of Iron-57 in Iron Metal and Very High Pressures", Science 141, 1035, 1963.
22. R.H. Herber and J. Spijkerman, "Mössbauer Line Broadening in SnO_2", J. Chem. Phys. 42, 4312, 1965.
23. P. Debrunner, R.W. Vaughan, A.R. Champion, J. Cohen, J. Moyzis and H.G. Drickamer, "Versatile High Pressure Mössbauer Apparatus", Rev. Sci. Instrum. 37, 1310, 1966.
24. H.S. Möller, "Pressure Dependence of the Isomer Shift of Sn^{119}", Z. Phys. 212, 107, 1968.
25. V.N. Panyushkin, "Shift of the Mössbauer Line of β-Sn During its Phase Transition Under Pressure", Sov. Phys. Solid State, 10, 1515, 1968.
26. V.N. Panyushkin, "Chamber for Studying Mössbauer Effect at Hydrostatic Pressures up to 10 kbar", Instrum. Exp. Tech. 1969, 478 , 1969.
27. C.W. Christoe and H.G. Drickamer, "Clamp Cell for High Pressure, Low Temperature X-ray and Mössbauer Resonance Studies," Rev. Sci. Instrum. 40, 169, 1969.
28. L.D. Roberts, D.O. Patterson, J.O. Thomson and R.P. Levey, "Solid State and Nuclear Results from a Measurement of the Pressure Dependence of the Energy of the Resonance gamma ray of ^{197}Au," Phys. Rev. 179, 656, 1969.
29. K. Syassen, C.W. Christoe, and W.B. Holzapfel, "Cryostat for X-ray Scattering and for Nuclear γ-resonance and Angular Correlation Studies at High Pressure", Z. Angew. Phys. 31, 261, 1971.
30. G.M. Kalvius and E. Kankeleit, "Recent Improvments in Instrumentation and Methods of Mössbauer Spectroscopy", p. 9, in Mössbauer Spectroscopy and its Application, International Atomic Energy Agency, Vienna, 1972.

31. D.L. Williamson, S. Bukshpan, R. Ingalls and H. Shechter, "Low Temperature Apparatus for High Pressure Mössbauer Studies", Rev. Sci. Instrum. 43, 194, 1972.
32. G. Wortmann, W.B. Holzapfel, G. Kaindl and G.M. Kalvius, "Pressure Dependence of the Isomer Shift of the 6,2 keV Gamma-rays of Ta-181", Bull. APS (II), 18, 546, 1972.
33. G. Kaindl, D. Salomon and G. Wortmann, "Mössbauer Isomer Shifts of the 6,2 keV Gamma-Rays of Tantalum-181,"p. 211 in Mössbauer Effect Methodology, Vol. 8, I.J. Gruverman and C.W. Seidell, Plenum Press, New York, 1973.
34. J.S. Schilling, U.F.G. Klein and W.B. Holzapfel, "Versatile High Pressure Cell for Use at Low Temperatures", Rev. Sci. Instrum. 45, 1353, 1974.
35. H.D. Bartunik, W.B. Holzapfel and R.L. Mössbauer, "Effect of Pressure on the Isomer Shift and Quadrupole Splitting in Au(III) Compounds", Phys. Lett. 33A, 469, 1970.
36. U.F.G. Klein, G. Wortmann, G.M. Kalvius, and W.B. Holzapfel, "Mössbauer-Effekt Messungen an zwei- und dreiwertigen Europium-Verbindungen unter hohen Drucken", Verh. Dtsch. Phys. Ges. (VI) 8, 350, 1973.
37. U.F.G. Klein, G. Wortmann, G.M. Kalvius, K. Syassen and W.B. Holzapfel, "Effect of Pressure on the Hyperfine Interactions in Mössbauer Studies on Eu^{2+} and Eu^{3+} Compounds", to be published.
38. A. Jayaraman, "Pressure-Volume Relationship and Pressure-induced Electronic and Structural Transformations in Eu and Yb Monochalcogenides", Phys. Rev. B9, 2513, 1973.
39. K. Syassen and W.B. Holzapfel, "Effect of Pressure and Temperature on the Magnetic Hyperfine Interaction of Fe (II) in CoO," Phys. Rev. B8, 1799, 1973.
40. H.N. Ok, and J.G. Mullen, "Magnetic Properties of Iron Ions in CoO(I) and CoO(II)", Phys. Rev. 168, 563, 1968.
41. C.J. Ballhausen and H.B. Gray, Molecular Orbital Theory, W.A. Benjamin Inc., New York, 1965.
42. R. Keller, J. Fenner and W.B. Holzapfel, "The Resistivity of Mixed-valence Compound $Cs_2Au_2Cl_6$ at High Pressure and Low Temperatures," Mater. Res. Bull. 9, 1363, 1974.
43. N. Elliot and L. Pauling, "The Crystal Structure of Cesium Aurous Auric Chloride, $Cs_2AgAuCl_6$, and Cesium Argentous Auric Chloride, $Cs_2AgAuCl_6$," J. Am. Chem. Soc.60,1846, 1938.
44. H. Prosser, Mössbauer Spektroskopie an Goldverbindungen unter hohem Druck, Diplomarbeit, Technische Universität, München, 1973.
45. H. Prosser, G. Wortmann, J. Fenner and W.B. Holzapfel, "Effect of Pressure on the Mössbauer Isomer Shift and Quadrupole Splitting on ^{197}Au in $Cs_2Au^{I}Au^{III}Cl_6$," to be published.
46. F.E. Wagner, U. Zahn, H.D. Bartunik, G. Kaindl and R.L. Cohen, "Mössbauer Shifts in a Selection of Elements and Systems", in Ref. 9.

47. H. Prosser, G. Wortmann, K. Syassen and W.B. Holzapfel, "Effect of Pressure on the Mössbauer Isomer Shift and Quadrupole Splitting in Au(I)-Cyanides," Z. Physik B24, 7, 1976.
48. W. Denner, Strukturanalyse an Einkristallen unter hydrostatischem Druck, Universität Karlsruhe, 1977.

X-RAY DIFFRACTION ON SOLIDS UNDER PRESSURE [+)]

Wilfried B. Holzapfel

Max-Planck-Institut für Festkörperforschung
Stuttgart, Federal Republic of Germany

1. Introduction

2. X-Ray Diffraction on Polycrystalline Samples

2.1 Angular Dispersive X-Ray Diffraction
2.2 Energy Dispersive X-Ray Diffraction

3. X-Ray Diffraction on Single Crystals

3.1 Buerger-Precession Technique
3.2 Automatic Four Circle X-Ray Diffractometer

4. Perspectives

+) This work has been supported in part by the Deutsche Forschungsgemeinschaft

Kelm, H. (Ed.) High Pressure Chemistry. 177–197. Copyright © 1978. D. Reidel Publishing Company.
All Rights Reserved.

1. INTRODUCTION

Changes of crystal structures and lattice parameters under pressure have been studied for many years by x-ray diffraction / 1 - 3 /. These studies resulted in a detailed knowledge about phase transitions and P-V-data for a wide variety of substances.

A number of new developments in the high pressure x-ray diffraction techniques have led to improvements in accuracy, to a better understanding of nonhydrostatic stress contributions, to an extension of the pressure range and to more sophisticated techniques for complete structure determinations on solids under pressure.

The intention of the present paper is therefore primarily to introduce the reader into this field and to point out recent developments and directions of future applications and progress.

2. X-RAY DIFFRACTION ON POLYCRYSTALLINE SAMPLES

Since the basic ideas of x-ray diffraction can be found in many text books, we want to recall here only that the Bragg equation

$$\sin\theta_{hkl} = \lambda/2\, d_{hkl}$$

relates the diffraction angles θ_{hkl} to the wave length λ of the diffracted x-rays and with the lattice spacings d_{hkl} of the diffracting lattice planes which are characterised by the Miller indices h k l .

With polycrystalline samples, one uses most commonly monochromatic x-rays, measures the diffraction pattern as a function of the diffraction angle, and determines from a set of diffraction angles θ_{hkl} a set of d_{hkl} values. This is the usual angular dispersive technique. On the other hand, one can use also white x-rays, a fixed diffraction angle θ and a solid state x-ray detector, which allows for an accurate analysis of the wave length λ or of the corresponding energy

$$E = (3.10\ \text{keV\AA})/\lambda$$

of the diffracted x-rays. Diffraction peaks on the x-ray energy spectrum appear in this case at energy values E_{hkl} which are related to the lattice spacings d_{hkl} again by the Bragg equation.

The advantages and disadvantages of these different techniques are discussed together with typical examples in the following subsections.

2.1 ANGULAR DISPERSIVE X-RAY DIFFRACTION

Detailed reviews about the use and applications of tungsten carbid anvil high pressure devices / 1,2 / as well as diamond anvil high pressure devices / 3 / in high pressure x-ray diffraction studies can be found in the literature. Therefore, only the basic ideas and recent improvements of these techniques are discussed in this section.

The typical geometry for x-ray diffraction on polycrystalline samples in tungsten-carbid anvil devices is shown in Fig. 1. The sample is usually mixed with a soft pressure transmitting medium and pressed into the center of a boron epoxy disk. This boron epoxy disc acts as a gasket which supports pressure in the center of this disc, when the sample is compressed by the two WC-anvils. Since boron-epoxy absorbs and scatters hard x-rays only weakly, x-ray scattering from the sample under pressure can be studied in the plane of the gasket as indicated in Fig. 1.

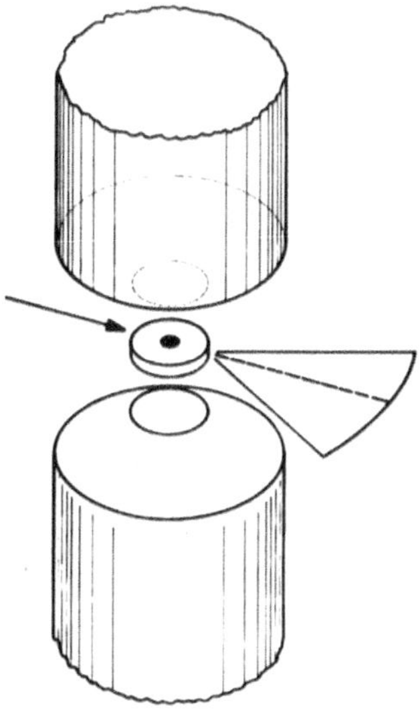

Fig. 1 Typical geometry for x-ray diffraction in tungsten carbid anvil devices

The force on the anvils can be generated in various ways / 1 - 9 /. The mechanism shown in Fig. 2 / 5,7 / offers the advantage, that it needs no hydraulic systems and is easily adaptable to the use in cryostats and with commercial x-ray diffractometers. The force is generated by turning the two screws at the top of the device. These screws push the wedge between the two knees which are straightened thereby and apply forces up to 20 tons (200 kN) to the lower piston. The force can be measured by calibrated strain gauges on these pistons. Usually,

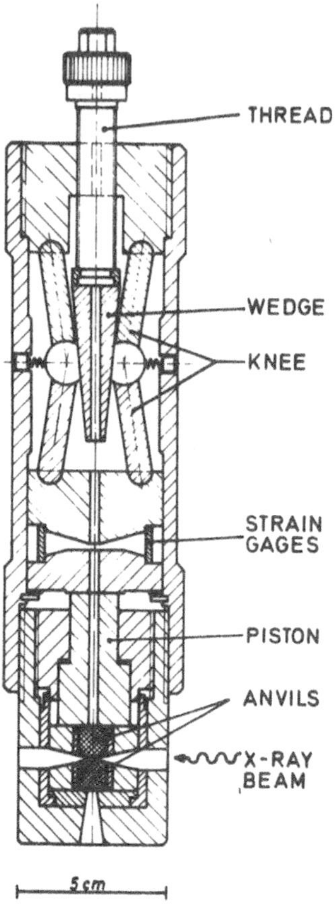

Fig. 2 Mechanical device for high pressure generation with tungsten carbid anvils / 5,7 /.

however, the pressure on the samples is determined from simultaneous x-ray measurements on a "marker" substances, for which the pressure dependence of the lattice parameters is well known. Two slits on the cell serve as windows for the incoming and diffracted x-ray with a free angular range of 100° on each side of the sample. The adaptation of this high pressure cell to a commercial x-ray diffractometer is shown in Fig. 3 / 5,10 /. Special collimators reduce the background scattering from the anvils and gasket material into the detector. Typical diffraction patterns from the chart recorder output of the diffractometer are shown in Fig. 4 / 5,10 / giving the countrate as a function of the scattering angle for a NaCl sample at 1 bar and at 103 kbar. The shifts of the individual lines under pressure are indicated in Fig. 4 by small arrows.

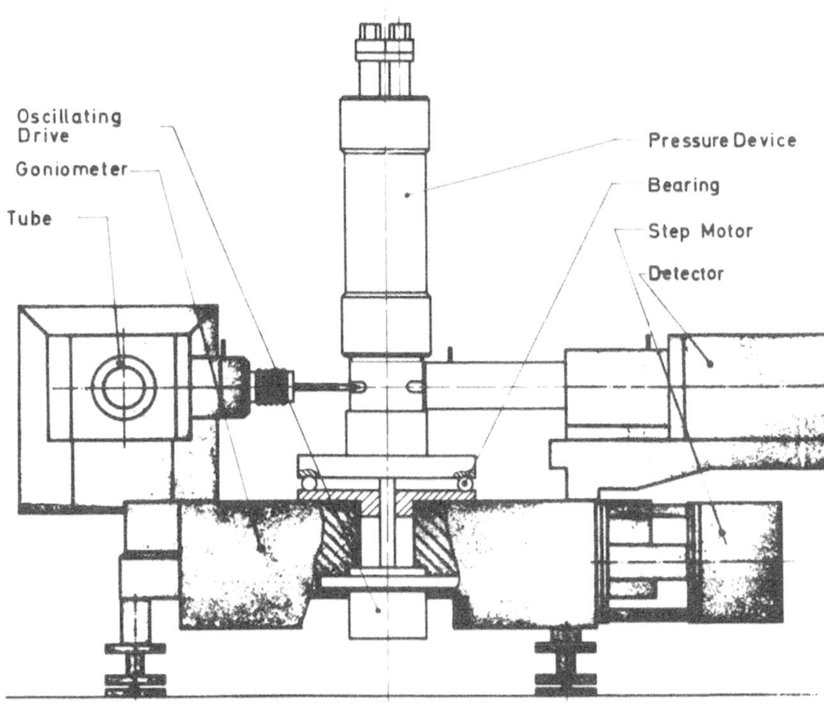

Fig. 3 Commercial x-ray diffractometer with tungsten carbid anvil high pressure cell / 5,10 /

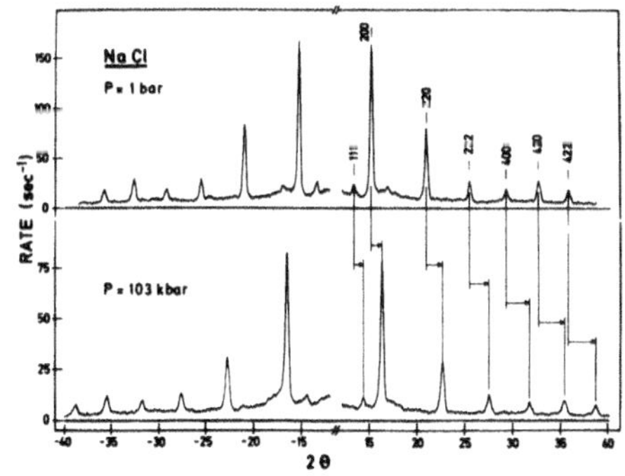

Fig. 4 Diffractometer recordings for NaCl at 1 bar and 103 kbar / 5,10 /

For a more accurate determination of the lattice parameters and their changes under pressure, the countrate is taken in a slow step scan procedure and registered digitally on punch tape.

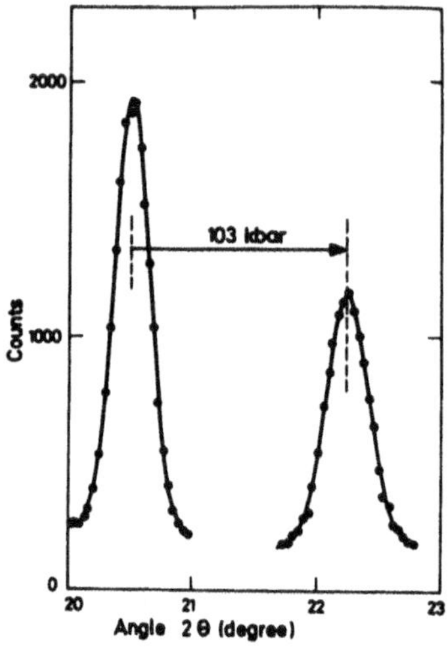

Fig. 5 Digital recordings of 220 reflections of NaCl at 1 bar and 103 kbar / 5,10 /

Fig. 5 shows data points obtained in this way for the 220 reflections of NaCl at 1 bar and 103 kbar together with least squares fitted Gaussian curves U(2θ). This procedure gives the line positions with an accuracy of typically 0.05% and the lattice parameters of cubic crystals at 1 bar with an accuracy of typically 0.02% / 5,10 /. The same accuracy is obtained under pressure only, if the pressure is truly hydrostatic. Otherwise, nonhydrostatic stresses lead to broadenings of the lines which can be noticed in Fig. 5, and which reduces the accuracy. Furthermore, the individual diffraction peaks are generally shifted in different ways by the uniaxial stresses. This effect has been analysed in detail / 11 /, and this analysis showed that the elastic anisotropy even of a cubic crystal leads in an ideal polycrystalline sample under uniaxial stress to different variations of the various lattice spacings d_{hkl} . If one calculates changes in the lattice parameter a under non-hydrostatic stresses from different x-ray reflections, one observes therefore a dependence of these a values on the Miller indices hkl as shown in Fig. 6 / 5,10 /, where the function Γ(hkl) is given by

$$\Gamma(hkl) = (h^2k^2+k^2l^2+l^2h^2)/(h^2+k^2+l^2).$$

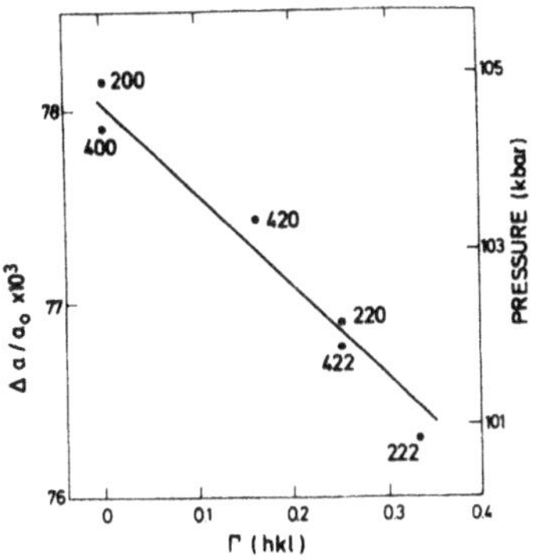

Fig. 6
Dependence of the change in the lattice parameter a in a polycrystalline sample of NaCl under nonhydrostatic stresses on the Miller indices hkl / 5,10 /

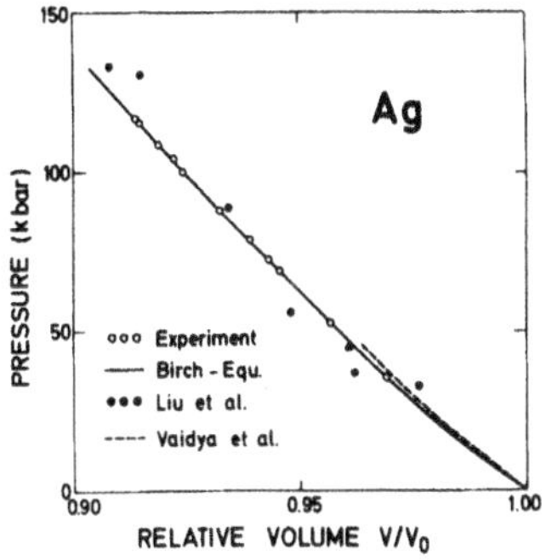

Fig. 7
Compression of silver as determined by different techniques / 7 /.

The slope of the line in Fig. 6 represents this effect,which depends not only on the macroscopic nonhydrostatic stresses contribution but also on the microscopic stress distribution and on the detailed stress and stain continuity at the grain boundaries as well as on the elastic anisotropy of the sample material / 11 /. Therefore, these uniaxial stresses lead to the dominant uncertainty in this kind of lattice parameter determinations under pressure. By careful sample preparation and intercomparison of measurements with different marker materials this uncertainty in the lattice parameter determination can be reduced to about 0.1% as shown in Fig. 7 / 7 /, which compares recent measurements on the compression of silver with earlier results /12,13/.

Higher accuracy in the pressure range up to 100 kbar can be obtained if one uses a truly hydrostatic pressure transmitting medium /14 / and either a liquid tight gasket ring made out of beryllium /15 / or specially shaped anvils of boron carbide as shown in Fig. 8 /5,16 /. The sample is filled together with the marker material and pressure transmitting liquid into the central cavity which is covered by a metal gasket. This gasket must not be transparent to the x-rays since they pass directly through the anvil. The cell has been used primarily for x-ray measurements at low temperatures and high pressures / 5,16 /. Fig. 9 / 16 / shows the cryostat which contains this high pressure cell and fits to the standard x-ray diffractometer shown in Fig. 3. First results with this set-up have been obtained for solid Xe at 85 K in the pressure range between 20 and 110 kbar /5,16 /. These data are shown in Fig. 10 together with data from piston cylinder compression measurements / 17 /.

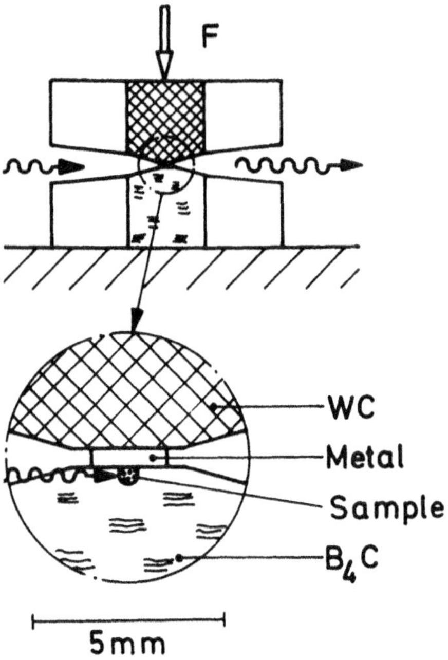

Fig. 8
High pressure cell with sample cavity for low temperature and hydrostatic high pressure x-ray diffraction / 5, 16 /.

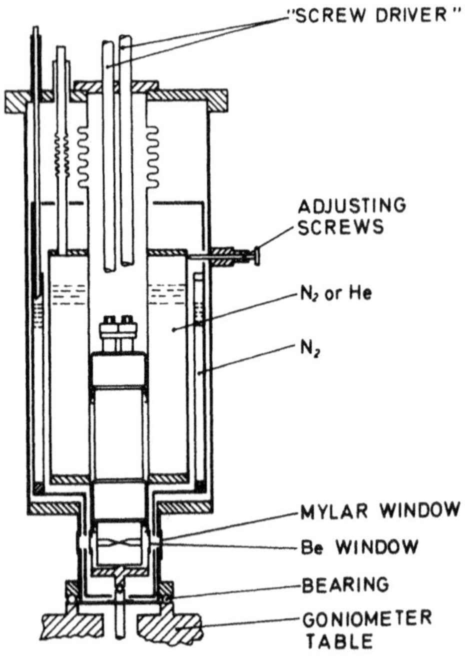

Fig. 9
Cryostat for high pressure x-ray diffraction / 5, 16 /.

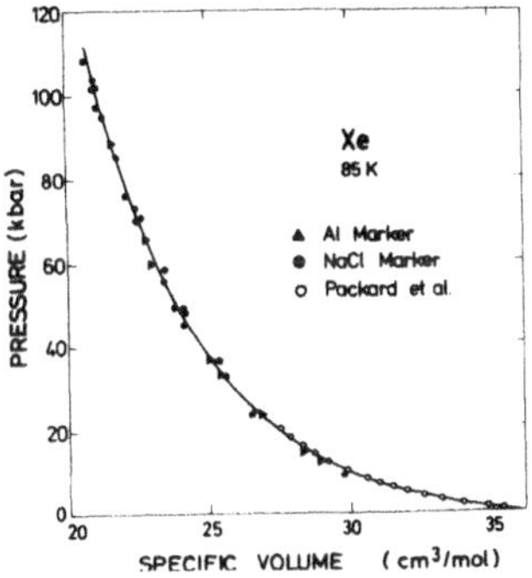

Fig. 10
Compression of solid Xenon at 85 K measured with different marker materials by powder x-ray diffraction / 5 / and compared with bulk compression measurements at low pressures / 17 /.

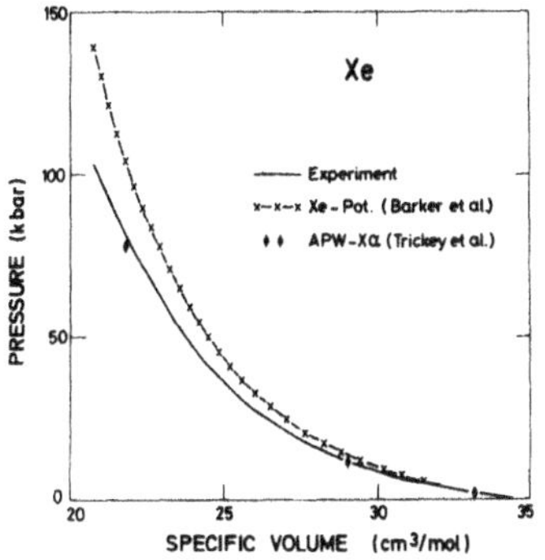

Fig. 11
Comparison of the experimentally determined compression of Xenon / 5,6,17 / with different theoretical predictions / 18, 19 / as given in reference / 20 / .

It is clear that these data are very interesting for the comparison with different theoretical predictions / 18 , 19 / as shown in Fig. 11. A detailed discussion of these results is given in Ref. / 20 / .

2.2 ENERGY DISPERSIVE X-RAY DIFFRACTION

Energy dispersive x-ray diffraction has already been used in the past together with various high pressure techniques / 21, 22 /.Only recently, however, this technique has been combined with diamond anvil high pressure devices / 6,28 /. Such a combination provides certain advantages. First of all, the diamond anvil technique allows for measurements under hydrostatic pressures of up to 100 kbar / 14 / and under quasihydrostatic pressures to about 1 Mbar / 23,24 /. The energy dispersive x-ray diffraction requires only small openings as they are typical for diamond anvil cells and offers the advantage of rather short counting times. These points will be illustrated in the next figures.

The heart of the gasketed diamond anvil cell is the high pressure cavity in the center of a metal gasket between the flat faces of the two diamond anvils as shown in Fig. 12.

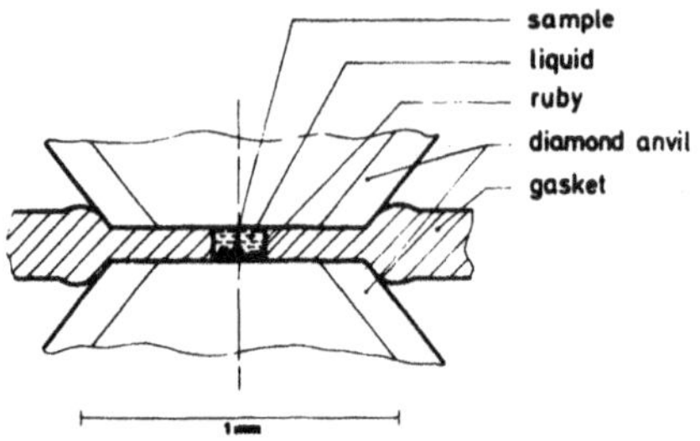

Fig. 12 High pressure cavity in the gasketed diamond anvil cell / 26 /

Usually, the sample is inserted into this cavity together with a ruby splinter which is used as manometer / 25 - 27 / and with a liquid which serves as hydrostatic pressure transmitting medium / 14 /. There are various different techniques available for the generation of the required forces on the diamond anvils / 3,6,9,16,14 /. One way /6,28 / is shown in Fig. 13. The diamonds are mounted in special backing plates, which allow for careful central and parallel alignment of the diamond anvil faces. Good guidance of the moving piston guarantees that the alignment is preserved also under pressure. The force on the piston is generated by the two brackets and levers when the two threaded rods are

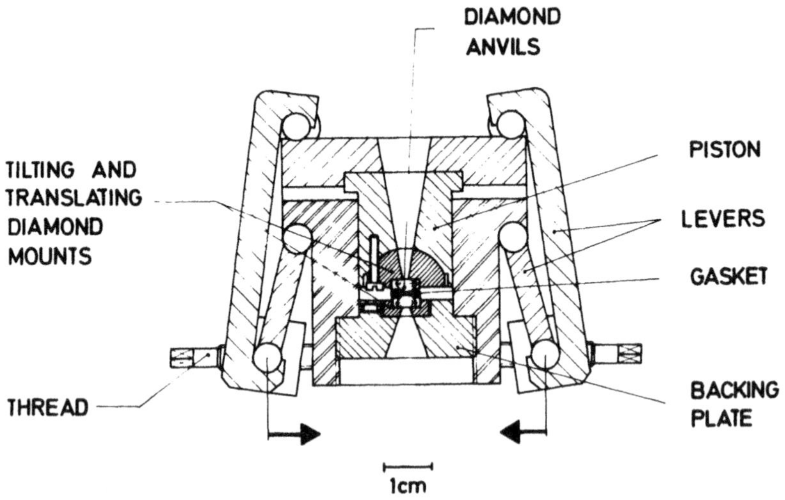

Fig. 13 Diamond anvil cell for optical and x-ray studies / 6 / .

turned synchronously by a special set of gears and pull thereby the two brackets together, forcing the piston with the upper diamond against the lower diamond and its backing plate.

The holes in the backing plates of the diamonds allow for visual observation of the samples, for luminescence measurements on ruby splinters in the samples cavity for pressure determinations / 25 - 27 / as well for the x-ray measurements on the sample. Most massive support of the diamond anvils is provided in this case by backing plates which have only a very small central hole. Since energy dispersive x-ray diffraction requires only one fixed diffraction angle of $2\theta \approx 10^{\circ}$, it is most easily adopted to these devices.

Other advantages of energy dispersive x-ray diffraction are the lower absorption of the x-rays due to their higher energies (20 to 60 keV) as well as higher counting rates due to the fact that all the detected x-rays from the white spectrum of the tungsten x-ray tube are energy analysed immediately and therefore lead to a nearly continuous growth of the whole diffraction pattern.

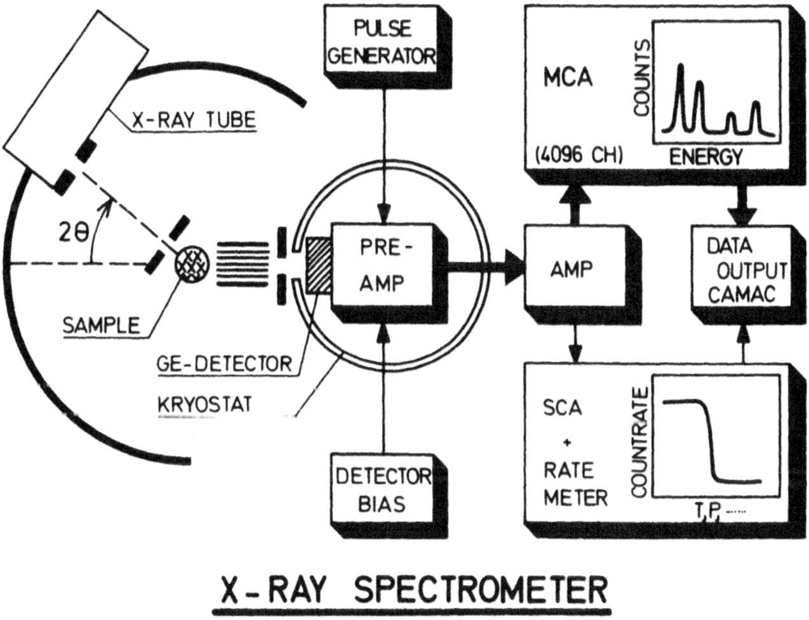

Fig. 14 Typical components of an energy dispersive x-ray diffractometer / 6 /.

The components of an energy dispersive x-ray diffraction system are illustrated in Fig. 14 / 6,28 /. The major components are the white x-ray tube, collimators for the x-ray beam in front and behind the sample defining the diffraction angle 2θ , a cooled solid state x-ray detector connected to the pulse preamplifier, amplifier and multichannel pulse height analyser (MCA), which shows finally the spectra on its display. For a detailed analysis of the line positions, the spectra are usually transferred to a computer. Spectra for NaCl are shown in Fig. 15 / 28 / for comparison with the angular dispersive measurements shown in Fig. 4. Spectra from a recent measurement on Yb metal under pressure / 29 / are given in Fig. 16. The 40 kbar spectrum is taken just at the phase transition between the low pressure fcc- and the high pressure bcc-phase of Yb and shows the lines from both phases. The shifts of the fcc-lines between 1 bar and 40 kbar as well as the shifts of the bcc-lines between 40 and 236 kbar are indicated by arrows. A detailed discussion of these measurements is given in ref. / 29 /.

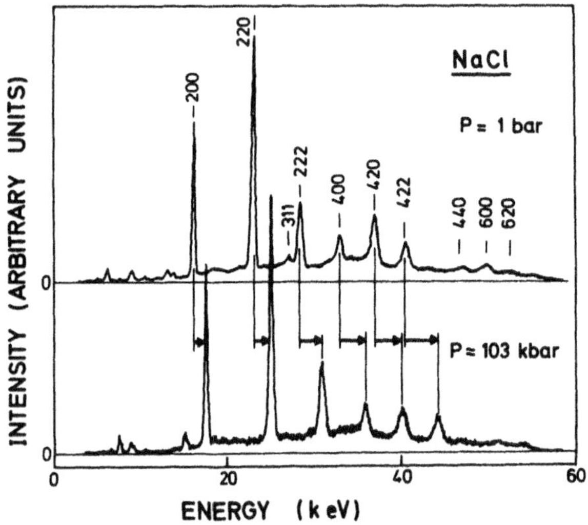

Fig. 15
Energy dispersive x-ray diffraction spectra for NaCl at 1 bar and 103 kbar / 28 /.

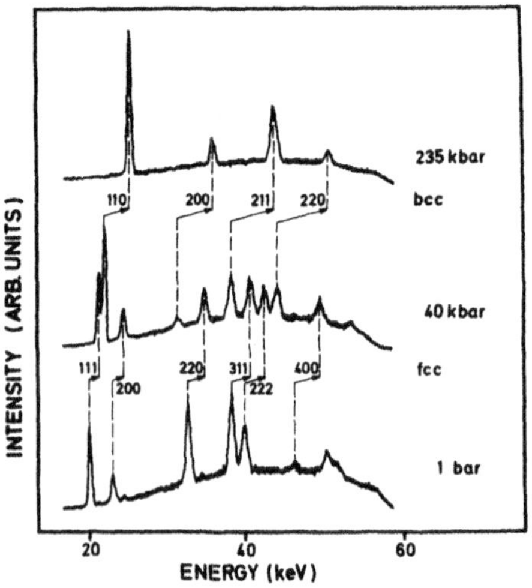

Fig. 16
Energy dispersive x-ray diffraction spectra of Yb at different pressures / 29 /.

3. X-RAY DIFFRACTION ON SINGLE CRYSTALS

The fascinating progress in the diamond anvil high pressure techniques / 14,26,27,30 / stimulated also further developments in high pressure single crystal x-ray diffractometry / 31 - 35,9 /. Two examples are given in the next sections.

3.1 Buerger Precession Technique

Various slightly different high pressure single crystal x-ray diffraction set-ups using the Buerger precession technique have been described in the literature / 36 - 38, 9 /. In the most recent application / 9 /, a gasketed diamond anvil cell was presented, which fits into commercial precession cameras and is capable of generating pressures up to 100 kbar and above. This cell is shown in Fig. 17 / 9 /. The backing plates of both diamonds are made out of beryllium and provide therefore large solid angles for the transmission and diffraction of the x-rays. On one side this solid angle is 90°, on the other side it is 100° .

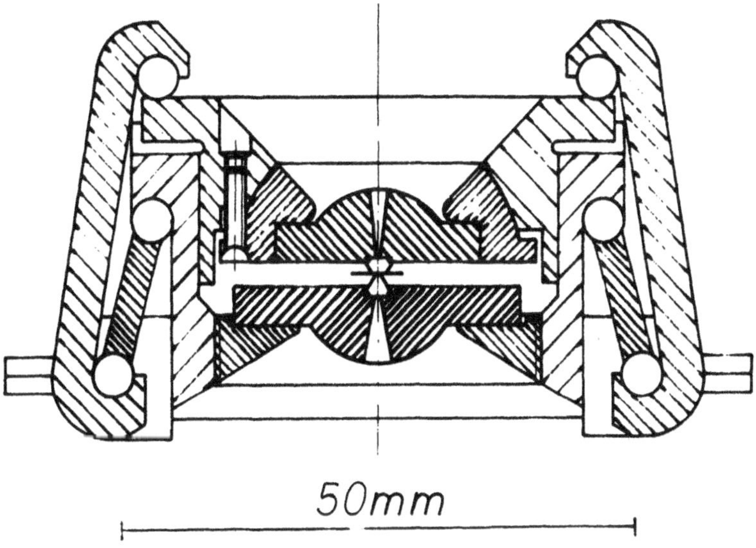

Fig. 17 Gasketed diamond anvil cell for single crystal x-ray diffraction up to 100 kbar and above /9/

The pressure is generated in this cell in the same way as in the device that was discussed before (see Fig. 13). The small holes in the beryllium backing plates allow for visual observation of the sample and for pressure measurements by the ruby fluorescence method / 25 - 27 /.

A photograph of this diamond anvil device mounted on a commercial precession camera is shown in Fig. 18 / 9 /.

Fig. 18. Precession camera with diamond anvil high pressure cell /9/

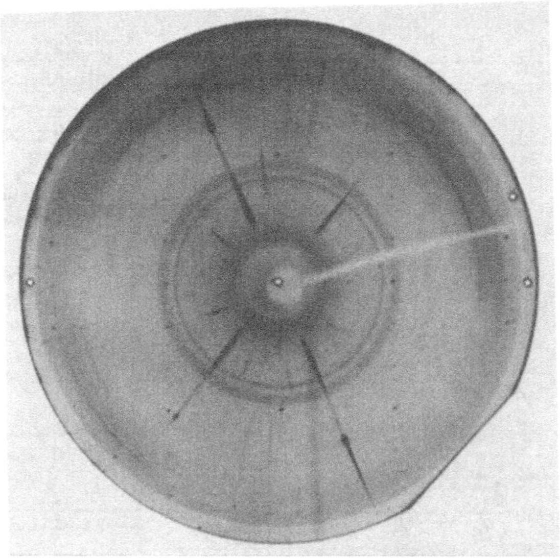

Fig. 19 Precession x-ray photograph of Se at 87 kbar showing streaks of the diamonds and small diffraction spot of the Se-(hhl)-plane / 9 /.

A typical precession photograph taken on this set-up with Zr-filtered MoKα radiation and an exposure time of 17 h is reproduced in Fig. 19 / 9 /. It shows small diffraction spots in the (hhl)-plane of trigonal Se at 87 kbar as well as strong streaks from the two diamonds, and diffuse rings from the beryllium backing. Lattice spacings can be determined directly from the distances between various reflections in this photograph. Atom position parameters are derived from the relative intensities of the various reflections with standard techniques / 9 /. Results for Se and Te under pressure are compared in Fig. 20 / 39 / with earlier data / 40,42 / measured on polycrystalline samples. In the case of Se, an attempt was made earlier / 41 / to derive the variation of the atom position parameter from changes in the relative intensities of the powder x-ray diffraction patterns. Strong effects from texture explain the discrepancy between these data and the more accurate results from the single crystal x-ray diffraction. More detailed discussions of these results as well as of the techniques are given in Ref. / 9, 39 /.

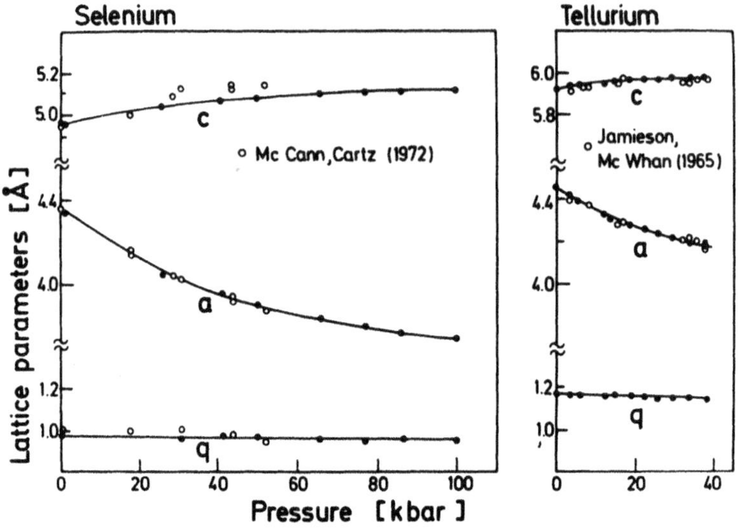

Fig. 20 Effect of pressure on lattice parameters a and c and atom position parameter q for trigonal Se and Te / 39 /.

3.2 Automatic Four Circle X-Ray Diffractometers

Diamond anvil devices have been adapted also to automatic four circle x-ray diffractometers / 31,33,9,35 /. A photograph of one of these adaptations / 35 / is shown in Fig. 21. Special care has been taken to allow for accurate centering of the sample together with the high pressure cell / 35 /.

One problem of this technique results from the fact that the measured x-ray intensities have to be correlated for absorption in the sample as well as in the diamonds, beryllium backings and in the gasket of the high pressure cell. In simple cases, one can use the low pressure absorption corrections / 39,42 / to scale the high pressure intensity data. This procedure was adopted in a first application of this technique where the atom positions in ruby were determined under pressures up to 85 kbar / 42,43 /. The results indicate / 42, 43 /, that atom position can be determined in simple cases with this procedure with an accuracy which is of the same order of magnitude as in similar measurements at atmospheric pressure. Typical data for the Al position parameter μ(Al) are shown in Fig. 22 which includes two different sets of data as well as the results of two different refinement procedures. The

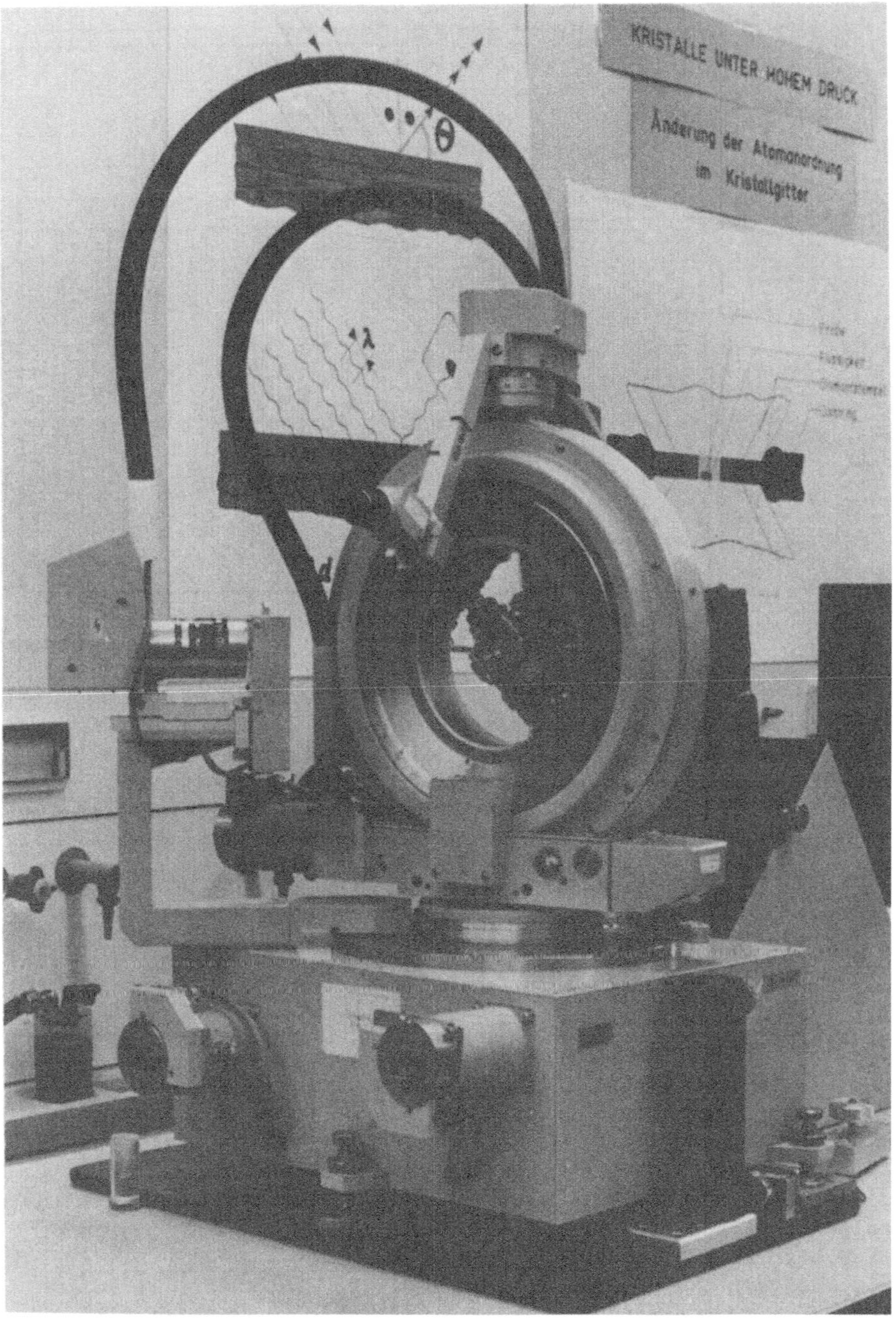

Fig. 21. Automatic four circle diffractometer with diamond anvil high pressure cell / 35/

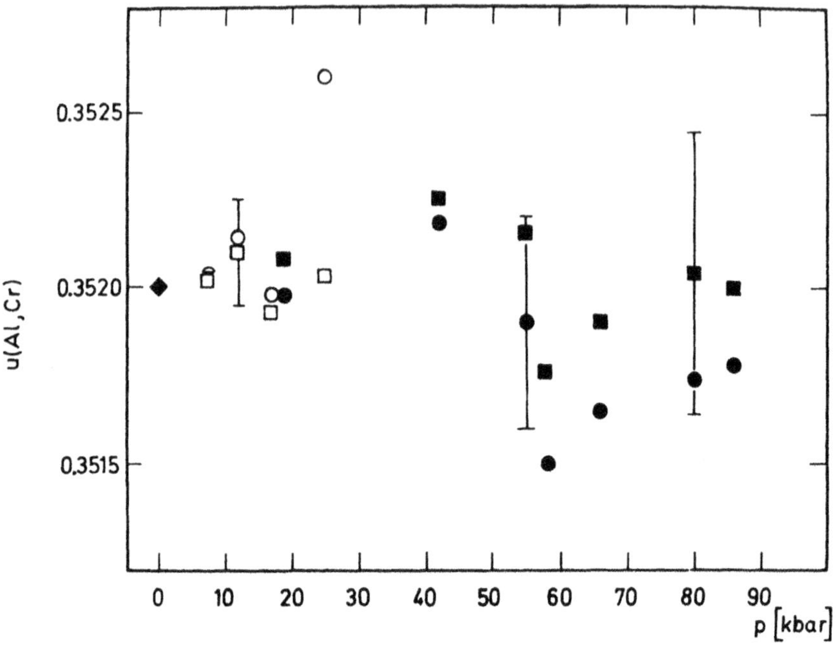

Fig. 22 Typical data for the effect of pressure on the atom position parameter of Al in Al_2O_3 derived from measurements with automatic four circle diffractometers. Open symbols are data from ref. 42, filled symbols and complete figure from ref. 43. Squares and circles represent results of different refinement procedures.

enlarged scale for μ(Al) emphasizes the purely statistical scattering of the results. Further details are given in the original literature / 42, 43 / .

4. PERSPECTIVES

Recent applications of the diamond anvil techniques indicate that there are good chances to extend the range of structural studies by these techniques to static pressures of 1 Mbar and above / 23,24 /. With the current developments in synchroton radiation x-ray sources, a white x-ray source has become available which has a spectral brightness about a factor of 1000 higher than conventional white x-ray sources. First high pressure energy dispersive x-ray diffraction studies with synchrotron radiation indicate in fact / 44 /, that one will be able to study also dynamical processes, diffusion and rapid phase changes with these sources. The possibility to use even smaller sample sizes with these sources will also help to extend the pressure range.

The extension of hydrostatic pressures to 100 kbar and above / 14 / increased the accuracy of x-ray measurements in this pressure range drastically. This has been demonstrated most strikingly by the adaptation of a gasketed diamond anvil cell to the Bond method / 32 /. This technique opens up the possibility to study also minor lattice distortions under pressure.

Recent software developments / 45 / are still improving the accuracy of the high pressure single crystal x-ray diffraction with automatic four circle diffractometers.

In conclusion, one can state, that the improvements in the diamond anvil high pressure technique have stimulated a number of recent developments in high pressure x-ray diffraction, and first experimental results of these improved techniques are just appearing in the literature.

ACKNOWLEDGEMENT

First of all I am very thankful to H.G. Drickamer for introducing me into this field some years ago. On the other hand it is also a great pleasure to acknowledge very stimulating cooperation and discussions with H. d'Amour, W. Dieterich, W. Denner, R. Keller, G.J. Piermarini, Heinz Schulz, D. Schiferl and K. Syassen.

REFERENCES

1. H.G.Drickamer, R.W.Lynch, R.L.Clendenen and E.A.Perez Albuerne Solid State Phys. 19, 135, 1966
2. D.B.McWhan, ed., Transactions of the American Crystallographic Association, Proceedings of the Symposium on Crystal Structure at High Pressure, Polycrystal Book Service, Pittsburgh, Pa, US, 1969
3. M.D.Banus, High Temp.-High Pressure 1, 483, 1969
4. K.Syassen, C.W.Christoe and W.B.Holzapfel, Z. angew. Physik 31, 261-262, 1971
5. K.Syassen, Thesis, University Stuttgart, 1974
6. K.Syassen and W.B. Holzapfel, europhys.conf.abstr. 1A, 75, 1975
7. K.Syassen and W.B.Holzapfel, europhys.conf.abstr. 1A, 85, 1975
8. R.Keller and W.B.Holzapfel, europhys.conf.abstr. 1A, 90, 1975
9. R.Keller and W.B.Holzapfel, Rev.Sci.Instrum.48, 517, 1977
10. K.Syassen and W.B.Holzapfel (to be published)
11. A.K.Singh and G.C.Kennedy J.Appl.Phys.45, 4686, 1974
12. S.N.Vaidya and G.C.Kennedy,J.Phys.Chem.Sol.31,2329, 1970
13. L.Liu, and W.A.Bassett,J.Appl.Phys.44, 1475, 1973
14. G.J.Piermarini,S.Block and J.D.Barnett, J.Appl.Phys.44, 5377, 1973
15. P.M.Halleck and B.Ollinger,Rev.Sci.Instrum.45, 1408, 1974
16. K.Syassen and W.B.Holzapfel europhys.conf.Abstr.1A, 87, 1975
17. J.R.Packard and C.A.Swenson J.Phys.Chem.Solids 24, 1405, 1963
18. Trickey,S.B.,Green Jr., F.R., and Averill,F.W.,Phys.Rev.B8, 4822, 1973
19. Barker,J.A., private communication
20. K.Syassen and W.B.Holzapfel (to be published)
21. P.J.Freud and P.N.LaMori, p.155 in ref. 2, 1969
22. L.M.Albritton and J.L.Margrave High Temp.-High Press. 4, 13, 1972
23. H.K.Mao and P.M.Bell, Science 191, 851, 1976
24. K.H.Mao and P.M.Bell, in High-Pressure Research ed. M.H.Manghna and S.Akimoto, Academic Press, New York, 1977
25. R.A.Forman, G.J.Piermarini, J.D.Barnett and S.Block, Science 17 284, 1972
26. J.D.Barnett, S.Block and L.J.Piermarini,Rev.Sci.Instrum.44, 1, 1973
27. G.J.Piermarini, S.Block,J.D.Barnett and R.A.Forman,J.Appl. Phys. 46, 2774, 1975
28. K.Syassen and W.B.Holzapfel (to be published)
29. K.Syassen and W.B.Holzapfel, Proceeding of the Sixth AIRAPT International High Pressure Conference, Boulder, 1977
30. S.Block and G.Piermarini,Physics Today p. 44, Sept. 1976
31. L.Merrill and W.A.Bassett, Rev.Sci.Instrum.45, 290, 1974
32. F.A.Mauer,C.R.Hubbard, G.J.Piermarini and S.Block, Adv.in x-ray Anal. 18, 437, 1975

33. D.Schiferl, Rev.Sci.Instrum.48, 24, 1977
34. L.W.Finger and H.King (to be published)
35. W.Denner, W.Dieterich, W.B.Holzapfel, R.Keller and H.Schulz (to be published)
36. C.E.Weir, S.Block and G.Piermarini, J.Res.Natl.Bur.Stand (US)69 C, 275, 1965
37. R.Fourme, J.Appl.Cryst.1, 23, 1968
38. H.Meyer, P.Bell and J.England, Carnegie Institution Yearbook 66, 541, 1967
39. R.Keller, W.B.Holzapfel and H.Schulz Phys.Rev.B 16, 4404, 1977
40. J.C.Jamieson and D.B.McWhan J.Chem.Phys.43, 1149, 1965
41. D.R.McCann and L.Lartz J.Appl.Phys.43, 4473, 1972
42. D.Schiferl and J.C.Jamieson (to be published)
43. H.d'Amour, D.Schiferl, W.Denner, H.Schulz and W.B.Holzapfel J.Appl.Phys. (to be published)
44. B.Buras, J.Staun Osen, L.Gerward, G.Will and W.Hinze, J.Appl.Phys., Crystallogr. 10, 431, 1977
45. W.Denner, Thesis, Universität Karlsruhe, 1977

SOLIDS UNDER PRESSURE

D. Bloch

Laboratoire Louis Néel, C.N.R.S., 166X,
38042 Grenoble-Cedex, France

ABSTRACT. In this lecture we discuss effects of pressure on order-disorder transitions and pressure induced phase transitions.

This lecture has only a modest goal, although its title is rather ambitious. Its purpose is to indicate to "high pressure chemists" tye type of problems on which are working their colleagues the "high pressure physicists". It will neither be a course on solid state physics |1,2| nor a general review on main results obtained by "high pressure physicists" |3|. It will rather be a list of examples chosen through their pedagogical character. These examples will mostly be selected in the field of phase transitions. It should be noticed however that in addition high pressure behaviour is often used as a supplementary parameter by solid state physicists to help them to choose between various theories which otherwise could not be separated using atmospheric pressure data. We will not discuss of these peculiar approaches since they are generally more complex to introduce than phase transition themselves. We will not discuss of experimental technics, since they are described in other Chapiters of this book. We shall simply mention that most usual characterization techniques, which are on use at atmospheric pressure can be used with an equal accuracy under high pressure conditions.

1. SOLIDS ARE COMPRESSIBLE

It is often tempting to separate the "lattice" properties of solids (compressibility, lattice motions, melting...) from the "electronic" ones (electrical resistivity, magnetic properties...)

Kelm, H. (Ed.) High Pressure Chemistry. 199–220. *Copyright © 1978. D. Reidel Publishing Company. All Rights Reserved.*

One of the major tools of solid state physics is the Born-Oppenheimer approximation : as the electrons are light particles, and have high speed, whereas the nuclei are heavy and move slowly, the properties of the electrons can be evaluated according to the Born-Oppenheimer approximation considering the lattice as "frozen". However, interactions between atomic motions and electrons are to be re-introduce to explain for instance superconductivity or various aspects of other "electronic" phase transitions.

Pressure changes the interatomic distances of these "frozen" lattices and can lead to major "electronic" modifications, for instance appearance or disappearance of metallic resistivity, of magnetic moments, modification of bonding... Major "electronic" changes are expected in those materials with the highest compressibility. However, hard materials in the vicinity of a continuous phase transition often present some kind of lattice softening and have therefore a high compressibility. It will then be difficult in many cases to clearly separate between "lattice" and "electronic" properties.

At the beginning of this lecture we will discuss some of the major parameters which govern the compressibility K of elements of the periodic table*. As can be seen from table I, the initial compressibility of solids for example (fig. 1), we observe that it possesses at 45 kbar, a volume which is 40 % of its P = 0 value. Furthermore, the compressibility, at 45 kbar, is reduced from a factor 70 from the initial compressibility. There is in fact no general rule from which to

Elements	He	Cs	U	Cu	C
$10^6 \times K$ |1| (bar^{-1})	1168	32	1.01	0.73	0.183
Z	2	55	92	29	66
Melting temperature (K)	-	302	1405	1356	4000

Table 1. Compressibility, atomic number and melting temperature of selected solids.

---

* $K = -\frac{1}{V}\left(\frac{dV}{dP}\right)_T$

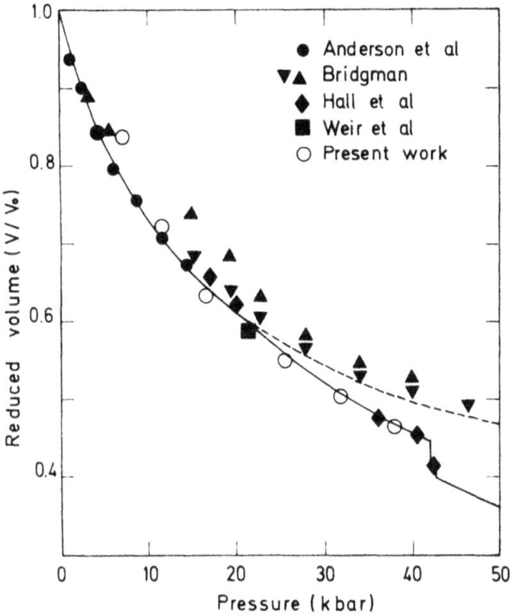

Fig. 1. P-V relationship for cesium |4|.

predict the compressibility of all elements. We will however indicate some general rules. The first one applies to matter at ultra-high pressure : it is the Thomas-Fermi approximation where the solid is considered as a lattice constructed from nuclei, plus a charge density cloud in which the Z electrons of each atom participate - without any consideration for energy level structure.

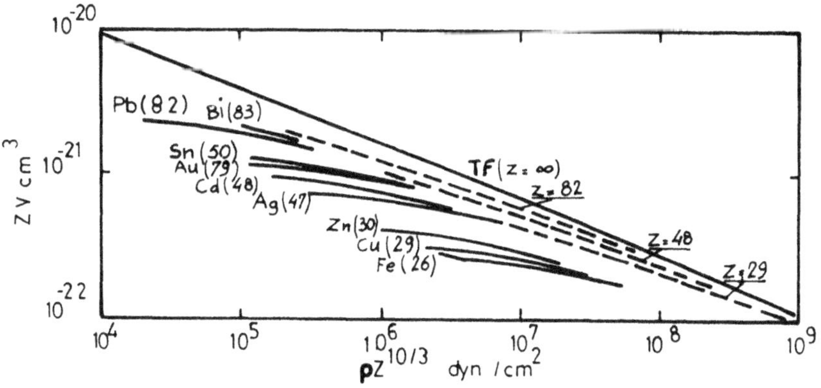

Fig. 2. Comparison of experimental P-V data with Thomas-Fermi-Dirac equation of state for comparable Z-values. The Thomas-Fermi equation of state corresponds to Z = ∞ |5|.

This gives the T-F curve, fig. 2, where the volume (per atom) V (xZ) is plotted versus applied pressure $P(xZ^{-10/3})$. Experimental results for few metallic element are plotted for comparison. Although the T-F curve does not give the experimental P-V relationship, it gives some indication of the limit for high Z or ultra-high pressure. An improvement on T-F equation consists in taking into account electronic exchange interactions : this gives rise to the Thomas-Fermi-Dirac (TFD) approximation, with results also indicated fig. 2. Although this approximation is useful for heavy elements or very high pressure, it does not give account of the P-V relationship for light elements, and laboratory pressures. The atomic number Z is not the appropriate parameter.

If we now proceed to real world, a few observations can be done :

a) Elements of similar outer configuration and similar bonding have similar compressibility. This compressibility increases with increasing interatomic distances (fig. 3).

b) Inner electrons have little influence on compressibility as is clear from an examination of the compressibility of the $4f^n$ metals. In fact Pr($4f^2$), Nd($4f^3$), Sm($4f^5$), Gd($4f^7$), Tb($4f^8$), Dy($4f^9$), Ho($4f^{10}$), Er($4f^{11}$), Tm($4f^{12}$) and Lu($4f^{14}$) have similar compressibilities, with Z varying from 59 (Pr) to 71 (Lu). This compressibility is $(2.9 \pm 0.5) 10^{-6}$ bar^{-1}. These metals can be considered as tripositive T^{3+} ions imbedded in a sea of collective electrons (6s 5d). However Eu($4f^6$) and Yb($4f^{13}$)

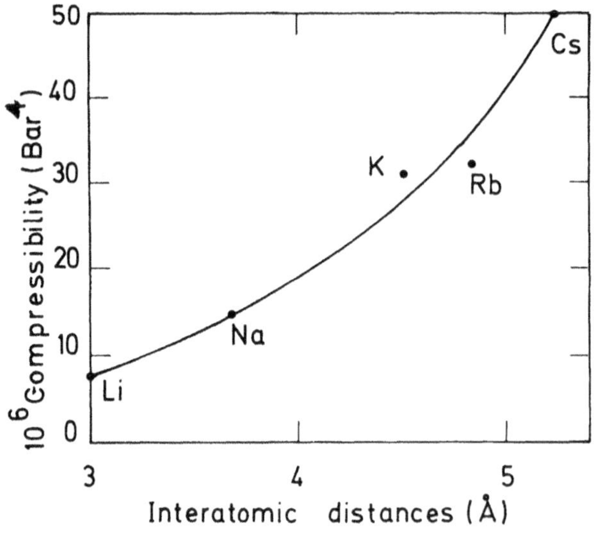

Fig. 3. Compressibility of alkaline metals versus interatomic distances.

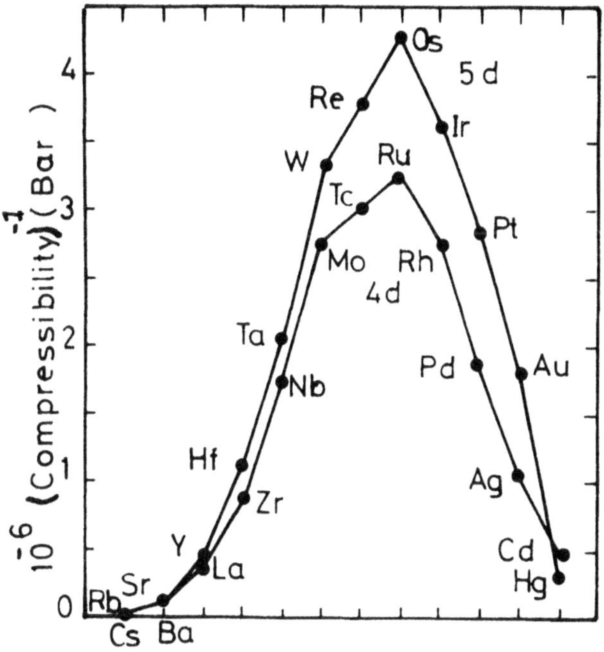

Fig. 4. Bulk moduli of 4d and 5d metals versus collective electrons number (4d - 5s ; 5d - 6s).

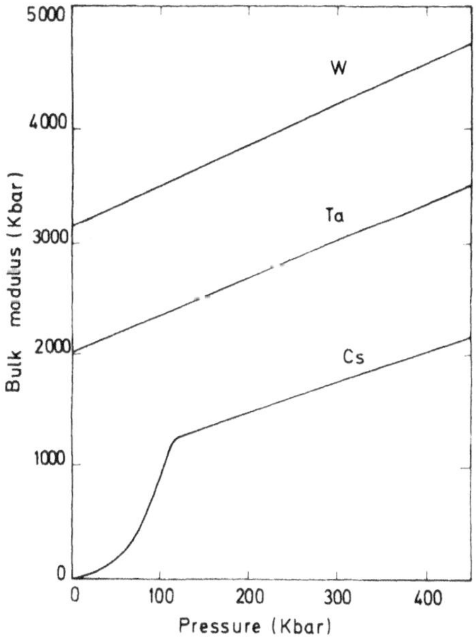

Fig. 5. Bulk moduli versus pressure : Cs and W |8|, Ta |9|.

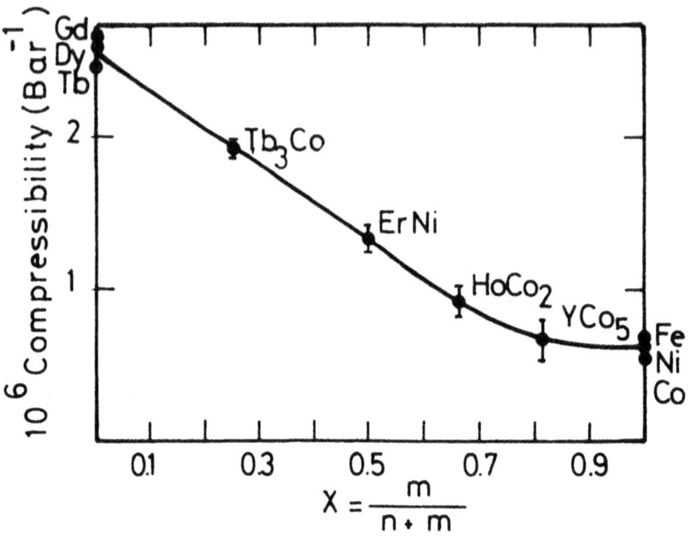

Fig. 6. Compressibility of A_nB_m compounds ; A is a rare earth and B a transition metal |10|.

and which possess respectively 2 and 4 collective electrons have quite different compressibility, respectively 6.80 x 10^{-6} and 7.52 x 10^{-6}.

c) The importance of the number of outer electrons on compressibility appears on a peculiarly clear manner for the transition metals (fig. 4). The bulk moduli (or inverse compressibility) of 4d or 5d elements varies approximately as Z(10 - Z), where Z is the number of d electrons |6,7|. The very large compressibility of Cs in the 10 - 100 kbar range (figs 1 and 5) is associated |4| with a continuous 6s → 5d transition. Above 100 kbar (fig. 5) Cs behaves like an usual 5d metal, and it presents superconductivity as Ta or W.

d) The last point we will mention is that for similar type of bonding, the compressibility is only weakly dependent on crystal structure ; for instance (fig. 6) Fe is body-centered cubic, Ni face-centered cubic, Co, Gd, Tb and Dy hexagonal compact and Tb_3Co and ErNi orthorhombic, YCo_5 hexagonal and $HoCo_2$ has the cubic Laves phase. All these elements and compounds are located on a smooth line as a function of rare earth (or transition metal) content.

2. ORDER-DISORDER TRANSITIONS

2.1 Phase diagram

If we consider a standard phase diagram, as given fig. 7,

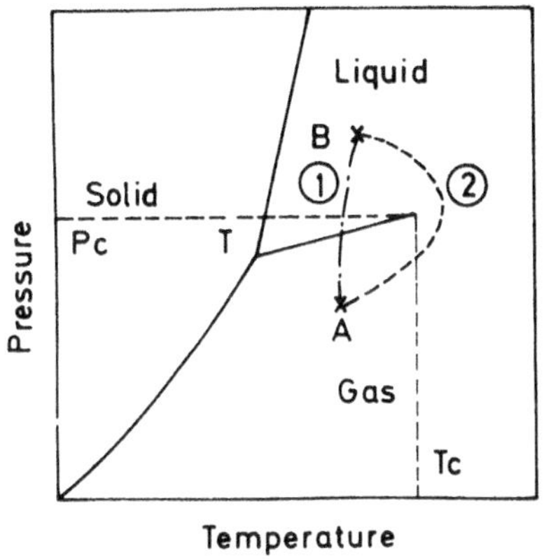

Fig. 7. Liquid, solid, gas phase diagram.

where the full lines represents discontinuous transitions (or first order transitions) between the liquid, solid or gaseous states, one notices, besides the triple point, at the junction of the three first order lines, a critical point which ends, at pressure P_c and temperature T_c, the liquid-gas separation line. Two points A and B of this phase diagram can be connected through a path (1) which intersects the first order line, where a discontinuous density change occurs. These two points can also be connected through a path (2) which turns around the critical point, and thus no discontinuous change occurs. Similar features

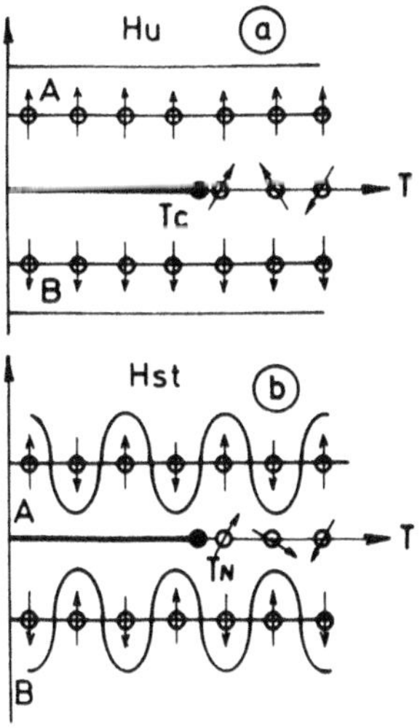

Fig. 8. Phase diagram of a ferromagnet (a) and of an antiferromagnet (b).

happen in a large variety of phase diagrams, which for instance describe magnetic, ferroelectric, superconducting... properties. Let us take for instance a magnetic crystal, which contains a regular array of atoms with identical magnetic moment. Interactions between these magnetic moments lead, at sufficiently low temperature, to a long range ordering of the moments. Ferromagnetism occurs when all the moments are parallel (fig. 8a) and antiferromagnetism when half the moments point upwards and half the moments point downwards (fig. 8b).

When a uniform magnetic field H_u is applied to a ferromagnet (fig. 8a), all the magnetic moments are directed along the field direction. When the magnetic field is reversed, the magnetic moments flip discontinuously from upwards (A) to downwards (B) direction. The crossing of the transition line ($H_u = 0$) is therefore associated with a discontinuous change of the overall magnetization. The transition line ends at a critical point, the so-called "Curie temperature". When turning around the critical temperature, the flipping is continuous and thus any point of the H_u - T diagram can be connected through a continuous transformation.

A similar feature can occur for an antiferromagnet if one considers a magnetic field whose direction is modulated so as to be aligned with the alternating magnetic moment (staggered field H_{st}) (fig. 8b). Reversing the staggered field leads to a discontinuous reversal of the atomic moment, for temperature lower than T_N, the "Néel temperature", or to a continuous one above T_N.

The usual role assigned to high pressure (or stress) is to change internal parameters, such as interactions between atomic moments. Thereby pressure (or stress) leads to a change in the values of the Curie - or Néel - temperatures which are tightly related to the magnitude of these interactions.

2.2 Order parameter

The interaction energy favours order, whereas the magnetic entropy favours disorder. The Curie (or Néel) temperature thus separates the ordered magnetic (low temperature) state from the disordered (high temperature) state. The degree of order is characterized by an order parameter σ which is zero in the disordered temperature range for a ferromagnet σ will be $M_A(T, H = 0^+)/M_A(T = 0, H = 0^+)$, whereas for an antiferromagnet, it will be $(M_A - M_B)(T)/(M_A - M_B)(T = 0)$, where $M_A - M_B$ is the difference between the magnetic moments of up and down sublattices.

In fact the order parameter can go to 0 (its value in the disordered temperature range) continuously (fig. 9c) or discontinuously (fig. 10d). The order-disorder transition is then said to occur continuously or discontinuously.

2.3 Landau theory - Critical and tricritical behaviour

In the calculation of thermodynamical quantities, a first approximation is to neglect all fluctuations in the magnetization. The difference between the actual free energy and the free energy for zero order parameter, can be expanded, near the critical point as a function of the order parameter which is then small compared to its T = 0 value

$$F(\sigma) - F(\sigma=0) = \frac{1}{2} A\sigma^2 + \frac{1}{4} B\sigma^4 + \frac{1}{6} C\sigma^6 + \ldots \quad (1)$$

for zero applied pressure (or stress).

The actual order parameter is that which minimizes the order parameter dependent part of the free energy, that is which satisfies to :

$$\frac{dF(\sigma)}{d\sigma} = A\sigma + B\sigma^3 + C\sigma^5 + \ldots = 0 \quad (2)$$

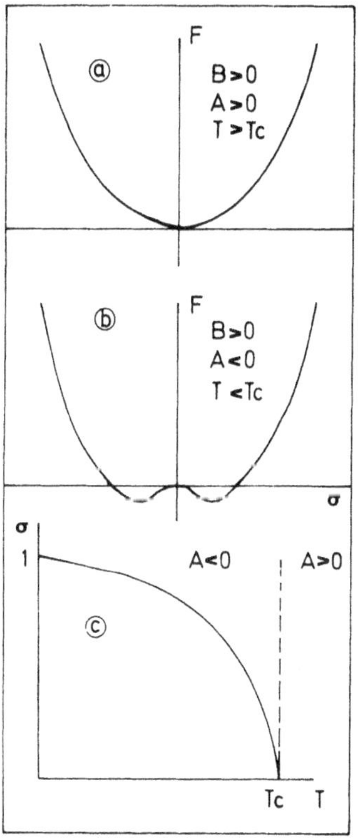

Fig. 9. Thermal dependence of the free energy (eq. 1) and order parameter (eq. 3) for B > 0.

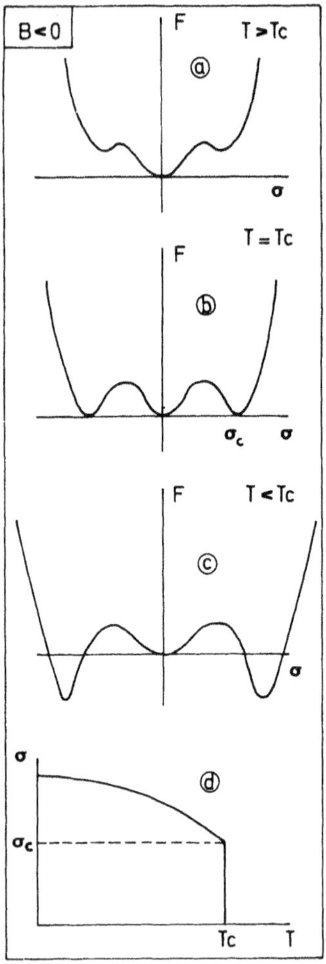

Fig. 10. Thermal dependence of the free energy (eq. 1) and order parameter for B < 0.

Neglecting term higher than 4th order in equation (1) leads to the non zero solution :

$$\sigma^2 = - A/B \tag{3}$$

We first take B as positive ; the order parameter goes to zero when A itself has a zero value (fig. 9a and b). If we take, following the classical Landau treatment A close to the critical temperature as a linear function of T, with B smoothly dependent on temperature then :

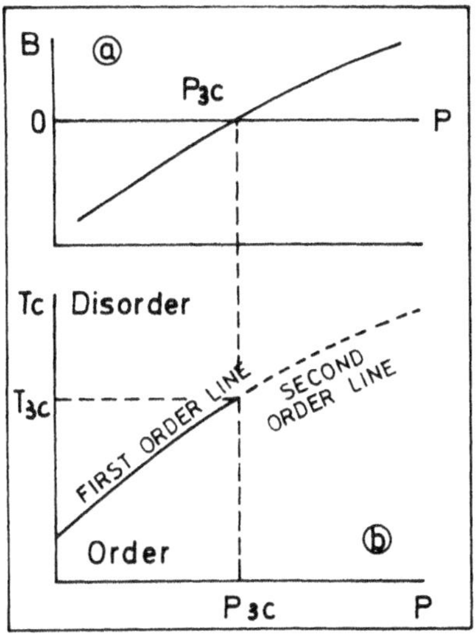

Fig. 11. Tricritical behaviour.

$$A = A'(T_c - T) \qquad (4)$$

leads to :

$$\sigma^2 = \frac{A'}{B}(T - T_c) \qquad (5)$$

The order parameter varies as the power 1/2 of the temperature (fig. 9c).

When $B < 0$ (fig. 10), order occurs discontinuously at temperature T_c. In this case, C should be taken as positive, and cannot be neglected, in order to insure that a phase transition exists.

An interesting situation occurs when B is pressure, or stress dependent, and changes its sign at pressure P_{3c}. Then simultaneously the ordering transition changes from continuous to discontinuous (fig. 11). Such a transition is called a tricritical transition ; T_{3c} and P_{3c} are the tricritical transition temperature and pressure. At the tricritical transition, when $B = 0$ then, from (2) :

$$\sigma^4 = \frac{A'}{C}(T - T_c) \qquad (6)$$

and the order parameter varies at the power 1/4 of the temperature.

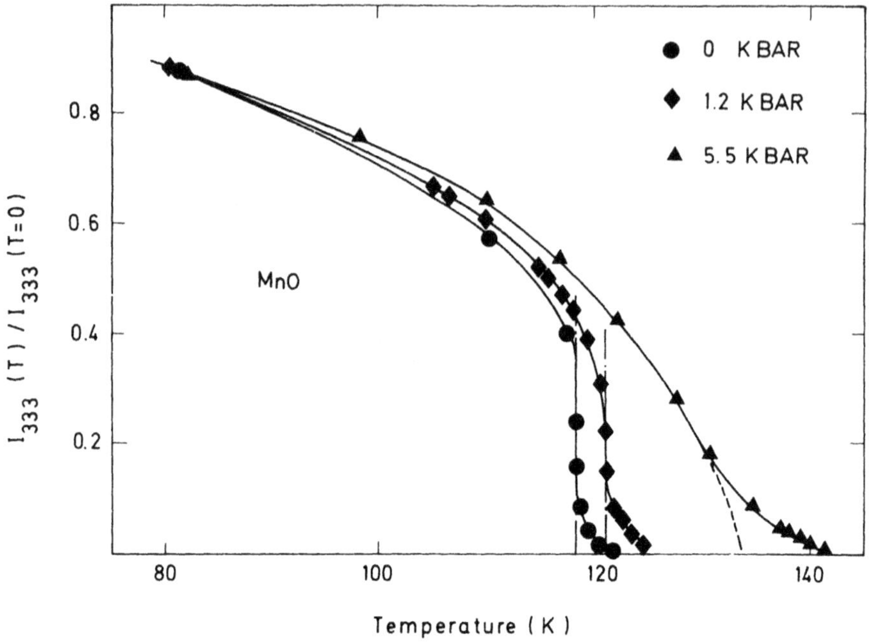

Fig. 12. Thermal dependence of the intensity of (333) magnetic reflections at various applied stresses.

2.4 Applications : tricritical properties

MnO : We first describe stress effects on manganese oxide, which is a simple antiferromagnet with the NaCl structure. In a neutron scattering experiment the intensity of the magnetic reflections is proportional to the square of the magnetic moments. Therefore, scattering experiments performed at various temperatures (fig. 12) can be used to obtain the temperature dependence of the magnetic moment, and as a next step the temperature dependence of the order parameter. The order-disorder transition is accompanied by a (111) rhombohedral distortion of the lattice, which is associated with the strongly distance dependent interaction between the moments. If we apply a stress along the (111) diagonal, we notice that the ordering parameter, which has a discontinuous behaviour at zero stress, becomes continuous at high stress |11|. As described above, this is a tricritical point where B changes sign. The ordering temperature increases with stress at a rate of $\sim$3 K/kbar and the tricritical pressure and temperature are approximately 4 kbar and 130 K.

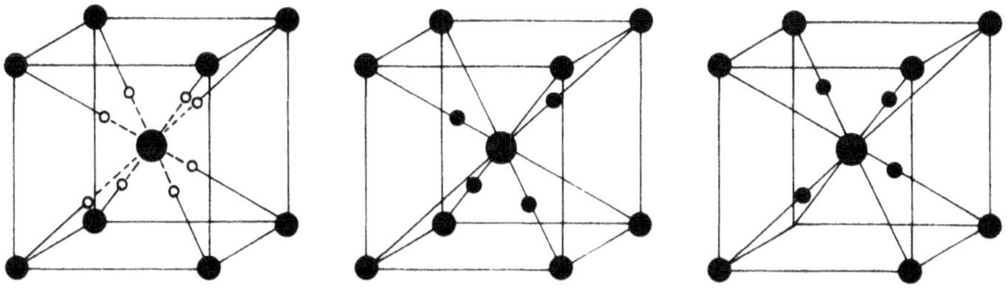

Fig. 13. Ordered and disordered NH_4Cl

NH_4Cl : We now consider another type of order-disorder transition which is encountered for instance in NH_4Cl (or ND_4Cl) (fig. 13). In the ordered phase the hydrogen (or deuterium) tetrahedra are oriented in one of the two equivalent position shown on the right. In the disordered phase the tetrahedra are distributed at random as indicated by the broken line |12|. The ordered phase results from an octupole-octupole nearest neighbour coupling between the hydrogen (or deuterium) tetrahedra. The long range order disappears at a temperature of ∿242 K (or 250 K for ND_4Cl). The ordering temperatures increase with pressure at a rate of 8.5 K/kbar, with tricritical pressure respectively at 128 bar and 1500 bar |12,13|.

K.D.P. : KH_2PO_4 (K.D.P.) is a ferroelectric material, which presents a spontaneous polarisation up to a critical temperature T_c ∿ 122 K. Tetrahedral phosphate (PO_4) groups are connected by O - H -- O hydrogen bonds.

The ferroelectric transition is associated with an order-disorder transition of the proton arrangement in the H-bonds. This order-disorder transition is accompanied with a deformation of the lattice, associated with a strong proton lattice interaction. This order-disorder transition is first order at zero pressure, and it occurs at T_c ∿ 122 K. It decreases with pressure at a rate of ∿4.5 K/kbar |14|, and the tricritical transition occurs at ∿2 kbar |15|. Large pressure dependences of the ordering temperatures are in fact common for most ferroelectrics. SbSI for instance, has a Curie temperature at ∿293 K, which decreases with pressure at a rate of -37 K/kbar |15|.It possesses a tricritical temperature at ∿235 K for P = 1.4 kbar |16|. This, in fact reflects the very intimate relationship which exists between ferroelectric ordering and spontaneous strain within the lattice |14|.

3. ELECTRONIC PHASE TRANSITION

High pressure experiments are of prime importance to understand electronic phase transitions, and among these transitions a special class is formed out of those transitions which can occur even at T = 0. In this paragraph, we will concentrate on these T = 0 effects.

3.1 Metal insulator transitions

Electrons in solids are submitted to a periodic lattice electrostatic potential. From this result a broadening of the atomic energy level in permitted energy bands (fig. 14). If we

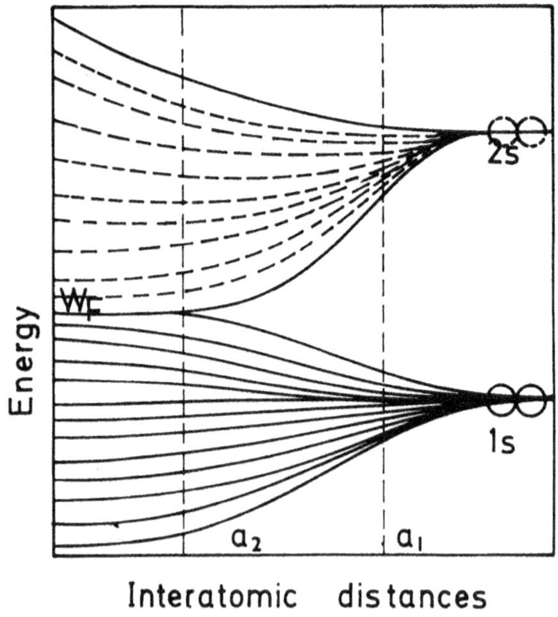

Fig. 14. Electronic energy bands versus interatomic distances.

consider an hypothetical helium crystal, at large interatomic distances (a_1) will exist a 1s energy band which can contain 2 electrons per atom. It will therefore be saturated from the 1s electrons of helium. The crystal is then an insulator since to insure electrical conductivity, electric field should modifies the electronic energies, but no authorized energy levels are available for the electrons. For large pressure (a_2), the 1s and 2s bands give rise to a common 1s - 2s band, which can contain up to 4 electrons per atom. Permitted energy levels exist just above the maximum energy level W_F and thus electrical conductivity can occur. The metal-insulator transition cannot be unambiguously predicted. Estimated values are for instance, for atoms with saturated outside shells : Xe : 700 kbar, Ar : 1300 kbar, Kr : 1400 kbar, Ne : 10000 kbar and He : 100000 kbar |17|.

However if we consider an hypothetical hydrogen crystal with only one outer electron, straigth-forward band theory would indicate a metallic conductivity at any interatomic distance. However, as indicated by N.L. Mott, due to electrostatic repulsion between electrons each electron stays on atomic site and no electrical conduction can occur. Metal-insulator transition in solid hydrogen has been predicted to occur in the Mbar range, when the high electron density screens the repulsive inter-electron Coulomb interactions.

Continuous or discontinuous metal-insulator transitions have been observed in a large variety of transition metals or rare earth compounds |18|. In fact modifications of the electronic conductivity is usually associated with a large number of modifications in other physical properties (interatomic distances, crystallographic structure, elastic properties, optical properties...) and the leading mechanism for these transformations is often difficult to assess.

A classical example is given from vanadium oxides such as V_2O_3 |19| (fig. 15). A pressure of $\sim$20 kbar permits, at T = 0, to observe an (antiferromagnetic) insulator-(paramagnetic) metal transition. In this system pressure has the same effect as decreasing atomic number and thus a generalized phase diagram can be constructed to compare pressure effects and concentration effects.

3.2 New superconductors

One of the major interest of high pressures is to give rise to high pressure modifications which, in numerous cases, are superconductive. Less than 30 out of the elements of the periodic tables have been recognized as superconductors at atmospheric pressure, but 14 more elements are superconductors in high pressure condition, below 200 kbar, although they are not superconductors

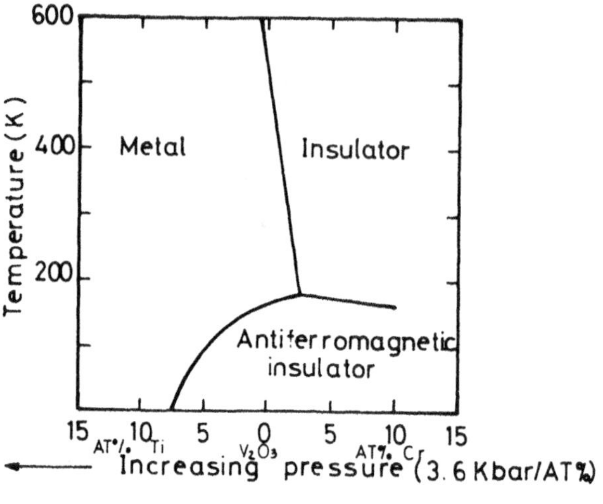

Fig. 15. Generalized phase diagram of transition temperature versus both pressure and atomic percents Cr and Ti in V_2O_3 |19|.

at atmospheric pressure. Among them are Si or Ge, Se or Te, P or As, Ce and Cs. The tetrahedrally bonded crystalline semiconductor Si and Ge transform for instance under pressure to a β-Sn structure, respectively at ∿130 and 100 kbar, which has metallic (and superconductive) properties. Furthermore high pressure modifications of elements which already are superconductors at atmospheric pressure can have higher transition temperature in the high pressure structure than in the zero pressure structure : this is the case for Sn and Ga.

3.3 Valence transitions

Moderate pressure can lead to drastic electronic phase transitions when the two electronic phases have, at room pressure, similar energies. This is the case for instance for cerium metal which presents two phase (α and γ) (fig. 16) of identical crystallographic structure (cfc) but with a large volume contraction (18 %). At zero pressure cerium in the γ-phase is trivalent (Ce^{3+}), whereas it has an intermediary state in the α-phase.

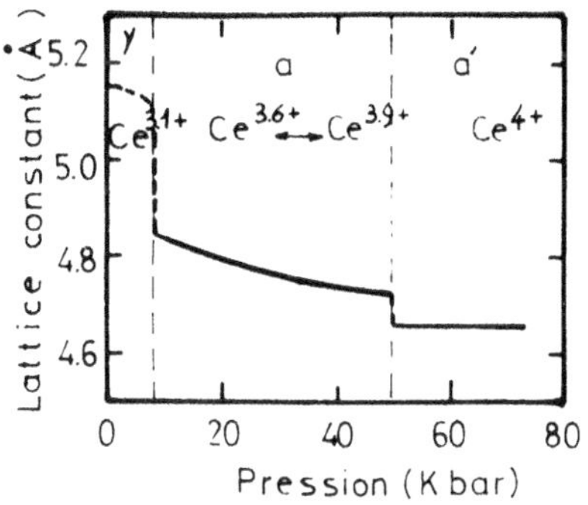

Fig. 16. Pressure dependence of the lattice constant of cerium |20|

Above 50 kbar, Ce transforms to an α' (superconductive) phase, which has a crystallographic structure which, at the present time, is not assessed. Recent experiments |21| have demonstrated that a new allotropic phase, of tetragonal body-centered type, exists at pressure above 122 kbar.

Our second example will be samarium sulfide, SmS, which presents, at 6.5 kbar, a discontinuous electronic transition. The phases on each side of the transition pressure possess identical crystallographic structure (NaCl) |22| but with a volume reduction of 12 %. This transition is associated with a valence change from Sm^{2+} to Sm^{3+}. That is to a 4f → 5d transition similar to that observed for pure cerium. It is accompanied from an insulator metal transition, and a subsequent change in color, from black to golden-yellow, which is associated with the different optical properties of the low pressure insulator and high pressure metallic phases. These transitions have in fact common features with the first order isostructural cfc - cfc transition observed (fig. 1) in Cs at 42.5 kbar and which is associated with a partial 6s → 5d transition, with a volume reduction of ∿10 %. Superconductivity has been detected in cesium, at high pressure, as it has been detected in cerium, as already mentioned. One can notice that in the three examples under consideration electronic rearrangement appears to be proceeding continuously before a discontinuous collapse in volume occurs.

3.4 Magnetic transition

Electronic transitions can be evidenced from the magnetic properties of the solids. We will indicate, using few examples, how ferromagnetism can disappear in transition metal alloys. In a d-band as given fig. 17a, the net overall magnetic moment results from the different number of electrons with magnetic moments up (↑) and down (↓). There is no magnetic moment (fig. 17b) when these two numbers are equal. The transition between state (a) and state (b) can occur continuously or discontinuously.

Fe : Iron is the most classical example of ferromagnetic transition metal. The magnetic moment of α-Fe, which has the body-centered cubic structure, is only weakly pressure dependent, with a relative variation 3×10^{-4} $kbar^{-1}$. At room temperature a crystallographic phase change to a hexagonal closed packed structure occurs at 130 kbar. Mössbauer experiments |23| have demonstrated that hcp iron is not magnetic at room temperature. In fact further experiments performed down to 4 K have not found any evidence of an ordered magnetic structure.

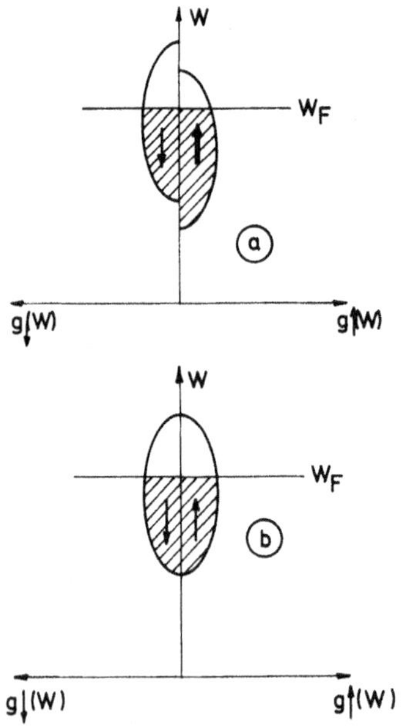

Fig. 17. d-electrons energy band of transition metal. The d-band can contain up to 10 electrons.

FeRh : The FeRh alloys have unusual magnetic properties. At room pressure the Fe magnetic moments have an antiferromagnetic arrangement at low temperature, below Θ_1, whereas they are ferromagnetic above Θ_1 (fig. 18). The antiferromagnetic-ferromagnetic transition observed at Θ_1 is discontinuous. (In fact Θ_1 increases from 143 to 403 K when rhodium concentrations increase from 50 to 53 %). The antiferromagnetic and ferromagnetic phases have an identical ordered CsCl structure but have different volume. In the antiferromagnetic phase each rhodium atom has 4↑ and 4↓ Fe moments in its first neighbourhood, whereas in the ferromagnetic phase each rhodium atom has 8 first neighbours with parallel magnetic moment. The influence of the first neighbours cancels out in the antiferromagnetic phase, where rhodium has no magnetic moment, whereas in the ferromagnetic range it possesses an induced magnetic moment ($\sim 1\ \mu_B$), The high pressure behaviour of FeRh (fig. 18) has been extensively studied |24|. The ferromagnetic phase disappears at pressure ∿60 kbar. In fact substitution of Pd to rhodium, which has a larger susceptibility increases the stability of the ferromagnetic phase, whereas Pt or Ir substitutions, which have a smaller magnetic susceptibility favours antiferromagnetism, and so does the substitution of Rh to Fe in the CsCl structure. In this case pressure thus favours a non-magnetic state and this is a general result for transition metal alloys.

$ZrZn_2$: Pressure induces a discontinuous dissappearance of magnetic moment on Fe in the case of pure iron, or on rhodium in FeRh alloy. We will now introduce examples where pressure leads to a progressive disappearance of the magnetic moment. In that case, close to the critical pressure for the disappearance of the magnetic moment M, the difference between the actual free energy

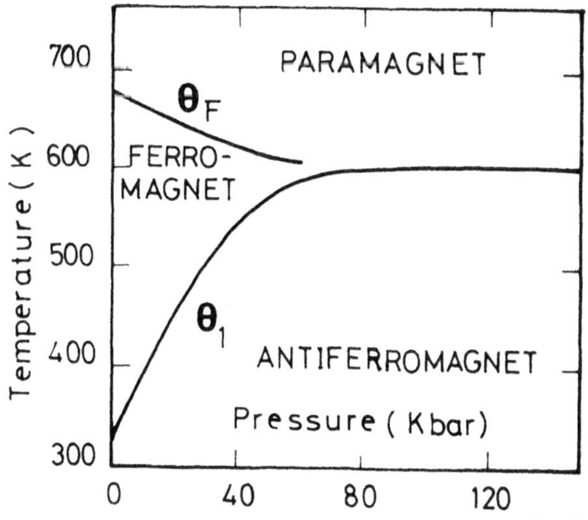

Fig. 18. Phase diagram of Fe-Rh alloys |24|.

and the energy without a magnetic moment can be developed, even at T = 0, using a Landau expansion similar to (1) :

$$F_{M,P} - P_{M=0,P} = \frac{1}{2} A(P)M^2 + \frac{1}{4} B(P)M^4 + \ldots \qquad (1')$$

where A(P) changes for instance linearly its sign at the "critical" concentration P_c, whereas B is a smooth function of P. Minimizing $F_{M,P}$ leads to the equation for M

$$M^2 = -\frac{A}{B} = \frac{A'}{B}(P - P_c) \qquad (5')$$

which indicates a parabolic dependence of M on P. This permits to understand, at least qualitatively, the high pressure behaviour of $ZrZn_2$ |25| (fig. 19). In fact a similar treatment, as applied in details to the Ni_xPt_{1-x} alloys |26| demonstrates that spontaneous magnetic moment and Curie temperatures have identical pressure dependences and therefore disappear simultaneously.

ACKNOWLEDGMENTS

The authors acknowledge D.B. McWhan for a critical reading of the manuscript.

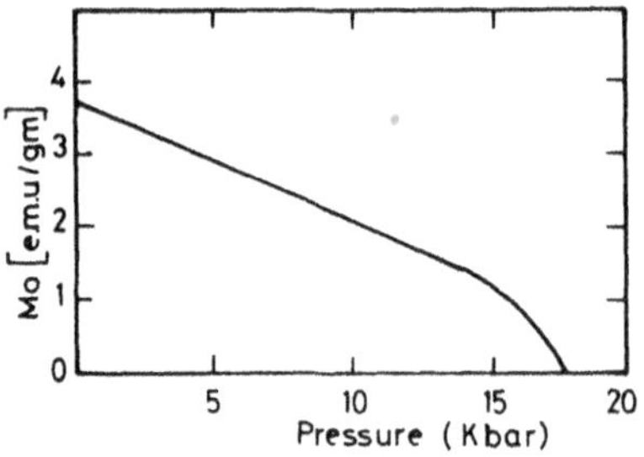

Fig. 19. Pressure dependence of the spontaneous magnetization of $ZrZn_2$ |25|.

REFERENCES

1. C. Kittel, Introduction to solid state physics, Wiley, 1976.
2. J.M. Ziman, Principles of the theory of solids, Cambridge Univ. Press, 1964.
3. For general references on High Pressure Solid State Physics, see for instance :
   a - Solids under pressure, W. Paul and D.M. Warschauer Edit., McGraw Hill,(1963.
   b - Physics of solids at high pressure, C.T. Tomizuka and R.M. Emrick Edit, Academic Press, 1965.
   c - Colloque International du C.N.R.S. sur les Propriétés physiques des solides sous pression, Grenoble, 1969, C.N.R.S., 188, 1970.
   d - High pressure and low temperature physics, C.W. Chu Edit., Plenum (to be published).
4. D.B. McWhan, G. Parisot and D. Bloch, J. Phys. F, Metal Phys., 4, L69, 1974.
5. L. Knopoff in High pressure physics and chemistry, R.S. Bradley Edit., Academic Press, 1, 247, 1963.
6. J. Friedel in Physics of metals I, Electrons, J. Ziman Edit., Cambridge, 1969.
7. J. Friedel and C.M. Sayers, J. de Phys. Lettres, 38, L263, 1977.
8. M.B. McWhan, Science, 176, 75, 1972 and references therein.
9. M.H. Rice, R.G. McQueen and J.M. Walsh in Solid state physics, F. Seitz and D. Turnbull Edit., 6, 1958.
10. R. Maury, C.R. Acad. Sc. Paris, 271, 950, 1970.
11. D. Bloch, D. Herrmann-Ronzaud, C. Vettier, W.B. Yelon and R. Alben, Phys. Rev. Letters, 35, 963, 1975.
12. W.B. Yelon, D.E. Cox, P.J. Kortman and W.B. Daniels, Phys. Rev., B9, 4843, 1974.
13. C.W. Garland and J.D. Baloga, Phys. Rev., B16, 331, 1977.
14. G. Samara in High pressure research, R.S. Bradley Edit., Academic Press, 3, 155, 1969.
15. P. Bastie, M. Vallade, C. Vettier and C.M.E. Zeyen, to be published.
16. P.S. Peercy, Phys. Rev. Letters, 1581, 1975.
17. N.H. March in Advances in high pressure research, R.S. Bradley Edit., Academic Press, 3, 241, 1969.
18. D. Adler in ref. 3c.
19. D.B. McWhan in ref. 3c.
20. E. Francheschi and G.L. Olcese, Phys. Rev. Letters, 22, 1299, 1969.
21. S. Endo, N. Fujioka,, N. Kawaï and H. Sasaki, to be published.
22. A. Jayaraman, A.K. Singh, A. Chatterjee and S. UshaDevi, Phys. Rev., 9B, 2513, 1974.
23. D.N. Pipkorn, C.K. Edge, P. Degrunner, G. de Pasquali, H.G. Drickamer and H. Frunenfelder, Phys. Rev., 135A, 1604, 1964.

24. a - D. Bloch, Annales de Physique, 1, 93, 1966.
b - E.G. Ponyatovskii, A.R. Kut-Sav and G.T. Dubovka, Sov. Phys. JETP Cryst., 12, 63, 1967.
c - J.M. Legrand, C. Susse in 3c.
d - G.T. Dubovka, Sov. Phys. JETP, 38, 1140, 1974.
25. J.G. Huber, M.B. Maple, D. Wohlleben and G.S. Knapp, Sol. Stat. Commun., 16, 211, 1975.
26. H.L. Alberts, J. Beille, D. Bloch and E.P. Wohlfarth, Phys. Rev., 9, 2233, 1974.

EXPERIMENTAL INVESTIGATIONS OF FLUIDS AT HIGH PRESSURES AND ELEVATED TEMPERATURES

E. U. Franck

Institute of Physical Chemistry, University of Karlsruhe, Federal Republic of Germany.

ABSTRACT. A survey is given of experimental methods, investigations and results on various properties of fluids and fluid mixtures at high pressures and high temperatures. Fluids of polar molecules and binary mixtures are emphasized. After a discussion of critical data and new PVT-results particularly on water, aqueous salt solutions and methanol, selected binary systems and their critical behaviour are considered. - Spectroscopic investigations and methods with fluids at high pressure and temperature are shown. The design of optical cells for special purposes is given and the quantitative investigation of ethylene - polyethylene polymerization and phase seperation by infrared techniques is described. The following section gives values for the static dielectric constant of dense subcritical and supercritical fluids, especially for water, acetonitrile and ammonia. The last chapter deals with ionic and metal states in high temperature fluids. Supercritical salt solutions, fused salts, mercury and alkali - alkali hydride mixtures are discussed. Finally a schematic picture of phenomena to be found with polar fluids in a wide range of temperatures and densities is presented.

1. INTRODUCTION

The application of high pressures to fluids can have several and quite different advantageous and technically useful effects in chemistry. This is particularly

Kelm, H. (Ed.) High Pressure Chemistry. 221–256. *Copyright © 1978. D. Reidel Publishing Company. All Rights Reserved.*

true, if high pressure can be combined with elevated temperatures. At supercritical conditions even moderate pressures of several kbar or only a few hundred bar can produce continuous density variations from gas-like to liquid-like values. Accordingly, the extent of intermolecular interaction varies considerably, especially if the fluid consists of polar molecules with strong and far-reaching interaction potentials. Such fluids have necessarily relatively high critical temperatures and are often corrosive. In order to investigate and utilize such interaction variations, suitable equipment and experimental procedures have to be designed. Selected examples and results of experimental investigations will be shown and discussed below. Some thermodynamic and transport properties will be dealt with at first. Apart from their importance as such, they are often essential for the understanding of other kinds of experimental results. The following two sections will be dedicated to spectroscopic and dielectric investigations at high pressures which reveal considerable information on intermolecular interaction and have also many application of a more practical nature. Finally several electrolytic solutions - mainly aqueous phases - at high pressures and temperatures will be discussed. Among other fields it is power plant operation and geochemistry,where the knowledge of such solutions is important.

2. THERMODYNAMIC AND TRANSPORT PROPERTIES

Because the critical temperatures and pressures give the lower limits of single-phase, homogeneous behavior, the knowledge of such "key" data serves to decide which pressures can be used to produce wide, continuous variations of density. New and extensive compilations of critical data are available in the literature (1). A few selected values are given in Table 1. Most of the "normal" fluids, which are of technical interest have critical temperatures and pressures lower than those of water. Since the critical density is usually about one third of that of the liquid at the triple point, pressures of a few hundred or thousand bar suffice to attain liquid-like densities at supercritical temperatures. Comparatively little is known at present of critical data of salts and fluid metals. Bismuth trichloride (2) has been investigated experimentally and also mercury and the alkali metals (3). Generally however, the values for these classes of substances have as yet to be estimated. It is note-

	T_c (K)	P_c (bar)
Ar	151	49
CH_4	191	46
C_2H_4	282	50
CO_2	304	74
NH_3	406	113
CH_3OH	513	81
C_6H_6	562	49
H_2O	647	220
$BiCl_3$	1178	119
Hg	1765	1510
K	2200	155
NaCl	3500?	250?
Cu	6000?	4000?
W	23000?	10000?

Table 1: Critical temperatures and pressures of selected substances.

worthy, that the range of the liquid state apparently extends over many thousands of degrees for some of the heavier metals.

Accurate equation of state data are available for many fluids, particularly for those of technical importance. Of this group, water is probably the best investigated. Existing international steam tables reach to about 1000 °C and 1000 bar. Precise, more recent data from static experiments are available to 1000 °C and 10 kbar (4) (5) (6). Shock wave experiments have been extended to several hundred kbar (7). Fig. 1 gives a number of isobars within a temperature-density diagram of water. Pressures of about 9 kbar are necessary to obtain the normal liquid density at 500°C,

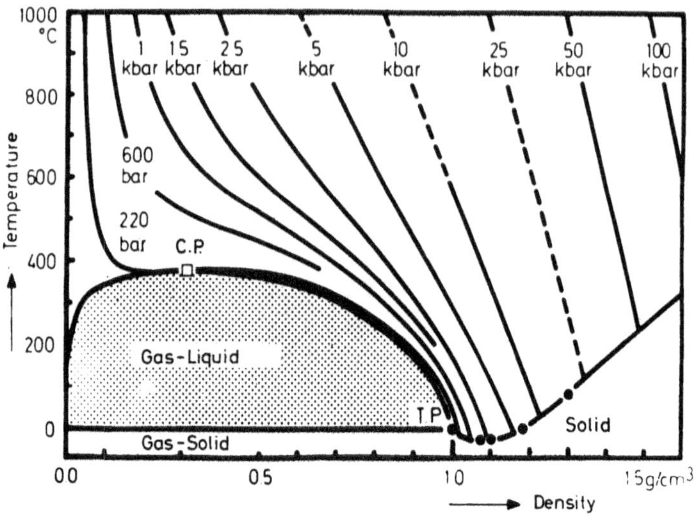

Fig. 1: Temperature-density diagram of water. The dotted isobars are interpolated.

and at 1000°C about 20 kbar would be needed.

Concentrated aqueous solutions of salts occur often in nature and in industry. The PVT-data of such solutions can be very different from those of pure water because of the contracting effect of ion hydration. As one example, Fig. 2 shows partial molar volumes of sodium chloride in water, which were calculated from a critical compilation of existing data (8). Another plot of new experimental values (9) of partial molar volumes of sodium chloride as a function of temperature as isobars is shown in Fig. 3. The expansion of the water with increasing temperature causes the partial molar volume of the salt to become negative, the less so, as the pressure increases. At constant total density the partial molar volume remains almost constant. It is expected, that measurements and diagrams of this kind can facilitate the prediction of PVT-data of other aqueous salt solutions for wide ranges of conditions.

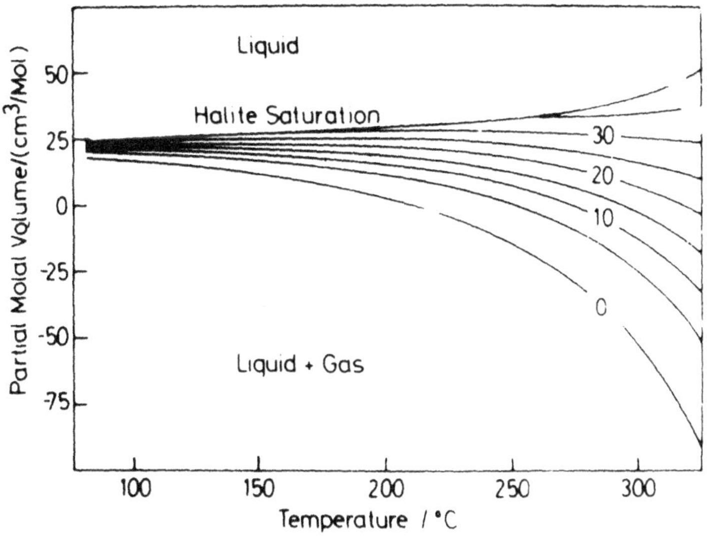

Fig. 2: The partial molar volume of NaCl in the liquid phase for vapor-saturated NaCl-solutions from zero weight percent to halite saturation calculated by J.L. Haas jr., U.S. Geological Survey Reston, Va. (8).

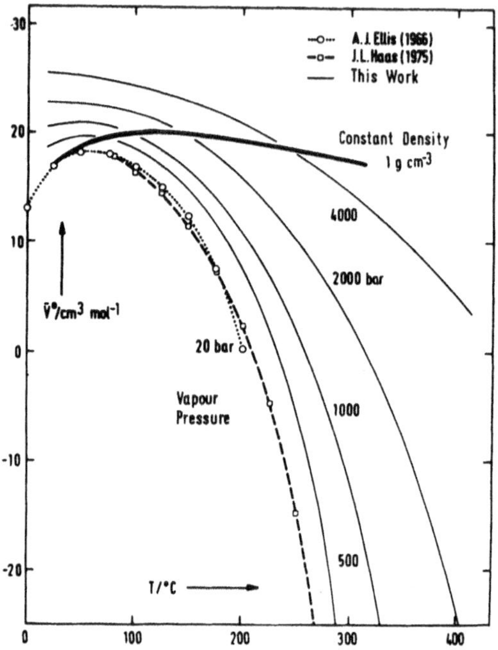

Fig. 3: The partial molar volume of NaCl in water at several constant high pressures and constant density of 1 g/cm^3 (9).

Precise PVT-measurements with fluids at high pressures can be made in different ways. The above mentioned salt solutions and also pure water to 600 °C and 4 kbar were investigated with an autoclave with internal heating (9). It is shown schematically in Fig. 4. Within the autoclave is a solution filled bellows, surrounded by pure water. Changes of the volume of the bellows are determined by the amount of pure water around it in the autoclave, which is derived from the position of a float on mercury in a second high pressure vessel at room temperature. The third region within the autoclave, which contains the heaters, is filled with pressurized argon.

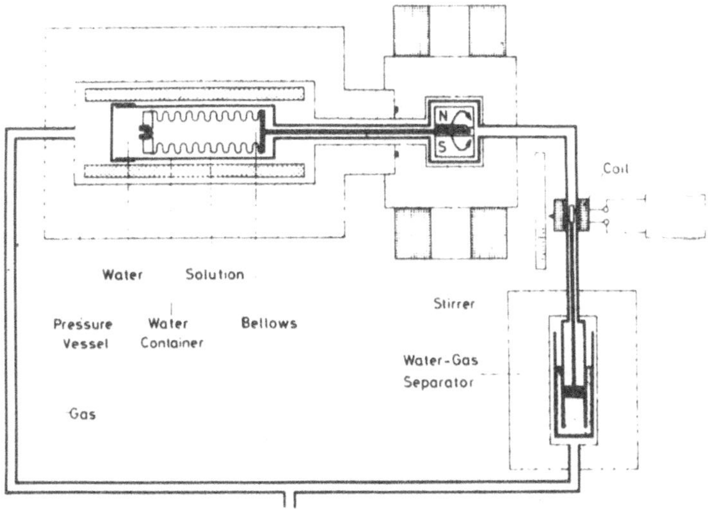

Fig. 4: Schematic diagram of an autoclave with internal heating and auxiliary equipment for PVT-measurements with water and aqueous solutions to 600 °C and 4 kbar (9).

It is interesting to compare the thermal properties of water with recent data for methanol which has a related molecular structure and also technical importance. The density of this compound was measured with an externally heated vessel to 8 kbar and 350 °C (10). Results are shown in Fig. 5. Although methanol is unstable at 350 °C, the decomposition in the monel-lined vessel was slow enough to permit the measurements. The PVT-data could be well represented by an extended Tait equation with temperature dependent coefficients. Compressibility coefficients, derived from the data by means of this equation, are shown in Fig. 6 in comparison with analogous numbers for hexane and water. At a "moderate" pressure of 1000 bar methanol and n-hexane are relatively close and water deviates considerably from these two. At 8000 bar, however, the data of water and methanol almost coincide. This demonstrates, that qualitative equation of state rules must be applied with caution for highly compressed fluids.

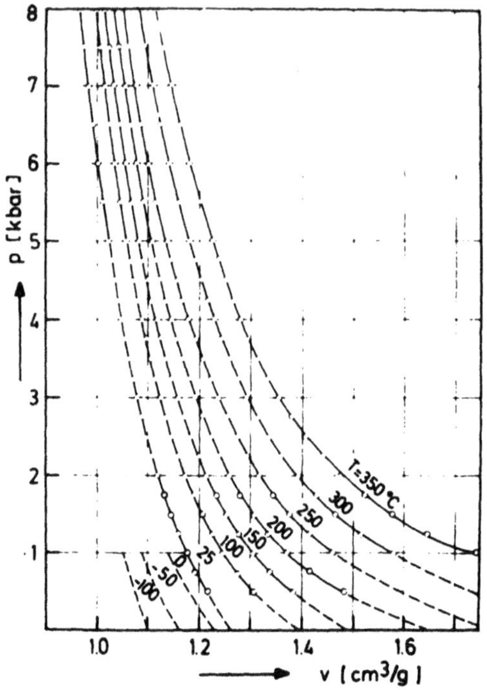

Fig. 5: Experimentally determined p-v-diagram of methanol

Of the three main transport properties, viscosity, thermal conductivity and diffusion, the viscosity - that is the momentum transfer - is probably the easiest to measure at high pressures and temperatures. Because diffusion coefficients and ion mobilities behave often roughly proportional to the inverse of the viscosity, this latter quantity is also of interest for chemistry. With increasing temperature the viscosity of a fluid at constant high density decreases, while it increases at constant low, gas like density. Accordingly, it is not too difficult, to roughly estimate viscosity for dense supercritical fluids if some reference data are available (11). The basis for this is the Enscog theory for fluids of simple, spherical molecules.

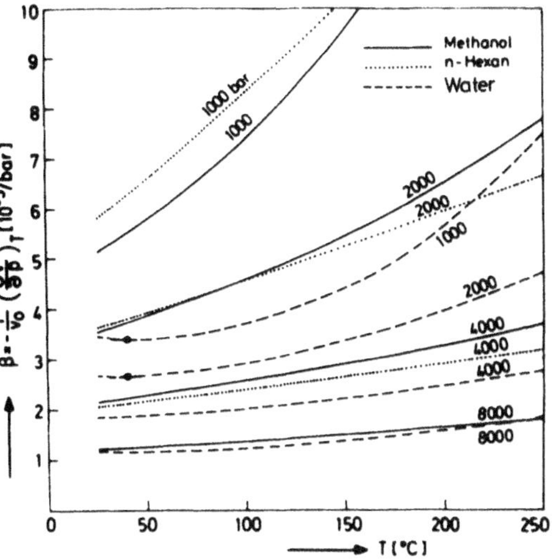

Fig. 6: Comparison of the compressibility of liquid methanol, n-hexane and water to 25o °C and 8 kbar.

These general priciples apply also to unusual fluids like metals approaching the critical region, as has been shown by recent measurements with mercury (12) (Fig. 7).

The data are for coexisting equilibrium liquid and gas states to 1200 °C. The extensions were calculated with a modified Enskog equation. The viscosity at th at the critical point is quite "normal", which underlines the conclusion, that the very high electric conductance in this region is indeed an electronic phenomenon.

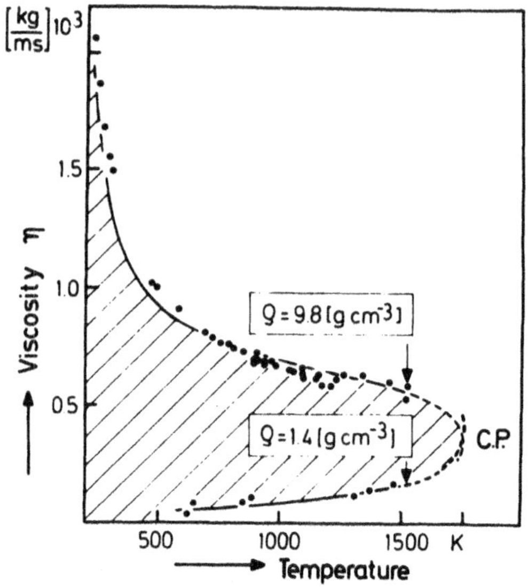

Fig. 7: The viscosity of liquid and gaseous mercury at saturation pressures. Values above 1500 K are calculated.

So far only thermodynamic and transport properties of one-component systems have been discussed. Obviously, binary, ternary and multicomponent systems are more elaborate to investigate, but nevertheless of considerable scientific and also of practical interest in several fields. Here, only a short discussion of fluid-fluid phase equilibria and critical properties will be given.

Binary systems have a critical curve in the pressure-temperature-composition diagram. The curve connects uninterrupted the two critical points of the pure components if those are relatively similar like

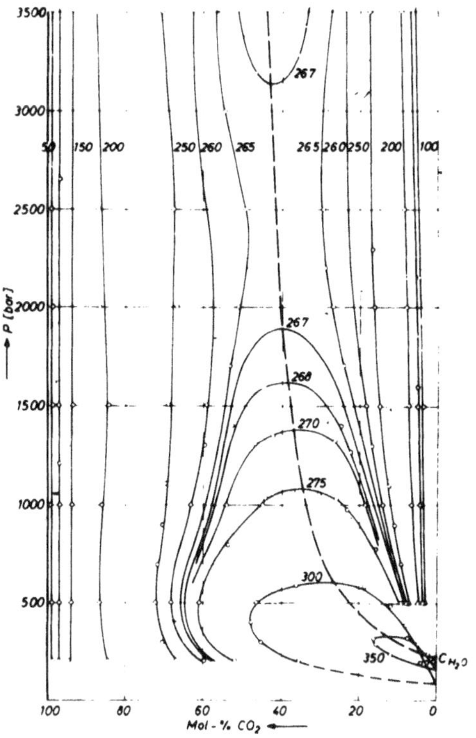

Fig. 8: Isotherms on the liquid-gas phase separation surface of the system H_2O-CO_2

ethane und hexane. If the molecules of the components differ strongly in polarity or size the critical curve may of the interrupted kind with an "upper" branch beginning at the critical point of the higher boiling component and extending in different ways to high pressures and temperatures (13). A particularly well investigated and important example is the system carbon dioxide-water. Fig. 8 gives isotherms on the fluid-fluid phase separation surface (14). Within the indicated isotherms is the region of two-phase behaviour. At 266 °C and 2500 bar the critical curve passes through a minimum. Such critical curves are a sort of envelopes of the region of partial immiscibility. A set of such critical curves for binary

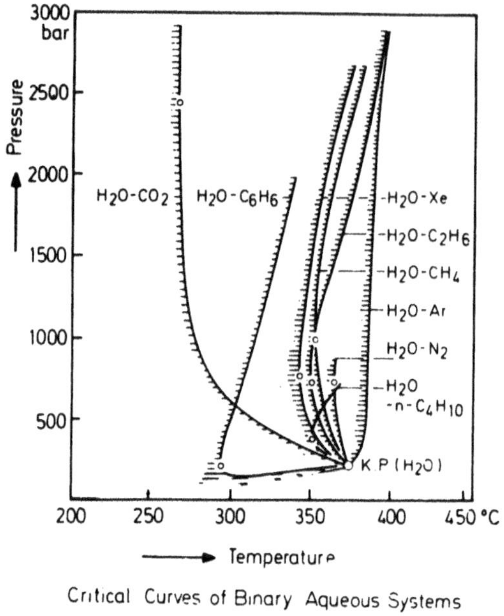

Fig. 9: Critical curves of binary aqueous systems.

systems (15) is given in Fig. 9. Above 400 °C all the eight nonpolar components are completely miscible with water - at least to pressures up to 3000 bar. Two critical curves extend to particular low temperatures: those for water-carbon dioxide and water-benzene. The first demonstrates the well-known interaction between the two partners. The second curve shows, that because of the relatively strong attractive interaction between the water dipoles and the aromatic electron system of the benzene, this compound is already completely miscible with water at 300 °C and 300 bar. Similar behaviour can be expected for other aromatic molecules. One might expect future use of such mixtures for separation processes and chemical or electrochemical reactions.

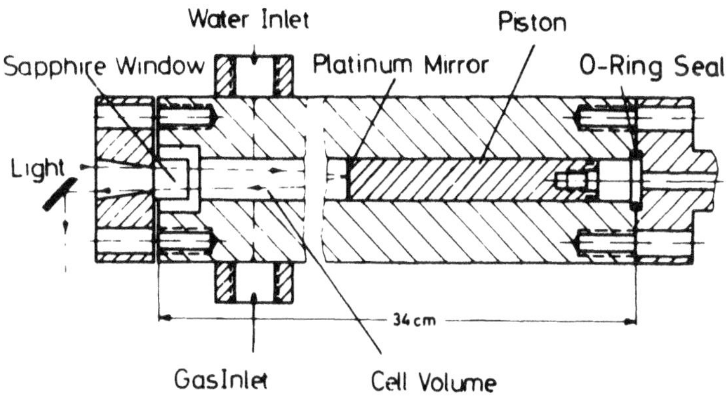

Fig. 10: Simple autoclave with variable cell volume to determine phase separations and PVT-data of fluid binary systems. Heating on cooling jackets not shown.

One type of apparatus (16) which has been very useful to determine fluid phase equilibria in binary systems and which is relatively simple to build and operate is shown in Fig. 10. The horizontal cylindrical autoclave has a volume of 5 to 25 cm^3. The size of this volume is given by the position of a movable piston, the position of which can be determined with a magnetic indicator. The piston is relatively long, so that the sample temperature, which is maintained by an adjustable external heater, can decrease to 100°C is applied. The O-ring separates the sample mixture from a second fluid at the back of the piston which has nearly the pressure of the sample. This fluid can be one of the components of the sample. At constant sample volume the pressure-temperature curves, "isochors" are recorded. Discontinuities on the isochores indicate points on the two-phase boundary surface in the P-T-composition diagram.

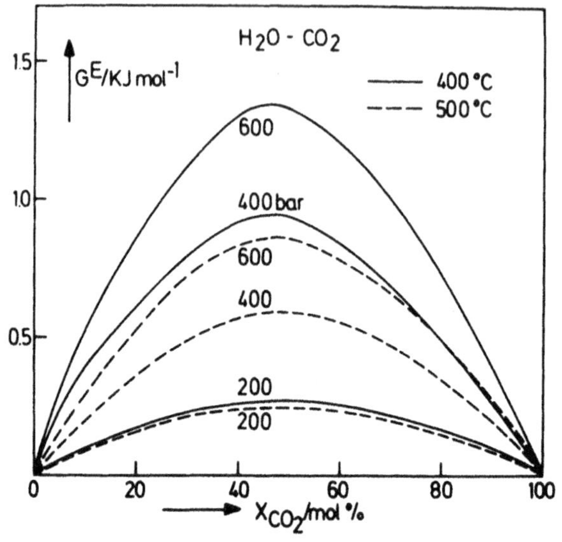

Fig. 11: Excess Gibbs energy at several constant pressures and temperatures for the binary system CO_2 - H_2O.

The apparatus described is also well suited to measure the density of the sample mixture in the homogeneous region. If enough accurate PVT-data are available, thermodynamic excess functions for supercritical conditions can be derived. One example is shown in Fig. 11, giving the excess Gibbs energy for water-carbon dioxide at several constant pressures and temperatures as a function of composition. Such functions are of use for the synthesis of minerals at hydrothermal conditions.

In connection with the investigation of natural hydrothermal fluids ternary systems are of particular interest. Experimental data for such system are still limited for obvious reasons. The system water-carbon dioxide-sodium chloride is relatively well investigated (17) (18). Fig. 12 gives a number of "isopleths", curves of constant composition, on a P-T-diagram. It can be seen, that the addition of less than two mole percent sodium chloride shifts such curves by a about one hundred degrees to higher temperatures (from B' to B, C, D respectively). The salt extends the two phase

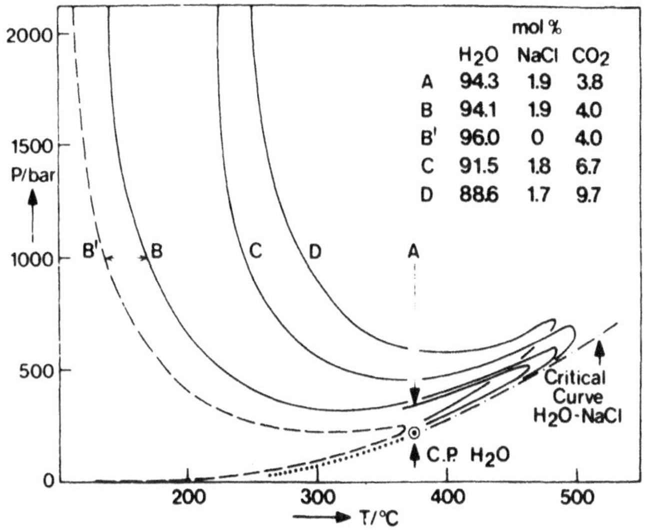

Fig. 12: Several isopleths (curves for constant composition) of the ternary system H_2O-CO_2-NaCl

region extensively - a particularly pronounced consequence of a "salting out" effect.

3. SPECTROSCOPIC INVESTIGATIONS

Thermodynamic and transport phenomena give only limited and unspecific information about structural properties of fluids under pressure. Spectroscopic investigations can be much more revealing. Since elevated pressures and temperatures can often be maintained only for limited periods, the short duration of many spectroscopic measurements is also quite desirable. Many different designs for optical cells, particularly for the study of solids at high pressure, are described in the literature (19). Research with fluids often requires elevated temperatures as well as high pressures which may cause special sealing and corrosion problems. In addition to glass also quartz, synthetic sapphire and occasionally pure silicon are useful window materials. Absorption measurements in the visible, UV- and IR-regions have been made as well as Raman and NMR-investigations.

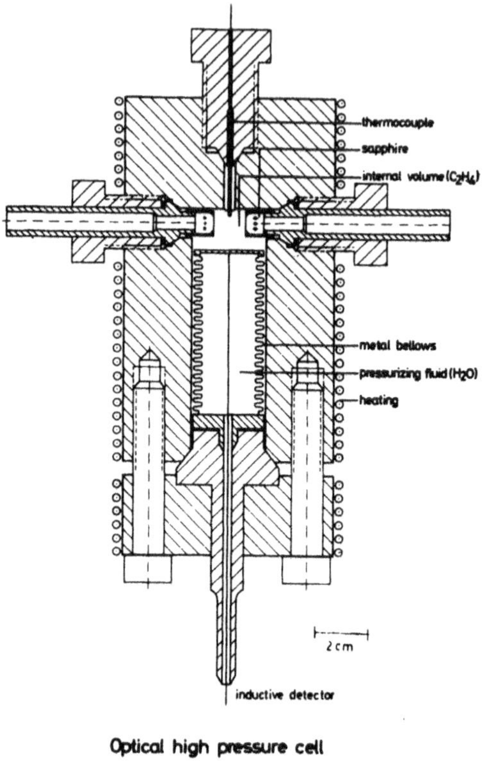

Fig. 13: Optical high pressure cell for infrared absorption measurements.

The latter are discussed elsewhere in this volume. Some examples of absorption and Raman spectroscopy will be given here.

Fig. 13 shows an infrared cell, in which absorption measurements and thermodynamic and kinetic determinations can be made with the same sample. The cell has two sapphire windows with so-called Poulter-type seals. The infrared light beam passes horizontally through windows and sample. Often it es desirable to vary temperature or pressure of the sample within the cell without the escape or introduction of fluid into the cell. For this purpose the cell in Fig. 13 contains a bellows, filled with an auxiliary pressurized fluid. The

The position of the front plate of the bellows and accordingly the free cell volume can be determined by means of a rod with a magnetic tip and an induction bridge from outside.

Infrared absorption at supercritical conditions is particular well suited to study intermolecular interaction by means of band shapes. Another, more practical example is the investigation of phase equilibria and reaction kinetics in high pressure ethylene polymerization. The position of the C-H-valence bands of ethylene and of - CH_2 - groups in the polymer are suffiently separated to permit quantitative determination of monomer and polymer concentrations in the optical cell which serves as a miniature reaction vessel. Preceeding measurements with several linear alkanes have shown, that the specific absorption intensity of -CH_2-groups is almost independent of temperature and pressure at least to 250 °C and 2000 bar. The monomer and polymer bands are even better separated in the first and second overton regions, where they have comparable intensities, so that concentrations can be determined equally from both kinds of bands. (20) (21) Fig. 14 gives an example for about 2000 bars.

In the near infrared region the turbidity caused by phase seperation can easily be observed. A combination of turbidity and concentration determinations at several temperatures and pressures could be used to find the monomer - polymer phase seperation surface in the temperature-pressure-composition space. The results (22), obtained recently, are shown in Fig. 15. Below the isotherms shown, the system is separated into two phases. The maxima of the isotherms on the left side could be connected by an "apparent" critical curve. This is not necessarily a true critical curve, since there is a molecular weight distribution of the polymer which is not iniform.

The rate of polymerization at constant temperature and pressure can be determined rather easily and accurately, using the optical cells as miniature reaction vessels. If the absorption in the region of the fundamental frequencies is used, and if infrared interferometry is applied, the measurements of very small degrees of polymerization - much below one percent - is possible, and this can be done at intervals of a few minutes. Thus initial rates of polymerization can be studied.

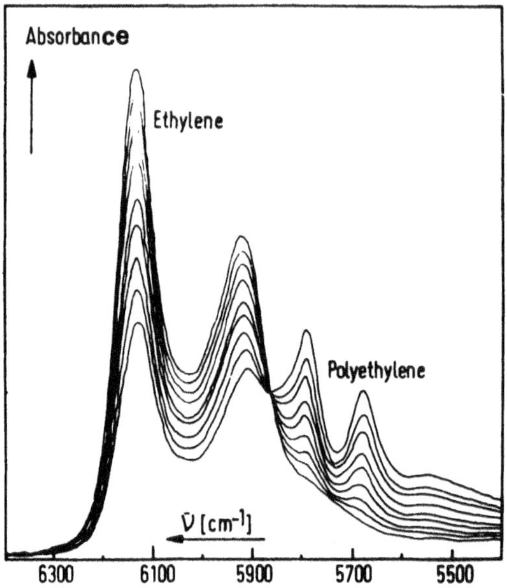

Fig. 14: Absorption of ethylene and polyethylene in the region of the first overtone of the C-H-volume vibrations at 235 °C and at about 2 kbar. The degree of polymerization approaches 50 %.

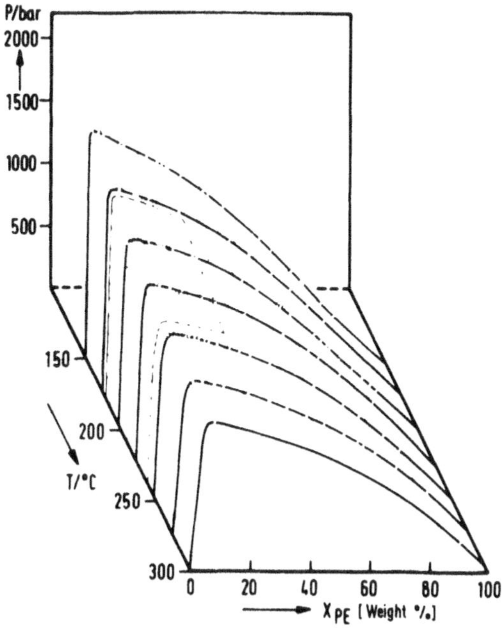

Fig. 15: Phase separation surface for ethylene - polyethylene determined by near-infrared absorption.

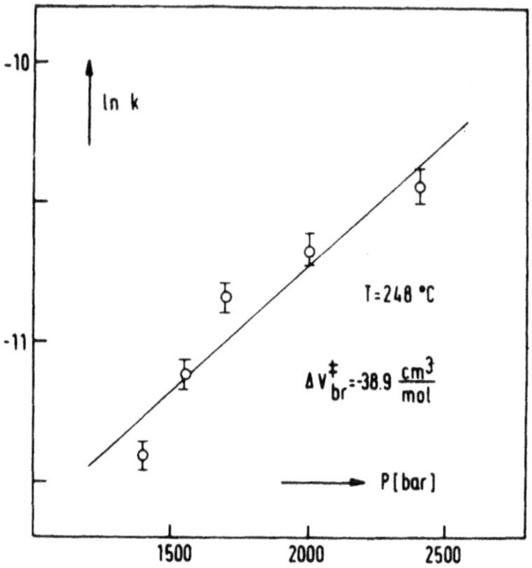

Fig. 16: Rate constant of observed "thermal polymerization" of ethylene as a function of pressure.

Recently the reaction rate of ethylene with not more than 1 ppm oxygen content was investigated. (22) The results give convincing evidence for a "thermal" polymerization without oxygen catalysis. Fig. 16, as an example, gives the polymerization rate constant at 248 °C as a function of pressure to 2500 bar. The high activation volume as well as the high activation energy of 155 kJ · mol^{-1} support the assumption of thermal polymerization. These results were obtained with optical cells similar to the one of Fig. 13.

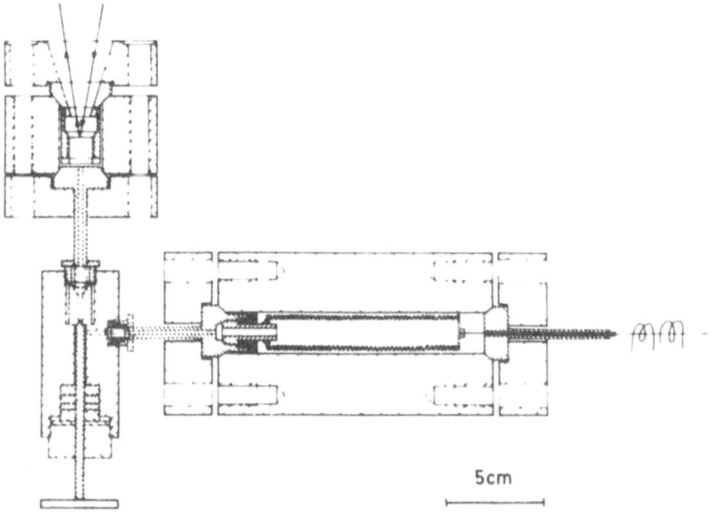

Fig. 17: Reflection type - one window - cell with auxiliary autoclave for premixing and storing corrosive fluid samples inside a non-corrosive bellows.

If very corrosive fluids are to be investigated, or if premixing of components at high temperature and pressure is necessary, it may be advisable, to use the optical cell with a connected auxiliary high temperature autoclave as whown in Fig.17. This is a one-window reflection type cell which has a metal mirror behind the sapphire window within the pressurized sample fluid. This sample fluid is prepared and stored inside the bellows of the auxiliary vessel. The sample in the bellows can be compressed from outside with an inert fluid, for example water. The length of the bellows is determined also from outside by means of a rod with a magnetic tip. Infrared investigations with H_2O-NaOH mixtures at all compositions to 350°C were made (23).

The absorption of high temperature, high pressure aqueous solutions has also been investigated in the visible and ultraviolet regions. Stability ranges of cobalt, nickel and copper complexes in concentrated solutions were studied in cells,which in part were

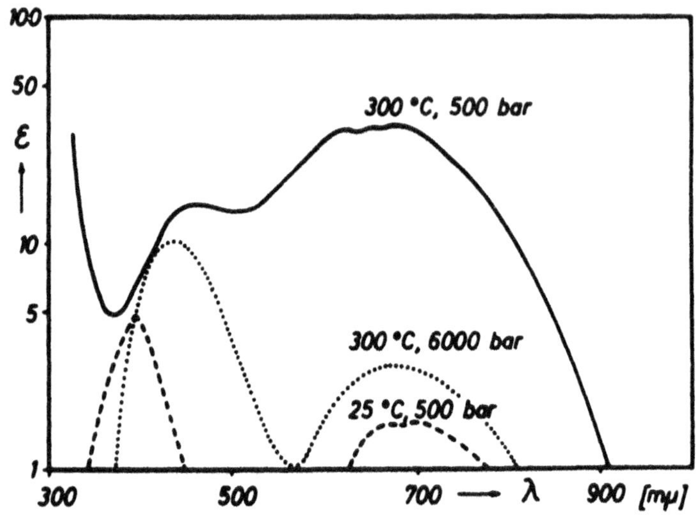

Fig. 18: Absorption spectrum of $NiCl_2$ (0.025 molal) in aqueous NaCl solutions (4.0 molal).

lined with teflon and which were equipped with sapphire windows (24) (25) (26). In Fig. 18 the absorption spectrum of $NiCl_2$ in concentrated aqueous NaCl- solutions is shown for 500 and 6000 bar. The higher absorption at the lower pressure is attributed to four-coordinated, nearly tetrahedral nickel complexes.

4. DIELECTRIC PROPERTIES

The dielectric phenomena are also among those, which can give information on structural properties of high temperature fluids and which also are often of considerable practical interest. This applies to the frequency dependent dielectric relaxation as well as to the "static" dielectric constant, observed at lower frequencies. Little is known as yet of relaxation phenomena for fluids at unusual conditions. Measurements of the static dielectric constant, however, have been extended with several fluids into the supercritical range. Cylindrical autoclaves of small dimensions were used with built-in condensers of a

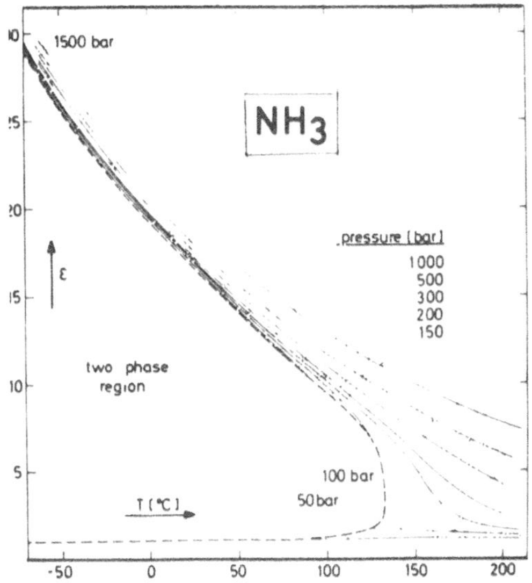

Fig. 19: Static dielectric constant ε of ammonia at constant pressures at subcritical and supercritical temperatures.

capacity between about 100 and 300 pF. The capacity could be varied at constant conditions of temperature and pressure by changing the condenser geometry. In some cases half cylinders were rotated around a common axis from outside. Thus the influence of the leads through the autoclave closures could be eliminated. Frequencies up to 1 MHz were used, which are below the range where relaxation occurs.

The static dielectric constant or permittivity, ε, was measured with fluids of small, polar molecules wich deviate not very much from spherical shape. Among these fluids were H_2O, HCl, NH_3, CH_3OH, CH_3CN and

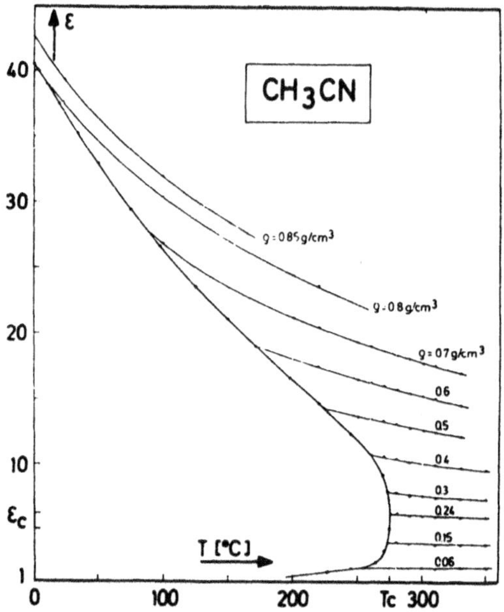

Fig. 20: Static dielectric constant of autonitrile at constant densities at subcritical and supercritical temperatures.

and several freons. (27) (28) (29) (30) (31) (32). As examples, Fig. 19 and 20 show the dielectric constants of ammonia and methyl cyanide as functions of temperature for constant values of pressure or density. To calculate the isochors of Fig. 20, the PVT-data of methyl cyanide had to be determined also. (33) The general appearance of both diagrams is similar and typical for the other compounds mentioned above. There is always a pronounced decrease of ε from the normal liquid region with increasing temperature not only at constant pressure but also at constant density. From the ε-data so called Kirkwood correlation factors, g, can be derived and discussed. These dimensionless

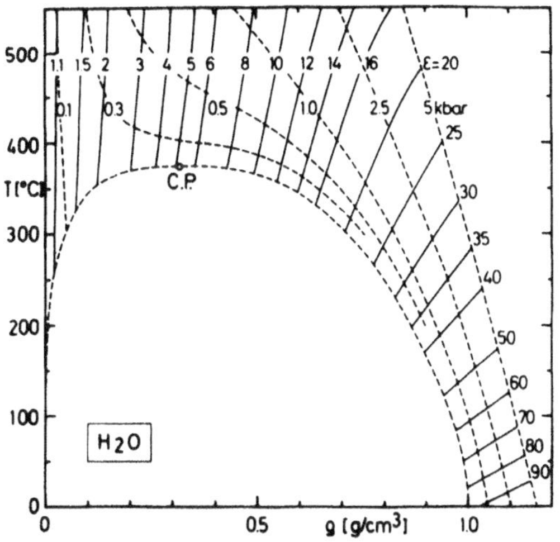

Fig. 21: Curves of constant values of the static dielectric constant of water within a temperature-density region to 550 °C and 1,2 g/cm³.

factors should be equal to unity at random distribution of the molecular dipoles. g-factors greater or smaller than unity indicate a tendency towards parallel or antiparallel orientation. The molecules with hydrogen bonds like water and methanol have g-factors considerably higher than 1.0 even at critical conditions. The dielectric g-factors can be related to expansion coefficients of the general pair correlation functions for fluids.

The dielectric constant of water has been studied for a wide range of conditions by many authors. A recent critical compilation (28) is shown in Fig. 21. It is obvious, that the unusually high ε-values occur only in a small range of low temperature and high density, but even at supercritical temperarure exists a wide range with ε-values between 10 and 25 where the aqueous phase is still a good electrolytic solvent.

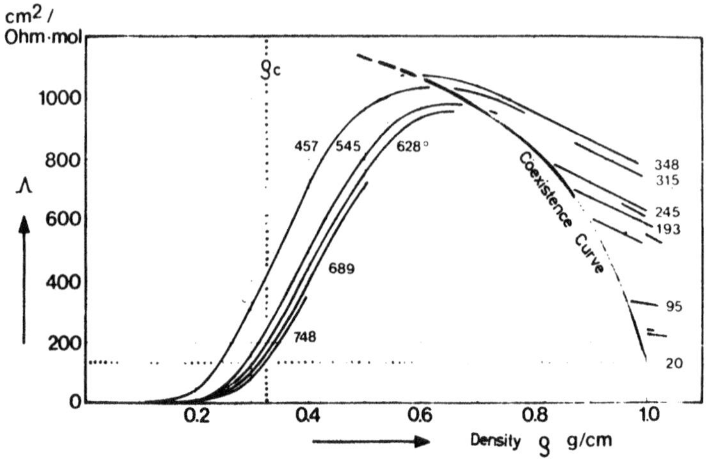

Fig. 22: Equivalent conductance of $KCl-H_2O$-solutions as a function of temperature and density. KCl-concentration: 0.01 m at 20 °C.

5. ION FORMATION AND ION CONDUCTIVITY.

Since supercritical water at high density must be a good electrolytic solvent as pointed out above, ionic solutions at these conditions should have a high conductivity, because the ion mobilities will be high in accordance with the low viscosity of the aqueous solvent. This is demonstrated by Fig. 22, which shows the equivalent conductance of KCl in dilute solution as a function of density by means of isotherms. (34) Below 0.2 g/cm^3 KCl is so weakly ionized, that the conductance is negligible. Around 0.6 g/cm^3 water density, the ion dissociation of the salt is almost complete and the equivalent conductance of the salt is about seven times as high as at normal conditions. With further density increase a slow decrease of the conductance is observed because of high viscosity. A survey of specific conductances of dilute KCl solutions up to 800 °C and 10 kbar is shown in Fig. 23. (25) The increase of the isobars below 400 °C is mainly due to the increase of ion mobilities. This influence, however, is more than compensated at higher temperature

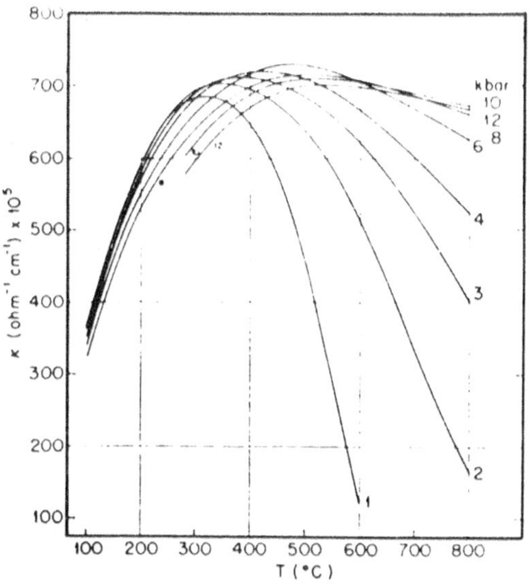

Fig. 23: Specific conductivity of 0.01 molal KCl-solutions at temperatures and pressures to 800 °C and 10 kbar.

by a decrease of concentration of charge carriers. To what extent this can be ascribed to an incomplete dissociation or to some kind of ion association is not yet known. In this respect it is interesting to investigate very concentrated aqueous salt solutions at high temperatures. An example is given in Fig. 24. The diagram is analogous to the one on Fig. 22. The equivalent conductance of sodium chloride in concentrated - 5 molal - solutions is plotted in the form of isotherms. The maximum of the conductance is only about 300 $cm^2Ohm^{-1}mol^{-1}$ and not 1000 $cm^2Ohm^{-1}mol^{-1}$ as in the dilute KCl solutions. The temperature dependence between 300 °C and 600 °C is very small. This be-

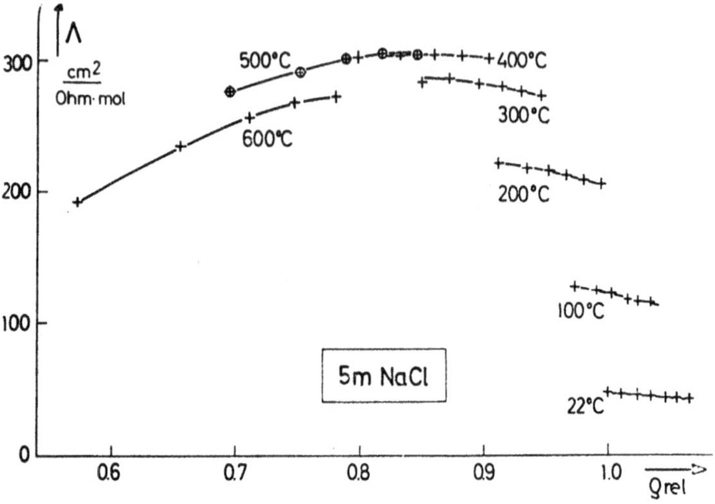

Fig. 24: Equivalent conductance of concentrated (5 molal) aqueous NaCl-solutions as a function of temperature and density to 600°C.

haviour is obviously again due to a reduction of charge carrier concentration, although a satisfactory quantitative description is not yet available. In certain respects do these high temperature concentrated solutions approach already the fluid salts.

Pure liquid salts have also been investigated at high pressure. Density and conductance measurements have been made. (36) (37) Of special interest are simple, stable salts such as alkali chlorides. These, however, have very high critical temperatures, so that wide variations of density cannot be obtained as yet. Bismuth trichloride has its critical point much lower at 905 °C and about 120 bar. (38) Density and conductivity measurements with this salt could be made to 1200 °C and nearly 4000 bar. (39) (40) With certain assuptions a degree of ionic dissociation, α, could be calculated, which is plotted for several temperatures as a function of density in Fig. 25. The diagram shows that at supercritical temperature almost complete ionization can be reached, but only at pressures of 3000 bar and higher. In the critical

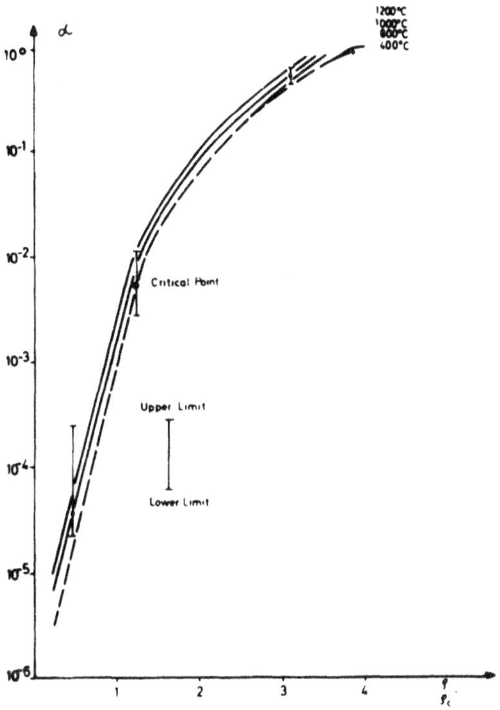

Fig. 25: Degree of dissociation of $BiCl_3$ as a function of reduced density (density /critical density) at several constant temperatures.

region the conductivity is very high, but α is only in the range of one percent. This behaviour reflects the fact, that $BiCl_3$ is not a very typical salt,since it contains a certain amount af covalent bonding.

It is possible, that ammonium chloride is an example for a salt which is completely or at least predominately dissociated to ions near the gas-liquid critical regien. Combined application of high temperature and high pressure permits to attain the critical region of NH_4Cl (T_c=882 °C,ρ_c=0.48 g/cm^3, P_c= 1635 bar) and perform measurements there. (41) Fig. 26 shows the coexistence curve of ammonium chloride.

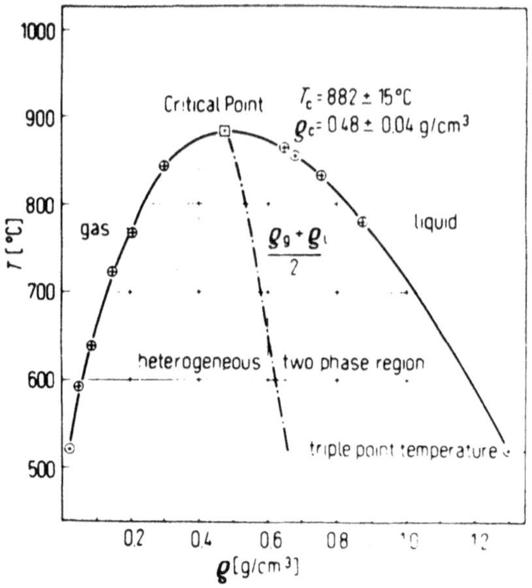

Fig. 26: Gas-liquid coexistence curve for fluid ammonium chloride.

Vapour pressure curve, equilibrium densities and the electrolytic conductance of the liquid have been measured. The high conductance suggests a high degree of ion formation at the critical temperature if the density is not lower than the critical density. This is in agreement with the known thermochemical data. The measurements were performed in several small high pressure cells, built in part from quartz, synthetic sapphire and niobium metal. - The complete transition of one fluid compound from insulating to ionic behaviour can probably be found with small polar substances as water and ammonia. It has been observed, that water, brought to about 1000 °C and 1oo kbar, is to about one percent ionized, and it has been suggested, that at 180 - 200 kbar water might exist as an ionic fluid: "hydronium hydroxide".

The study of the continuous nonmetal-metal transition is even more interesting and of more fundamental importance than the above mentioned transitions to ionic fluids. Certain nonmetal-metal transitions can be observed at high pressures. Two examples shall be given here. The first is from the

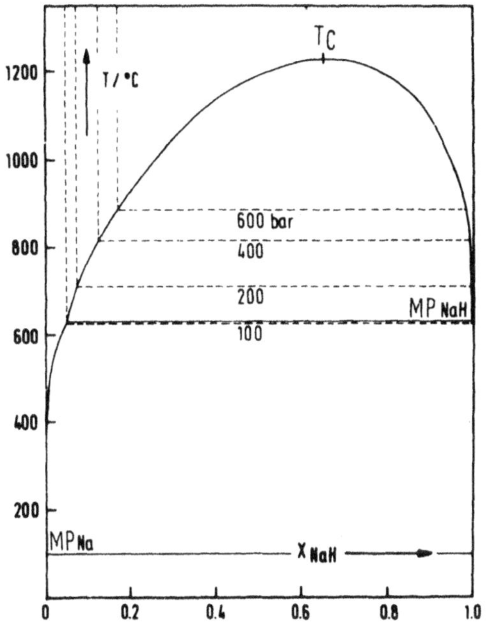

Fig. 27: Part of the experimentally determined sodium-sodium hydride phase diagram. MP: melting points. Critical temperature, T_c, calculated.

field of fused salts. A number of molten salts are miscible with their pure liquid metal. In some cases such a miscibility exists for the whole range of concentrations. This phenomen has been found and studied for many alkali halides (43) and for lithium hydride (44) at normal pressures. Since the alkali hydrides with their simple anion are particularly interesting for possible future theoretical interpretation, the systems sodium-sodium hydride and cesium-cesium hydride habe been studied. (45) Because of the insufficient stability of these hydrides, hydrogen pressures up to 800 bar at temperatures to 900 °C had to be applied.

The autoclaves were made of nickel-base superalloys, which did not show hydrogen embrittleness. This material is, however, very permeable for hydrogen at these conditions. Therefore, the autoclave had two inner compartments, one inside of the other. Hydrogen was constantly lost from the outer compartment at high pressures and temperatures, but this loss was replaced by pumping new hydrogen into this part at rate, by which equality of hydrogen pressures in both compartments was maintained. No loss occured from the inner part, which contained a small capsule of very pure iron in which the alkali metal and its hydride was enclosed. Thus the equilibrium hydrogen pressure over metal-metal hydride mixtures of different compositions at different temperatures could be measured. Since these measurements could not be made above 900°C, the critical mixing temperature of the Na-NaH-system could not be reached. It was estimated to be 1230 ±70°C. The hydrogen pressure at this point would be around 2500 bar. For the Cs-CsH-system the critical mixing temperature could be determined. It was 675°C (46). In Fig.27 a part of the Na-NaH phase diagram is shown as a temperature-composition plot with several isobars.

Another well investigated group of examples for nonmetal-metal transitions at high pressures are pure supercritical fluid metals. These are discussed and explained in detail by F. Hensel (47). One very well studied case is supercritical mercury, the specific conductance of which is shown in Fig. 28. This figure shows mercury gas at the supercritical temperature of 1550°C. If it is compressed to 2000 bar or higher, its conductance increases by many orders of magnitude and reaches metal-like values in the dense, disordered state. A detailed discussion proves that this dense fluid can indeed be called a metal.

As pointed out in the introduction, the combined application of high pressures and high temperatures opens up a wide field of phenomena which can be exhibited by polar fluids at different conditions. This is demonstrated by a schematic diagram by Tödheide (Fig. 29). (48) All these phenomena have been found and investigated in one fluid or another. It is hoped, that high pressured studies of this kind will contribute to the efforts to obtain in the future a very general theoretical description and understanding of fluids of strongly interacting particles over a wide range of density and temperature.

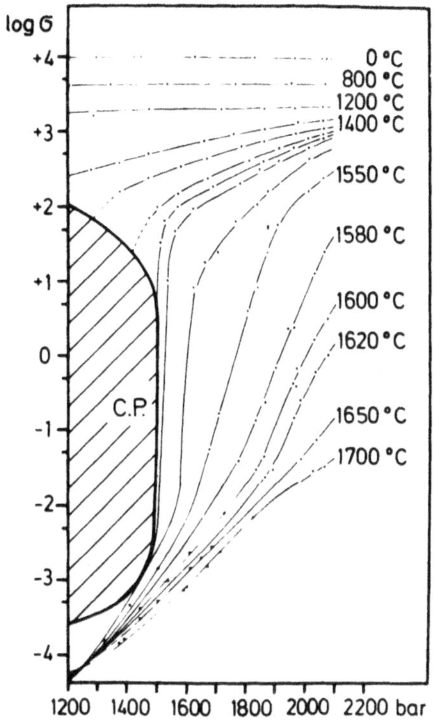

Fig. 28: Specific conductance σ of fluid mercury to 1700 °C and 2200 bar.

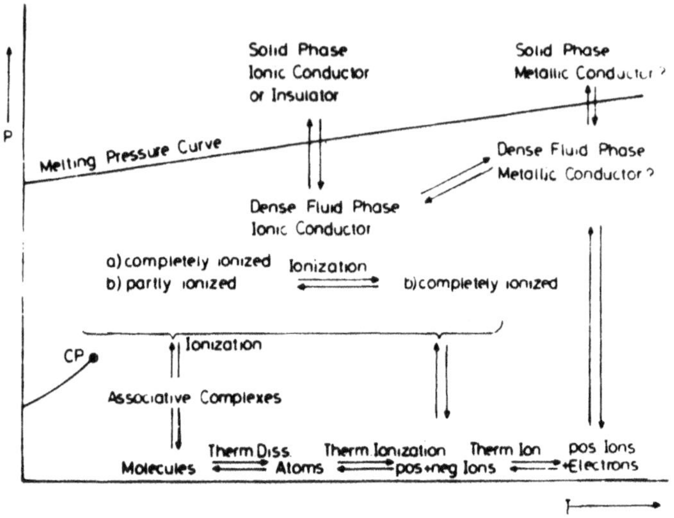

Fig. 29: Schematic compilation of phenomena to be expected with polar fluids within a wide range of temperatures and pressures. CP: Critical point.

REFERENCES

1. J.F. Mathews, Chem.Rev. 72, 71 (1972)
A.P. Kudchadker, G.H. Alani, B.I. Zwolinski, Chem.Rev. 68, 659(1968)
2. G. Treiber, K. Tödheide, Ber.Bunsenges.phys. Chem. 77, 1079(1973)
3. F. Hensel, E.U. Franck: "Thermodynamic Properties of Fluid Metals of High Temperatures and High Pressures", in "Experimental Thermodynamics, Vol. II", Editors B. LeNeindre, B. Vodar, IUPAC, Pure and Applied Chemistry, Butterworths, London, 1975
4. K. Tödheide: "Water at High Temperatures and High Pressures" in "Water, Vol.I" p.463, Editor F. Franks, Plenum Press, N.Y., London,1972
5. C. C.W. Burnham, I.R. Holloway, N.F. Davies, Amer.J.Sci. 267 A, 70 (1969)
6. S. Maier, E.U. Franck, Ber.Bunsenges.phys.Chem. 70, 639(1966)
H. Köster, E.U. Franck, Ber.Bunsenges.phys.Chem. 73,716(1969)
R. Hilbert, Thesis, Institute of Physical Chemistry, Karlsruhe University, 1978
7. I.M. Walsh, M.H. Rice, J.Chem.Phys. 26, 815 (1957)
A.C. Mitchell, W.L. Nellis, Preprint UCRL 79022 Lawrence Livermore Laboratory, USA 1977
8. I.L. Haas jr., U.S. Geological Survey, Report 75-615 (1975)
9. R. Hilbert, Thesis, Institute of Physical Chemistry, Karlsruhe University, 1978
10. R. Taani, Ph.D. Thesis, Institute of Physical Chemistry, Karlsruhe University, 1976
11. T.M. Reed, K.E. Gubbins, "Applied Statistical Mechanics" McGraw Hill, N.Y.,1973
12. H.v.Tippelskrich, E.U. Franck, F. Hensel, J. Kestin, Ber.Bunsenges.phys.Chem. 79,889(1975)
13. I.S. Rowlinson, "Liquids and Liquid Mixtures", 2nd edition, Butterworths, London,1969
14. K. Tödheide, E.U. Franck, Z.Physik.Chemie, N.F. 22, 232(1959), 37, 387(1963)
15. E.U. Franck, Pure and Appl.Chem., 38,449(1974)
16. H. Lentz, Rev.Sci.Instruments, 40, 371(1969)
E.U. Franck, H. Lentz, H.Welsch, Z.Physik. Chemie N.F. 93, 95(1974)
17. S. Takenouchi, G.C. Kennedy, Amer.J.Sci. 262, 1055(1964)
18. M. Gehrig, Thesis, Institute of Physical Chem. Karlsruhe University, 1975

I.R. Ferraro, L.I. Basile, Appl. Spectroscop. 28, 505(1974)
20. L.D. Naake, Thesis, Institute of Physical Chemistry, Karlsruhe University, 1976
21. S.v.Tapavicza, M. Buback, E.U. Franck, High Temperatures - High Pressures, 7, 535(1975)
M. Buback, E.U. Franck, J.Chim.Physique, 72, 601(1975)
F.W. Nees, M.Buback, Z.Naturforschg. 31a,1690 (1976)
M.Buback, F.W. Nees, Ber.Bunsenges.phys.Chem. 80, 650(1976)
22. F.W. Nees, Thesis, Institute of Physical Chem. Karlsruhe University, 1978
23. M. Charuel, Thesis, Institute of Physical Chem. Karlsruhe University, 1974
24. H.D. Lüdemann, E.U. Franck, Ber.Bunsenges. phys.Chem. 71, 455(1967)
25. H.D. Lüdemann, E.U. Franck, Ber.Bunsenges. phys.Chem. 72, 514(1968)
26. B. Scholz, H.D. Lüdemann, E.U. Franck, Ber. Bunsenges.phys.Chem. 76, 406(1972)
27. K. Heger, Thesis, Institut of Physical Chem. Karlsruhe University, 1969
M. Uematsu, W. Harder, E.U. Franck, "The Static Dielectric Constant of Water to 550°C and 5 kbar". Report of the "International Association on the Properties of Steam " IAPS, 1977
29. W. Harder, Thesis, Institute of Physical Chem. Karlsruhe University, 1972
30. M. Buback, W. Harder, Ber.Bunsenges.phys.Chem. 81, 603, 609(1977)
31. K. Reuter, Thesis, Institute of Physical Chem. Karlsruhe University, 1974
32. W. Hill, Thesis, Institute of Physical Chemistry Karlsruhe University, 1974
E.U. Franck, R. Deul, British Chem.Soc.Faraday Division, General Discussion No. 66, (1978) in press
33. A.Z. Francesconi, E.U. Franck, H. Lentz, Ber. Bunsenges.phys.Chem. 79, 897(1975)
34. E.U. Franck, Z.physikal.Chem. N.F. 8, 107, 192 (1956)
35. A.S. Quist, W.L. Marshell, E.U. Franck, W.v. Osten, J.Physical.Chem. 74, 2241(1970)
36. G. Goldmann, K. Tödheide, Z.Naturforschg. 31a, 656(1976)
37. G. Goldmann, K. Tödheide, Z.Naturforschg. 31a, 769(1976)
38. I.W. Johnson, D. Cubicciotti, J.physical.Chem. 68, 2275(1964)

39. G. Treiber, K. Tödheide, Ber.Bunsenges.phys. Chem. 77, 540(1973)
40. G. Treiber, K. Tödheide, Ber.Bunsenges.phys. Chem. 77, 1079(1973)
41. M. Buback, E.U. Franck, Ber.Bunsenges.phys.Chem. 76, 350(1972); 77, 1074(1973)
42. W. Holzapfel, E.U. Franck, Ber.Bunsenges.phys. Chem. 70, 1105(1966)
S.D. Hamann, M. Linton, Trans.Far.Soc. 65, 2186(1969)
43. M.A. Bredig in "Molten Salt Chemistry" p.367, Editor: M. Blander, Wiley, N.Y. 1964
44. E. Veleckis, E.H. Deventer, M. Blander, J. Physical Chem. 78, 1933(1974)
45. W. Klostermeier, Thesis, Institute of Physical Chemistry, Karlsruhe University, 1978
46. A.W. Szafranski, W. Klostermeier, E.U. Franck, To be submitted to Ber.Bunsenges.phys.Chem.
47. F. Hensel, E.U. Franck, Ber.Bunsenges.phys.Chem. 70, 1154(1966); Rev. Mod. Physics, 40, 697(1968)
Experimental Thermodynamics, Vol.II. Editors: B. LeNeindre, B. Vodar, Pure and Applied Chemistry, Butterworths, London, 1975
F. Hensel, Angewandte Chemie, 86, 459(1974)
48. K. Tödheide, Naturwissenschaften 57, 72(1970)

THE EFFECT OF PRESSURE ON THE ELECTRONIC STRUCTURE OF FLUIDS

F. Hensel

Institute of Physical Chemistry of the University of Marburg, 3550 Marburg, Germany

ABSTRACT. Selected experimental results for properties of metallic and semiconducting fluids at high temperatures and pressures particularly in the supercritical regions are presented and discussed. The results are discussed with special emphasis on their relationship to some theoretical models for the density induced nonmetal to metal transitions in fluid metals and the temperature induced nonmetal to metal transition in fluid Se and Te.

1. INTRODUCTION

The electronic and molecular structure of fluids depend strongly on density and temperature. Since with pressure the density can be varied at constant temperature, pressure is obviously an important variable for the understanding of the electronic structure and the related electrical transport and thermodynamic properties of fluids. A common effect of increasing density is to broaden the bands of allowed energy and to decrease the gaps of forbidden energy between bands irrespective of whether the material is crystalline or fluid. In the latter case, however, the density of states is expected to tail into the gap due to the lack of long-range order. The broadening of the bands can ultimately lead to a continuous transformation from an insulating or semiconducting to a metallic state resulting from an overlap of the highest occupied and the first excited band [1]. Besides this

Kelm, H. (Ed.) High Pressure Chemistry. 259–279. Copyright © 1978. D. Reidel Publishing Company. All Rights Reserved.

band crossing transition a second kind of density induced metal-insulator transition has been discussed by Mott for some arrays of one-electron atoms in which the interaction between the electrons is essential [1], [2]. For large interatomic distances an antiferromagnetic insulator is predicted and the Mott transition can be described by the crossing of two Hubbard bands [1]. For disordered systems including fluids an additional other form of metal-insulator transition can occur according to whether the Fermi energy of a degenerate electron gas lies above or below a mobility edge. This Anderson-transition [3] has been extensively discussed in numerous recent papers [cp. 2].
A unique possibility for the experimental verification of these transitions in fluids is the observation of the appearance of metallic properties in a metal vapour, which occurs if the vapour density is continuously increased to sufficiently high values [4], [5]. Such continuous variation of density is only possible, however, if the vapour is compressed above the critical temperature T_c which terminates the vapour pressure curve. A list of critical data for a few selected typical metals together with some chalcogenic elements is assembled into fig.1.

Substance	T_c (K)	p_c (bar)
Hg	1760	1510
Cs	2020	110
Rb	2100	130
K	2200	155
Te	2500*	270*
Se	1760	380
S	1315	180

* estimated

Fig. 1.

Another very interesting transformation from semiconducting to metallic behaviour with increasing temperature and pressure is exhibited by the liquid chalcogenic elements Se and Te [6], [7], [8]. Both liquids are nonmetallic near their melting points because of well developed molecular species such as Se_8-rings and polymeric chains. As the temperature is increased the rings and chains dissociate or decrease in length, with the process of dissociation into a completely random atomic configuration typical for fluid metals. Again application of very high temperatures implies that the heating up of the liquid takes place at pressures greater than the liquid-vapour critical pressure p_c.
The aim of the present paper is to review some recent experimental results with particular emphasis on their relevance to the metal-nonmetal transitions mentioned above.

2. EXPANDED MERCURY

Because liquid mercury has a relatively low critical temperature, extensive experimental results including d_c conductivity [5], [9], [10], thermoelectric power [5], [11], Hall effect [12], Knight shift [13] and optical absorption [14], [15] are available at sub- and supercritical conditions. All these properties establish the occurrence of a metal-nonmetal transition induced by a density change in the range between 11 and 8 g/cm^3. There mercury exhibits a rapid variation of many of its thermodynamic and transport properties [16].
The following noteworthy observations are specifically important for the present understanding of the electronic structure of expanded fluid mercury.

a) The electrical conductivity, σ, decreases by eight orders of magnitude as the density ρ is reduced from 13.6 g/cm^3 to 2 g/cm^3. At high densities between 13.6 g/cm^3 and 11 g/cm^3 σ varies from 10^{+4} to $3\times10^{+3}$ ohm^{-1}cm^{-1}. Here the electron mean free path, L, exceeds the reciprocal Fermi wavevector, i.e. $k_F \cdot L > 1$, and σ is properly described by Ziman's "Nearly Free Electron" (NFE) model [17]. For ρ = 11 g/cm^3 L becomes comparable with $1/k_F$. It is not likely that the NFE-model will hold here where the electron ion interaction is so strong. Electronic transport in this strong scattering (s.s.) regime has been discussed in numerous theoretical papers [cp. 2].

The essential result is, that as the density of states $N(E)$ changes with some parameter, e.g. density ρ or temperature T, σ will change in proportion to the square of $(N(E_F)$. Between 11 and 9 g/cm^3 σ varies from 3000 to 300 ohm^{-1} cm^{-1} corresponding to a decrease in $g = N(E_F)/N(E_F)_{free}$, the ratio of $N(E)$ at the Fermi energy E_F to the corresponding free electron density of states, from 1 to 0.3. For densities smaller than 9 g/cm^3 σ becomes smaller than 300 ohm^{-1}cm^{-1}, which is the value for minimum metallic conductivity σ_{min} as indicated by Mott [2]. For σ smaller than σ_{min} the main current should be carried by thermally excited electrons as in fluid semiconductors. This, in fact, is consistent with the observed more rapid increase of $(\frac{\partial \ln \sigma}{\partial \ln V})_T$ and $(\frac{\partial \ln \sigma}{\partial T})_V$ for σ smaller than 9 g/cm^3 [11].

b) The Hall coefficient, R, [12] of expanded mercury in the density range between 13.6 and 11 g/cm^3 exhibits the free electron value $R_{free} = (ne)^{-1}$, whereas it shows marked positive deviations from the free electron value $(1 < R/R_{free} < 3)$ for the range between 11 and 9 g/cm^3. The latter is consistent with the strong scattering theory of Friedman [18], which predicts a functional dependence of $R/R_{free} \propto g^{-1}$.

c) In the density range between 13.5 g/cm^3 and 11 g/cm^3 the Hall mobility, $\mu = R\sigma$, [12] exhibits a relatively large decrease from 0.71 to 0.24 cm^{-2}volt^{-1} sec^{-1}, whereas it shows only a moderate density dependence from 0.24 to 0.1 cm^{-2}volt^{-1}sec^{-1} between 11 and 9 g/cm^3. For densities between 9 and 9.5 g/cm^3 the mobility is practically constant.

d) The thermoelectric power, S, [11] is negative, revealing a slow decrease between 13.6 to 9 g/cm^3 with a typical nonmetallic value of -86 μV/K at 9 g/cm^3. For densities smaller than 9 g/cm^3 a more rapid increase in $(\frac{\partial S}{\partial T})_V$ and $(\frac{\partial S}{\partial V})_T$ is observed [11]. Furthermore for densities smaller than 8 g/cm^3 σ and S are consistent with thermal excitation of electrons from localized states to extended states across a real gap or a mobility gap. From the correlation between S and σ for the semiconducting range $\ln \sigma = ek^{-1} \cdot S + \text{const}$ with $e \cdot k^{-1} = 1.16 \cdot 10^{+4}$ deg$\cdot$volt^{-1} it has been concluded [11] that the onset of semiconducting behaviour occurs for a density of about 8 g/cm^3.

e) The Knight shift, K, [13] decreases roughly linearly with decreasing density in the density range between 13.6 to 11 g/cm^3 and becomes density independent in the density range between 11 and 9 g/cm^3. An abrupt drop to a nearly zero shift is observed between 9 and 8 g/cm^3.

A comparison of all the data described above leads to the conclusion that a gradual transformation from a metallic to a semiconducting state occurs in fluid mercury at a density of about 9 g/cm^3 (ρ_c, the critical density, is 5.3 g/cm^3). As mentioned above, this type of band crossing metal-semiconductor transition is expected for a low density fluid or hypothetical crystalline divalent metal like mercury. In the case of a fluid, however, the situation is more complicated, because in disordered systems the singularities in the density-of-states function at the band edges are smeared out; that is the band edges exhibit tails within the region which is forbidden in the corresponding crystalline solid [2]. Fig.2 shows a schematic illu-

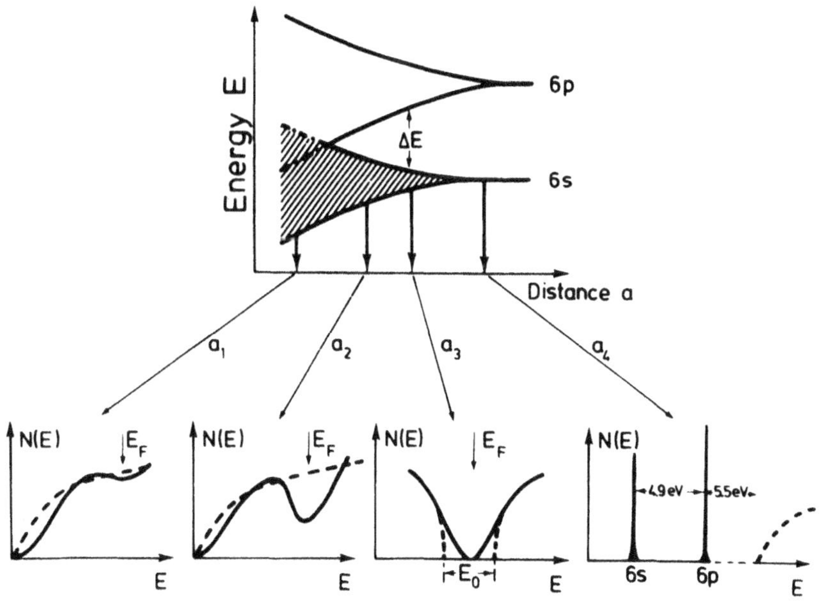

Fig.2 Density of states in expanded fluid mercury (schematic). E_F is the Fermi energy. The dotted line gives the free electron density of states.

stration for the situation in an expanded Hg-crystal and the corresponding fluids for four different densities. In the limit of a very small density the distance between the 6s-valence band and the 6p-conduction band must tend to the distance of the discrete energy levels in the isolated noninteracting atoms, i.e. 4.9 eV. To avoid confusion the higher energy levels between the 6p state and the conduction band above the ionization limit of 10.4 eV are not shown in fig.2. For this distance the great majority of the valence electrons are still attached to their parent atoms and only a small fraction of them are split by thermal excitation into ions and free electrons. On the other hand at high liquid like densities (ϱ between 13.6 and 11 g/cm^3) σ , S, R and K are consistent with the assumption that Hg is a nearly free electron metal; i.e. $N(E_F)$ should not deviate largely from $N(E_F)_{free}$ (the dotted line in fig.2 indicates $N(E)_{free} \propto \sqrt{E}$). Between these two extreme density limits the intermediate state indicated in fig.2 must occur. The gap is replaced by a pseudogap, i.e. a minimum in N(E). For divalent Hg E_F lies in or near the minimum.

The spectral dependence of the absorption coefficient of mercury at the constant supercritical temperature of 1640°C and selected densities between 2.5 g/cm^3 and 4.0 g/cm^3 is shown in fig.3 [15]. The absorption edges exhibit the expected red shift with increasing density and are found to be exponential in form at values of K up to about $2 \cdot 10^3 cm^{-1}$. The exponential shape of the absorption edge reminds us of the Urbach rule which is obeyed in many materials, e.g. in alkali halides, many semiconductors and glasses [19]. It is supposed that the similar absorption behaviour in all these different substances has the same physical reason. As a general physical model Dow and Redfield [20] propose that the Urbach rule arises from an electric-field broadening of an exciton line. In amorphous and fluid materials one must consider the fluctuations in composition or density as possible sources of local fields. However there is an additional possibility, which has been suggested for disordered semiconductors by Tauc [19], namely that the Urbach tail arises from electronic transitions between states in the band edge tails, the density of which is assumed to be an exponential function of energy.

In the strong absorption region of the edge ($K \approx 2 \times 10^{+3} cm^{-1}$) the absorption constant K increases less steeply with increasing $\hbar\omega$ and can be described by the relation

$$\sqrt{K \cdot \hbar\omega} = \text{const} \cdot (\hbar\omega - E_g)$$

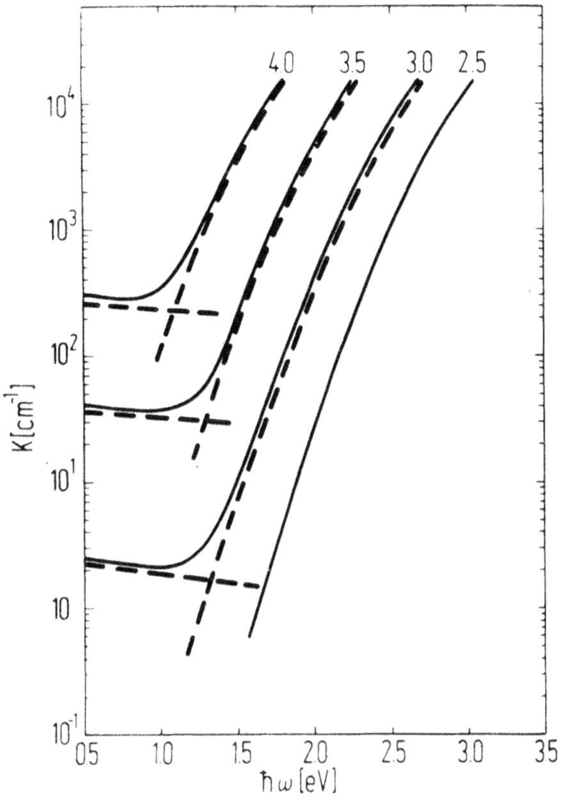

Fig.3. Optical absorption edges of fluid mercury at different densities; experimental results (full line) compared with the results of model calculation (dotted line). The temperature is constant 1640°C.

This relation is of a form similar to that for direct interband electronic transitions in crystalline semiconductors, and was derived by Tauc under the assumption that transitions take place between pairs of states in parabolic valence and parabolic conduction bands separated by an energy $\hbar\omega$ and that the optical matrix element is constant in the energy range under consideration [19]. The constant E_g is used to define an optical gap, which is shown as a function of the density in fig.4. It should be emphasized that E_g is an extrapolated rather than a real zero in the density of states. At densities larger than 3 g/cm^3 the absorp-

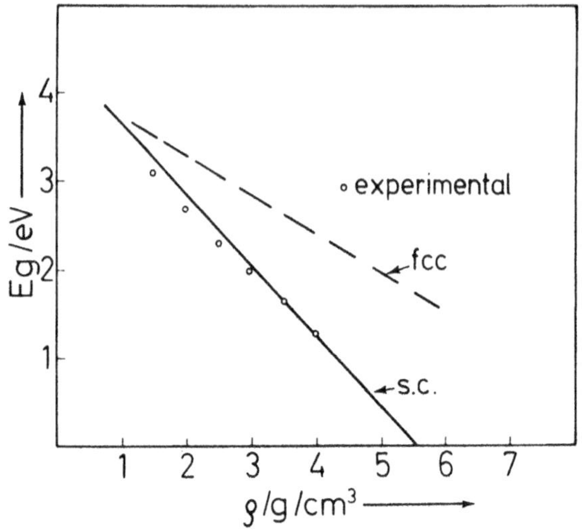

Fig.4 Optical gap of fluid mercury as a function of density (T = 1640°C) compared with results of a band structure calculation for expanded fcc and s.c. mercury crystals

tion curves flatten out to approximately constant values of the absorption coefficient at long wavelengths up to the infrared region. The level of these long wavelength tails increase very rapidly with increasing density and the density dependence is the same as observed for the d.c. conductivity.
It is surely tempting to compare the density dependence of the optical gap with recent band structure calculations carried out for hypothetical crystalline Hg in the low density [21], [22], [23] and the high density limit [24], [25]. The most significat result of these calculations is that the energy gap is very sensitive to changes in structure. As can be seen from fig.4 the data for the expanded simple cubic (s.c.) crystal structure deviate significantly from the fcc band gap [21]. This can be understood in terms of coordination numbers: the fcc structure is highly coordinated as appropriate for a dense system whereas the s.c. structure has a coordination number of 6 only. Since it can be inferred from very recent measurements of the radial distribution function for expanded Rb [26], that the coordination number decreases rapidly upon expan-

sion the low coordinated s.c. structure is more likely a reasonable model for the vapour. However, the agreement between the measured gap and the calculated for the s.c. structure evident from fig.4 is extremely close and should be regarded to be incidental. The dotted curves in fig.3 are the result of an attempt to describe the optical properties on the basis of such a band model [27]. First it is assumed that N(E) of the conduction and the valence band not to close to the band edges is of a parabolic form and that the optical matrix element M is constant in the energy range under consideration. In addition it is assumed that as a consequence of the density fluctuations at the high temperatures N(E) exhibits tails near the band edges in the form of a Gaussian distribition. Fitting the value of M and the constant in the Gaussian distribution to match the experimental curve for each density separately gives fair agreement with experimental curves as shown in fig.3. The form of the long wavelength absorption tails is compared with the theory of free-carrier (intraband) absorption in the strong scattering range of transport developed by Mott [2] and Hindly [28]. Again fitting of the value of M leads to excellent agreement between theory and measurement (fig.3).

However it should be pointed out that the value for the low density limit for a meaningful band structure calculation is unknown. From very recent unpublished optical absorption and emission measurements it is suggested that extended states in the upper 6p-band exist down to 1.5 g/cm^3. This suggestion is based on the observation of discontinuities in the first derivative of the intensities of absorption and emission of thermal radiation as a function of density at constant wavelength. In fig.5 the plot of density versus the photon energy for constant emission of thermal radiation corresponding to a constant absorption-constant reveals the discontinuity in the slope at 1.5 g/cm^3. Since the optical transition matrix element should depend strongly on the nature of the initial and final states it is believed that the discontinuity in fig.5 indicates the transition from localized to extended states in the 6p-band of Hg.

Both the optical measurements and the band model indicate that the band gap is closed at densities between 5 and 6 g/cm^3. This is in constrast to all the electrical and the Knight shift data which show nonmetallic behaviour at densities below 9 g/cm^3. Such a discrepancy is very often observed in disordered semiconduc-

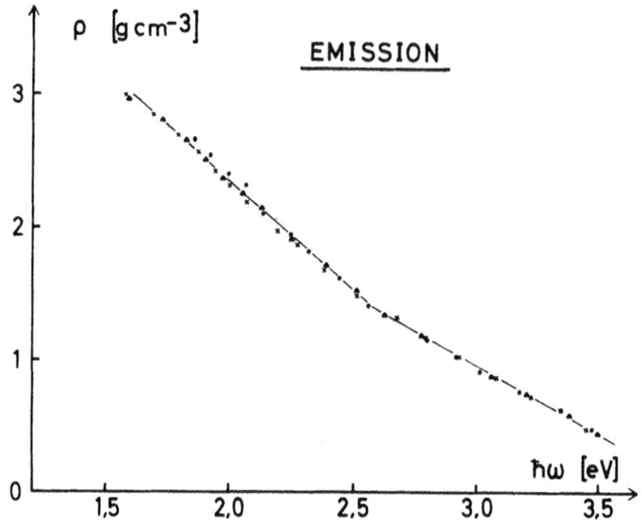

Fig.5 Plot of pairs of density and photon energy for constant emission intensity of thermal radiation.

tors where localized states are proposed to increase the apparent electrical band gap [2], and there is no reason not to assume a similar explaination for fluid Hg. This leads to the conclusion that the observed metal-nonmetal transition is essentially related to the disorder of high temperature Hg and is therefore an Anderson transition [3].
An alternative explaination has been proposed by Cohen and Jortner [29]. They suggested the existence of an inhomogeneous transport regime in mercury between 9 g/cm^3 and 8 g/cm^3. They assumed that the microscopic inhomogenities arise from density fluctuations. The metal-nonmetal transition in such a system with metallic and semiconducting clusters is then discussed within the framework of percolation theory. It is necessary to point out that from existing experimental results it cannot be decided whether the Cohen-Jortner proposal is right. More accurate experimental work on expanded mercury is necessary to make it possible to formulate a theory which is tested experimentally.

However, recent measurements of the isothermal compressibility χ and the thermal expansion coefficient α for expanded mercury as a function of the density by Schönherr [30] indicate an enormous increase of χ and

α at densities smaller than 9 g/cm^3 as shown for χ in fig.6.

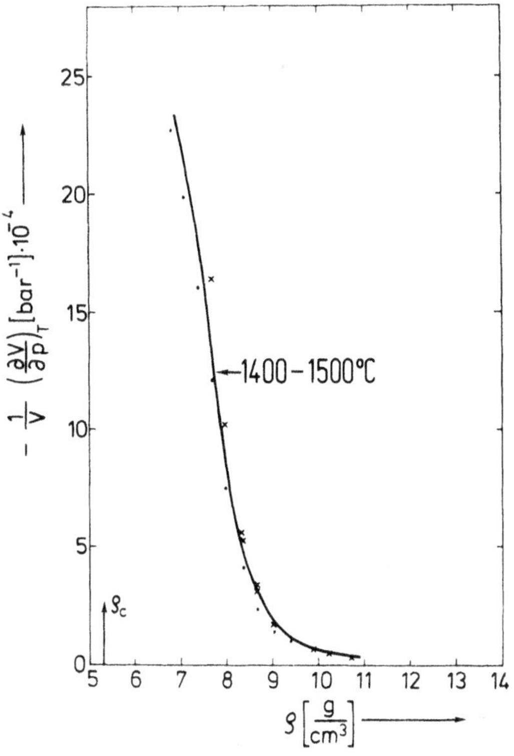

Fig.6 Thermal compressibility of fluid mercury as a function of the density

It is certainly interesting to speculate on whether the mutual interdependence of electron-electron screening and interatomic forces necessarily leads to density fluctuations in the metal-nonmetal transition range.
Finally it should be pointed out that there is an apparent contradiction between the Knight shift and the electrical transport in the strong scattering regime between 11 und 9 g/cm^3. As mentioned above conductivity σ and Hall constant R are consistent with the assumption that $N(E_F)$ decreases rapidly in this range, whereas the Knight shift is found to be constant. This suggests that the S component of N(E) is constant. A detailed discussion of this problem has recently been

published by Mattheis and Warren [25].

3. EXPANDED ALKALI METALS

For an array of one-electron atoms, whether crystalline or not, a metal-nonmetal transition can only occur as a consequence of the interaction between the electrons [31]. The formal treatment is due to Hubbard [32]. He finds that if the distances between the atoms exceed a critical value, a splitting of the conduction band into two bands occurs. The lower band is occupied and the upper band is empty and the material is a nonmetal with an activation energy for conduction. In numerous papers it has been pointed out by Mott [cp.2], that in a disordered expanded system of monovalent atoms with a mean interatomic distance near that for the metal-nonmetal transition, the density of states can be represented by two overlapping Hubbard bands, so that the position of the Fermi energy is in the minimum of the density of states. If the depth of the minimum is large enough, the states around the Fermi energy should become localized; i.e. a mobility gap should be formed. This gives a gradual transition to a nonmetallic state which experimentally should be very similar to the transition observed for divalent mercury.

From recent measurements of the electrical conductivity σ and the absolute thermoelectric power S for fluid Cs [33], [34], Rb [35], [36] and K [37] it can be suggested that in the fluid alkali metals NFE-behaviour exists over a wide density range down to about 50% of the normal melting point density. At this density the electron mean free path L is approximately equal to the reciprocal wavevector k_F at the Fermi edge, the temperature coefficient of σ changes sign from negative to positive and the Hall coefficient R [38] exhibits nearly the free electron value down to this density. Therefore the Ziman (NFE)-model [17] should apply for the electrical transport properties of the fluid alkali metals in this density range. This has very recently been proved experimentally for fluid rubidium by measurements of the density [25], [36], the electrical conductivity [35], [36] and the structure factor S(Q) [39], [26], [40] as a function of temperature and pressure. The results are shown in fig.7 for different temperatures and pressures. It is found [40] that the density dependence of σ can be well represented by the (NFE)-model in the density range between 1.4 g/cm^3 and 0.9 g/cm^3, i.e. that σ is mainly determined by the number of free valence electrons, the configuration of the scatters S(Q) and the scattering properties of the screened positively charged ions.

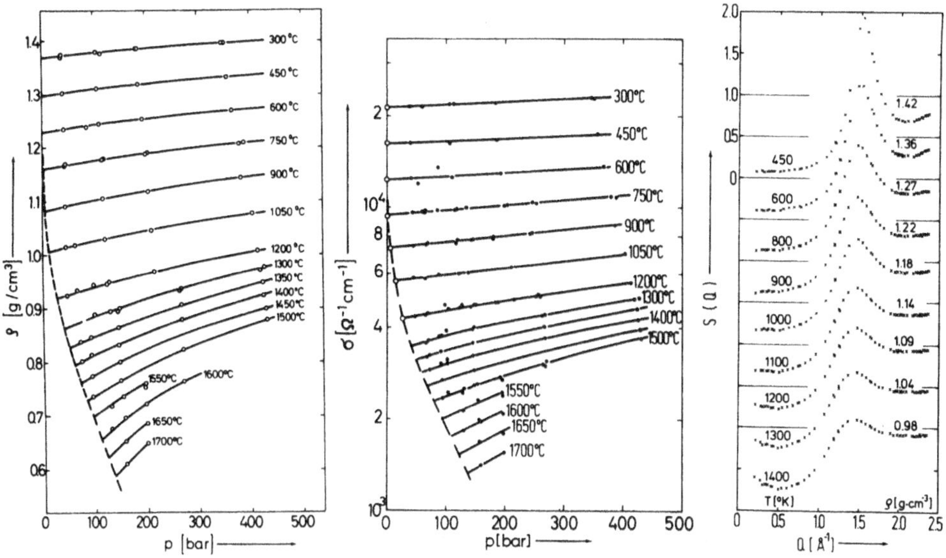

Fig.7 Density ρ, conductivity σ and structure factor S(Q) for fluid Rb at different temperatures and different pressures or densitites.

In the S(Q) curves mainly three particular features may be distinguished. The first peak, which is relatively sharp at low temperatures, broadens appreciably with increasing temperature and its height decreases drastically. This fact illustrates that the "shell of nearest neighbours" is less well defined and smears out with increasing temperature or decreasing density. At the same time the position of the first maximum is shifted only slightly to lower Q values between 450 K and 1400 K, which simply shows that expansion of the fluid leads to a large decrease in the coordination number, whereas the nearest neighbour-distance remains nearly unchanged [26]. The main overall feature for Q values above 1 $Å^{-1}$ is a considerable flattening of the structure factor.
At Q values below 0.5 $Å^{-1}$ two interesting observations can be made. Whereas for temperatures below 1100 K S(Q) remains constant below $Q \leq 0.6$ $Å^{-1}$, a slight increase of S(Q) with decreasing $Q \leq 0.4$ $Å^{-1}$ starts to show up at 1200 K, and is well pronounced at 1400 K. The second

interesting finding is that at the smallest Q reached in this experiment, S(Q) increases by almost a factor of five between 450 K and 1400 K. This is consistent with the observed increase in the isothermal compressibility on expansion of fluid rubidium which is related to the structure factor at Q = 0 $Å^{-1}$ by the well-known equation

$$S(0) = k_B T \chi_T n$$

where k_B is the Boltzmann constant and n the number density per cm^3 [39].

In many theoretical treatments of simple liquid metals, especially the alkali metals close to their melting points, it has been shown that the structure factors S(Q) can be modelled, in a crude approximation, by hard-spheres of an appropriately chosen diameter [cp. 41]. However since screening of the ionic charges by the electron gas leads to effective pair interactions in metals which are always implicit functions of the electron density, i.e. of the ion number density, the hard-sphere diameters have to be considered to be density dependent. A least squares fit of the Percus-Yevick equation [42] with a hard core potential to the S(Q)-data of fig.7 shows a slight nearly linear decrease for the hard core diameter of Rb of less than 10% in the density range between 1.4 and 1.04 g/cm^3 [40]. The temperature dependence is believed to be negligible. It has recently been shown by Pfeifer [36] that the hard-sphere model can also successfully be applied to p-V-T data of expanded metals at higher temperatures. He showed that the pressure and compressibility of fluid Rb in the density range between 1.4 and 0.7 g/cm^3 (fig.8) can be separated in a temperature and an only volume dependent contribution. The slopes of the linear form of the isochors in fig.8 can be well described by the classical van der Waals equation $(\partial p/\partial T)_V = R/(V - b)$. Since the form of the volume dependence of the cohesive energy in fluid metals is not known accurately enough, Pfeifer simply used the functional form $E = -a \cdot V^{-n}$ fitting the value of a and n to match the experimental p-V-T data of fig.7. He got a surprisingly good description of the p-V-T data of liquid Rb in the large density range between 1.4 to 0.7 g/cm^3 at temperatures between 200 and 1700°C with the simple equation of state with only three constants a, b and n

$$p = \frac{RT}{V - b} - \frac{a}{V^{n+1}}$$

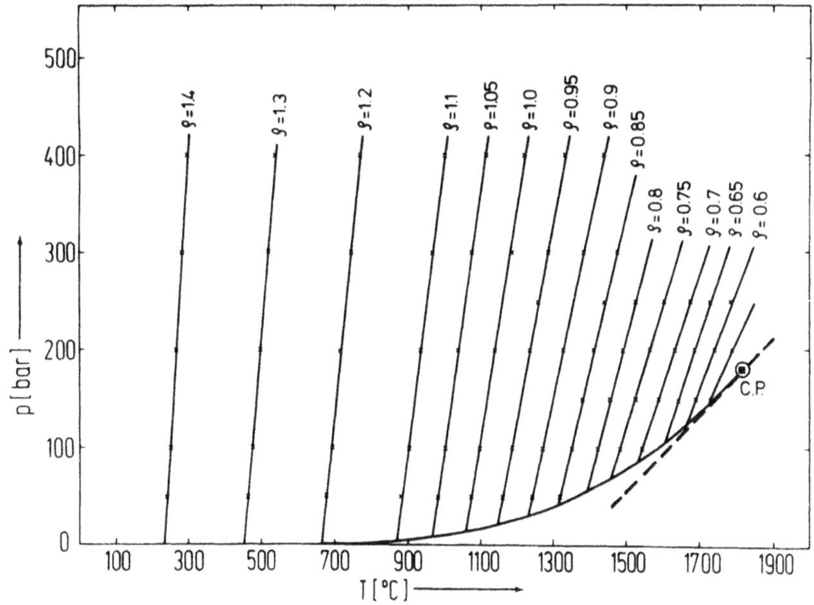

Fig.8 Isochors of liquid Rb

4. ELEMENTAL SEMICONDUCTORS

A progressive variation of the electrical properties from insulating to metallic is exhibited by the liquid elements of group VI A with increasing atomic weight. Liquid sulphur is an insulator, selenium is a liquid semiconductor, tellurium appears to be a borderline case exhibiting both metallic and semiconducting behaviour, while polonium is a liquid metal. Especially liquid selenium and tellurium and their liquid mixtures has been studied at moderate temperatures by many investigators both in regard to their electrical properties and their molecular and electronic structure [8], [43]. Their behaviour is consistent with the assumption that the chain-like structure, i.e. the covalent bonds in two-fold coordination, of solid tellurium, selenium and of the solid solutions of both consisting of mixed chains [4] persists to a large extent in the melt at temperatures close to the melting point [45], [46], [6]. This assumption leads to a well-developed energy gap and semiconducting properties. However if the temperature is increased, the chains should dissociate and decrease in length; the decrease of the chain length goes in the direction of a completely random atomic configuration which is typical for metals i.e. a semiconductor to metal transformation is expected to occur with increasing temperature. This proposal is consistent with recent measurements of the electrical conductivity of fluid tellurium [47], selenium [8] and Se-Te mixtures [6], [48] up to very high temperatures at moderate pressures. Selected data are shown in fig. 9. The limitations of the "semiconductor model" for

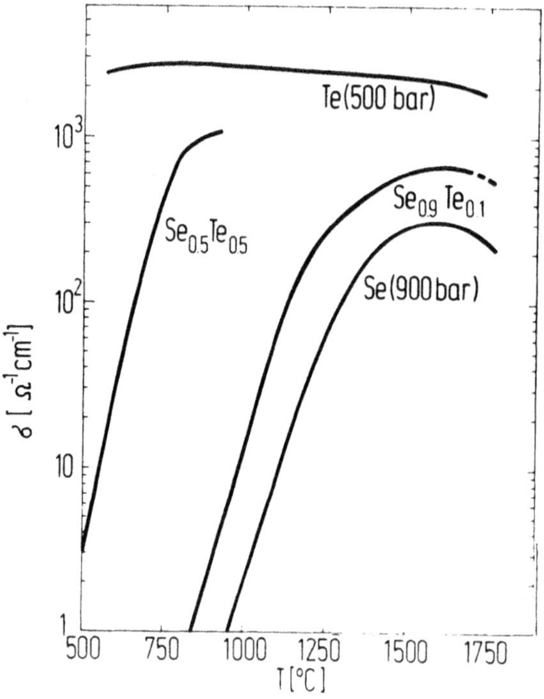

Fig.9 Electrical conductivity of fluid Se, Te and Se-Te mixtures at moderate pressures as a function of temperature.

the considered liquids become apparent at high temperatures. The dominant feature of the conductivity isobars is a more or less large increase from "semiconducting conductivities" (except for Te) at low temperature to a maximum in the nearly-metallic range 100-3000 $ohm^{-1}cm^{-1}$. Above the maximum σ decreases at the highest temperatures. It is reasonable to suppose that the temperature effect in the liquid results from the combined effects of thermal activation of carriers, increasing coordination number, and decreasing density from thermal expansion. The former two effects should increase the conductivity while the density decrease works in the opposite direction. The maxima in the isobars can then be understood in terms of the dominance of the density reduction at high temperatures. There is a characeristic shift of the temperature, at which the transformation from semiconducting to metallic behaviour occurs with increasing bond strength of the covalently bonded atoms within the chains. Mössbauer data indicate that the binding of a covalent Se-Te bond is stronger than the average of the Se-Se and Te-Te bond [49]. As a consequence addition of selenium to tellurium tends to stabilize the covalent bonds between atoms in two-fold coordination or semiconductor configuration.

Structural data over a large temperature range are only available for fluid Te [45]. Cabane and Friedel [43] conclude from a detailed analysis of these data that the number of atoms in the first coordination shell is about two near the melting point and changes to about three at about 900°C. They described the structure of liquid tellurium in this temperature range in terms of a covalently bonded network in which the coordination number of the first shell is between 2 and 3. They then consider in detail the band structure which is to be expected from this model and propose that the density of conduction electrons increases with temperature due to the increase of ternary sites. The hypothesis of an increase in the coordination number in fluid selenium, tellurium and Se-Te mixtures with increasing temperature is consistent with recent measurement of the density of these systems by Thurn and Ruska [50]. The anomalous temperature dependence of the density is demonstrated by fig.10, which shows the thermal expansion coefficient α derived from the density data of Thurn and Ruska and Fischer and Schmutzler [51]. The results clearly demonstrate the change of the short range order of the fluids. Obviously the average coordination number increases with increasing temperature leading to a more dense packing despite of the competing thermal expansion. Negative values of α for liquid Te are only observed in the supercooled melt, whereas α for Se starts to decrease at temperatures larger than about 800°C. This difference is certainly related to the larger stability of the bonds within the selenium chain-molecules. The transformation to a more metallic state occurs at higher temperatures for selenium than for tellurium. For selenium the maximum conductivity is observed near the critical tem-

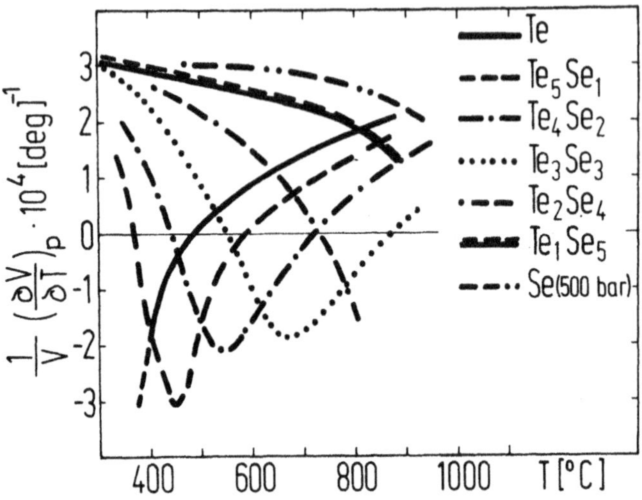

Fig.10 Thermal expansion coefficient of fluid Se-Te mixtures as a function of temperature

perature of the liquid-vapour phase equilibrium, whereas for tellurium the maxima occur far beyond the critical temperature. The binding of the covalent S-S bond in sulfur ring-or chain-molecules is much stronger than the Se-Se bond. Thus indications for metallic conduction in sulfur have been reported only at very high pressures of more than 100 kbars at high temperatures. Recent optical absorption measurements for fluid sulfur show a well developed absorption edge even at supercritical temperatures of 1100°C at supercritical pressures [52]. From the pressure dependence of the spectrum of sulfur at 1100°C shown in fig.11 it is evident that the edge at higher densities is formed by an overlap of absorption bands of at least 3 sulfur species with absorption peaks or shoulders at 265 nm and 285 nm belonging to cyclo-S_8, at 400 nm belonging to S_4 and at 530 nm belonging to S_3.

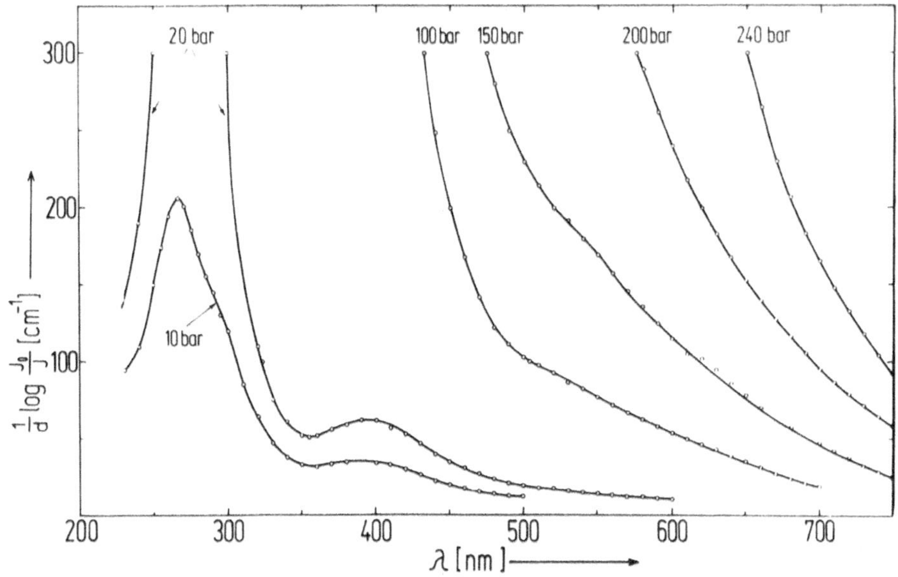

Fig.11 Absorption spectrum of a 100 μ thick sulfur film at the supercritical temperature of 1100°C at different pressures.

REFERENCES

1. N.F.Mott, Metal-Insulator Transitions, Taylor and Francis LTD, London, 1974.
2. N.F.Mott and E.A.Davis, Electronic Processes in Noncrystalline Materials, Clarendron Press, Oxford, 1971.
3. P.W.Anderson, Physic.Rev. 109, 1492 (1958).
4. F.Hensel, Angew.Chem.internat.Edit. 13, 446 (1974).
5. R.W.Schmutzler and F.Hensel, J.Non-Cryst.Solids 8, 718 (1972).
6. J.C.Perron, Advances in Physics 16, 657 (1967).
7. H.Hoshino, R.W.Schmutzler, F.Hensel, Ber.Bunsenges.phys.Chem. 79, 1186 (1976).
8. H.Hoshino, R.W.Schmutzler, W.Warren, F.Hensel, Phil.Mag. 33, 255 (1976).
9. F.Hensel and E.U.Franck, Ber.Bunsenges.phys.Chem. 70, 1154 (1966).
10. I.K.Kikoin and A.R.Sechenkov, Physics Metals Metallogr. 24, 5 (1967).
11. R.W.Schmutzler and F.Hensel, Ber.Bunsenges.phys. Chem. 76, 531 (1972).

12. U.Even and J.Jortner, Physic.Rev.Letters 28, 31 (1972).
13. V.El-Hanany and W.W.Warren, Physic.Rev.Letters 34, 1276 (1975).
14. F.Hensel, Ber.Bunsenges.phys.Chem. 75, 847 (1971).
15. H.Uchtmann, F.Hensel, Phys.Letters 53A, 239 (1975).
16. F.Hensel, Liquid Metals, 1976 ed.by Evans and Greenwood, Bristol, 1977.
17. T.E.Faber, An Introduction to the Theory of Liquid Metals, University Press, Cambridge, 1972.
18. L.Friedman, J.Non-cryst.Solids 6, 329 (1971).
19. J.Tauc, Amorphous and Liquid Semiconductors, Plenum Press, New York, 1974.
20. J.D.Dow and D.Redfield, Phys.Rev. 5B, 594 (1972).
21. H.Overhof, H.Uchtmann and F.Hensel, J.Phys.F. 6, 523 (1976).
22. P.Fritzson and K.-F.Berggren, Solid State Commun. 19, 385 (1976).
23. F.Yonezawa and F.Martino, Solid State Commun. 18, 1471 (1976).
24. M.A.C.Devillers and R.G.Ross, J.Phys.F. 5, 73 (1975).
25. L.F.Mattheis and W.W.Warren Jr., Phys.Rev. B16, 624 (1977).
26. W.Freyland, W.Gläser and F.Hensel, German Bunsentagung 1978.
27. H.Uchtmann, Thesis Marburg 1978.
28. N.K.Hindley, J.non-cryst.Solids 5, 17, 31 (1970).
29. M.H.Cohen and J.Jortner, Proc.of the Fifth Int. Conf.on Amorphous and Liquid Semiconductors, eds. J.Stuke and W.Bredig, Taylor and Francis, London, 1974.
30. G.Schönherr, Thesis Marburg 1978.
31. N.F.Mott, Philos.Mag. 6, 287 (1961).
32. J.Hubbard, Proc.Roy.Soc.(London) A281, 401 (1964).
33. H.P.Pfeifer, W.Freyland, F.Hensel, Phys.Letters 43, 111 (1973).
34. G.Franz, Diplomarbeit 1977, Marburg.
35. H.P.Pfeifer, W.Freyland and F.Hensel, Ber.Bunsenges.phys.Chem. 80, 716 (1976).
36. H.P.Pfeifer, Thesis Marburg 1977.
37. W.Freyland and F.Hensel, Ber.Bunsenges.phys.Chem. 76, 347 (1972).
38. U.Even and W.Freyland, J.Phys.F. 5, L104 (1975).
39. R.Block, J.B.Suck, W.Gläser, W.Freyland, F.Hensel, Ber.Bunsenges.phys.Chem. 80, 718 (1976).
40. R.Block, J.B.Suck, W.Gläser, W.Freyland and F. Hensel, Liquid Metals 1976, ed.R.Evans and D.A. Greenwood, The Institute of Physics, Bristol and London 1976.

41. N.W.Ashcroft and J.Lekner, Phys.Rev. 145, 83 (1966).
42. P.A.Egelstaff, An Introduction to the Liquid State, Academic Press, London and New York, 1967.
43. B.Cabane and J.Friedel, J.Phys.(Paris) 32, 73 (1971).
44. E.Grison, Journ.Chem.Phys. 19, 1109 (1951).
45. G.Tourand, B.Cabane and M.Breuil, J.non-cryst.Solids 8-10, 676 (1972).
46. K.Suzuki and M.Misawa, Liquid Metals 1976, ed. by Evans and Greenwood, The Institute of Physics, Bristol 1976.
47. H.Endo, H.Hoshino, R.W.Schmutzler and F.Hensel, Liquid Metals 1976, p.412, ed.by Evans and Greenwood, The Institute of Physics, Bristol 1976.
48. A.A.Andreev, T.Turgunov, V.A.Alekseev, Fiz.Tverd. Tela 16, 12, 3360 (1974).
49. P.Boolchand and P.Suromyi, Phys.Rev. B7, 57 (1973).
50. H.Thurn and J.Ruska, J.non-cryst.Solids 22, 331 (1976).
51. R.Fischer and R.W.Schmutzler, private communication.
52. W.W.Warren, G.Weser and F.Hensel, Ber.Bunsenges. phys.Chem. 1978 to be published.

DETERMINATION AND INTERPRETATION OF VOLUMES OF ACTIVATION

H. Kelm and D.A. Palmer

Institute for Physical Chemistry, University of Frankfurt, Robert Mayer Str. 11, D 6000 Frankfurt am Main, West Germany.

ABSTRACT. The volume of activation, as derived from the pressure dependence of the observed rate constant of a chemical reaction, is discussed in general terms, rather than a systematic treatment of various classes of reactions, which are in fact reported elsewhere (1-7). Emphasis is given to the formal evaluation of the volume of activation as well as to the numerical procedures. An attempt is also made to interpret the volume of activation in terms of a molecular concept.

I. INTRODUCTION

Using the framework of the transition state theory (8) for the description of kinetic processes, a series of quasi-thermodynamic correlations can be derived. Thus the logarithm of the rate constant, k_x, of a chemical reaction can be considered to be proportional to the free enthalpy of activation, $\Delta G^{\neq}$:

$$\ln k_x = \ln k_B T/h - \Delta G^{\neq}/RT \qquad (1)$$

The pressure derivative leads to the volume of activation, $\Delta V^{\neq}$:

$$(d \ln k_x)/dP = - \Delta V^{\neq}/RT \qquad (2)$$

If the concentrations of the reactants are not measured in mole fractions, but in molarities, Eq. (2) must be modified:

$$k_x = k_c (RT/P)^{\Delta\nu} \qquad (3)$$

$$d (\ln k_c + \Delta\nu \ln V)/dP = - \Delta V^{\neq}/RT \qquad (4)$$

Kelm, H. (Ed.) High Pressure Chemistry. 281-309. *Copyright* © 1978. *D. Reidel Publishing Company.*
All Rights Reserved.

such that:

$$(d \ln k_c)/dP = - (\Delta V^{\neq}/RT) - \Delta\nu\kappa RT \tag{5}$$

where $\Delta\nu$ is the difference in the stoichiometric indicies between the transition state and reactants and κ is the compressibility of the solvent. The product $\Delta\nu\kappa RT$ does not appear in truly first-order reactions. For second-order reactions in water this correction term amounts to 1 to 2 $cm^3\ mol^{-1}$, which in many cases represents the accuracy reached in the determination of the volume of activation.

It should be mentioned that analogous to the volume of activation, the pressure of activation, $\Delta P^{\neq}$, is derived by:

$$(d \ln k_x)/dV = - \Delta P^{\neq}/RT \tag{6}$$

The relationship between the two activation parameters is given by the following equation:

$$\Delta P^{\neq} = - \Delta V^{\neq}(dP/dV) = \Delta V^{\neq}/V\kappa \tag{7}$$

According to the fundamental equations of Gibbs and Helmholtz the following correlations between the activation parameters can be established:

$$(d\ \Delta G^{\neq}/dP)_T = (d\ \Delta H^{\neq}/dP)_T - T(d\ \Delta S^{\neq}/dP)_T \tag{8}$$

$$(d\ \Delta V^{\neq}/dT)_P = (d\ \Delta S^{\neq}/dP)_T \tag{9}$$

$$(d\ \Delta H^{\neq}/dP)_T = \Delta V^{\neq} - T(d\ \Delta V^{\neq}/dT)_P \tag{10}$$

Eqs. (9) and (10), which are commonly known as Maxwell Correlations, are also prevalent to a number of systems studied over restricted pressure and temperature ranges (9). The applicability of these correlations is often taken as an indication that the same mechanism is operating over the ranges studied.

Analogous to the temperature derivative, the second pressure derivative of the free enthalpy of activation is defined as follows:

$$\Delta \beta^{\neq} = - (d\ \Delta V^{\neq}/dP)_T = - \{(dV_{\neq}/dP)_T - \Sigma_i(dV_i/dP)_T\} \tag{11}$$

where $\Delta\beta^{\neq}$ represents the pressure dependence of the volume of activation, or the compressibility coefficient of activation. From this quantity the compressibility of activation, $\Delta\kappa^{\neq}$, can be defined:

$$\Delta\kappa^{\neq} = - (d\ \Delta V^{\neq}/dP)_T/\Delta V^{\neq} = \Delta\beta^{\neq}/\Delta V^{\neq} \tag{12}$$

Both functions involve the difference in the compressibilities of the transition state and reactants. The molecular significance of these quantities is described in section IV.

II. VOLUMES OF ACTIVATION OF COMPLEX REACTIONS

The quasi-thermodynamic quantity, $\Delta V^{\neq}$, cannot always be derived from simple rate expressions, i.e. from rate constants of essentially unidirectional processes reaching completion. In many cases the observed rate constant, k_{obs}, is a function of several constants. Various procedures can be applied to separate the individual contributions to the overall $\Delta V^{\neq}_{exp}$. In the following sub-sections some of the most common complications will be discussed and an evaluation of the volumes of activation will also be described.

1) First-Order Consecutive Reactions

For the sake of simplicity, the following series of first-order reactions has been selected:

$$A \xrightarrow{k_1} B \xrightarrow{k_2} C$$

(As long as the individual steps remain unidirectional, any second-order contribution can be converted into a pseudo-first-order one by using an excess concentration of one of the reactants). Consecutive reactions of this type are frequently encountered in the multiple hydrolysis of coordination compounds. The time dependencies of educt, intermediate, and product concentrations are given by the rate expressions:

$$- d[A]/dt = k_1[A] \; ; \; d[B]/dt = k_1[A] - k_2[B] \; ;$$

$$d[C]/dt = k_2[B] \qquad (13)$$

In order to determine the individual rate constants a knowledge of at least two of these time dependencies is necessary. These can in turn be determined at different pressures and permit the application of Eqs. (3) and (4) for the evaluation of the corresponding volumes of activation.

Several procedures have been applied to determine the individual rate constants. Among these, a graphical separation can be used in those cases where the two rate constants are numerically quite different. Such a system is shown in Fig. 1, where the logarithm of a function proportional to the concentrations of A, B and C is plotted against time. A photometric method can be

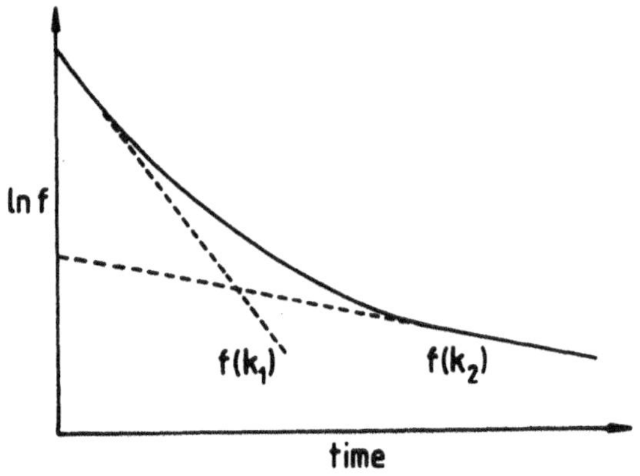

Fig. 1. The variation in the logarithm of a function, f, which is proportional to concentration, with time during a two-step, first-order consecutive reaction.

applied for the separation if the spectra of the involved species show one or more isosbestic points between either A and B, and/or B and C. Time dependencies measured at such isosbestic points enable the individual rate constants to be evaluated. An example of this spectrophotometric analysis is presented in Fig. 2.

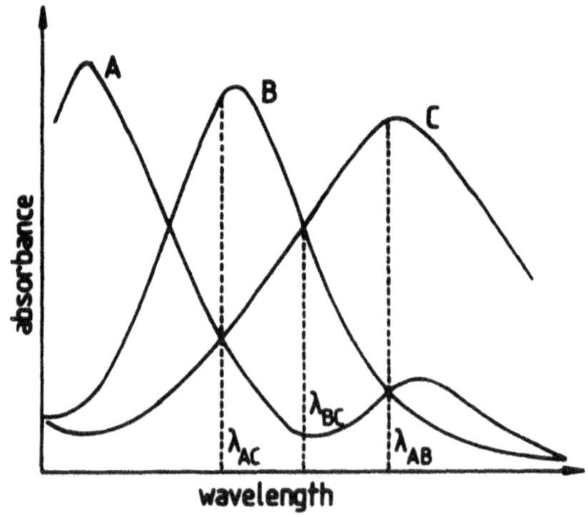

Fig. 2. Spectra of the species A, B and C, showing isosbestic points at the wavelengths, λ_{AC}, λ_{BC} and λ_{AB}.

2) First-Order Reversible Reactions

The phenomenon of a reaction not going to completion is frequently encountered. The simplest system in this respect is one which envolves opposing first-order reactions:

$$A \underset{k_{-1}}{\overset{k_1}{\rightleftharpoons}} B$$

The time dependence of the concentrations of A and B are represented by:

$$- d[A]/dt = d[B]/dt = k_1[A] - k_{-1}[B] \quad (14)$$

The observed rate constant consists of the sum of the rate constants k_1 and k_{-1}.

$$k_{obs} = k_1 + k_{-1} \quad (15)$$

Either the determination of the initial slopes of ln [A] or ln [B] versus time plots, or an independent determination of the actual equilibrium constant, $K = k_1/k_{-1}$, is necessary to evaluate the individual rate constants and from there, their pressure dependencies.

3) Second- / First-Order Reversible Reactions

These conditions prevail in systems representing association equilibria:

$$A + B \underset{k_{-1}}{\overset{k_1}{\rightleftharpoons}} C$$

$$- d[A]/dt = - d[B]/dt = d[C]/dt = k_1[A][B] - k_{-1}[C] \quad (16)$$

As can be seen from Eq. (16), the use of an excess of one reactant (e.g. $[B]_o >> [A]$) virtually reduces the rate equation to the form shown above in Eq. (15):

$$k_{obs} = k' + k_{-1} = k_1[B]_o + k_{-1} \quad (17)$$

By varing the excess concentrations of B, a plot of k_{obs} versus $[B]_o$ will yield both rate constants as illustrated in Fig. 3. A series of these plots at different pressures finally leads to the calculation of the corresponding volumes of activation according to Eq. (5).

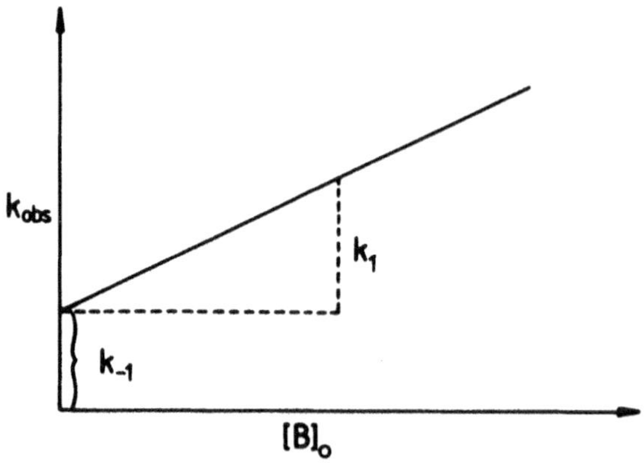

Fig. 3. The concentration dependence of the observed rate constant for a reversible first- / pseudo-first-order reaction.

4) First-Order Consecutive Reactions with a Reversible Step

This rather common situation is presented in the following scheme:

$$A \underset{k_{-1}}{\overset{k_1}{\rightleftharpoons}} B \xrightarrow{k_2} C$$

in which the initial step may also be pseudo-first-order. The time dependencies of the concentrations of these species can be written as:

$$- d\,[A]/dt = k_1[A] - k_{-1}[B] \;;\; d\,[C]/dt = k_2[B]$$

$$d\,[B]/dt = k_1[A] - k_{-1}[B] - k_2[B] \qquad (18)$$

Under Steady State Conditions involving [B] , i.e. $d\,[B]/dt \cong 0$, one obtains:

$$- d\,[A]/dt = k_{obs}[A] \text{ with } k_{obs} = k_1k_2/(k_{-1} + k_2) \qquad (19)$$

Several cases can be discussed for this system:

i) Where the rate of the reverse reaction, k_{-1}, is much slower than the forward reaction, represented by k_2. Under these conditions, which are assumed to exist at high pressure, the measured rate-determining step and thus the volume of activation can be exclusively ascribed to the k_1 process.

$$k_{-1} << k_2 \;;\; k_{obs} = k_1 \;;$$

$$(d \ln k_{obs})/dP = - \Delta V_1^{\neq}/RT \qquad (20)$$

ii) In the case when the rates between A and B are fast compared to the consecutive reaction, a typical pre-equilibrium is established and the experimentally observed rate constant is the product of a rate constant and an equilibrium constant.

$$k_1 \cong k_{-1} >> k_2 \;;\; k_{obs} = k_1 k_2/k_{-1} = k_2 K \;;$$

$$(d \ln k_{obs})/dP = - \Delta V_2^{\neq}/RT - \Delta \bar{V}/RT \qquad (21)$$

iii) Should the rates of the forward and reverse reactions be comparable, a composite expression for $\Delta V^{\neq}$ is obtained and must be treated in detail. An example of this are radical decay-recombination processes (10).

$$k_1 \cong k_{-1} \cong k_2 \;;\; k_{obs} = k_1 k_2/(k_{-1} + k_2) \;;$$

$$(d \ln k_{obs})/dP = - \Delta V_1^{\neq}/RT - \Delta V_2^{\neq}/RT - (d \ln(k_{-1} + k_2))/dP \qquad (22)$$

5) Parallel Reactions

Systems in which a substance reacts via two parallel, unidirectional pathways may be characterized by the following scheme:

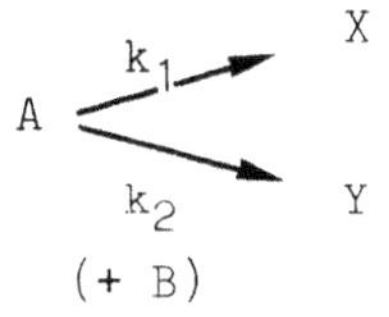

The variation in the concentration of A with time results in the typical equation for k_{obs}:

$$- d[A]/dt = k_1[A] + k_2[A][B] = (k_1 + k_2[B])[A]$$

$$k_{obs} = k_1 + k_2[B] \qquad (23)$$

Here again, plots of k_{obs} versus [B] yield the individual rate constants. A variation of pressure provides the data for the evaluation of the respective volumes of activation, as shown in Fig. 4.

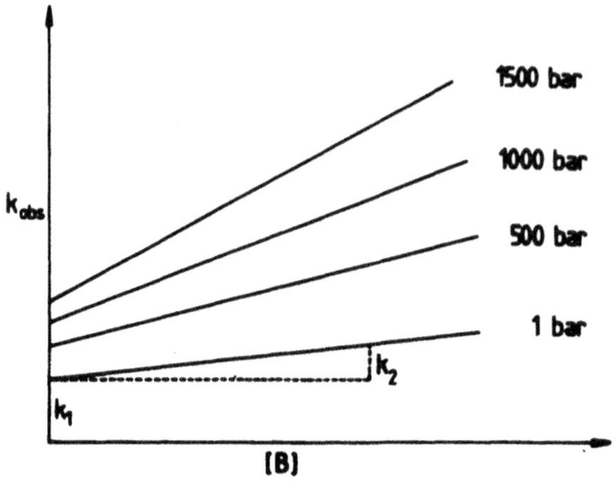

Fig. 4. The concentration dependence of the observed rate constant for two parallel first- and second-order reactions.

6) Chain Reactions

Understandably, composite rate constants are obtained for chain reactions. Examples of these are polymerization reactions which are treated in the article by Jenner in this volume.

The experimentally obtained volume of activation consequently contains several contributions which can sometimes be individually determined, as in the case of the production of initiator radicals.

7) General Remarks

The general shape of the ln k_{obs} versus pressure plot may provide valuable information concerning the nature of the prevailing rate processes and their pressure dependencies. Within this context a minimum, as shown in Fig. 5, is a strong indication of two parallel reactions with counteracting pressure dependencies. On the other hand a maximum could well be caused by additional effects, such as viscosity. This typical behaviour is presented in Fig. 6 where the rate of a second-order reaction becomes limited by diffusion in an increasingly more viscous medium.

III. DATA PROCESSING IN HIGH PRESSURE KINETICS

Pressure dependencies of rate constants may be linear or nonlinear with either positive or negative slopes. Some generalized examples are shown in Fig. 7.

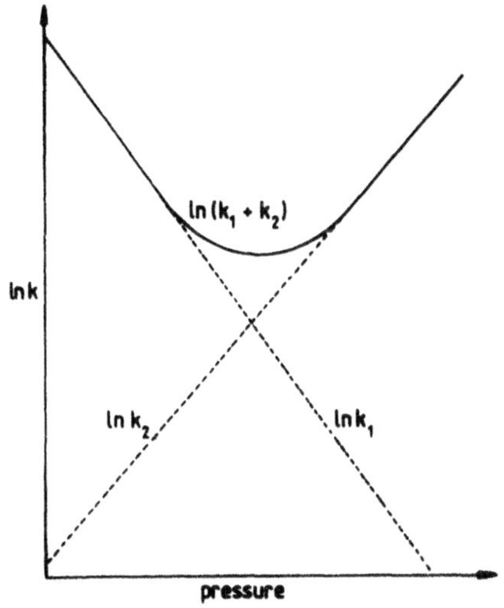

Fig. 5. The pressure dependence of the logarithm of an observed rate constant indicating the presence of two competing rate constants k_1 and k_2.

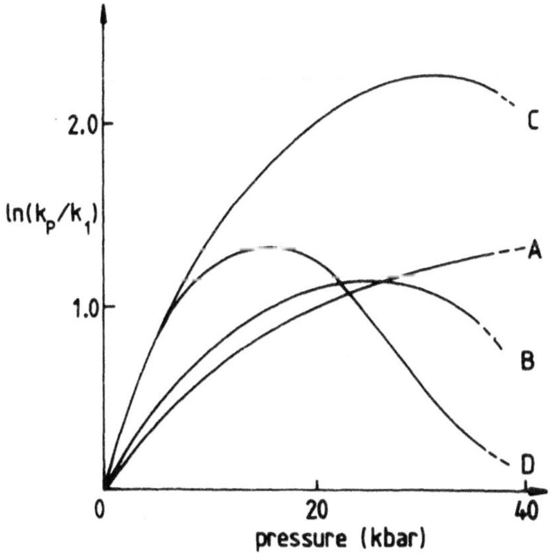

Fig. 6. The effect of pressure on the rate of reaction between ethyl bromide and alkoxide ions, where: A ≡ MeO^- in MeOH; B ≡ iso-PrO^- in iso-PrOH; C ≡ Eugenoxide ions in iso-PrOH; D ≡ Eugenoxide ions in 1:1 Eugenol and iso-PrOH.

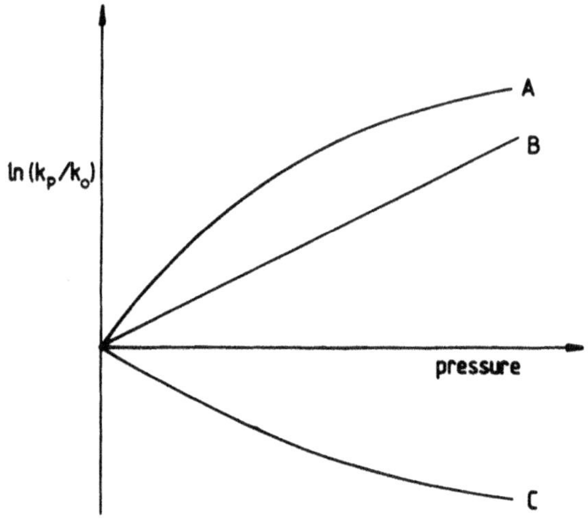

Fig. 7. Various features of the pressure dependencies of the logarithm of rate constants as exemplified by: A, Cylcoaddition Reactions; B, Ligand Substitution Reactions; C, Homolytic Scission Reactions.

A completely theoretical description of the dependence of ln k on pressure is not yet available. Therefore a number of empirical procedures have been proposed (2,12,13) for the determination of the slopes of these curves. Some of these procedures will be discussed and compared in this chapter.

1) Linear Function

Assuming the function F, for the description of ln k, is linear with pressure:

$$F = \ln k = A + BP \tag{24}$$

$$F = \ln k = \ln k_o + BP \tag{24a}$$

Here it is implied that the volume of activation is pressure independent, i.e. the compressibilities of the reactants and transition state are either zero or simply cancel out. Eq. (24a) requires a forced intercept of the straight line equal to the rate constant at normal pressure, represented as $\ln k_o$, thereby emphasizing this value. Justification of such a treatment must be made individually.

In many cases where linear relationships appear to occur, the pressure range normally used should be extended to ensure that linearity persists. Nevertheless, volumes of activation calculated on the assumption of a linear dependence represent minimum absolute quantities which are often sufficient for mechanistic discussions.

2) Quadratic Function

Two very similar functions can again be used:

$$F = \ln k = A + BP + CP^2 \tag{25}$$

$$F = \ln k = \ln k_o + BP + CP^2 \tag{25a}$$

Here again they differ only in the derivation of the intercept, which retains the above mentioned consequences.

This treatment recognizes that $d\,\Delta V^{\neq}/dP \neq 0$, but does assume that $d^2\,\Delta V^{\neq}/dP^2 = 0$.

$$(d \ln k_p/k_o)/dP = B + 2CP = -\Delta V^{\neq}/RT \tag{26}$$

$$-d\,\Delta V^{\neq}/dP = \Delta\beta^{\neq} = 2RTC \tag{27}$$

Such a treatment either implies that the isothermal compressibilities of the reactants and transition state are pressure independent, or that the difference between these quantities for the sum of the educts and transition state must be zero. The first alternative is generally known not to be the case and the second would be fortuitous, especially for reactions in which the polarity increases substantially during the development of the transition state.

There is no physical justification for the use of the quadratic function. Nevertheless, it is quite often applied and considered by many to best describe the data obtained. Its mathematical treatment is certainly simple enough. The volume of activation at zero pressure equals -RTB and its pressure derivative, - 2RTC.

The linearization of the quadratic function is accomplished as follows:

$$1/P\,(\ln (k_p/k_o)) = B + CP \tag{28}$$

and often shows the weakness of this function, especially at higher pressures. Eckert (5) could impressively demonstrate this for a Menschutkin reaction, as shown in Fig. 8.

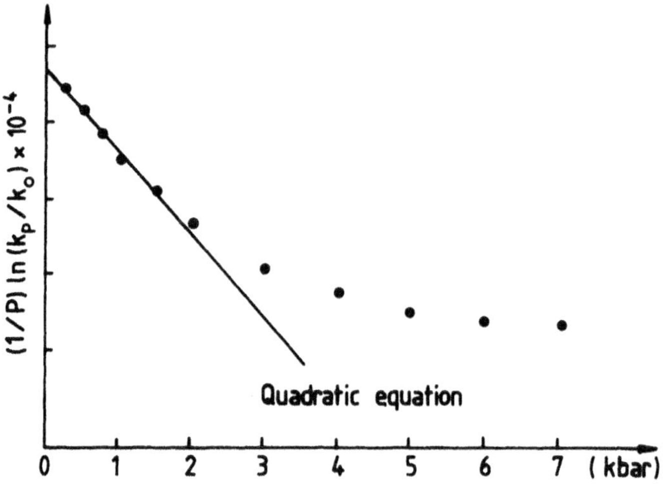

Fig. 8. The pressure dependence of the reaction between methyl iodide and pyridine in acetone at 50°C. (Ref. (5))

Another drawback of the quadratic equation appears to be its changing emphasis and influence on the slope at zero pressure. With increasing pressure and the inclusion of more points at higher pressures, the slope at zero pressure decreases, as can be seen from the results given in Table 1. A systematic reduction in the relative weight of points near the interesting region of $P = 0$ tends to lead to a decrease in the absolute value of $\Delta V^{\neq}$. It should be mentioned that several workers (2-4) in the field of

Table 1. Effect of the pressure range on $\Delta V^{\neq}$ for the cycloaddition reaction between cyclopentadiene and dimethylacetylenedicarboxylate in ethyl acetate at 10°C. (Ref. 5)

N° of Points	Upper Pressure kbar	$\Delta V_o^{\neq}$ $cm^3\ mol^{-1}$	Standard Deviation
6	1.04	-31.3	1.2%
7	1.38	-30.2	1.5%
8	2.07	-28.2	2.2%
9	3.10	-27.0	2.5%
10	4.14	-26.9	2.3%
11	5.17	-25.2	3.8%
12	6.21	-23.3	5.9%

pressure kinetics have recommended that more emphasis be given to data collection at the low pressure region.

Some authors have used a third-order polynomial equation to represent their data. Apart from the lack of justification of this treatment (extremum behaviour), it does not bring any significant improvement over the use of the quadratic equation.

3) Fractional Function

A fractional function may also be used for the presentation of data.

$$F = \ln k = A + BP/(C + P) \qquad (29)$$

The intercept A can again be substituted by the experimental value of ln k_o or taken as an adjustable quantity. Similar to isothermal adsorption curves, the parameter B affects the amplification while C modifies the pressure saturation. It is safe to assume that at pressures larger than 100C the function becomes essentially independent of pressure, which means that (d ln k/dP) becomes virtually zero. On the basis of a molecular interpretation of $\Delta V^{\neq}$ this is not acceptable. However at moderate pressures (i.e. in the curved region) the function often quite adequately describes the measured data and presents a convenient way of calculating the volume of activation and the compressibility of activation through computer adjustment of B and C:

$$(d \ln k)/dP = -\Delta V^{\neq}/RT = CB/(C + P)^2 \qquad (30)$$

$$d\,\Delta V^{\neq}/dP = -\Delta \beta^{\neq} = 2B/C^2 \qquad (31)$$

A similar equation was proposed by El'yanov and Hamann (14) for the pressure dependence of equilibrium constants of weak acids. It can also be applied for the estimation of the pressure dependence of $\Delta V^{\neq}$:

$$F = \ln k = \ln k_o - \Phi \Delta V_o^{\neq}/T \qquad (32)$$

$$F = \ln k = \ln k_o - (\Delta V_o^{\neq}/RT)P/(1 + BP) \qquad (33)$$

where $\Delta V_o^{\neq}$ is the volume of activation at zero pressure.

4) Exponential Function

Another description of the asymptotic behaviour of ln k versus P plots has been used in the form of equation (34)

$$F = \ln k = A + B[1 - \exp(-CP)] \tag{34}$$

Again B acts as an amplification and A shifts the pressure range of saturation. At high pressures F approaches a constant value thereby involving the same problems already outlined for the fractional function.

5) Tait-like Function

As early as 1888 Tait proposed an empirical equation for the compressibilities, κ, of organic liquids:

$$\kappa = (V_o - V_p)/V_o = C \log[(B + P)/P] \tag{35}$$

$$V_p = V_o - V_oC \log[(B + P)/(B + 1)] \tag{36}$$

Here V_o and V_p represent the molar volumes of liquids at zero pressure and pressure P, respectively, while B and C are empirical constants. Thus the second term in Eq. (36) describes the change in molar volume with pressure. Applied to reactants, R, and transition state, TS, one obtains an equation for the volume of activation which separates the effect at zero pressure from the compressibility effects at higher pressures.

$$\begin{aligned} F = \Delta V_p^{\neq} &= V_p^{\neq} - \Sigma V_p^R \\ &= \Delta V_o^{\neq} - \{V_o^{TS}C^{TS}\log[(B^{TS} + P)/(B^{TS} + 1)] \\ &\qquad - \Sigma V_o^R C^R \log[(B^R + P)/(B^R + 1)]\} \end{aligned} \tag{37}$$

Application of this concept implies the following:

i) If the reaction is affected by pressure (i.e. the molar volume of the transition state is different from the sum of the molar volumes of the reactants), then the volume of activation must vary with pressure. A plot of ln k versus P will not be linear.

ii) The mathematical description of this pressure dependence predicts negative volumes of activation at high pressures.

iii) It is assumed that the Tait equation, found to be a good approximation for pure liquids, also applies to the partial molar volumes of the transition state and reactants.

When the Tait equation was applied to pure organic liquids, a nearly constant value for C of ca. 0.215 emerged. Benson and Berson (16) introduced this constant C value into Eq. (37) assuming that solutions of reactants and transition state also show this phenomenon. In order to approximate the function

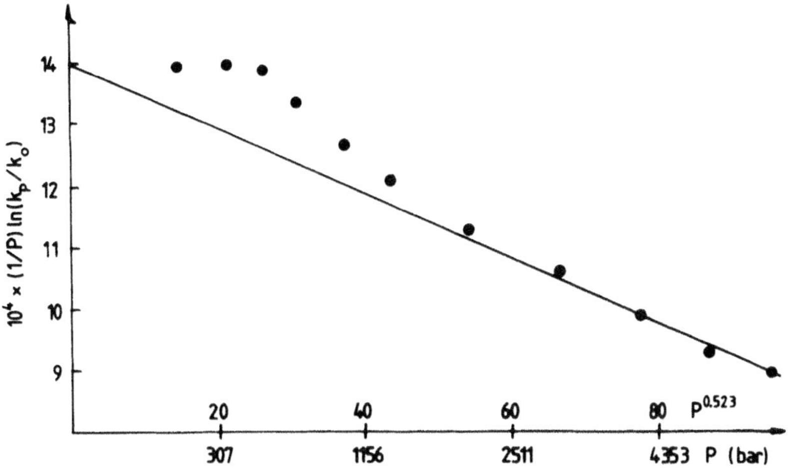

Fig. 9. Benson and Berson treatment of the pressure dependent rate data for the reaction of isoprene with maleic anhydride in ethyl acetate at 35°C taken from Ref. (17).

F = {VC log [(B + P)/B + 1)]} for reactants and transition state in Eq. (37), these authors used an exponential expression, $F(P) = (\mathrm{const})P^n + Q$. Such a function is virtually linear in the range, $0.3 < \log F(P) < 1.5$, which in turn corresponds to the pressure range, $2 < P < 10$ kbar. The best value for n was found to be 0.523. As a result, they proposed the following equation as a possible description of non-polar reactions:

$$F = (1/P) \ln (k_p/k_o) = A + BP^{0.523} \tag{38}$$

This is known as the Benson-Berson Equation. For an estimation of the volume of activation at zero pressure (i.e. the value of A), a rather long extrapolation is necessary, as can be clearly seen from Fig. 9.

6) Graphical Methods

In order to avoid that any unjustified mathematical preconception be imposed on the data, graphical methods can be applied to evaluate the volumes of activation. Two typical approaches used in the literature are demonstrated in Fig. 10 and 11.

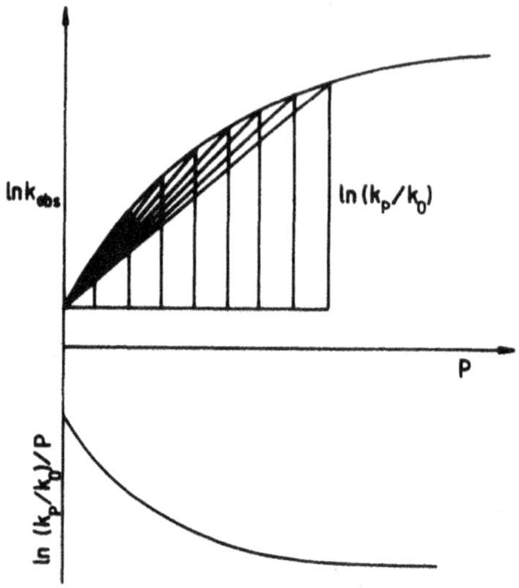

Fig. 10. Graphical slope analysis of ln k_{obs} versus pressure data. An extrapolation of the slope increments to P = 0 yields $\Delta V_0^{\neq}$.

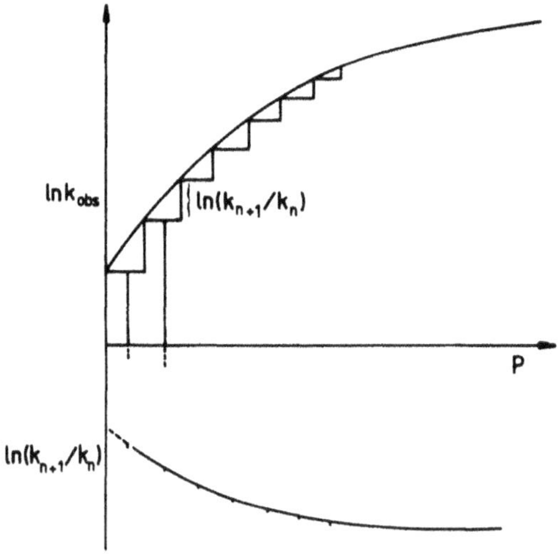

Fig. 11. A graphical analysis of ln k_{obs} versus pressure data using an incremental slope analysis. Using small pressure increments the initial slope of ln (k_{n+1}/k_n) versus pressure gives $\Delta V_0^{\neq}$.

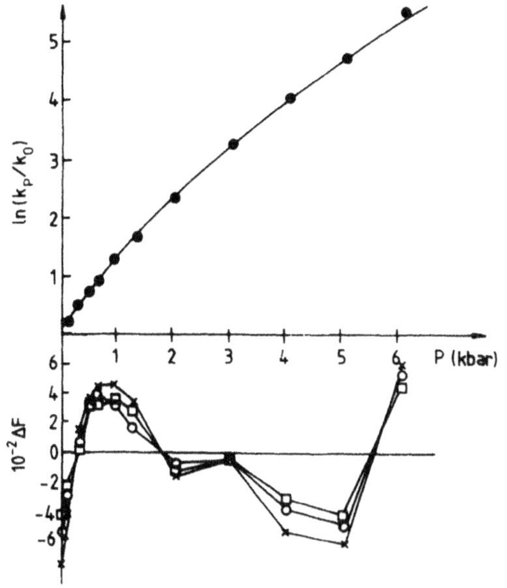

Fig. 12. Cycloaddition reaction between isoprene and maleic anhydride in ethyl acetate at 35°C (taken from Ref. (17)).

×	$F = A + BP + CP^2$	$\Delta V_0^{\neq} = -30.3\ cm^3\ mol^{-1}$
O	$F = A + B[1 - \exp(-CP)]$	$= -31.7$ " "
□	$F = A + BP/(C + P)$	$= -32.6$ " "
	Benson-Berson	$= -36.4$ " "

7) Comparison of Methods

For a comparison of the different methods two examples have been chosen which cover different ranges of pressure. Firstly, a cycloaddition reaction which was studied up to 6 kbar by Grieger and Eckert (17) is presented in Fig. 12. Since the fitted curves which result from the various functions are rather similar, a hand-drawn curve is shown. The deviations between the experimental and calculated points for the individual curve-fitting processes are also plotted as a function of pressure. The resulting volumes of activation are listed and are essentially similar. It can be seen that for the different processes similar deviations, ΔF, appear which would tend to indicate a systematic error in the data points.

The second example involves a ligand substitution reaction of a platinum(II) complex (18). These data were treated in the same manner and the results are shown in Fig. 13. Similar systematic trends can be observed, although to a lesser extent.

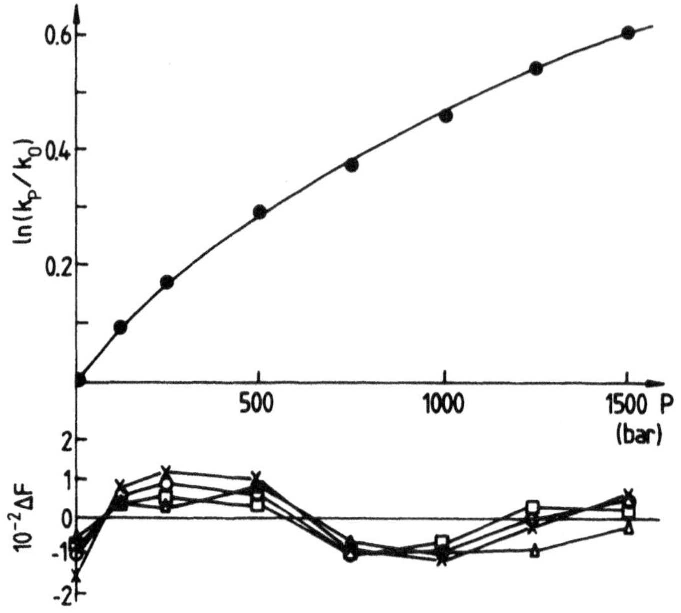

Fig. 13. Ligand substitution reaction involving $Pt(dien)Cl^{+}$ and OH^{-} in aqueous solution (taken from Ref. (18)), where dien = diethenetriamine .

X	$F = A + BP + CP^2$	$\Delta V_0^{\neq} = -14.7 \pm 0.8\ cm^3\ mol^{-1}$
O	$F = A + BP + CP^2 + DP^3$	$= -17.9 \pm 1.0$ " "
□	$F = A + B[1 - \exp(-CP)]$	$= -16.2 \pm 2.0$ " "
Δ	$F = A + BP/(C + P)$	$= -17.3 \pm 0.5$ " "

As a rule none of the applied functions appear to be superior, although it can be said that the quadratic function generally underestimates the value of $\Delta V_0^{\neq}$. Eckert (13) has recently stressed this point and shown that except for the quadratic function, application of the remaining functions is justified up to ca. 10 kbar (i.e. corresponding to a compression of most organic solvents of less than 30%).

IV. INTERPRETATION OF VOLUMES OF ACTIVATION

1) General Description of the Volume of Activation

The quantity, $\Delta V_0^{\neq}$, which is derived from the pressure dependence of the rate constant, documents, within its error limits, the difference between the partial molar volume of the transition state and the sum of the partial molar volumes of the educts. Since the partial molar volumes of the educts and - if so desired -

of the products can generally be determined by precision density measurements, one can assign a numerical value to the volume of the transition state and place it within a volume profile of the reaction. It is therefore understandable that one is able to explain such volume quantities in terms of the individual molecular processes and the consequences thereof.

Contrary to the thermodynamic treatment, where $\Delta V^{\neq}_{exp}$ is thought of in terms of the pressure derivatives of $\Delta H^{\neq}$ and $\Delta S^{\neq}$ (Eqs. (8-10)), one is here inclined to conceive $\Delta V^{\neq}_{exp}$ as a combination of an intrinsic and a solvation part (19).

$$\Delta V^{\neq}_{exp} = \Delta V^{\neq}_{intr} + \Delta V^{\neq}_{solv} \qquad (39)$$

The intrinsic part results from nuclear displacements within the reactants during the formation of the transition state. Simply stated, it reflects the stretching of a bond during a dissociative process or the formation of a new bond during an associative step. If the change in overlap of the van der Waal's spheres is taken as a first approximation of the intrinsic contribution, then the latter can be estimated on a molar basis, see Fig. 14. Thus bond formation between two spheres of radius 3 Å results in a volume of activation of ca. $-12\ cm^3\ mol^{-1}$.

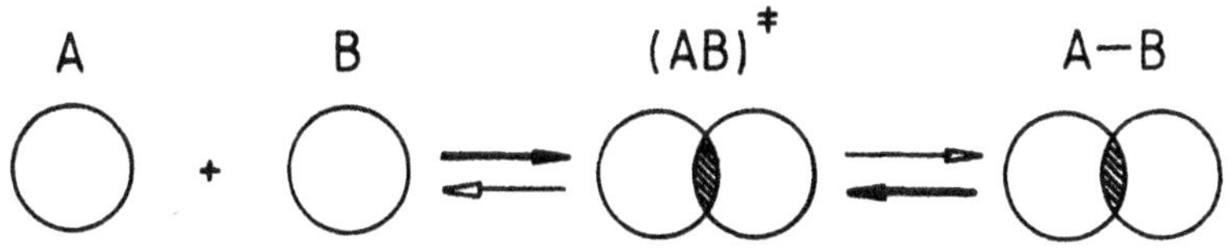

Forward reaction - Pressure acceleration: $-\Delta V^{\neq}_{intr}$.

Reverse reaction - Pressure retardation: $+\Delta V^{\neq}_{intr}$.

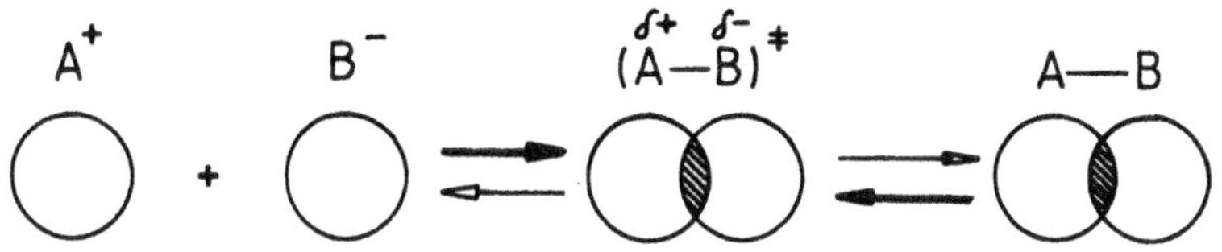

Forward reaction - Pressure retardation: $+\Delta V^{\neq}_{solv}$.

Reverse reaction - Pressure acceleration: $-\Delta V^{\neq}_{solv}$.

Fig. 14. Simplified hard-sphere models which illustrate the contributions to $\Delta V^{\neq}_{exp}$.

Experience has shown that bond formation and bond cleavage involving the same kind of atoms, studied for instance in an equilibrium, does not always result in the same absolute value for the volume of activation. This can be taken as an indication that the transition state must not necessarily lie in the center of the reaction coordinate.

Not only reduced or increased overlap of the van der Waals radii of the reaction centers are involved in determining the magnitude of $\Delta V^{\neq}_{exp}$. Steric crowding by neighbouring groups can also contribute to $\Delta V^{\neq}_{intr}$. The typical nonpolar hydrogen-transfer reaction (20) between diphenyl-picryl-hydracyl radicals (DPPH·) and thiol derivatives (HSR) to form diphenyl-picryl-hydrazine (DPPHH) and a thioxyl radical (RS·) demonstrates that increased steric crowding is manifested in more negative $\Delta V^{\neq}_{exp}$ values, as shown in Table 2.

Table 2. Volumes of activation of hydrogen-transfer reactions in toluene. (Ref. 20)

$$\text{DPPH}\cdot + \text{HSR} \longrightarrow \text{DPPHH} + \text{RS}\cdot$$

HSR	Temperature °C	$\Delta V^{\neq}_{exp}$ $cm^3\ mol^{-1}$
C_4H_9SH	35	- 13
$C_6H_{13}SH$	30	- 20
tert-$C_8H_{17}SH$	35	- 21
tert-C_4H_9SH	35	- 29

Changes in polarity during the formation of the transition state give rise to volume effects which are represented by the $\Delta V^{\neq}_{solv}$ term. The development of charges or polar moments during the reaction increase the strength of the interactions between reactants and solvent, thereby resulting in a negative contribution to $\Delta V^{\neq}_{solv}$. Naturally, a quenching of charge or polar moments leads to the opposite effect. A typical reaction in which charge is partially developed is the Menschutkin reaction, i.e. the quaternization of tertiary amines. A reaction of this type has been investigated (21) as a function of pressure in a number of solvents. These results, which are listed in Table 3, show that $\Delta V^{\neq}_{exp}$ is very sensitive to the nature of the solvent (i.e. $\Delta V^{\neq}_{solv}$ represents a very significant contribution to $\Delta V^{\neq}_{exp}$).

Table 3. Solvent dependence of the volume of activation for the reaction of methyl iodide with pyridine at 50°C. (Ref. 21)

$$C_5H_5N + CH_3I \longrightarrow C_5H_5NCH_3^+ + I^-$$

Solvent	$\Delta V^{\neq}_{exp}$ cm^3 mol^{-1}
Carbon Tetrachloride	- 37.5
Benzene	- 35.1
Acetone	- 33.2
Chlorobenzene	- 30.7
Methanol	- 26.6
Nitrobenzene	- 21.3

For a discussion of reaction mechanisms the intrinsic part of the volume of activation is of primary interest. It is therefore desirable to describe the solvation term quantitatively, if possible, in order to establish the intrinsic term.

2) Estimations of the Solvation Contribution

The task of estimating the solvation contribution to the volume of activation is equivalent to solving the problem of quantitatively describing the solvent effect on rates of reactions. The latter has not yet been accomplished and as a consequence, an estimation of $\Delta V^{\neq}_{solv}$ must remain a crude approximation.

If one accepts the existence in solution of dispersion, repulsion and electrostatic forces, as well as specific interactions caused by functional groups (e.g. hydrogen bonding), then these approximations can be made.

The volume change incurred on mixing two nonpolar liquids, which do not undergo any specific interactions, is usually less than 1 to 2 cm^3 mol^{-1}. Thus neglecting the effect of dispersion forces is justified, since the accuracy normally reached in high pressure kinetics is of the same order of magnitude.

Interactions stemming from special functional groups, such as hydrogen bonding, donor-acceptor complex formation or hydrophobic interactions, are difficult to assess in terms of volume changes. For instance, hydrogen bonding is expected to result in a negative contribution due to bond formation. However, should this be an intramolecular hydrogen bond, then ring formation will occur, which is normally accompanied by a positive volume change. Both contributions counteract so that in many cases the overall effect is unpredictable.

Repulsive forces are usually taken into account by using finite radii for the involved species in which charges or electric moments are embedded. In this connection, Whalley (22) postulates cavities which have the same properties as the bulk solvent (i.e. they are compressible and expandable).

The most significant contribution to $\Delta V^{\neq}_{solv}$ stems from electrostatic interactions involving ionic charges, dipole or quadrupole moments. Considering the solvent medium as a continuum characterized by a dielectric constant and the solute species as spheres or spherical cavities with a dielectric constant of unity contained within them, also carrying point-like charges and/or moments at their centers, an expression for the free enthalpy of the electrostatic interactions can be derived:

$$\Delta G_{el} = -\underset{\text{(ionic)}}{(Ne^2z^2/2a_o)(1 - 1/D)} - \underset{\text{(dipolar)}}{(N\mu^2/a_o^3)[(D - 1)/(2D + 1)]}$$

$$- \underset{\text{(quadrupolar)}}{(3Nq^2/2a_o^5)[(D - 1)/(3D + 2)]} \qquad (40)$$

The corresponding pressure derivative amounts to:

$$\begin{aligned}\Delta V_{el} = &-(Ne^2z^2/2a_o)\{[d(1/D)/dP] + (1 - 1/D)(d\ \ln a_o/dP)\}\\ &-(3N\mu^2/2a_o^3)\{(1/2)[d(1/D)/dP] + (1 - 3/2D)(d\ \ln a_o/dP)\}\\ &-(5Nq^2/2a_o^5)\{(1/3)[d(1/D)/dP] + (1 - 5/3D)(d\ \ln a_o/dP)\}\end{aligned} \qquad \text{..........(41)}$$

This procedure assumes that charges and moments are independent of pressure. There is little information available concerning this point. However, a study in our laboratory showed that the dipole moment of acetonitrile did not change up to 1.6 kbar.

Model calculations by Whalley (22), which assume a cavity radius, a_o, of 3 Å; unit charges; dipole moments of 10 D; and quadrupole moments of (2×10^{-25}) e.s.u., gave the volume quantities shown in Table 4.

Neglecting the pressure dependence of a_o, one can use the remaining functions, including the pressure dependencies of the dielectric constant, as descriptions of the $\Delta V^{\neq}_{solv}$ term.

Table 4. Volumes of solvation due to electrostatic interactions. (volumes in cm^3 mol^{-1})

Solvent	Charge (1e)	Dipole (10 D)	Quadrupole Moment (2×10^{-25} e.s.u.)
Water	1.3	1.8	1.4
Methanol	4	6	4
Acetone	7	10	7
Benzene	27	39	28

Certainly this treatment can only be applied to strongly polar reactions in which electrostatic interactions predominate.

For the Menschutkin reaction mentioned earlier in Table 3, the function for the dipolar interactions can be used as a solvent-characterizing parameter. A plot of $\Delta V^{\neq}_{exp}$ versus this parameter yields a fairly good straight line, as can be seen from Fig. 15.

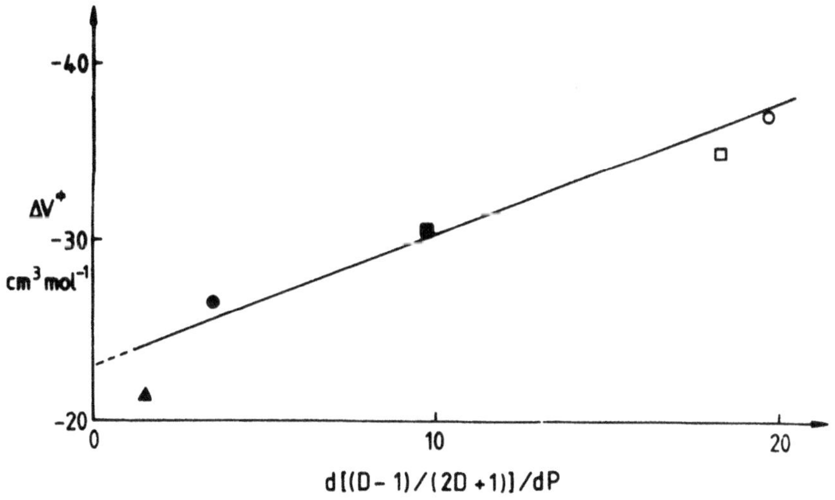

Fig. 15. Solvent dependence of $\Delta V^{\neq}_{exp}$ for the Menschutkin reaction between methyl iodide and pyridine at 50°C, where ○ carbon tetrachloride; □ benzene; ■ chlorobenzene; ● methanol; ▲ nitrobenzene.

The intercept in such a plot can be taken as an estimation of the intrinsic contribution to the volume of activation, $\Delta V^{\neq}_{intr}$, and in this case indicates that bond making is rate determining. The quantities then obtained by subtracting $\Delta V^{\neq}_{intr}$ (here ca. -22 $cm^3\ mol^{-1}$) from the measured $\Delta V^{\neq}_{exp}$ values originate from electrostriction of the respective solvents. It is worth mentioning that solvents of lower polarity show larger electrostriction effects. Pushing this treatment to the limit, one could use the slope of the line to estimate the dipole moments involved. Indeed, in the present example a rather reasonable value of ca. 7 D is obtained for the transition state.

This admittedly crude treatment of the volume of activation data has been shown to be useful in other cases in which relatively drastic changes in polarity could be assumed, e.g. the two-step cycloaddition reactions between tetracyanoethylene and enol ethers.

There have also been attempts made to describe the solvent interactions by physical properties of model compounds (probes). For highly polar transition states the charge transfer transition energy, E_T, of a betaine molecule was used (23,24).

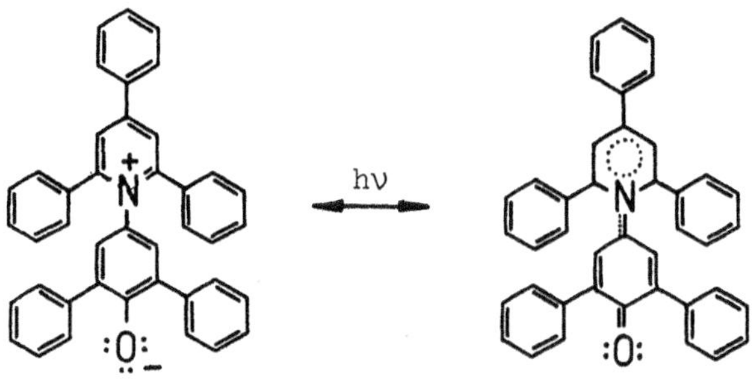

Due to the varying degrees of stabilization of the ground state of the betaine in different solvents, a solvatochromic shift of E_T is observed at normal pressure. It is therefore conceivable that a further shift of E_T can be induced by the application of pressure (i.e. a dν/dP parameter, also referred to as E_P). These E_P parameters could then be considered proportional to $\Delta V^{\neq}_{solv}$, with the advantage that E_P should reflect solvational perturbations in the neighbourhood of the probe rather than in the bulk solvent. Unfortunately, the effect of pressure on E_T proved to be very small (25) and generally unsystematic. One possible explanation of this is that the internal pressure created by the interaction of the polar betaine molecule in its ground state with the solvent is sufficiently large at 1 bar that additional hydrostatic pressure makes little influence on the overall energetics of the system. The applied pressures were within the range 1 to 3000 bar.

3) Pressure Susceptibility of Dissolved Species

The electrostrictive approach used above implies that a_o is pressure independent. In other words, it is assumed that molecular moities are not influenced by moderate pressures, with respect to their bond lengths and angles. The observation that the dipole moment of a dissolved species is apparently not influenced by pressure partially establishes this fact.

From its basic principles, IR spectroscopy is one of the most informative experimental techniques concerning bond energies and angles. Therefore further valuable information could be obtained from IR pressure studies of solutions. With this aim in mind we studied (26) the pressure dependence of the OH stretching frequencies of a series of substituted phenols in carbon disulphide. In fact the first overtone of the stretching frequency was investigated as it should be even more sensitive to the effects of applied pressure. Here again, the inherent problem in such an investigation is the separation of the solvent effects, which are possibly due to hydrogen bonding, from the actual effect on the molecule of interest. However, by increasing the steric hindrance around the OH-group it was hoped that this would reduce its accessibility to solvent molecules. It was found that ortho-substitution by tertiary butyl groups did indeed reduce the pressure shift by a factor of about 2. This reduced value, ca. 10 Å $kbar^{-1}$, towards longer wavelengths, can well be interpreted as a residual solvent influence rather than a distortion of the molecule as such a distortion would be expected to show a shift to higher energies. Therefore it was again concluded that no significant change in frequency occurred at moderate pressures which could be related to distortion of the solute.

While pressure effects on the infrared spectra of reactants can be, in principle, obtained and interpreted, those of transition states are inaccessable. Only differences between transition state properties and those of the reactants can be obtained by kinetic measurements. From the conclusion drawn above that the reactants are not affected intrinsically by moderate pressure, as exemplified in the case of phenols, pressure effects on the transition state should be observable in the kinetic parameters when they originate from different vibrational levels. Such a situation is provided by studying the primary isotope effect. In other words, if pressure distorts the energy profile (i.e. the Morse function) of the transition state in a different manner from that of the ground state, then the zero point energy levels for the hydrogen species will be more affected than for deuterium. As a result, a pressure dependence of the primary isotope effect obtained and, for a suitable reaction, $\Delta V^{\neq}_{exp}$(H) should differ from $\Delta V^{\neq}_{exp}$(D). Using the hydrogen transfer reaction between diphenyl-picryl-hydracyl, as acceptor, and, as donors, those substituted phenols which were shown to be pressure independent with respect to their OH vibrations, no influence of pressure on the isotope effect could be detected (26). For the reactions of DPPH with 2,4,6-tri-tert-butyl-, 2,6-di-tert-butyl-, and 2,6-di-tert-butyl-4-methyl-phenol, the $\Delta V^{\neq}_{exp}$(H) and $\Delta V^{\neq}_{exp}$(D) values turned out to be - 13.5 ± 1 cm^3 mol^{-1}). Thus the conclusion can be drawn that no detectable distortion of either the reactants or transition states in solution occurs at moderate pressures.

In several cases the compressibilities of solid materials were used as approximate guide lines for the compressibilities of dissolved species. Needless to say, the compressibilities which result from lattice compressions, phase transitions, etc., do not give the desired information. On the other hand, spectral data originating from transitions which are determined by bond lengths and bond angles could well render useful information. Next to IR spectroscopy, the d-d transitions of coordination compouns can serve as a suitable probe. It is well known that these ligand field bands depend on the metal-ligand bond length. Thus a compression or shortening of these distances will manifest itself in the spectrum. With this in mind, Stranks (6) calculated the degree of compression of a dipositively charged octahedral complex and found it to be << 1 cm^3 mol^{-1} (i.e. well within the accepted error limits for volumes of activation).

At the present time it seems fairly safe to assume that dissolved species are only slightly, or not at all, affected by moderate pressures. However, this does not include the solvation sheeths and near-neighbour configurations. Consequently, the assumptions that ionic charges, electrostatic moments and intrinsic radii are pressure independent in the equations discussed in the preceeding section, appear to be justified. This results bring a substantial simplification of those correlations, thereby making them more widely applicable.

4) Pressure dependence of volumes of activation.

In a thermodynamic treatment, a pressure dependence of the volume of activation indicates that the difference in partial molar volumes of the transition state and reactants changes with pressure. The difference may be caused by the transition state, the reactants, or both. Only a combined determination of the partial molar volumes of the educts at different pressures and high pressure kinetics will allow one to establish the exact origin of the effect. The major problem is to find a technique to determine partial molar volumes, or changes therein, with sufficient accuracy at high pressures. Therefore, only a few such extensive studies have been undertaken.

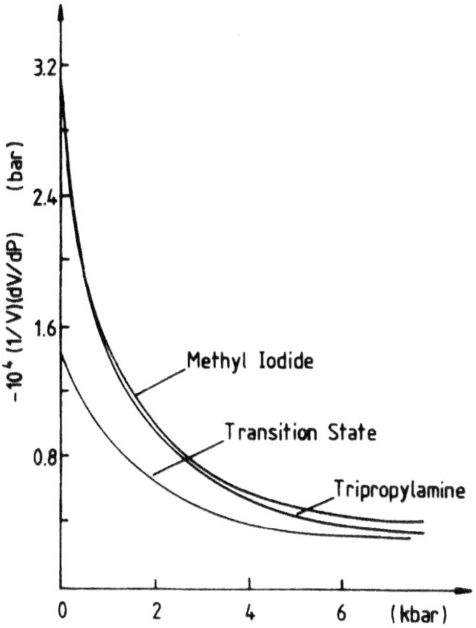

Fig. 16. Compressibilities of the reactants and transition state of a Menschutkin reaction as a function of pressure

Eckert (13) was able to show that for a Menschutkin reaction that both reactant and transition state partial molar volumes change in a similar way with increasing pressure, as depicted in Fig. 16. This suggests that the pressure dependence is caused by the solvent contribution. Here again, each partial molar volume can be considered to consist of an intrinsic part and a solvation part. Taking for granted that the intrinsic contribution is incompressible, the observed compression must be ascribed to solvation effects. For this reason the pressure dependence of the volume of activation can often be correlated with the compressibility (or molar volume changes) of the solvent.

Even the application of such approximate relationships as the Kirkwood equation seems to hold at high pressures. Such a test was also made for a Menschutkin reaction (21).

The situation is more complicated when a solvent molecule acts as a reactant as in solvolysis reactions. In cases of metal ions in solution (6,27), the situation can be described with the help of the model shown in Fig. 17, which shows the centrosymmetric solvation shells surrounding a trivalent cation in water. In the immediate neighourhood of the ion is a well defined coordination sphere of ligands L which may also be solvent molecules. Second

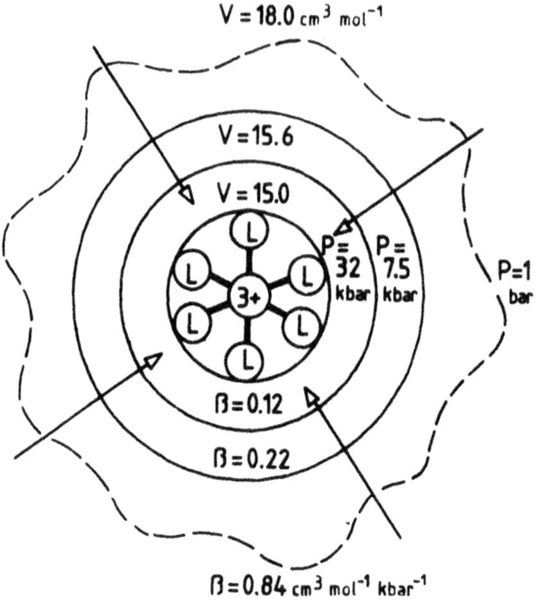

Fig. 17. Solvation of a trivalent metal cation in water, where V represents the partial molar volume of water; ß is the compressibility coefficient of water; and P is the internal pressure exerted on the solvent.

and third coordination spheres of solvent molecules orientated towards the charge center, also possibly containing other species such as anions, are also thought to exist. Beyond these ordered regions lies the bulk solvent. In the coordination spheres different compressions and compressibilities prevail which can be the cause of the pressure dependence of $\Delta V^{\neq}_{exp}$, depending on the underlying mechanism (i.e. whether a solvent molecule moves from the bulk water into the first coordination sphere or from the already compressed second or third coordination spheres).

ACKNOWLEDGMENTS

The authors wish to thank several colleagues and the publishers for permission to reproduce Figures 5,6,8,16,17 and the data in Tables 1,2,4. The results from our laboratory discussed here were supported by various grants of the Deutsche Forschungsgemeinschaft.

REFERENCES

1. S.D. Hamann, "Physico-Chemical Effects of Pressure", Butterworths, London, (1957).
2. E. Whalley, Adv. Phys. Org. Chem., 2, 92 (1964).
3. W.J. le Noble, Prog. Phys. Org. Chem., 5, 207 (1967).
4. G. Kohnstam, Prog. React. Kinetics, 5, 335 (1970).
5. C.A. Eckert, Ann. Rev. Phys. Chem., 239 (1972).
6. D.R. Stranks, Pure and Appl. Chem., 38, 303 (1974).
7. G. Jenner, Angew. Chem., 87, 186 (1975).
8. H. Eyring, J. Chem. Phys., 3, 107 (1935).
9. a) H.-D. Brauer and H. Kelm, Z. Physikal. Chem. (N.F.), 76, 98 (1971).
   b) H. Tiltscher and Y.G. Wang, Z. Physikal. Chem. (N.F.), 99, 307 (1976).
10. a) R.C. Neuman, Jr. and J.V. Behar, J. Am. Chem. Soc., 91, 6024 (1969).
    b) R.C. Neuman, Jr. and R.J. Bussey, J. Am. Chem. Soc., 92, 2440 (1970).
    c) R.C. Neuman, Jr. and J.V. Behar, J. Org. Chem., 36, 657 (1971).
11. S.D. Hamann, Trans. Faraday Soc., 54, 507 (1958).
12. H.S. Golinskin, W.G. Laidlaw and J.B. Hyne, Can. J. Soc., 44, 2193 (1966).
13. C.A. Eckert, Proc. VI.AIRAPT Conference on High Pressure, Boulder, Col., (1977), in print.
14. B.S. El'yanov and S.D. Hamann, Aust. J. Chem., 28, 945 (1975).
15. P.G. Tait, "Report on some of the Physical Properties of Fresh-Water and of Sea-Water" from "Physics and Chemistry", H.M.S. Challenger, Vol. II, Part IV, (1888).
16. S.W. Benson and J.A. Berson, J. Am. Chem. Soc., 84, 152 (1961).
17. R.A. Grieger and C.A. Eckert, Trans. Faraday Soc., 66, 2579 (1970).
18. D.A. Palmer and H. Kelm, Inorg. Chim. Acta, 19, 117 (1976).
19. M.G. Evans and M. Polanyi, Trans. Faraday Soc., 31, 875 (1935).
20. K.E. Russell, J. Phys. Chem., 58, 437 (1954).
21. H. Hartmann, H.-D. Brauer, H. Kelm and G. Rinck, Z. Physikal. Chem. (N.F.), 61, 53 (1968).
22. E. Whalley, J. Chem. Phys., 38, 1400 (1963).
23. K. Tamura, Y. Ogo and T. Imoto, Bull. Chem. Soc. Japan, 46, 2988 (1973).
24. T. Tamura and T. Imoto, Bull. Chem. Soc. Japan, 48, 369 (1975).
25. J. v. Jouanne, D.A. Palmer and H. Kelm, Bull. Chem. Soc. Japan, in print.
26. D.A. Palmer and H. Kelm, Aust. J. Chem., 30, 1229 (1977).
27. T.W. Swaddle, Proc. VI.AIRAPT Conference on High Pressure, Boulder, Col., (1977), in print.

FAST REACTIONS IN SOLUTION UNDER HIGH PRESSURE

K. Heremans

Department of Chemistry,
Katholieke Universiteit te Leuven, Belgium.

ABSTRACT. A review is given of the relaxation methods used to study fast reactions in solution under pressure. Special attention is given to a newly developed stopped-flow capable of measuring reaction rates in the msec range up to 1.2 kbar.

1. INTRODUCTION

A primary task of chemistry is to describe the behaviour of molecules in terms of reactions with other molecules. Classical chemical kinetics has shown that most chemical reactions are composed of elementary step reactions which themselves are to fast to study by conventional means. The introduction of rapid reaction techniques has therefore greatly improved our understanding of the reactive behaviour of molecules.

In principle there are two possibilities to study fast reactions. One can introduce rapid mixing techniques coupled to fast detection to follow the reaction after mixing is completed. This approach has been used by Hartridge and Roughton.

A completely different approach is to perturb the equilibrium position of the reaction by one of the parameters that control the equilibrium, i.e. temperature, pressure, electric field or concentration. This approach has been pioneered by Eigen and De Maeyer (1,2).

The use of pressure to obtain activation volumes of fast reactions goes back to Brower (3). He was the first to use pressure-jump techniques at high pressure. The more popular temperature-jump technique is now used by several groups to study fast reactions under pressure (4-7). The stopped-flow technique has proven to be more resistant against high pressure. Recently, our group

Kelm, H. (Ed.) High Pressure Chemistry. 311-324. Copyright © 1978. D. Reidel Publishing Company.
All Rights Reserved.

in Leuven has taken the most simple approach by putting the whole mechanism into a high pressure bomb (8).

In this paper we discuss the information that can be obtained from the study of fast reactions under pressure. A few examples will be given mainly from the biological field, the authors main interest.

There are several good textbooks for students who want to go into the field of fast reaction kinetics (9-11). Biological applications can be studied also (12-13). For a more advanced approach we refer to ref (1,2).

2. PHYSICALLY AND CHEMICALLY CONTROLLED REACTIONS

In this discussion we limit ourselves to reactions in solution. We refer to Dr. Troes paper in this volume for a discussion of reactions in the gas phase.

We consider a bimolecular process

$$A^{+} + B^{-} \rightleftharpoons AB$$

Smoluchowski (14) applied Fick's law to obtain the rate of coagulation of particles in solution. Debye (15) extended his equation to the interaction of charged molecules. For the second order rate constant he obtained

$$\vec{k}_d = \frac{4\pi N_a\, a(D_a+D_b)(z_a z_b e_o^2/\varepsilon akT)}{\exp(z_a z_b e_o^2/\varepsilon akT) - 1}$$

In this equation the symbols have the following meaning : N_a : Avogadros number; a : distance of closest approach of the ions; D : diffusion coefficient; z : charge of the ions; ε : dielectric constant of the solvent. One can calculate $\vec{k}_d$ to be 10^9 to 10^{11} $M^{-1}sec^{-1}$. The upper limit being obtained when protons or molecules with large D values or of opposite charges are considered.

The factor 4π in the equation should be modified if allowance is made for the fact that not all encounters between molecules are successful. For the interaction of an enzyme with its substrate where only a small spot on the molecule is active for interaction, the factor reduces to $2\pi(1-\cos\frac{\alpha}{2})$. This expression takes into account the fact that the interaction site on the enzyme is a cleft with opening angle α. If one further allows for orientation effects, one comes down to 3×10^7 $M^{-1}sec^{-1}$ as a highest rate constant for the interaction of a small molecule with a protein molecule. A few experimental values are given in Table I.

A basic assumption of the model is that every encounter between molecules is chemically effective and very rapid compared to the diffusion process.

A more general scheme is therefore

$$A + B \underset{\overleftarrow{k}_d}{\overset{\overrightarrow{k}_d}{\rightleftharpoons}} A\text{-}B \underset{\overleftarrow{k}_{tr}}{\overset{\overrightarrow{k}_{tr}}{\rightleftharpoons}} AB$$

The overall rate constant for the formation of AB is, assuming A-B to be a stationary intermediate

$$\overrightarrow{k} = \frac{\overrightarrow{k}_d \cdot \overrightarrow{k}_{tr}}{\overleftarrow{k}_d + \overrightarrow{k}_{tr}}$$

From this equation the meaning of the term diffusion control can be seen. If $\overleftarrow{k}_d \ll \overrightarrow{k}_{tr}$ then $\overrightarrow{k} = \overrightarrow{k}_d$. For the reaction to be diffusion controlled, the chemical transformation ($\overrightarrow{k}_{tr}$) must occur in less than 10^{-9}–10^{-11} sec. These processes are usually not considered as chemical reactions.

If on the other hand $\overrightarrow{k}_{tr} \ll \overleftarrow{k}_d$ then

$$\overrightarrow{k} = \overrightarrow{k}_{tr} \cdot \frac{\overrightarrow{k}_d}{\overleftarrow{k}_d}$$

We now have a chemically controlled reaction and the measured rate constant ($\overrightarrow{k}$) is a product of the chemical transformation and the equilibrium constant for complex formation without involving bond formation

$$K_F = \frac{\overrightarrow{k}_d}{\overleftarrow{k}_d} = \frac{4\Pi N_a}{3} a^3 \exp(z_a z_b e_o^2/\varepsilon a k T)$$

This is the familiar Fuoss equation for the formation of ion pairs (16).

With this admittedly crude model, we can calculate activation volumes and overall volume changes when we differentiate the equations with respect to pressure. When the reaction is entirely controlled by the diffusion of the ions we calculate $\Delta V^{\#} = 1$ ml in aqueous solution. An experimental observation which would fit in this picture was reported by Carnevale and Litovitz (17). See Table I.

When the chemical transformation is rate limiting we obtain

$$\Delta V^{\#}_{obs} = \Delta V^{\#}_{tr} + \Delta V_F$$

Here $\Delta V^{\#}_{tr}$ contains volume change contributions from bond making and breaking, specific solvent effects, sterical hindrance, etc. For considerations of these factors we refer to the contributions by H. Kelm and W. le Noble in this volume. Except for the solvent,

Table I

Examples of data obtained with fast reaction techniques under pressure

Reaction	Rate constant (M^{-1} sec^{-1})	$\Delta V^{\#}$ (ml)	Method[a] (ref)
$NH_4^+ + OH^-$	4.7×10^{10}	0.5	US(17)
Bu_4N^+ + Picrate$^-$	6×10^{11}	25[b]	EFM(26)
FMN + AMP	1.1×10^{9}	2.4	FL(39)
$Fe^{3+} + NCS^-$		6[c]	PJ(3)
$Fe^{3+} + NCS^-$		0	TJ(27)
$(FeOH)^{2+} + NCS^-$		7.1	TJ(27)
$Fe^{3+} + NCS^-$		-12	SF(8)
$(FeOH)^{2+} + NCS^-$		7.8	SF(8)
Haem + CO	2.8×10^{7}	2[d]	LF(25)
Haem + CO	1.9×10^{6}	14[e]	LF(25)
Haemoglobin + O_2	3.2×10^{7}	5	LF(25)
Haemoglobin + CO (fast)	2.8×10^{6}	-3.1	LF(25)
Haemoglobin + CO (slow)	6×10^{4}	-21.2	LF(25)
β-Lactoglobulin + Bromoplenolblue	6.4×10^{5}	+34	TJ(6)
Trypsin + Proflavin	4.6×10^{5}	+17.5	TJ(7)
Chymotrypsin + Proflavin	4.6×10^{7}	+8.7	TJ(7)

a. U.S. : ultrasonics; EFM : electric field modulation; FL : fluorescence lifetime; PJ : pressure-jump; SF : stopped-flow; LF : laser flash photolysis; TJ : temperature-jump
b. solvent : diethylether
c. direct and base catalyzed pathway
d. solvent : ethyleneglycol + water (80 % v/v)
e. solvent : glycerol

which is always water or aqueous salt solutions, the situation is not different for biological reactions. ΔV_F can be calculated when assumptions are made about a the distance of closest approach of the ions. In water for $z_a z_b = -1$ and a = 0.5 nm, ΔV_F = 1.5 ml. Hemmes (18) has considered volume changes of ion association in relation to the interpretation of ultrasonic studies.

From an inspection of the Fuoss equation one might be tempted to conclude that for reactions between noncharged molecules or a charged and a noncharged molecule, $\Delta V_F = 0$. Although this might be true, it is not a priori so. The problem is to write an equation for the interactionpotential which then can be integrated in the equation

$$k_d = \frac{4\pi N_a(D_a + D_b)}{\int_{\alpha}^{\infty} \exp(-U/kT)\,dr/r^2}$$

For overall volume changes it seems also necessary to make a distinction between aliphatic and aromatic systems in aqueous solutions (19). At present it is not entirely possible to give a theoretical model to account for the experimental observations of these systems.

3. EXPERIMENTAL APPROACH TO THE STUDY OF FAST REACTIONS

Fig. 1 gives a summary of the available techniques for the study of rapid reactions with their time scale of application. We restrict ourselves to the chemical reactions in the sence described in the previous sections.

Formally flow methods belong to the classical techniques although special techniques are used for the detection of changes in the concentration of the reactants. A considerable advantage of the flow methods is that they are not restricted to equilibrium reactions. Modern stopped-flow methods can measure half lives of reactions in the order of msecs depending to a large extent on the changes in extinction coefficient occuring in the reaction.

A completely different approach to the study of fast reactions was introduced by Eigen and De Maeyer (1,2).

Suppose that the reaction

$$A + B \rightleftharpoons AB$$

has a finite ΔH, ΔV or ΔM, i.e. change in enthalpy, volume or dipole moment, then it is possible to perturb the equilibrium since the following equations hold

$$\frac{d \ln K}{dT} = \frac{\Delta H}{RT^2} \qquad \frac{d \ln K}{dP} = \frac{-\Delta V}{RT} \qquad \frac{d \ln K}{dE} = \frac{\Delta M}{RT}$$

Two possibilities can be used to perturb chemical equilibrat

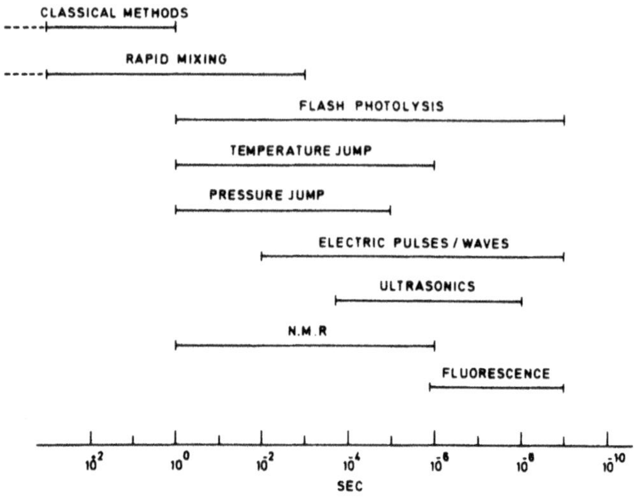

Fig. 1. Fast reaction techniques and their time range of application expressed as half-life in seconds. Adapted from ref. (9).

Two possibilities can be used to perturb chemical equilibria : a stepwise perturbation or a periodic perturbation.

In the most popular technique, the temperature-jump method, the chemical equilibrium responds to a stepwise change in temperature through a relaxation to the new equilibrium position defined by the new temperature of the medium. The same principle applies to the pressure-jump and electricfield-jump method. When the rate equations can be linearized, i.e. when the perturbation is not to large, it can be shown that the relaxation time for the reaction is given by

$$1/\tau = k_1(\bar{A} + \bar{B}) + k_{-1}$$

$\bar{A}$ and $\bar{B}$ being the equilibrium concentrations of A and B. From the dependence of the reverse relaxation time against the sum of the equilibrium concentrations, both rate constants can be obtained. An example is given in Fig. 2.

For periodic perturbations one can easily see that the response of the chemical equilibrium will also be periodic but lagging behind the perturbation with a phase difference which is a measure of the rate of the chemical reaction (dispersion methods). Alter-

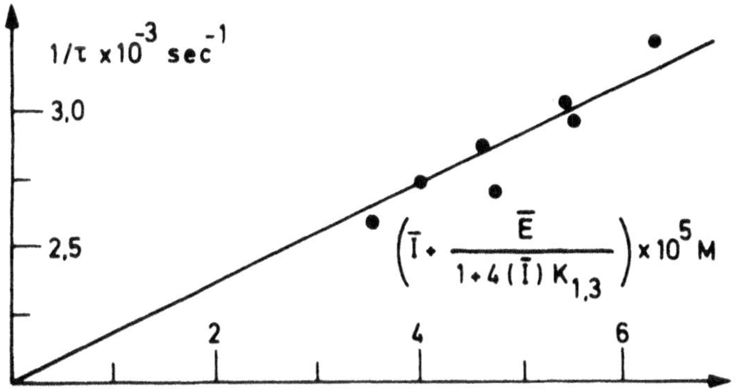

Fig. 2. The interaction of the dye Biebrich Scarlet with the active site of Chymotrypsin. Dependence of the relaxation time on the equilibrium concentrations. Corrections are introduced for the concentration dependent association of the dye.

natively one can measure the amplitude loss (absorption methods). Ultrasonic (20) and electricfield techniques (21-23) are examples of this method of perturbation.

From the above discussed techniques only stopped-flow, temperature-jump and, to lesser extent, pressure-jump have gained popularity.

4. FAST REACTION TECHNIQUES UNDER HIGH PRESSURE

The first relaxation method to be used under pressure was pressure-jump. Brower (3) already in 1968 described a method to produce pressure jumps in the kilobar range. With cut-off valves it was possible to make jumps from any desired pressure to higher or lower pressures. The method has been used recently by Whisnia (24) to study the association of ribosome subunit.

Several groups have built temperature-jump machines for work under high pressure.

Caldin has described an apparatus which uses a laser as a

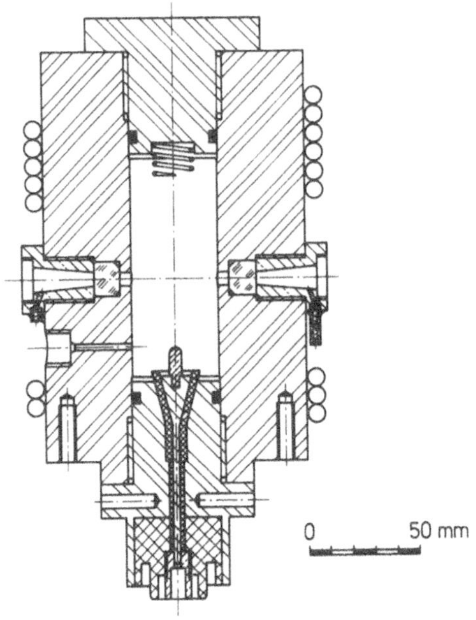

Fig. 3. Outline of a high pressure bomb for chemical relaxation spectrometry under pressure.

heating source (4). This simplifies the design of the apparatus since heat is supplied from outside through a window. Using the absorption band at 1060 nm of water, jumps of about 0.1 K can be obtained. Jumps of 5 K can be obtained when dyes are used as energy absorber. The method has several advantages but the presence of dyes implies the study of reactions which do not interact with the dyes. Another advantage of the apparatus is that it can be used to do flash-photolysis work under pressure (25).

Grieger et al. (6) have described an instrument where the discharge of a capacitor through the observation cell is used to heat the solution. Heating times of the order or microseconds can be obtained with this method.

Jost (5) has described a similar instrument using gas as a pressure transmitting medium.

Our group in Leuven has developed a machine for work up to 3 kbar using a high voltage capacitor as heating source (7). Oil is used as a pressure transmitting medium. The cell which contains the solution under investigation can be filled and cleaned outside the high pressure bomb. Fig. 3 shows a simple version of this bomb for work up to 1.5 kbar. A special feature is the high voltage feed-trough. The isolation is made of nylon. The windows

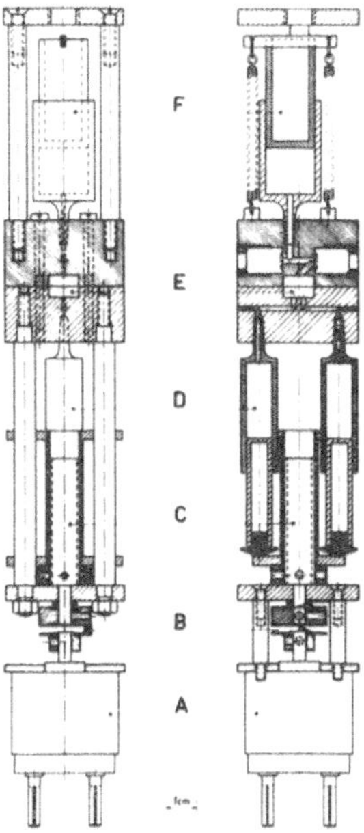

Fig. 4. Outline of stopped-flow mechanism for experiments under pressure. A,B,C : syringe driving mechanism; D : feed syringes; E : mixing chamber; F : waste syringe. The whole unit is placed into a high pressure chamber.

are in plexiglas while the cell (not shown) is a conventional temperature-jump cell provided with a teflon membrane to transduce the pressure. The temperature-jump instrument was purchased from Messanlagen studiengesellschaft mbH, Göttingen, Germany.

Periodic perturbation techniques have also been used under high pressure. Carnevale and Litovitz (17) have studied a diffusion controlled reaction under pressure with ultrasonic techniques. Recently, stationary electric field methods have been applied to study ion pair formation in low polarity solvent (26).

The development of a high pressure stopped-flow has proven to be more difficult (8). We have chosen the most simple approach : using a single step motor as a syringe driving mechanism, the stopped-flow unit i.e. syringe driving mechanism, mixing chamber, observation chamber and stopping syringe can easily be put into a high pressure bomb. An outline of the stopped-flow unit is given in fig. 4. The high pressure bomb is mounted on a support which allows introduction of the stopped-flow unit in the vertical position. After closure of the bomb the measurements can be made in the horizontal position. The course of the reaction is followed optically. The dead time of the instrument is of the order of 20 msec.

5. APPLICATIONS

The method described have been used to study inorganic reactions (27), proton transfer in organic solvents (28), biological reactions (29) and reactions in micelles (30). Possible applications of the high pressure stopped flow method have been given elsewhere (8). A selection of data obtained with fast reaction techniques under pressure is given in table I.

In this section we give a detailed account of a biological redox reaction where the use of the temperature-jump method and stopped-flow experiments have given complementary information.

Vandersypen (31) studied the redox reaction of cytochrome C with Ironcyanides.

$$\text{Ferri-cyt} + Fe(CN)_6^{4-} \overset{k_r}{\rightleftharpoons} \text{Ferro-cyt} + Fe(CN)_6^{3-}$$

With a large excess of $Fe(CN)_6^{4-}$, the expression for the relaxation time is as follows

$$1/_\tau = k_r\ (Fe_o) + k_o\ [a\ (\text{cyt } c_o) + 2\ (Fe(CN)_6^{3-}]$$

where (Fe_o) and (cyt c_o) stand for the total concentration of Fe and Cytochrome. a gives the fraction of reduced cyt in the preparation. Fig. 5 gives a plot of the measured relaxation times as a function of pressure.

The following results are obtained for the activation volumes :

$$\Delta V_r^\# = +\ 13 \text{ ml} \qquad \Delta V_o^\# = -\ 24 \text{ ml}$$

The activation volume for the oxidation can be compared with the values calculated for outersphere redox-reactions in inorganic chemistry (32). The value for the reduction is less well understood, except when one assumes a preequilibrium complex between cyt and $Fe(CN)_6^{4-}$.

Stopped-flow experiments have been performed by Snauwaert (8) for the reduction of Cyt with Na-ascorbate. The activation volume for the reduction is then - 21 ml. This is in agreement with the oxidation value indicating that the reduction is also outer-sphere. The positive activation volume observed with $Fe(CN)_6^{4-}$ can therefore be ascribed to ion pair formation between the positive charges on the protein surrounding the heme group and the negative charges of the reductant.

Examples of the use of temperature jump and stopped-flow under high pressure to study enzyme systems will be described in the

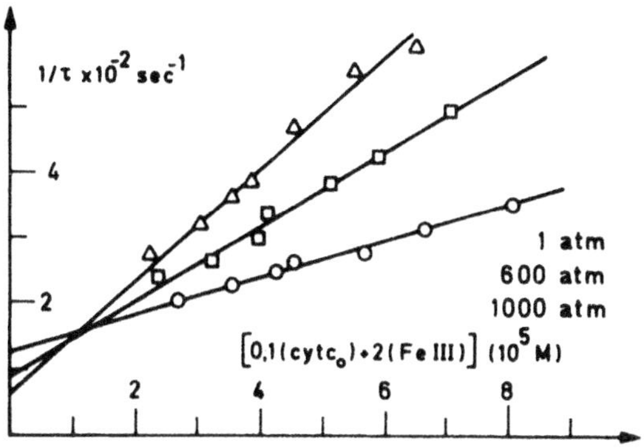

Fig. 5. Redox reaction of cytochrome c with iron-cyanides. Dependence of relaxation time on concentration and pressure.

accompanying paper in this volume.

6. FUTURE DEVELOPMENTS

Outside the high pressure field, the stopped-flow method is the most popular especially in inorganic and biological chemistry. It is to be expected that it will gain the same popularity in the high pressure field. The time range of the reactions that can be studied is considerably extended over the present sampling techniques used in high pressure chemistry. The fact that the technique is not restricted to equilibrium reactions is of considerable importance.

The student who wants to study the kinetics of fast equilibrium reactions has the choice between temperature- and pressure-jump techniques. Before going into it, it is useful to study the advantages and disadvantages of both techniques. Knoche (33) and Davis and Gutfreund (34) have discussed this matters in great detail.

In this paper no attention has been given to the use of High pressure high resolution NMR as a technique for the study of chemical reactions. Merbach et al. (35) as well as Lüdemann et al. (36) have recently demonstrated the possibilities of these methods.

Under certain conditions, fluorescence yield and lifetimes can be used to measure rate constants of diffusion controlled reactions. We refer to the paper by Brauer et al. in this volume. Weber et al. (39) have applied the method under pressure to obtain activation volumes for the complex formation between isoalloxazine and adenine.

The study of cooperative phenomena i.e. changes in conformation of biological macromolecules and structures, is of considerable interest for an understanding of biological phenomena. Because of their cooperative nature, even small perturbations are to large to give a linear response of the system. Clegg et al. (37) have described an apparatus for the study of these phenomena.

A further step would of course be to look to the fluctuations of the system in equilibrium. This possibility has been considered but in practice it is not very easy to extract the information concerning the chemical reaction out of the signal which largely reflect diffusion processes. Recently the method was used to look at changes in the size and the shape of phospholipid vesicles under pressure (38).

ACKNOWLEDGEMENTS

I am grateful to Professor Leo De Maeyer for introduction in the rapid reaction field. I thank my colleagues A. Persoons, L. Hellemans and Y. Engelborghs for stimulating discussions.

The work described from our group is supported by the Belgian National Science Foundation.

REFERENCES

1. M. Eigen and L. De Maeyer, Relaxation Methods, in : Investigations of Rates and Mechanisms of Reactions, ed. by S.L. Friess, E.S. Lewis and A. Weissberger, Vol. VIII, Part II, Interscience, New York, 1963.
2. M. Eigen and Leo De Maeyer, Theoretical basis of relaxation spectrometry, in : Investigations of Rates and Mechanisms of Reactions, ed. by A. Weissberger, Vol. VI, Part II, Interscience, New York, 1974.
3. K.R. Brower, J. Amer. Chem. Soc., 90, 5401, 1968
4. E.F. Caldin, M.W. Grant, B.B. Hasinoff and P.A. Tregloan J. Phys. E., 6, 349, 1973.
5. A. Jost, Ber. Bunsenges. physik. Chem., 78, 300, 1974.
6. A.D. Yu, M.D. Waissbluth and R.A. Grieger, Rev. Sci. Instrum., 44, 1390, 1973.
7. K. Heremans, J. Snauwaert, H. Vandersypen and Y. Van Nuland, Proc. 4th Int. Conf. High Pressure, Kyoto, 623, 1974.

8. K. Heremans, J. Snauwaert and J. Rijckenberg, Proc. 6th Int. Conf. High Pressure, Boulder, in press, 1977.
9. D.N. Hague, Fast Reactions, Wiley-Interscience, London, 1971.
10. C.F. Bernasconi, Relaxation Kinetics, Academic Press, New York, 1976.
11. H. Strehlow and W. Knoche, Fundamentals of Chemical Relaxation, Verlag Chemie, Weinheim, 1977.
12. H. Gutfreund, Enzymes : Physical Principles, Wiley-Interscience, London, 1972.
13. I. Pecht and R. Rigler, eds. Chemical Relaxation in Molecular Biology, Springer-Verlag, Berlin, 1977.
14. M.V. Smoluchowski, Z. phys. Chem., 92, 129, 1917.
15. P. Debye, Trans. Electrochem. Soc., 82, 265, 1942.
16. R.M. Fuoss, J. Am. Chem. Soc., 80, 5059, 1958.
17. E.H. Carnevale and T.A. Litovitz, J. Acoust. Soc. Am., 30, 610, 1958.
18. P. Hemmes, J. Phys. Chem., 76, 895, 1972.
19. K. Heremans, Proc. 6th Int. Conf. High Pressure, Boulder, in press, 1977.
20. J. Stuehr, Ultrasonic Methods, in ref. 2.
21. Leo De Maeyer and Andre Persoons, Electric Field Methods, in ref. 2.
22. A. Persoons, J. Phys. Chem., 78, 1210, 1974.
23. L. Hellemans and L. De Maeyer, J. Chem. Phys., 63, 3490, 1975.
24. A.A. Wishnia et al., Proc. 6th Int. Conf. High Pressure, Boulder, in press, 1977.
25. B.B. Hasinoff, Biochemistry, 13, 3111, 1974; E.F. Caldin and B.B. Hasinoff, J.C.S. Faraday I, 71, 515, 1975.
26. J. Everaert, F. Nauwelaers and A. Persoons, in preparation.
27. E.F. Caldin, M.W. Grant and B.B. Hasinoff, J.C.S. Faraday I, 68, 2247, 1972; A. Jost, Ber. Bunsenges. physisk. Chem., 80, 316, 1976.
28. C.D. Hubbard, C.J. Wilson and E.F. Caldin, J. Am. Chem. Soc., 98, 1870, 1976.
29. M.D. Waisbluth and R.A. Grieger, Arch. Biochem. Biophys., 159, 639, 1973; K. Heremans et al. ref. 7.
30. R.S. Seright, R.A. Grieger-Block and D. Thusius, Proc. 6th Int. Conf. High Pressure, Boulder, in press, 1977.
31. H. Vandersypen and K. Heremans, Arch. Int. Physiol. Biochim., 82, 792, 1974.
32. D.R. Stranks, Pure Applied Chemistry, 38, 303, 1974.
33. W. Knoch and G. Wiese, Rev. Sci. Instrum., 47, 220, 1976.
34. J.S. Davis and H. Gutfreund, FEBS Letters, 72, 199, 1976.
35. A.E. Merbach and H. Vanni, Helvetica Chim. Act., 60, 1124, 1977.
36. H.D. Lüdemann, R. Rauchschwalbe and E. Lang, Angew. Chem. Int. Ed. Engl., 16, 331, 1977.
37. R.M. Clegg, E.L. Elson and B.W. Maxfield, Biopolymers, 14, 883, 1975.
38. F. Ceuterick, K. Heremans, H. Desmedt, P. Nieuwenhuysen and J. Clauwaert, Meeting of Belgian, Dutch and German Biophysics

societies, Liege, 1977.

39. G. Weber, F. Tanaka, B.Y. Okamoto and H.G. Drickamer, Proc. Nat. Acad. Sci. USA, 71, 1264, 1974.

ORGANIC MODEL REACTIONS UNDER PRESSURE

W. J. le Noble

Department of Chemistry, State University of New York, Stony Brook, New York 11794

It is probably fair to say that the lectures given here to date have been concerned primarily with physics, and with the physical properties of compounds and sometimes mixtures; in contrast, my remarks and those of my colleagues to follow are almost exclusively directed at chemistry and chemical properties. It goes without saying, however, that the chemist who uses high pressure methodology depends very much indeed on the knowledge that has been gathered by the physicists and physical chemists.

The pedagogical nature of these sessions makes it desirable that some attention be given to some fairly elementary features normally glossed over or omitted altogether in research reports. By the same token, it seems unnecessary for the present purpose to support all results with quotations of the original literature; however, those who wish to do further studies in this field might begin by reading the review literature quoted at the end of this lecture (from which all the data to be discussed here were taken) and in Professor Kelm's contributions to this Conference.

One of the advantages that chemists have in this area is that their apparatus is generally extremely simple. As should be obvious from the preceding lectures, it is easy indeed to

Kelm, H. (Ed.) High Pressure Chemistry. 325-343. *Copyright © 1978. D. Reidel Publishing Company. All Rights Reserved.*

generate accurately known pressures up to a few kbar in a cylindrical space of a few cm diameter and a dm or two in length, and in fact, equipment to achieve just that is commercially available. So are window vessels, enabling one to carry out studies of solutions under pressure virtually as easily as under atmospheric conditions. The installation of electrically insulated leads into high pressure vessels is also standard, and hence the appraisal of rate and equilibrium constants via conductance measurements is a routine operation. The cost of the high pressure vessels and the pressure generating equipment is quite modest, and often negligible compared to that of the analytical equipment used.

A few aspects of chemistry at high pressure are perhaps not so obvious from the earlier presentations, and may be dwelt on in a bit more detail here. One of these is the need for excellent temperature control: rate and equilibrium constants are usually very sensitive to small temperature changes, and if one is going to deduce from the much smaller pressure induced changes in rate or equilibrium composition how large the corresponding volume differences might be, then there is obviously no room for major uncertainties about the temperature. And there are plenty of ways in which such uncertainties may arise. First of all, the high pressure vessel inside the thermostat is connected by heavy-walled, heat conducting pipes to the generating equipment outside, and hence it is not safe to assume without the support of measurement that the temperature is simply that of the bath. Secondly, when the pressure is raised from atmospheric to one of several kbars, the heat of compression, which may amount to several tens of degrees, is produced in more or less adiabatic fashion. This makes it necessary to either wait for the dissipation of this heat (which fortunately doesn't take very long - perhaps some minutes, especially if appropriately precooled solutions are employed) or to use an arrangement allowing the reagents to be mixed after temperature equilibrium has been re-established. I stress this point because the popularization of high pressure chemistry in recent years has to some extent been characterized by the enthusiastic publication of many improbable data produced under conditions in which temperature control was almost certainly inadequate. Thus, it should be noted that good temperature control becomes difficult outside the range of 0-200°C with the all metal apparatus. Another limiting circumstance is that many common solvents (benzene, t-butanol, dimethyl sulfoxide, cyclohexane, and others) are solids at elevated pressures at room temperature.

In part for the same reason it is highly desirable that dilute solutions be used in such research. Solutions much more concentrated than millimolar in reagents producing an exothermic reaction may give rise to unforeseen temperature rises that may well exaggerate pressure induced accelerations. Unwanted pressure changes could also occur. But even apart from these temperature and pressure factors, there is also the question of activity coefficients and their pressure dependence - a problem which is best minimized by the use of dilute solutions. The possibility of pressure induced solubility decreases can also be countered in this way. The need for dilute solutions is especially severe in reactions producing rather insoluble gases. While these will usually remain in solution during the experiment, this condition changes upon the release of pressure; in this instance at least, nature informs the failing experimenter of the error of his ways. It should perhaps be added here that these requirements are of less importance when pressure is applied for purely synthetic reasons.

Some thought should also be given to the rate, when activation volumes are the objective. In view of the compression heat problem, it used to be necessary to work under such conditions of temperature, concentration, solvent and catalyst as would produce reaction times of the order of some hours. Nowadays high pressure mixing equipment is available and as we shall hear during the remainder of this Conference, virtually the entire galaxy of rapid reaction techniques (jump and flash methods, stopped flow, and so on) has now been adapted for use at high pressures as well. However, the type of equipment needed is obviously linked to this question of rates, and the chemist considering his own entry into the field should devote his attention to it.

The actual vessels used for the solutions of interest must obviously scrupulously avoid any contact between these solutions and the hydraulic fluid. Mercury can be used in many instances when it does not in any way enter into the reaction (this rules out much photochemistry, for example). My own preference runs to the syringe principle; they can be adapted to virtually any special circumstances, and if they are well made, no problem is encountered because of diffusion through the annular space between jacket and plunger.

The mathematics underlying the chemist's operations at high pressure are even more simple-minded than the technology. The fundamental relation is

$$\partial\overline{G}/\partial p = \overline{V}$$

In view of the stress on dilute solutions, it should be understood that partiality of the thermodynamic properties applies in all of our considerations concerning those properties; however, for the sake of convenience and habit, I will omit the term partial most of the time. Since the equilibrium constant is given by

$$K = e^{-\Delta G/RT}$$

it follows that

$$\Delta V = \partial\Delta G/\partial p = -RT\partial \ln K/\partial p$$

Likewise, in kinetics we have the Eyring equation for the rate constant

$$k = (RT/Nh)e^{-\Delta G^{\ddagger}/RT}$$

and accordingly

$$\Delta V^{\ddagger} = \partial \Delta G^{\ddagger}/\partial p = -RT\partial \ln k/\partial p$$

In the case of equilibria, one can approach the equation from either side: one can measure the pressure dependence of an equilibrium constant and calculate the reaction volume, or vice versa. For kineticists, however, it is a fundamental fact of life that the transition state is not available for direct scrutiny, and hence if its volume is of interest, this can be determined only from the pressure dependence of the rate constant; in fact, the presence of many of us at this Conference can be traced to solely this condition!

Several points need to be made concerning the quantities $\partial \ln K/\partial p$ and $\partial \ln k/\partial p$. First of all, these derivatives may have either sign; volume changes may of course be either in the one direction or the other. Secondly, these pressure coefficients are not constants. They sometimes appear to be, at least over small pressure intervals, and when that is the case, the determination of the corresponding volume changes is simple indeed. More often, however, the experimental data show clearly that ΔV and $\Delta V^{\ddagger}$ are pressure dependent, and the questions then arise how to determine the slopes, and for which pressure to report them. The first of these questions will be considered in some detail in Professor Kelm's lecture; suffice it here to say that graph fitting, the least squares fitting of various empirical equations, and the straightjacketing of data in the framework of pseudo rigorous equations have all been used. With careful work, the pressure can be measured to a few bars and the rate or equilibrium constant to a few tenths of one percent; if sufficient data are available over sufficiently small intervals, the volume changes can be determined to within a few tenths of one

cm^3/mol.

The second point, the pressure for which to report, is usually dictated by the fact that virtually any other information about the reaction one might seek to correlate with is available only at atmospheric pressure, and most chemists have therefore adopted the habit of reporting ΔV_o or $\Delta V_o^{\ddagger}$ (liquid compressibilities being what they are, the volume changes at zero pressure differ of course only negligibly from those at one atmosphere). This choice is not without drawbacks, however. Thus, one ignores the information hidden in the curvature (the compressibility of activation), although fortunately at least the inorganic chemists have recently begun to find applications for this parameter. But the habit of reporting ΔV_o and $\Delta V_o^{\ddagger}$ also raises the twin questions about accuracy and precision. Accuracy because ordinarily no data are available at negative pressures. On this point I believe that the audience may be reassured. For equilibria at least, there are no significant discrepancies between reaction volumes calculated from the pressure data and those determined directly from density data. For chemical reactions one attempt has been recorded to measure the rate constants over a range of applied pressures as well as in the gas phase, and a smooth curve could be drawn through all of these data by assuming that the pressure to be plotted should be the externally applied pressure plus the internal pressure of the solvent. Such a result should of course be expected only if one considers a reaction the rate of which is virtually independent of the nature of the solvent; if this is not the case and special solvent-solute are indicated at any point along the reaction coordinate, it would no longer be safe to assume the simple pressure addition just mentioned. The problem of precision is again one to be discussed in more detail by Professor Kelm; I would only like to emphasize the special need for precise data at pressures near zero. While this seems obvious, it is a point that needs to be made again and again, especially to enthusiastic though inexperienced students eager to collect data only at the highest pressures the apparatus is capable of, if not beyond.

Let us now consider a simple equilibrium, namely the auto-ionization of water:

$$H_2O \rightleftharpoons H^+ + OH^-$$

for which at 1 atm and 25°, $K_w = 10^{-14}$. In this instance the need to work with conductivity water makes the measurement of Kw under pressure difficult; however, as noted, we have the alternative of measuring the volume change directly. This can be done by calculating the partial volumes of the participants in the reaction from density data as a function of concentration. At each concentration C one can compute the apparent molal volume ϕ_v from

$$\phi_v = \frac{M}{d} - \frac{d - d_o}{d_o} \frac{1000}{C}$$

where M is the molecular weight, d the density of the solution and d_o that of the solvent. A plot of ϕ_v vs C (or $\sqrt{C}$ for 1:1 electrolytes) then by extrapolation gives the partial molal volume at infinite dilution. If we have these data for HCl, NaOH, and NaCl in water, then the ionization volume for water is given by

$$\Delta V_o(i) = \bar{V}_{HCl} + \bar{V}_{NaOH} - \bar{V}_{NaCl} - V_{H_2O}$$

The result of -22.1 ± 0.1 cm^3/mol is of interest, for several reasons. First of all, so far as the absolute value is concerned, it is reassuring to note that this result is of the right order of magnitude; thus, molecular dimensions being of the order of Å (10^{-8} cm), and since 1 Å^3/molecule = $N(10^{-8})^3$ = 0.61 cm^3/mol, we should expect that molar volume changes are of the order of cm^3. Secondly, although the process considered is formally a dissociation, it is characterized by a volume decrease. Thirdly, since the molar volume of water itself is only 18 cm^3, the decrease cannot be ascribed to the ionizing water molecules themselves; at least some must be attributed to interactions of the ions with their neighbors. We now know, in fact, that the interactions of ions with solvents are the most important ones there are in terms of volume effects. The ions orient, polarize, attract, and thereby compress the neighboring molecules, and the resulting decrease in overall volume is known as electrostriction. The Born equation, based on a model of hard sphere ions and a structureless, but compressible and polarizable dielectric, provides the well known free energy change for the immersal of the ion of radius r and charge q in a medium of dielectric constant D; the pressure derivative of this equation , known as the Drude-Nernst equation, gives the corresponding volume effect

$$V_e = -\frac{q^2}{2r}\frac{\partial \ln D}{D\partial p}$$

As you know, organic chemists long ago adopted the custom of studying new phenomena and techniques by means of model equilibria, and this almost always means: acid-base equilibria. Enormous numbers of acids and bases with a wide variety of structural features are available, and the main structural effects on the equilibrium constants are large and well understood. The effect of pressure on ionization is no exception: thus, each of the terms affecting this electrostriction can be studied by means of this standard reaction. It may be said at the outset that many of the ΔVi -values quoted below were measured by both approaches (use of $\overline{V}$ or $\partial pK/\partial p$ data); all apply to water at 25° unless otherwise stated.

First, there are no major structural effects. Thus, ΔVi for the ionization of ammonia is about -28 cm^3/mol, hence not so different from that of water. Studies of other simple amines in most cases give results near -28 cm^3/mol, and families of acids similarly tend to cluster about a common value of ΔVi.

The radial dependence is clear from a comparison of water and hydrogen sulfide: the latter has a ΔVi of -16 cm^3/mol, and similar differences show up between alcohols and mercaptans. Agreement with theory is not quantitative, but that should perhaps not be expected. Another word of caution is in order here: the radius r is that of the atom(s) bearing the charge, and not some average calculated for the entire molecule.

The charge dependence in the Drude-Nernst equation predicts that charge delocalization from 1 to n sites will reduce the electrostriction volume by n. Again the agreement is qualitative (note that part of Ve in each case is due to the common ion) but strong. Thus, compared to water with -22 cm^3/mol, phenol has a ΔVi of -18 cm^3/mol, reflecting delocalization of the charge into the benzene ring; carboxylic acids have values near -13 cm^3/mol, with the charge now distributed equally between the two oxygen atoms; p-nitrophenol, with even more extensive opportunities for delocalization, is characterized by -10 cm^3/mol.

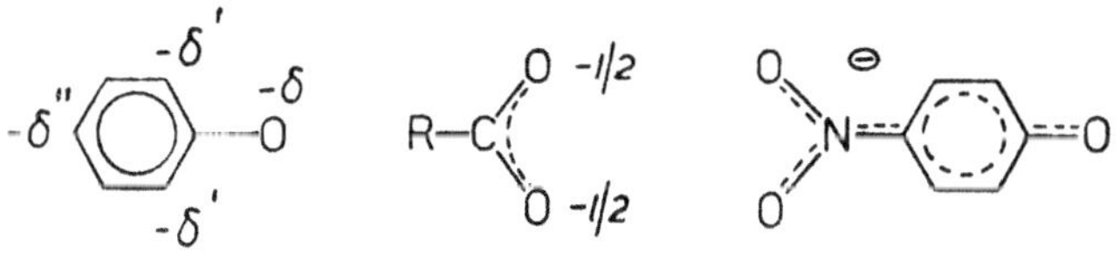

Similar observations can be quoted for bases. Compared to ammonia with -28 cm^3/mol, guanidine has a value of -21 cm^3/mol. Equally striking is the fact that when ionization must occur at

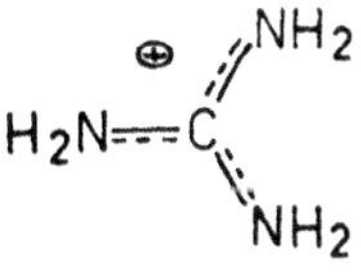

or near a site already bearing a charge of the same sign, then ΔVi becomes larger. Thus, the ammonium ion

has a ΔVi(2) which is -35 cm^3/mol; the same effect is noted in the many dibasic acids when the ionization volumes for the first and second steps are compared.

The solvent dependence of the electrostriction volume is hidden in the term $\partial D/D^2\partial p$. The nature of this term suggests that the effect should be large, and that the volume decreases should be largest in the non-polar media. This prediction often seems to startle chemists; however, it is perfectly reasonable even on intuitive grounds. Thus, polar solvents are usually already fairly dense because of interactions between the molecules themselves, and hence the introduction of an ionic charge cannot magnify the effect that much further; also, a low dielectric constant means that the ionic forces operate over long distances. In any case, the predictions are in excellent accord with laboratory experience. The ionization volume of phenol, quoted above at -18 cm^3/mol in water, equals -38 cm^3/mol in methanol, and long lists of similar solvent effects could be recorded. It is a pity that it is so difficult to study ionic equilibria in truly non-polar media; however, some hints in that direction are now available. It is known, for example, that ion-pair dissociation of divalent ionic species such as $MnSO_4$ in water have reaction volumes of -7 cm^3/mol; if it were possible to study monovalent ion pairs in this medium, -2 cm^3/mol ($-7/2^2$) would be expected. One such pair has been studied in ether: the dissociation of tetra-*n*-butylammonium picrate pairs has a volume decrease greater than 100 cm^3/mol!

There are additional effects not predicted by the Drude-Nernst treatment that have emerged from the acid-base studies. Thus, a nearby charge of opposite sign reduces electrostriction; glycine anion, for example, has a ΔVi of -23 cm^3/mol *vs* -28 for methylamine.

$$^{-}OOCCH_2NH_2 \quad \underline{vs} \quad CH_3NH_2$$

When the ionization of acetic acid in water is compared with that of carbon dioxide, the former has a ΔVi of -15 cm^3/mol and the latter -27 cm^3/mol: this large difference is due to the bond that is first formed when carbon dioxide is converted into carbonic acid.

$$CH_3COOH \rightleftharpoons H^+ + CH_3COO^-$$

$$H_2O + CO_2 \rightleftharpoons HOCOOH \rightleftharpoons H^+ + HOCOO^-$$

The bonding effect on volume is also clearly visible in other equilibria; thus, pressure favors N_2O_4 over NO_2, and hydrogen-bonded dimers over monomers; it promotes the hydration of carbonyl compounds, and charge transfer complexation. The volume contractions for these processes vary from 5 to 20 cm^3/mol, depending primarily on the strength of the interaction.

When we turn our attention to kinetics, the following new questions crop up. First of all, as noted already, we cannot measure $\overline{V}^{\ddagger}$ directly and hence we are forced to calculate $\Delta V^{\ddagger}$ from the pressure coefficient of the rate constant; however, once this is done we can turn the tables and combine the result with partial volume data of the reactants to determine $\overline{V}^{\ddagger}$ indirectly. Thus, we have facile access to one of the most important properties of the transition state, not as a difference, but in absolute terms. It is in fact the only transition state property that can be routinely evaluated.

A second point I wish to touch on here is the applicability of transition state theory. It is known to most chemists and frequently stressed by some, that the derivation of Eyring's equation requires assumptions the validity of which is strongest in slow reactions with high barriers. How closely the barrier may approach kT before the activation treatment becomes untenable is a question of much debate. My own attitude is that the concept of the activation parameter (whether $\Delta G^{\ddagger}$, $\Delta S^{\ddagger}$, $\Delta H^{\ddagger}$, or $\Delta V^{\ddagger}$) has so successfully correlated such enormous numbers of data of such diversity, that no chemist aware of this success is going to give up this treatment because of a dialectic impurity unless another of comparable simplicity can take its place; nevertheless, caution should certainly be the watchword for those colleagues who measure pressure effects on extremely fast reactions such as diffusion controlled chemistry, or photochemistry with its myriad of ultrafast steps.

Thirdly, a word about corrections. Many papers in this field, perhaps even most of them, contain a statement to the effect that compressibility corrections were made. It seems reassuring that authors take the trouble to correct their data for any inaccuracies they may contain; however, in the present instance it must be feared that most of these "corrections" are not merely unnecessary, but wrong. If we are going to compare rate constants at various pressures, it is of course necessary that these constants be calculated on the basis of pressure independent concentration units, and these may be mole fractions, moles per kg solvent (molalility units), or moles per liter <u>at one atmosphere</u> (molarity units). Since the solutions used are invariably made up at atmospheric pressure, no corrections are necessary; such a requirement would exist only if the solutions had been mixed at high pressure, and if only the concentration at that pressure were known. This point has been made especially clearly and convincingly by Hamann in one of the review articles listed at the end of this paper.

Still another important consideration is this. The lengths of bonds in transition is of course much more variable than that of bonds in stable molecules; that and a corresponding variability of dipole moments are the major sources of the range of values one may observe in a series of otherwise similar reactions. This must be expected, of course; the point of interest here is that this range is generally much smaller than that of other activation parameters. Thus, the volume changes are often compared with the entropies, the factors underlying both of them being largely the same; however, inspection of lists of $\Delta V^{\ddagger}$ and $\Delta S^{\ddagger}$ values generally shows the latter to show the wider and more capricious variations. The further differences that $S^{\ddagger}$ itself cannot be readily determined, and that entropy does not yield as easily as does volume to intuitive insight are additional major reasons why the volume in the reaction profile is more useful than the entropy.

Reactions producing two identical radicals by a simple bond fission should be characterized by a volume increase in the transition state, and this is the case; thus, $\Delta V^{\ddagger} \approx + 10$ cm^3/mol in:

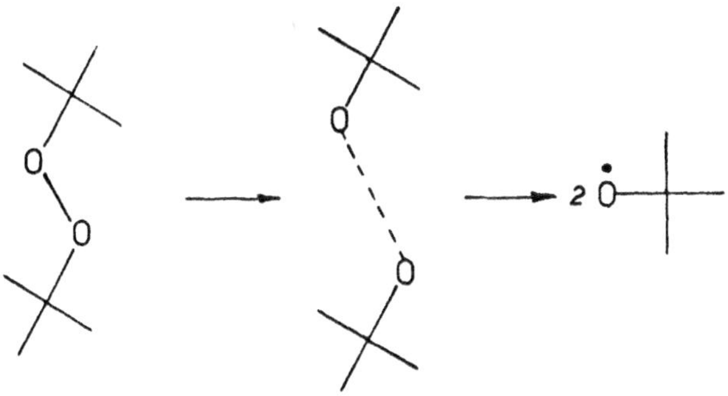

The bond extension calculated from this value is about 0.5 Å, which seems reasonable. In fact, however, the picture is often somewhat complicated by a factor called internal return. If the kinetic experiment is designed to count only the escaped radicals and not the pairs that are the immediate result of bond cleavage, then $\Delta V^{\ddagger}$ is the sum of the reaction volume for pair formation and the activation volume for escape from the cage. Conclusions about bond extensions should therefore not be drawn unless these mechanistic details have been carefully unravelled.

Bond fission producing fragments other than radicals are also pressure inhibited; the reactions:

$$ArCON_3 \longrightarrow N_2 + ArNCO$$

$$Ar\text{-}N_2^{+} \longrightarrow N_2 + Ar^{+}$$

$$CCl_3CO_2^{-} \longrightarrow CO_2 + CCl_3^{-}$$

have activation volumes of + 3, + 10, and + 6 cm^3/mol, respectively.

Recombination of caged radicals should be favored by pressure, $\Delta V^{\ddagger}$ should be negative, and this has been demonstrated in one or two difficult instances. Bond formation of radicals separated by greater distances must first diffuse toward one another. This is the case in the termination step in polymerization. In this instance, fairly rare in bimolecular reactions, the diffusion step is rate controlling, and accordingly, $\Delta V^{\ddagger}$ equals + 20 cm^3/mol.

Displacements are reactions in which one bond is substituted for another; for example:

$$R \;+\; Br\text{—}CCl_3 \longrightarrow RBr \;+\; CCl_3$$

$$EtBr + MeO^{-} \longrightarrow MeOEt + Br^{-}$$

In such reactions, $\Delta V^{\ddagger}$ is generally found to be between -5 and -10 cm^3/mol; evidently bond making in the transition state has progressed further than bond breaking:

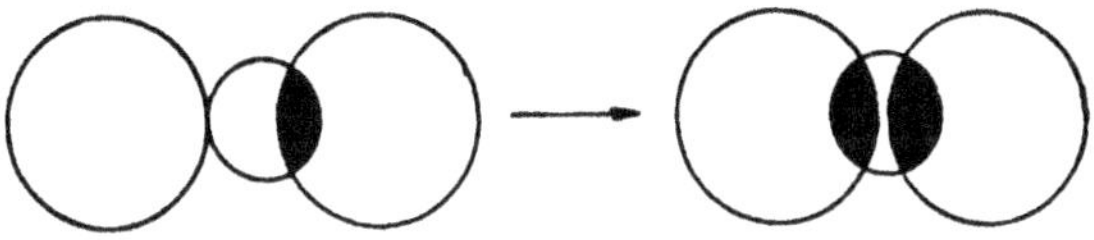

Ionization reactions are strongly accelerated by pressure, as should be expected on the basis of the arguments made earlier for electrostriction. The Drude-Nernst model explains the kinetic data very well, too. Again, there are no major structural effects, and a wide variety of tosylate esters, for example, can be described by the general conclusion that $\Delta V^{\ddagger}$ at 25° in 80% aqueous ethanol equals -21 ± 2 cm^3/mol. Again, the radial dependence of the ions plays a role; the ethanolysis of ethyl chloride, bromide, and iodide have activation volumes of -40, -30, and -25 cm^3/mol, respectively. Again, charge dispersal is in the picture; the solvolysis of allylic (and benzylic) substrates is accelerated by pressure, but less than that of saturated analogs.

And finally, again the solvent polarity is important; the hydrolysis and methanolysis of *t*-butyl chloride are characterized by activation volumes of -10 and -20 cm^3/mol, respectively. One can in fact show that there is a correlation between pressure effect and degree of charge separation: just as pressure favors the concentration of loose ion pairs in equilibrium with tight ones, so it favors products resulting from separate ions over those resulting from paired ions. The difference in $\Delta V^{\ddagger}$ between the two competing reactions shown below is 4 cm^3/mol.

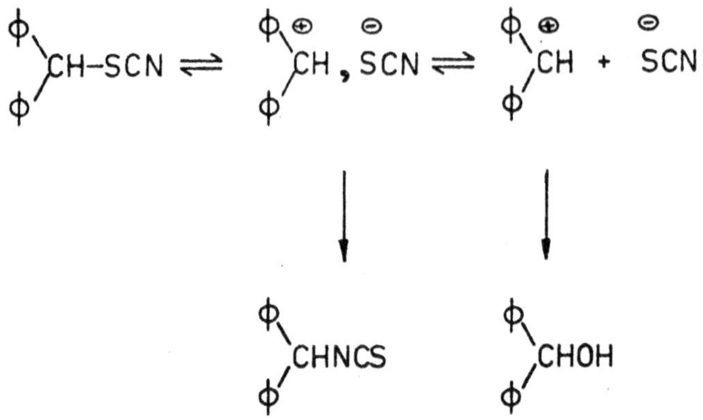

Neutralizations are inhibited, as expected. The well-known and historical Wöhler synthesis of urea has a $\Delta V^{\ddagger}$ of +7 cm^3/mol; the base catalyzed elimination reaction

$$Et_3S^+ + OH^- \longrightarrow C_2H_4 + Et_2S + H_2O$$

has an activation volume of +11 cm^3/mol in water. Solvent dependence now of course works in the opposite direction: the use of non-polar solvents in neutralization reactions should be more severely retarded. The following reaction, which is a close analog of the one just shown, has a $\Delta V^{\ddagger}$ of +32 cm^3/mol in nitrobenzene:

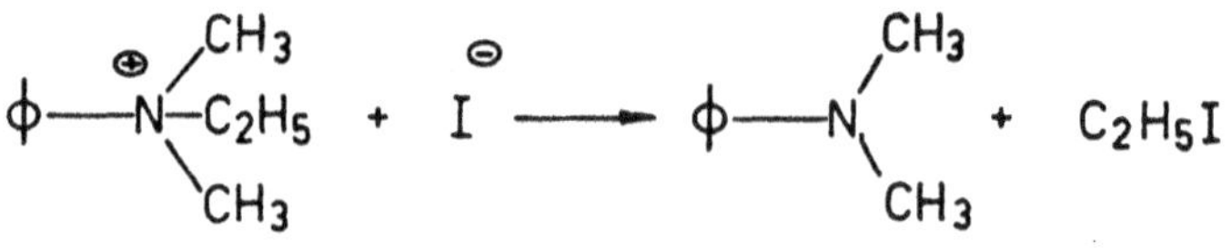

In isomerizations, $\Delta V^{\ddagger}$ may be large or small, depending on the severity of the bond stretching or polarization that may be involved. Cyclohexane inversion, for instance, is a reaction in which the only major feature is carbon-carbon bond rotation; $\Delta V^{\ddagger} = -1\ cm^3/mol$.

Similarly, biphenyl racemization depends on relatively minor bond bending, $\Delta V^{\ddagger} \approx -2\ cm^3/mol$ in that case.

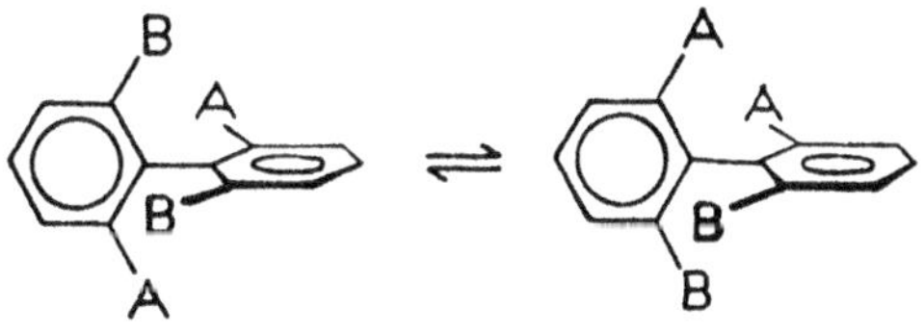

The bond rotation of dimethylformamide has the much larger activation volume of + 10 cm^3/mol. In this instance, rotation destroys the overlap of the CON system; in the transition state the opportunity for resonance is lost and the dipole moment accordingly reduced.

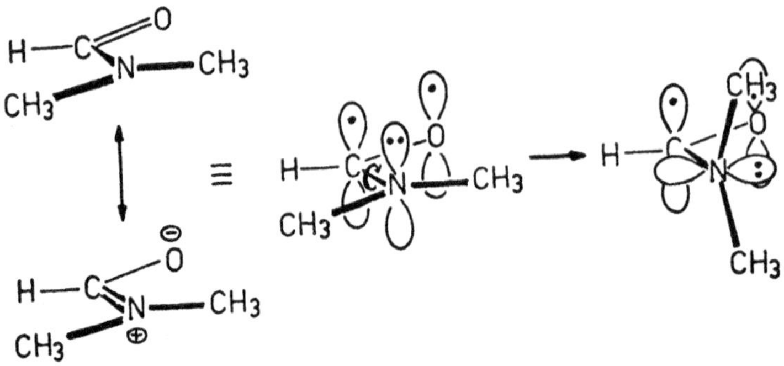

Non-degenerate rearrangements such as the Claisen rearrangement obviously involve more drastic electron relocations, and larger $\Delta V^{\ddagger}$ values would not be unexpected. In this reaction $\Delta V^{\ddagger} \approx -10\ cm^3/mol$.

Since my remarks here constitute a lecture more than a review, it would serve no important purpose to do more than hint at the hundreds, if not thousands, of measurements with reactions in which more than one of the factors described above is operating. It is surely well-known to the audience that most organic reactions do possess a mix of such factors; that diversity in fact is what makes organic chemistry so enticing or exasperating, depending on whether you're in the field or out of it. Perhaps a single example may suffice. For this purpose I chose the Menshutkin reaction; it is chemistry which has been much studied and which is still conceptually simple:

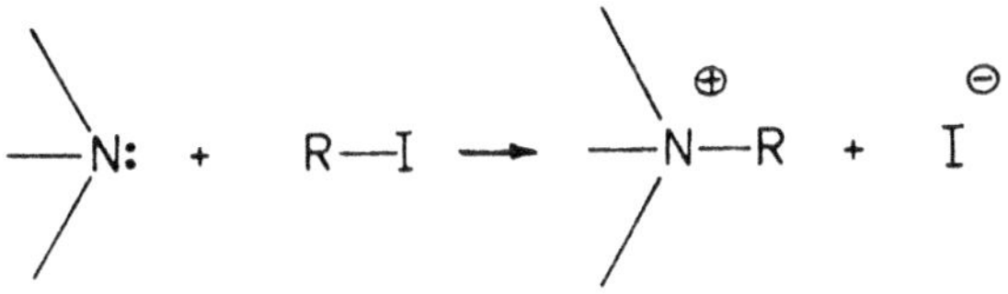

It combines two features: it is both a displacement and an ionization reaction. Both of these lead to volume diminution, and the latter part should be solvent sensitive. This combination then should lead one to expect that the reaction should be greatly accelerated by pressure, and especially so in non-polar solvents. The literature on this subject is voluminous, if I may say so, but the message is clear: $\Delta V^{\ddagger}$ -values of -20 to -50 cm^3/mol are common for this reaction.

Those interested in further details may wish to consult the following literature:

(1) W. le Noble, Progr. Phys. Org. Chem., 5, 207 (1967); this paper listed all $\Delta V^{\ddagger}$ -values reported through 1966 (~ 500).

(2) S. D. Hamann, Mod. Aspects Electrochem., 9, 47 (1974); this publication reports all $\Delta V^{\ddagger}$ -values known through then (~ 250).

(3) T. Asano and W. le Noble, Chem. Rev., 78, 0000 (1978). This article will have about 1500 new data, and list earlier reviews.

ORGANIC PROBLEM REACTIONS UNDER PRESSURE

W. J. le Noble

Department of Chemistry, State University of New York,
Stony Brook, N. Y. 11794

No technique or new approach worth its salt survives for long in the public's interest if all that it can do is to show that it could have solved old problems had it been on the scene on time. High pressure chemistry is no exception to this harsh rule; to pay its way it must be applicable to current problems. In my first lecture I attempted to show that volume changes can be very nicely correlated with structural changes when the latter have already been well defined; now it is time to see whether this correlation is strong enough to be capable of extrapolation into less well reconnoitered territory. As in my first lecture, detailed references to the original work to be discussed now may be found in the review literature quoted there.

To Whalley belongs the honor of being the first chemist to solve a mechanistic puzzle by means of high pressure chemistry. This puzzle was the question of whether carbonium ions generated in acidic, solvolytic media from such precursors as ethers, esters, amides, acetals, and so on, were free, or bound to solvent. For example:

Kelm, H. (Ed.) High Pressure Chemistry. 345-363. *Copyright* © 1978. *D. Reidel Publishing Company. All Rights Reserved.*

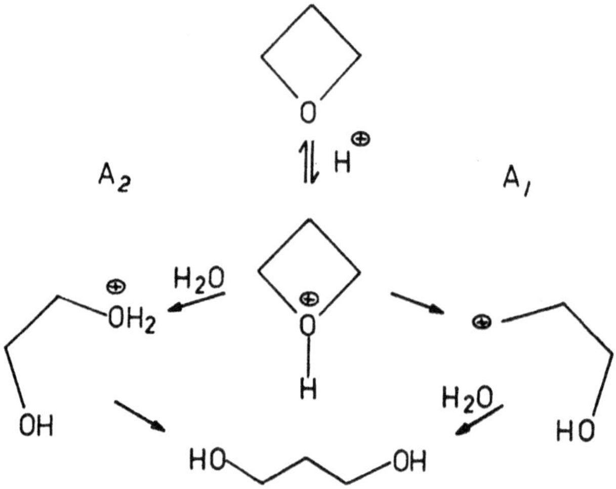

If the first alternative is true, the rate controlling step in this acid catalyzed reaction is unimolecular, the carbonium ion is free, and the reaction would be, in the chemical jargon, an A_1 reaction; if the solvent molecule is involved at the start, the reaction is bimolecular (A_2) and the carbonium ion is not free. Arguments used in the past have included especially effects of isotopic substitution in the solvent (D_2O _vs_ H_2O) and of acidity, with correlation of k vs either h_o or pH being used as a basis for assignment to the A_1 and A_2 categories, respectively. The difficulty here has been the fact that the conclusions so drawn do not correlate with intuition. The sample reaction shown, for example, by the use of the Hammett acidity function, is concluded to be an A_1 reaction; yet, as any organic chemist knows, primary carbonium ions and nucleophilic media are an improbable combination.

The prediction that one makes on the basis of pressure effects is conceptually very much simpler than either one of the older criteria: if we assume no large volume change associated with the preequilibrium (very reasonable in view of the absence of structural effects in model acid-base equilibria!), the first pathway will have a positive activation volume, and the second a negative one. Experimentally, one obtains the latter result: the reaction is accelerated by pressure, and hence the reaction is in fact an A_2 reaction. I should add at once that the same is true for other solvolytic reactions involving simple ethers, esters, and amides; positive activation volumes are obtained when stable cations such as tertiary or alkoxy substituted carbonium ions are potential products. Thus, a much better correspondence resulted between these assignments and other sources of chemical folklore. As an interesting sidelight, perhaps, the new assignments have not been adopted by everyone: fruitless discussion of the old conclusions still continues in some literature. Evidently, this approach is still suspect to some chemists, evidently under the impression that the solutions are in touch with rusty autoclaves and greasy parts. Let us hope that these sessions here will contribute to the demise of those superstitions.

We may perhaps more briefly discuss this question of the freeness of intermediates in other types of reactions. One prominent among these is the carbene, which forms in B_1 reactions if it is free, or via an intermediate which is nothing more than an oxygen ylid if the B_2 mechanism is operating:

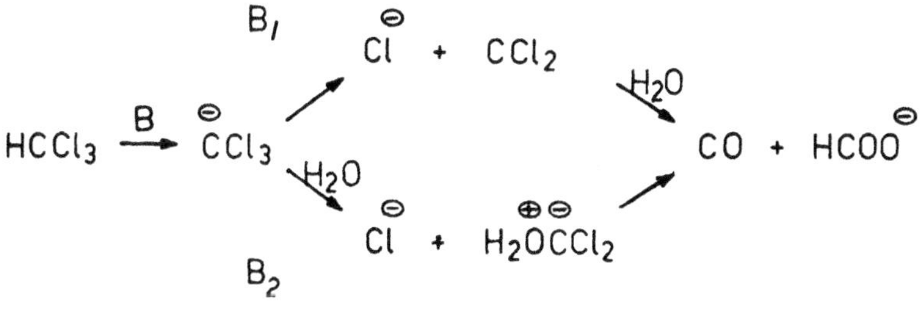

The activation volume is + 16 cm^3/mol; thus, the carbene is free, the mechanism is B_1 , and a question left dangling ever since Hine postulated this pathway in the early fifties is answered. Similar conclusions have been reached for other carbenes, and for the benzyne intermediate. In this series, there is an exception to point to as well: the allenylidene drawn below is formed solvolytically in a reaction which is virtually totally insensitive to pressure. The interpretation offered for this

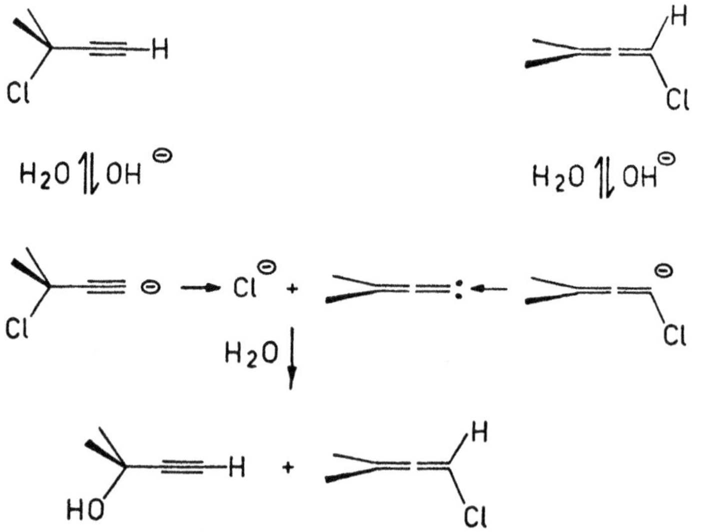

curious exception was that the incipient carbene is polarized by the leaving anion:

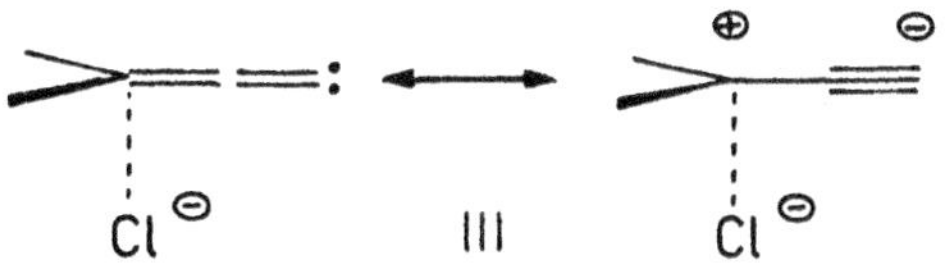

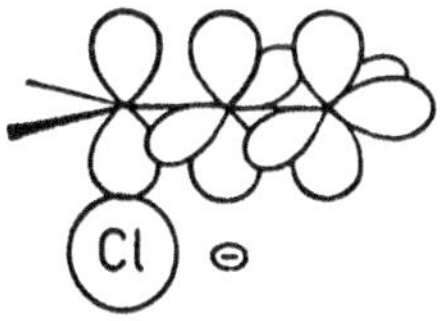

The sense of polarization is in accord with what we know about the stability of tertiary cations and acetylenic anions, and the dipole would by virtue of electrostriction account for the lack of a volume increase. Subsequent conventional investigations have supported this picture. Thus, it would be but a small extrapolation to say that the carbene is paired with the leaving group, much as ion pairs form in the first step of solvolysis. Such pairing can be detected if internal return occurs. Evidently such return does take place: when the reaction of the acetylene is carried out in the presence of a large excess of chloride ion, the amount of isomeric allene produced can be made almost independent of the concentration of the added ion (by proper choice of medium). This independence can only mean that the carbene and the ion did not go separate ways before the return. The organic chemists in the audience can probably appreciate the possibility that these contentions can also be supported in stereochemical ways; thus, if internal return occurs, the following scheme should apply. Appropriate experiments are in progress; the conclusion at this point is nevertheless already inescapable that high pressure kinetics can not only be helpful in classifying individual reactions, but exceptional results may point the way to features previously completely unsuspected.

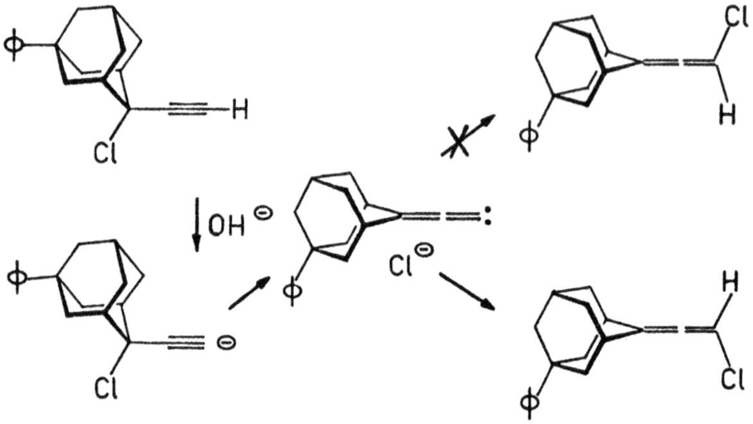

The nitrenes constitute still another group of intermediates capable of study in this way. Fluoronitrene is an instructive example. Earlier experiments had strongly hinted that when difluoramine is hydrolyzed in aqueous base, the nitrene is an intermediate; thus, the reaction is exceptionally fast compared to that of difluoramine with other simple anions, and the rate law - first order in HNF_2 - demands an intermediate containing one nitrogen atom although the product (N_2F_2) has two. D-exchange is rapid, excluding NF_2^- as the possible intermediate; yet no definite assignment could be made because all attempts to trap NF in the way nitrenes can usually be caught (with olefins to give aziridines) led to failure.

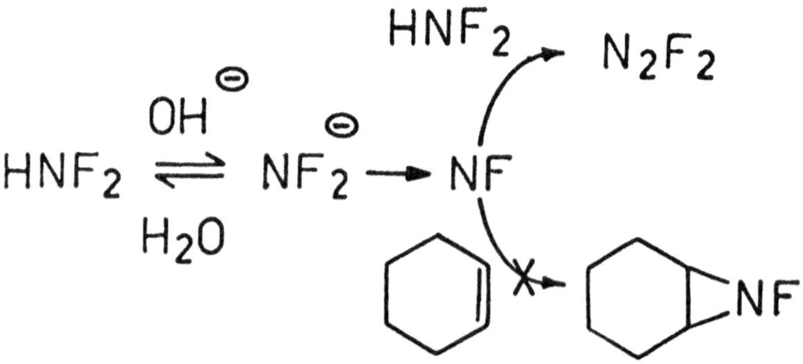

High pressure study then showed that the reaction had a positive activation volume in spite of the small radius of the leaving fluoride anion. Totally convinced by this observation, we made a renewed study of the trapping reactions, and while time precludes my getting into this work very deeply, it may be said now that the reasons for the failure of the trapping efforts have become clear. Fluoronitrene is namely extraordinarily efficient in abstracting hydrogen atoms; thus, even benzene suffers this fate and is converted into biphenyl. On the other hand, use of nitrosocompounds did eventually lead to successful trapping: the product shown below was the result.

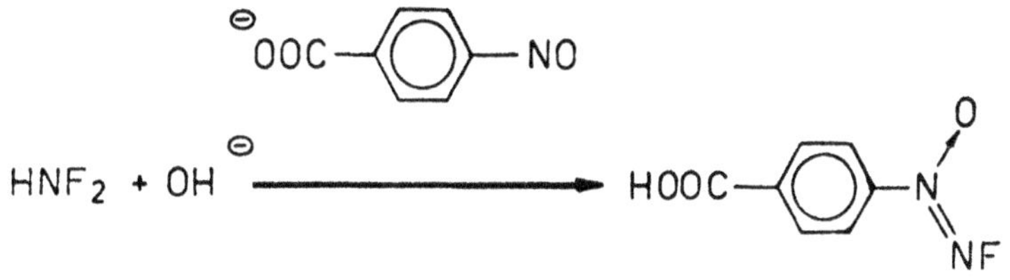

Perhaps the conclusion warranted here is that activation volume data can be sufficiently persuasive, if only to its practitioners, that they become willing to deny the emperor's clothes, all opinion to the contrary notwithstanding.

The examples all illustrate results obtained in pursuit of questions regarding the free vs bound states of intermediates; however, these are not the only questions one can raise. Another question often arising occurs when a bond as yet unformed results from the closure of radical or ionic sites; in other words, whether the intermediate is a diradical or a zwitterion. One neat example is the following. In the cycloaddition of tetracyanoethylene (TCNE to insiders) to enol ethers, the stereochemistry has clearly implicated an intermediate which closes to product in a final step:

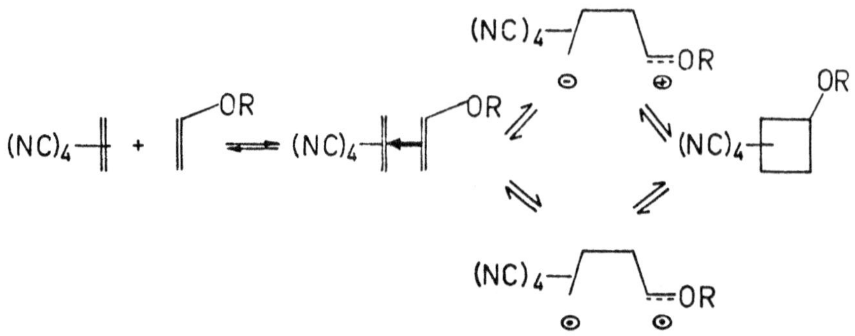

Two high pressure groups have looked at this reaction; one in the forward direction, the other in the reverse. In this rare instance, the backward study yielded the more direct information: that reaction is accelerated virtually as strongly as is solvolysis, proving the ionic nature of the intermediate.

Without a doubt, the most controversial problem that has ever arisen in mechanistic organic chemistry is that of the so-called non-classical ions. It is not remotely possible - or desirable - to recount even a fraction of the literature and laboratory effort that has been devoted to this problem; all that can be done here is to indicate with one example what the problem is all about, and why it so stirred the emotions.

It was proposed in the early fifties that when the two epimeric 2-norbornyl esters solvolyze, the <u>exo</u>-isomer is formed with assistance of the 1,6- bonding sigma electrons giving a non-classical intermediate, whereas the <u>endo</u>-analog ionizes without such assistance to a classical cation.

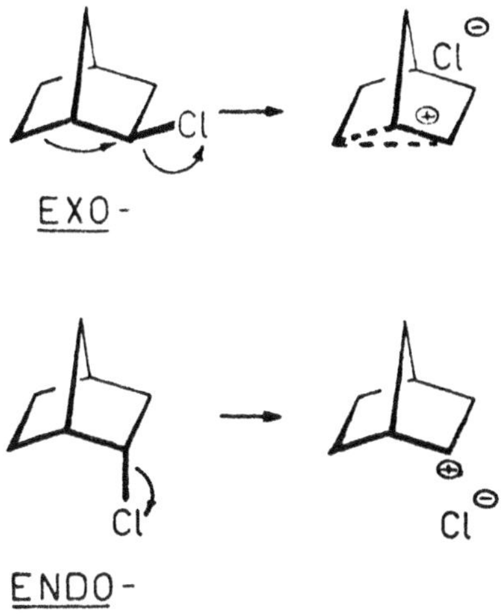

The non-classical intermediate is pictured by those who do not accept its existence as rather a pair of rapidly equilibrating classical ions; this view can indeed account for virtually all of the results just as well as the symmetrical structure:

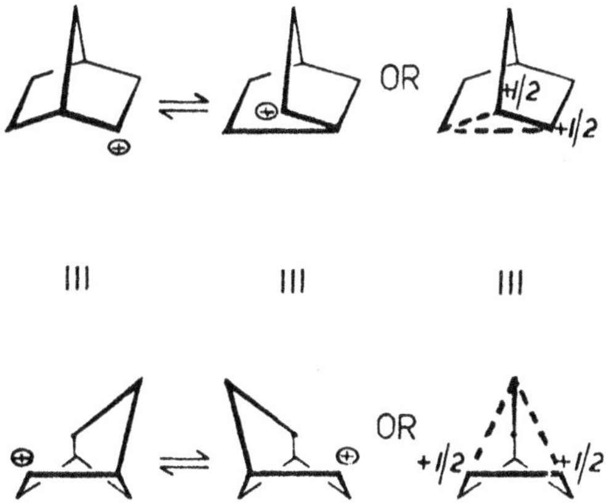

Perhaps the main reason for the reluctance of many organic chemists to accept this proposal stems from the century of success that followed Kekulé's suggestion of a tetravalent carbon atom. One doesn't depart lightly from a habit which hasn't been faulted for that long! The non-classical ions, however, propose exactly that; thus, the 6- carbon atom would in fact be bound to five other atoms:

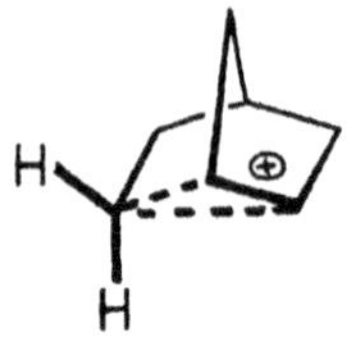

The high pressure kinetics results support the proposal. The exo-isomer solvolysis is pressure accelerated, as ionizations always are, but less so than the endo-isomer because of the delocalization of the positive charge. Since the anions are the same and the subdivision of the plus charge is by 2, $\Delta V^{\ddagger}_{exo}$ should be 25% less negative than $\Delta V^{\ddagger}_{endo}$. The results are -14.3 and -17.8 cm^3/mol, respectively. The agreement in a number of other cases has been equally good.

The most recent major contribution to mechanistic organic chemistry resulting from work by high pressure kineticists involves concerted reactions; i.e., reactions that do not traverse any intermediate stages at all, but rather pass smoothly from reactant to product in one fell swoop, with all atoms involved moving from initial to final positions without stopping. As an example, consider the ionization of γ-haloamines. This reaction leads to three fragments

and the question may be raised: does this happen concertedly, or in two stages?

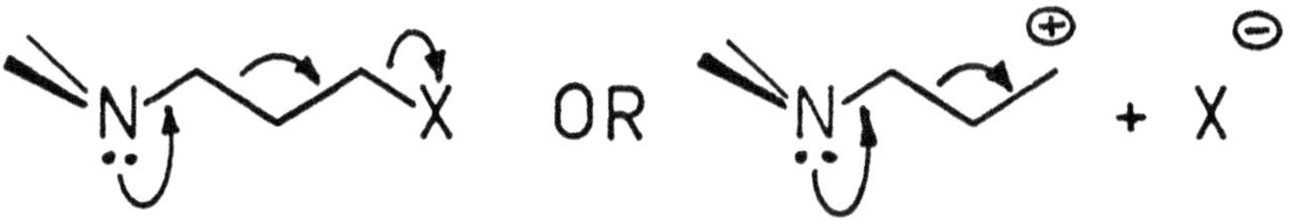

One important argument has been advanced: the rate of many of these reactions is faster than that of the carbon analogs capable of only the first step, although the greater electronegativity of the nitrogen atom should lead to a slower reaction

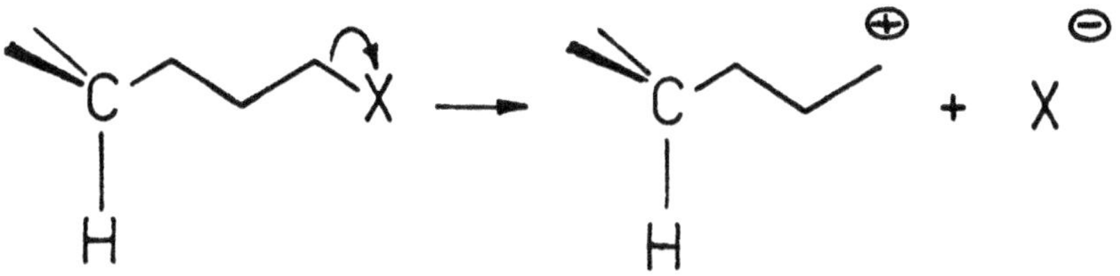

(inductive effect). The weakness of that argument is that it is not clear how much slower these haloamines should be.

The high pressure argument is that the extra breaking bond should show up as a positive contribution to $\Delta V^{\ddagger}$, and indeed this is what happens. Thus, one finds that

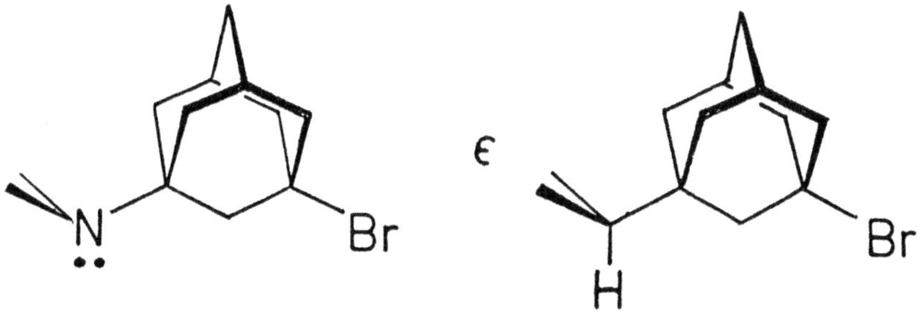

react in aqueous alcohol with activation volumes of -11.9 and -19.5 cm^3/mol, respectively, and the same difference appears consistently in a set of similar pairs in which the rate ratio N to C varies from 0.2 to 50,000; but interestingly, at the low end of this scale with a rate ratio of 0.12, there's a change: $\Delta V^{\ddagger}_N$ is then -23.8 cm^3/mol. We have then reached the limit of the inductive effect, which at this distance is evidently a factor of eight. Equally interesting, the products in that reaction are still fragmentation products: clearly the nature of the products in that case would have been misleading.

The decomposition of certain anions into multiple fragments is perhaps an even simpler example, since the volume changes are then not complicated by the appearance of new charges as is the case in solvolytic fragmentations. If we consider the reaction:

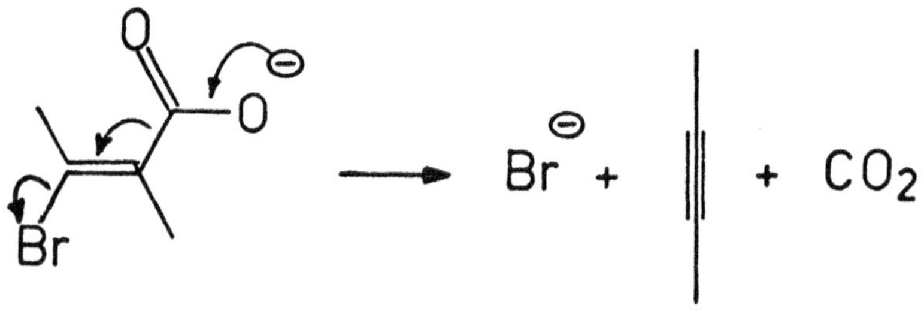

concertedness is a safe conclusion if it realized that $\Delta V^{\ddagger}$, which equals +18 cm^3/mol in this case, may be compared to the range of +8 ± 2 cm^3/mol normally observed in simple one bond decarboxylations such as

$$CCl_3COO^- \longrightarrow CCl_3^- + CO_2$$

An exceedingly instructive case is that of the apparently complex reaction

$$CH_2ClCONHNH_2 + OH^- \longrightarrow H_2O + Cl^- + CH_3COOH + N_2 + N_2H_4$$

The mechanism of this reaction has been studied, and was concluded to be

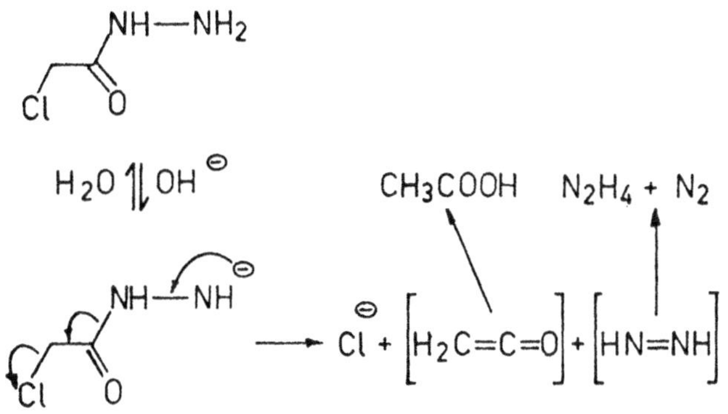

In support it may be noted that ketene is indeed known to be rapidly hydrated in water, and that diazene does indeed disproportionate to nitrogen and hydrazine. However, the reaction is pressure accelerated and $\Delta V^{\ddagger}$ is -5 cm^3/mol. With a discrepancy of that magnitude (expected: +18, found: -5 cm^3/mol), it may be considered certain that the mechanism on which the expected value was based is wrong, and indeed was indicated by subsequent investigation. Methyl labelling of the hydrazino group as well as basicity effects at high pH in stopped flow experiments revealed that the α-NH group rather than the β-NH_2 donates a proton to yield the first anionic intermediates. Internal displacement then occurs as the rate controlling step; it is this feature that produces the negative activation volume. The reaction then proceeds in rapid steps that cannot all be traced, but one of the intermediates, the acetyldiazene ($CH_3CON{=}NH$) is sufficiently stable that its presence can be visually detected.

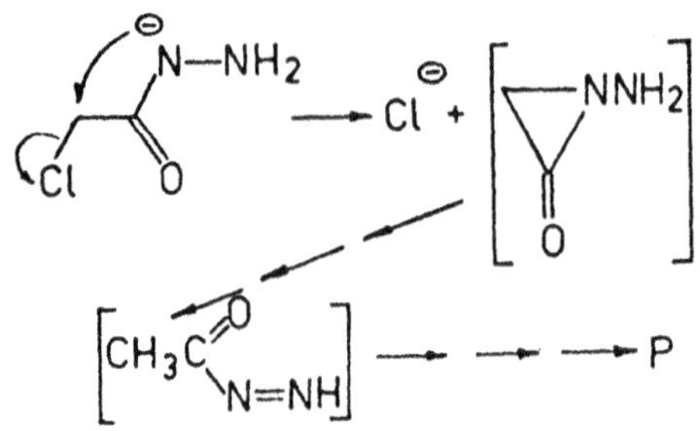

The moral of this story is that a discrepancy between predicted and observed activation volume should not be dismissed since it may very well reveal shortcomings in the mechanism adopted for the prediction.

Beside the fragmentations, there is a second major group of reactions in which the question of concertedness plays an important, indeed crucial role: i.e., the pericyclic reactions. In these reactions the Woodward-Hoffmann rules predict the stereochemistry provided that the reactions are concerted and involve no intermediates. As an example, we may consider the well-known Diels-Alder reaction; in the nomenclature used in this field, it is a [4 + 2] cycloaddition.

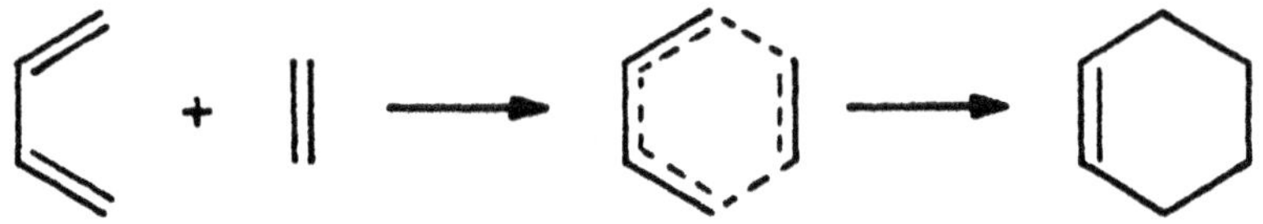

The Diels-Alder reaction is characterized by remarkable stereospecificity, and accordingly, there is every reason to believe that the reaction is concerted. This means that the two bonds form simultaneously, and hence that the nuclei in the transition state are already near their final positions. The activation volumes of the many Diels-Alder reactions which have been studied under pressure by now bear this out: in all instances the activation volumes are nearly comparable to the reaction volumes, the latter amounting to about -40 cm^3/mol in most cases.

These results are in sharp contrast to those [2 + 2] cyclo-additions which occur via intermediates. In simple dimerizations diradicals are strongly implicated by a host of observations; in such cases $\Delta V^{\ddagger}$ is far less negative, and when direct competition occurs between the two, the Diels-Alder reaction is greatly favored by the application of pressure. Thus, in:

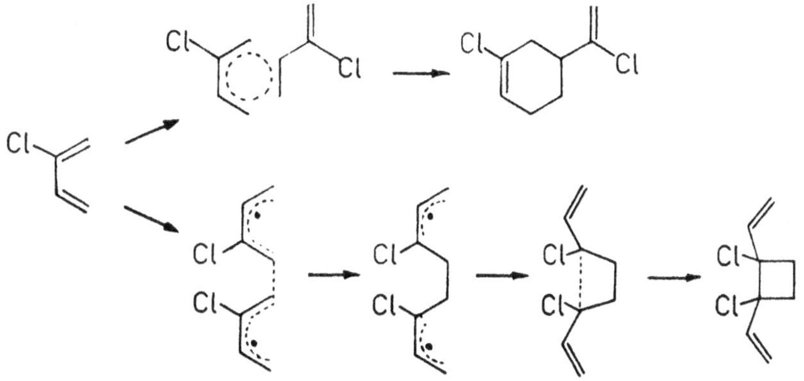

$$\Delta\Delta V^{\ddagger} = V^{\ddagger}_{[2+4]} - V^{\ddagger}_{[2+2]} = -10\ cm^3/mol.$$

A number of instances have come to light in which the activation volume is in fact more negative than the reaction volumes, meaning that some of the atoms in the transition state are more closely together than in the final state! These cases have been observed in reactions in which secondary orbital interactions have been postulated to account for stereochemistry. For example:

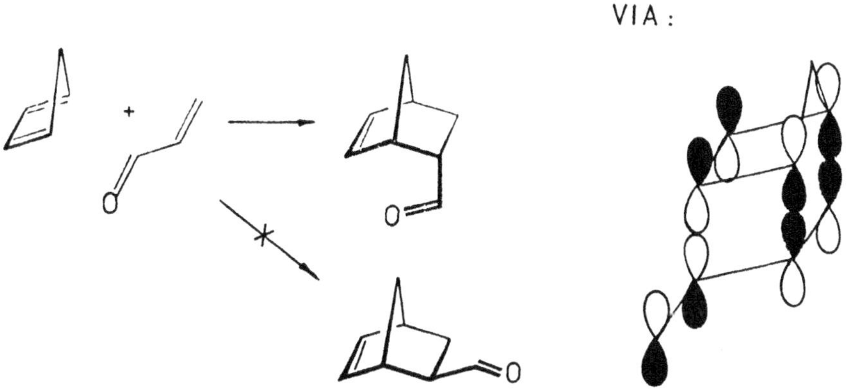

The following example is again an especially instructive one. We studied the cycloaddition of tropone to cyclopentadiene, which is known as a [6 + 4] cycloaddition with the prerequisite number of electrons for an allowed, concerted reaction; we expected $\Delta V^{\ddagger}$ to be close to -30 cm^3/mol and ΔV near -40 cm^3/mol as in the case of the Diels-Alder reactions.

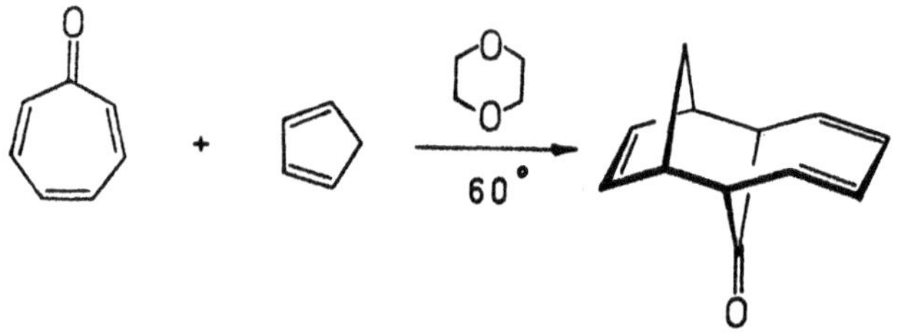

Much to our initial surprise, reaction was accelerated only very little by pressure; the activation volume was only -7 cm^3/mol; the conclusion seemed inescapable that a stepwise reaction was taking place in spite of all earlier opinions to the contrary. In order to make sure that we were not being misled by an equally exceptional value value of the reaction volume, we measured the latter also, in a simple dilatometric experiment. ΔV turned out to be only -5 cm^3/mol! This solved the problem in a sense: with $\Delta V^{\ddagger}$ and ΔV being so similar, the reaction was indeed concerted after all. But instead we have a new puzzle: why are $\Delta V^{\ddagger}$ and ΔV so small? This question can also readily be answered since it is so straightforward a matter to measure individual partial volumes. The results were that tropone, cyclopentadiene, and the adduct have partial volumes under the conditions chosen of 85.6, 97.2, and 177.2 cm^3/mol. These data yield a reaction volume of 177.2 - 97.2 - 85.6 = -5.6 cm^3/mol, in agreement with the dilatometric experiment; but more important, it reveals that tropone is the culprit in our mistaken assumptions regarding ΔV and $\Delta V^{\ddagger}$. Thus, it should be noted that it is some 10 cm^3/mol smaller than cyclopentadiene, even though it has three additional atoms, two carbon and one oxygen. Estimates on the basis of parachor suggest that about 30 cm^3/mol is "missing" in the case of tropone. The reason is that tropone is polar because of resonance; the dipole and the resulting electrostriction are apparently responsible. In accord with this, other physical

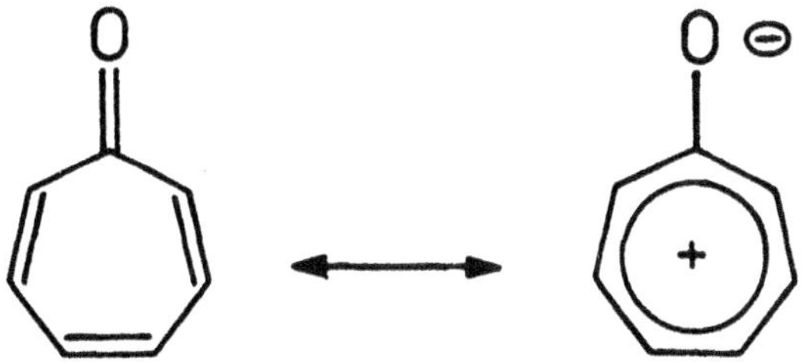

properties are known to be exceptional as well: total miscibility with water, a dipole moment of 4.2 D, a boiling point of 105° at 10 torr, and a $\nu_{C=O}$ of 1650 cm^{-1} all testify to the polar nature of the carbonyl group. What all this teaches us is that the opportunity to measure individual partial volumes is of exceptional value in making predictions of $\Delta V^{\ddagger}$ and ΔV; no counterpart exists in this respect for those who prefer $\Delta S^{\ddagger}$ as a probe in mechanistic chemistry. In other pericyclic reactions, $\Delta V^{\ddagger}$ should be of similar use, but this has not yet been demonstrated.

I remarked earlier that the reaction coordinate is an important variable in consideration of the activation volume: clearly the transition state volume is dependent on the length of bonds in transition, and on the magnitude of charges in formation or disappearance. It's therefore perhaps no wonder that the well-known Hammond postulate should figure prominently in high pressure investigations.

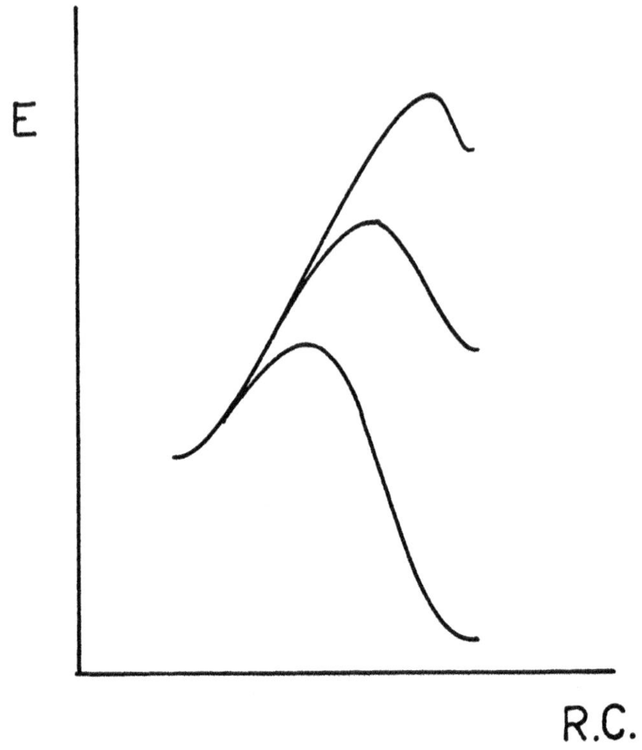

The Hammond postulate: greater exothermicity implies an earlier transition state.

Thus, the following data were determined:

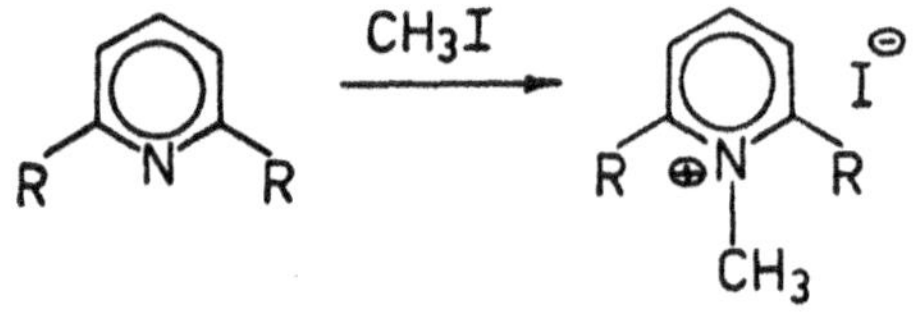

R	10^7k	$\Delta V^{\ddagger}$	$\Delta V^{\ddagger}/\Delta V$
H	2100	-22	0.47
Me	60	-24	0.50
Et	8	-27	0.55
i-Pr	0.3	-30	0.59
t-Bu	<0.001	<-50	>0.90

The large rate decreases as the size of substituents is raised are indicative of decreases in exothermicity as well; at the same time, $\Delta V^{\ddagger}/\Delta V$ shows that the lateness of the transition state increases accordingly. We have found additional evidence for this proposition in that there is a measurable increase in the Cl^{35}/Cl^{37} isotope effect between pyridine and 2,6-lutidine.

Perhaps one of the more promising near future applications is in synthesis. At 10 kbar, 10 cm^3 of space represents 2.5 kcal/mol, enough to change a 9:1 product ratio to one of 1:9. Such changes, and even better, have been realized, and there is surely more to come. Pressure induced improvements in yield and in stereospecificity remain closely tied to our ability to make intelligent guesses about volume changes, however, and the prospective users of high pressure synthesis would do well to understand these.

After listing all these benefits and advantages of high pressure research, I would probably be wise to finish with a few caveats. While it may seem obvious even to students of chemistry, high pressure kinetics is certainly not always the only or the best means of investigating a mechanism, and in many instances knowledge of $\Delta V^{\ddagger}$ is not helpful at all in reaching a decision between alternative pathways. And when it can be, it is most valuable if there is information not only about the unknown reaction, but also of analogs of well-known mechanism. Even then, a line must be drawn somewhere between what is a significant deviation from expected values and what constitutes a random type of difference outside of our ability to interpret. I have no wish to make a pronouncement of what this interpretability limit is; it depends on the data available, on the similarity of the unknown reaction and its analogs, and above all, on the care that was lavished on the precision and accuracy of the measurements.

EFFECTS OF PRESSURE ON THE FORMATION AND PROPERTIES OF POLYMERS

G. Jenner

Laboratoire de Chimie Organique Appliquée,
Institut de Chimie, Université Louis Pasteur,
Strasbourg, France

FOREWORD. Temperature and pressure constitute the two fundamental parameters in nature. However until the beginning of the century and despite the work of pioneers such as Amagat, Boyle and Bridgman, temperature has been the priviledged parameter for the obvious reason that it is simpler and less expensive to heat a substance than to compress it.

However with the help of multiple high pressure experimental techniques developed in the last twenty years, investigations under pressure increase yearly by a rhythm of 3 to 7% (Statistics from the Stanford Research Institute). This growing interest reflects the deep modifications of the physical and chemical properties induced by pressure.

My lecture will be devoted to the high pressure synthesis of polymers and the pressure effect upon some of their properties.

Kelm, H. (Ed.) High Pressure Chemistry. 365–420. *Copyright* © 1978. *D. Reidel Publishing Company.*
All Rights Reserved.

I. INTRODUCTION AND THEORY

The rate of chemical reactions in gaseous and condensed phases is influenced by pressure in following way :
- compressing the reaction medium increases the concentration of the reacting molecules (concentration effect). It appears mainly in the 0-1000 bar region.
- pressure changes the rate of intermolecular diffusion
- pressure compresses the molecules themselves, deforming the electron clouds and affecting the collision rate (kinetic effect).

In dense liquid phase, the free space required for the formation of the transition state is restricted and the influence of pressure on the reaction rate overshadows all other effects.

The application of high pressures to the polymerization reactions presents a potential interest, for practical and theoretical reasons. First, the successive addition of molecules of monomer leads to a decrease in volume and therefore is thermodynamically favored by an increase in pressure. Secondly, pressure affects the various elemental processes of the polymerization and hence permits the synthesis of polymers with possible structure and properties different of those for polymers prepared in usual pressure conditions.

Before giving some features of the high pressure homopolymerization, it seems necessary to recall briefly the main physical effects of pressure able to affect the course of the polymerization reactions in the liquid phase.

I.1 Physical effects of pressure

I.1.1 Volumetric effects. Two parameters are commonly used to describe the decrease in volume of a compressed substance :
- compressibility

$$\beta = -\frac{1}{V}\left(\frac{\partial V}{\partial P}\right)_T = -\left[\frac{\partial \mathrm{Ln}\ V}{\partial P}\right]_T$$

- compression coefficient

$$\kappa = \frac{V_o - V}{V}$$

Although the decrease in volume for liquids is much less than for gases, it is not negligible under high pressures and reaches at room temperature 4 to 10 % at 1000 bar and 20 to 30 % at 10 000 bar. β increases with temperature and diminishes with increasing pressure, but the more rigid is the monomer structure, the less the compressibility. For example, the existence of cycles or aromatic rings leads to an attenuation of the volumetric effect induced by pressure. To calculate the compressibility of monomers one may use an empirical relationship, the so-called Tait equation :

$$\kappa = C\ \mathrm{Ln}\left(\frac{B + P}{B}\right)$$

where B and C are constants independent on pressure.

The compressibility of polymers is of course smaller than that of liquids. For polystyrene, polyethylene, at 20°C, the compression coefficient has a value of 2 to 3 % at 1000 bar and 4 to 6 % at 2000 bar.

I.1.2. Viscosity. The viscosity η of liquids increases considerably under pressure. At 12000 bar, the ratio η_p/η_o can reach a value of 1000. The important increase of the viscosity of the reaction medium constitutes the physical effect of pressure whose consequences are the most determining in polymerization reactions. This effect is particularly reflected in the gel effect : the termination rate due to recombination governed by the diffusion rate of the macroradicals decreases, when the viscosity of the medium increases. Thus, as soon as the bulk polymerization carried out at a given pressure starts, the diffusion rate of the macroradicals in the compressed monomer (with viscosity η_p) is already identical with the rate which would be reached at atmospheric pressure for acertain conversion for which the viscosity should be equal to η_p. Therefore the gel effect appears at a yield the lower as the pressure is higher.

I.1.3 Phase equilibria. Pressure is able to modify the nature of the reaction medium : phase change (solidification) or phase separation.

- Transition liquid-solid of the monomer

The thermodynamic relationship between pressure and melting temperature T_m of a pure compound is given by the Clapeyron equation :

$$\frac{dT_m}{dP} = T_m\frac{\Delta V}{\Delta H}$$

ΔV, ΔH : variations in volume and enthalpy during the solidification

For all organic liquids, T_m increases with increasing pressure. Table I shows two examples : for a monomer (styrene) [1] and a solvent (benzene) [2]

Table 1. Solidification of styrene and benzene as function of pressure

Pressure (bar)	T_m (°C) styrene	benzene
1	- 30.6	5.5
500	- 20	20.3
1000	- 11	33.7
2000	6	56
3000	20	79
3700	30	93
6000	55	134
10000	98	192

- Solubility of the polymer

In an ideal solution, the solubility of a solid in a liquid is related with pressure according to :

$$\left[\frac{\partial \mathrm{Ln}\, f_s}{\partial P}\right]_T = \frac{V_s - \bar{V}_s}{RT}$$

f_s : molar fraction of the solute in the saturated solution
V^s_s : molar volume of the solid
$\bar{V}_s$: partial molar volume of the solid in the saturated solution

The above relation is only an approximation for real solutions and for polymers which cannot be assimilated to an unique molecular entity when their polydispersity is important.

The problem of the solubility of the polymer in the reactional phase under pressure has been carefully investigated for the system ethylene-polyethylene [3] for which the phenomenon of phase separation is essential for the application of industrial processes.

I.2. Kinetic aspects

The high pressure thermodynamics is well known and is given in detail by other authors during the present ASI. In the transition state theory, the fundamental relationships governing the condensed phase equilibria may be written as functions of the activation volume $\Delta V^{\ddagger}$:

$$\left[\frac{\partial \mathrm{Ln}\, k}{\partial P}\right]_T = -\left[\frac{\partial \Delta G^{\ddagger}}{RT\partial P}\right]_T = -\frac{\Delta V^{\ddagger}}{RT}$$

$$\left[\frac{\partial \Delta S^{\neq}}{\partial P}\right]_T = \left[\frac{\partial \Delta V^{\neq}}{\partial T}\right]_P$$

$$\left[\frac{\partial \Delta H^{\neq}}{\partial P}\right]_T = \Delta V^{\neq} - T\left[\frac{\partial \Delta V^{\neq}}{\partial T}\right]_P$$

In these expressions, the activation parameters (free energy $\Delta G^{\neq}$, enthalpy $\Delta H^{\neq}$, entropy $\Delta S^{\neq}$) characterizing the transition state have their usual meaning.

I.2.1 Pressure dependence of the rate constant.
Rigorously speaking, the pressure dependence of the rate constant k for a reaction :

$$\nu_1 A_1 + \ldots\ldots + \nu_j A_j \rightleftharpoons X^{\neq} \longrightarrow \nu_1' B_1 + \ldots\ldots + \nu_j' B_j$$

is written as :

$$\left[\frac{\partial \mathrm{Ln}\, k}{\partial P}\right]_T = -\frac{\Delta V^{\neq}}{RT} + \beta \Sigma \nu_i$$

with $\Sigma \nu_i = 1 - \Sigma \nu_j$
It must be noticed that, except for unimolecular reactions such as the decomposition of the initiator :

$$A \rightleftharpoons X^{\neq} \longrightarrow D \qquad \text{where } \Sigma \nu_i = 0$$

the compressibility of the solution must be taken into account in the final expression for k. This is for instance the case of the propagation, termination or transfer reactions where $\Sigma \nu_i = -1$.

Now if we introduce the yield τ instead the reaction rate constant, we obtain :

$$v = k\, (A_1)^{\alpha_1} \ldots\ldots (A_j)^{\alpha_j}$$

where α_j is the reaction order versus reactant j, (A_j) being the molar concentration.

$$\left[\frac{\partial \mathrm{Ln}\, v}{\partial P}\right]_T = \left[\frac{\partial \mathrm{Ln}\, k}{\partial P}\right]_T + \beta \Sigma \alpha_j$$

In this expression $\beta \Sigma \alpha_j$ represents the rate increase due to the "concentration effect" of pressure

$$\left[\frac{\partial \mathrm{Ln}\, v}{\partial P}\right]_T = -\frac{\Delta V^{\neq}}{RT} + \beta\left[\Sigma \alpha_j + 1 - \Sigma \nu_j\right]$$

If v is defined by the decrease of the concentration of teactant j, we write :

$$v = -\frac{d(A_j)}{dt}$$

$$v = (A_j)\frac{d\tau}{dt} = (A_j)\,\tau^{*}$$

$$\left[\frac{\partial \mathrm{Ln}\,\tau^{*}}{\partial P}\right]_T = -\frac{\Delta V^{\ddagger}}{RT} + \beta\left[\Sigma\alpha_j - \Sigma\nu_j\right]$$

In these expressions, it can be seen that the compression of the solution influences the value of the reaction rate through the stoechiometric coefficients $\Sigma\nu_j$ of the reactants and the orders $\Sigma\alpha_j$.

I.2.2 Application to polymerization reactions

I.2.2.1. Free radical polymerizations. The general features of free radical polymerizations have been known and understood for at least 30 years. The acceptance of a simple chain mechanism is universal and the traditional kinetic scheme for free radical polymerization in which the chain reaction is initiated by the thermal or photochemical decomposition of a suitable molecule I, is as follows :

- initiation

$$I \xrightarrow{k_d} 2\,R^{\cdot} \qquad \text{with} \quad v_d = 2\,f\,k_d\,[I]$$

(f : efficiency of the initiator)

$$R^{\cdot} + M \xrightarrow{k_i} P_1^{\cdot} \qquad \text{with} \quad v_i = k_i[R^{\cdot}][M]$$

- propagation

$$P_m^{\cdot} + M \xrightarrow{k_p} P_{m+1}^{\cdot} \quad \text{with} \quad v_p = k_p[M][P_m^{\cdot}]$$

- transfer

$$P_m^{\cdot} + S \xrightarrow{k_s} P_m + S^{\cdot} \quad \text{with} \quad v = k_s[S][P_m^{\cdot}]$$

$$M + S^{\cdot} \longrightarrow P_1^{\cdot}$$

- termination

$$P_m^{\cdot} + P_n^{\cdot} \xrightarrow{k_t} \text{polymer} \qquad \text{with} \quad v_t = 2\,k_t\,[P^{\cdot}]^2$$

where $[M]$ and $[R^{\cdot}]$, as well as $[P^{\cdot}] = \Sigma\,[P_m^{\cdot}]$ are the concentrations of monomer and primary radicals and the total concentration of polymer radicals, respectively. S represents a chain transfer agent (monomer, initiator or solvent).

Assuming radical reactivity is independent of radical size and any transfer derived radicals, $S^{\cdot}$, rapidly add monomer and are not consumed in any other process, application of the steady state hypothesis leads to the following expression for the overall polymerization rate :

$$v = k_p [M] \left[\frac{f k_d [I]}{k_t} \right]^{1/2}$$

In polymerization systems, as described above, the elemental reactions may be regarded as competing reactions. Propagation favours high polymer production, while transfer and side reactions lead to low molecular products. High pressure investigation of polymerizations is a feasible method for determining which of the competing reactions are most influenced by pressure and hence which have the greater volume decreases in the transition complex.

Extensive studies have been made on radical polymerization and copolymerization in liquid phase and it has been reported that the kinetic changes caused by pressure often alter the molecular weight and the structure of the polymer which is produced. To understand, correlate and predict the chemical effect of pressure on polymerization reactions, its effect on elementary reactions must be determined.

Let us consider the pressure effect on the reaction rate v :

$$\frac{\partial \mathrm{Ln}\, v}{\partial P} = \frac{\partial \mathrm{Ln}\, k_p}{\partial P} + \frac{1}{2}\frac{\partial \mathrm{Ln}\, k_d}{\partial P} - \frac{1}{2}\frac{\partial \mathrm{Ln}\, k_t}{\partial P} + \beta(\Sigma \alpha_j + 1 - \Sigma \nu_j)$$

Introducing the activation volume, we write :

$$\frac{\partial \mathrm{Ln}\, v}{\partial P} = - \frac{\Delta V_p^{\ddagger}}{RT} - \beta - \frac{1}{2}\frac{\Delta V_d^{\ddagger}}{RT} - \frac{1}{2}\left(- \frac{\Delta V_t^{\ddagger}}{RT} - \beta\right) + \beta + \frac{1}{2}\beta$$

or $$\frac{\partial \mathrm{Ln}\, v}{\partial P} = - \frac{\Delta V_o^{\ddagger}}{RT} + \beta$$

The overall activation volume $\Delta V_o^{\ddagger}$ is related with the activation volumes of the elemental steps :

- of propagation $\Delta V_p^{\ddagger}$
- of termination $\Delta V_t^{\ddagger}$
- of decomposition of the initiator $\Delta V_d^{\ddagger}$

The last term $\Delta V_d^{\ddagger}$ is in fact a composite quantity including the pressure dependence of the efficiency f :

$$\Delta V_d^{\ddagger} = \Delta V_{do}^{\ddagger} + \frac{\partial \mathrm{Ln}\, f}{\partial P}$$

$\Delta V_d^{\ddagger}$ may be easily determined from a pressure study of the decomposition of the initiator :

$$\Delta V_d^{\ddagger} \simeq V_{do}^{\ddagger} = - RT \frac{\partial \mathrm{Ln}\, v_d}{\partial P}$$

It is a bit more difficult to calculate $\Delta V_p^{\neq}$ and $\Delta V_t^{\neq}$ which understands the knowledge of the quantity

$$\xi = \frac{\partial \mathrm{Ln}\, R^{\cdot}}{\partial P}$$

since :

$$\frac{\partial \mathrm{Ln}\, v_p}{\partial P} = -\frac{\Delta V_p^{\neq}}{RT} + \xi$$

$$\frac{\partial \mathrm{Ln}\, v_t}{\partial P} = -\frac{\Delta V_t^{\neq}}{RT} - \beta + 2\xi$$

but the determination of ξ

$$\xi = -\left(\frac{\Delta V_o^{\neq} - \Delta V_p^{\neq}}{RT}\right) + \beta = -\frac{1}{2}\left(\frac{\Delta V_d^{\neq} - \Delta V_t^{\neq}}{RT}\right) + \beta$$

involves experimental techniques allowing the separate measurement of k_p and k_t (rotating sector method).

I.2.2.2. Ionic polymerizations. In a general sense, ionic polymerizations are reactions in which chain propagation proceeds through true ions or ionic aggregates (not always well defined) or only polarized molecules. Active centers may be anions, cations, radical-ions or other complex ionic species.

In this case, the value of the activation volume depends on volume change not only of the reactants, but also of the surroundings molecules of solvent. The overall activation volume $\Delta V^{\neq}$ is necessarly split into two terms :

$$\Delta V^{\neq} = \Delta V_s^{\neq} + \Delta V_e^{\neq}$$

$\Delta V_s^{\neq}$ is the change in volume resulting from structural modifications in the geometry of the molecule, when the transition state is formed (in free radical polymerizations $\Delta V^{\neq} \simeq \Delta V_s^{\neq}$).
$\Delta V_e^{\neq}$ is the accompanying volume change due to the modified interaction between the molecules of solvent and reacting molecules and characterizes the polarity of the transition state.
In reactions developping electrical charges in the activation step, the electrostatic forces due to charges and high electrostatic moments are strong and may sometimes be so important that these forces as well as the repulsive forces dominate. In that case, the pressure effect is very important. This may occur in ionic polymerizations.

I.3 Conclusion

Pressure on bulk liquid monomer influences the polymerization process only in so far as it affects the individual processes. Homogeneous liquid polymerization is favored by :

- the acceleration of the bimolecular propagation or growth step
- in some situations, retardation of the termination step as a consequence of increased viscosity
- the elevation of the ceiling temperature
- in some cases the elimination of inhibition by impurities

In the very high pressure range (particularly above 10 000 bar), phase changes may determine the polymerization process, through diffusion and morphological factors.

From the multitude of investigations, pressure appears as a powerful tool for :

- the increase of the yield
- the preparation of modified polymers (change in structure)
- the synthesis of new polymers
- the determination of mechanisms

II. ADDITION HOMOPOLYMERIZATIONS AT HIGH PRESSURES

II.1 Radical homopolymerizations in the liquid phase

II.1.1 Kinetic scheme. The general equation describing vinyl polymerizations predicts orders of 0.5 and 1 for initiator and monomer respectively. Non-ideality in vinyl polymerization is thus reflected in orders of initiator and monomer other than 0.5 and 1. Such cases are generally encountered in polymerizations carried out in presence of additives and high concentrations of initiator. It is now established that primary radical termination and degradative chain transfer are probably the most important causes of non-ideality in vinyl polymerizations. Scott and Senogles [4] have summarized the different possibilities which may occur in vinyl polymerizations according to the values of the reaction orders versus monomer (α) and initiator (β) respectively (Table II)

Table II. Features of vinyl polymerizations according to the values of α and β

α	β	Conclusions
1	0.5	ideal polymerization
$1 \leq \alpha \leq 1.5$	0.5	initiation with cage effect
>1	<0.5	termination by primary radicals
$0 \leq \alpha < 1$	> 0.5	degradative chain transfer (to monomer)
>1	0 - 0.5	ib. (to initiator)
>1	0.5 - 1	ib. (to solvent)

Table III. Kinetic scheme in free radical homopolymerization

Reaction	Rate	Process
$I \xrightarrow{k_d} 2R°$	$2fk_d [I]$	production of primary radicals
$R° + M \xrightarrow{k_i} P_1°$	$k_i [P°][M]$	chain initiation
$P_1° + M \xrightarrow{k_p} P_2°$ $P_m° + M \xrightarrow{k_p} P°_{m+1}$	$k_p [P°][M]$	propagation
$P_m° + P_n° \xrightarrow{k_t} P_{m+n}$ $\xrightarrow{k_t} P_m + P_n$	$k_t [P°]^2$	termination of growing chains by mutual deactivation
$P° + R° \xrightarrow{k} PR$	$k [P°][R°]$	primary radical termination
$P_m° + X \xrightarrow{k_{tr}} P_m + X°$	$k_{tr} [P°][X]$	chain transfer (X = monomer, solvent, initiator)
$X° + M \xrightarrow{k_{i'}} P_1°$	$k_{i'} [M][X°]$	re-initiation
$P° + X° \xrightarrow{k_{t'}} PX$	$k_{t'} [P°][X°]$	termination involving radical formed due to transfer and a macroradical

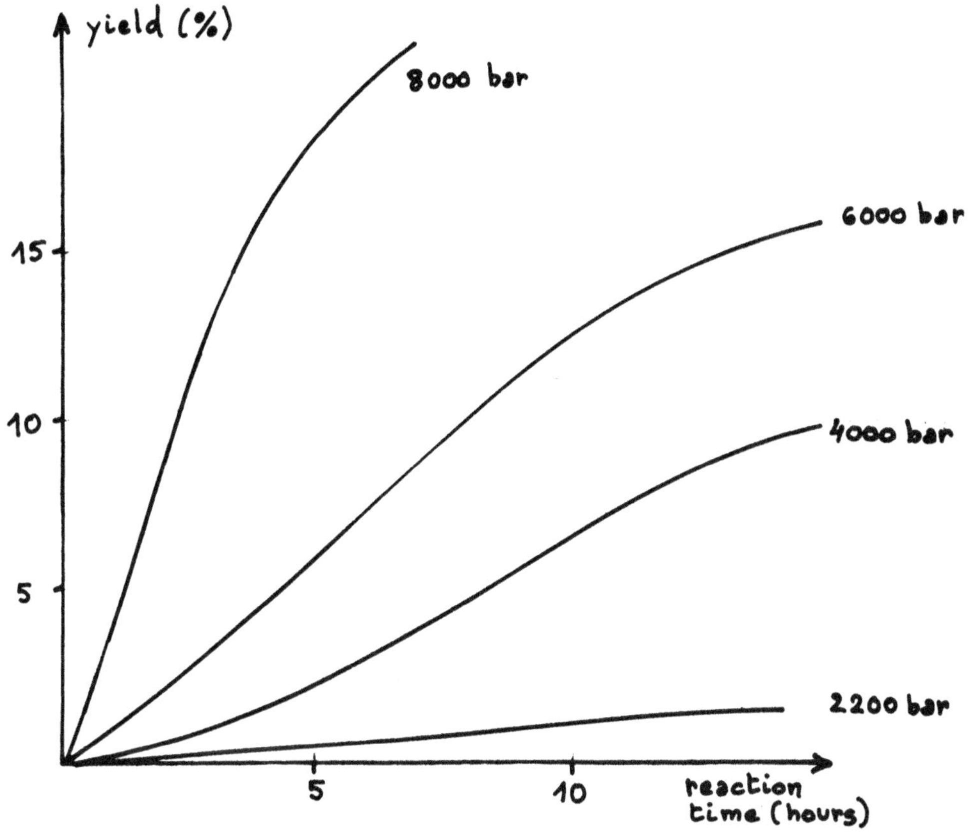

Fig. 1. High pressure polymerization of indene (kinetics)

The kinetic scheme must take into account the reactions that may occur in initiated vinyl polymerization. However the recombination reactions of primary radicals and of radicals formed as a result of chain transfer or the cross-combination have been omitted, because all of which are likely to be of negligible importance.

The pressure dependence of the reaction rate and hence the determination of the activation volume can give substantial information on the features of the addition free-radical polymerization.

II.1.2 High pressure ideal homopolymerization. The ideal behaviour in vinyl polymerizations is encountered for many monomers. Let us take the example of indene and styrene. (table IV) Figure I shows typical kinetic curves obtained in the homopolymerization of indene.

Table IV. Variation of the ratio k_p/k_1 with pressure for two homopolymerizations

Pressure (bar)	k_p/k_1 (indene) (ref. 5)	k_p/k_1 (styrene) (ref.6)
1	1	1
1000	2.1	1.5
2000	4.3	4.6
4000	14.9	14.7
6000	32.3	-
8000	64.2	-
$\Delta V^{\neq}$ (cc/mole)	- 21.0	- 17.0
Conditions	T = 64°C Solvent : heptane	T = 100°C bulk

We have seen in the introductory part of this lecture that, for strictly ideal homopolymerizations :

$$\Delta V^{\neq} = \Delta V_p^{\neq} + \frac{\Delta V_d^{\neq}}{2} - \frac{\Delta V_t^{\neq}}{2}$$

Using the rotating-sector method, Ogo et al. have been able to separate properly the pressure effect on the individual reactions of some polymerizations and found following values [7,8](table V)

Table V. Activation volumes for various monomers at 30°C (cc/mole)

Monomer	$\Delta V^{\neq}$	$\Delta V_p^{\neq}$	$\Delta V_t^{\neq}$	$\Delta V_i^{\neq}$
styrene	- 17.1	- 17.9	+ 13.3	+ 13.3
vinyl acetate	- 17.2	- 24.0	+ 16.3	+ 19.9
n-butyl acrylate	- 26.3	- 22.5	+ 20.8	+ 13.2
butyl methacrylate	- 17.4	- 23.2	+ 17.8	+ 29.4
methyl methacrylate	- 16.0	- 19.0	+ 25.0	+ 31.0
octyl methacrylate	- 20.4	- 24.7	-	-

It can be observed that the values of the overall activation volume $\Delta V^{\neq}$ have almost the same magnitude, except surprisingly for butyl acrylate. This feature seems to be a good criterion to define ideal behaviour for free-radical polymerization.

Several efforts to evaluate $\Delta V^{\neq}$ have been made from the transition state model, but the adoption of different arbitrary assumptions about the configuration and interatomic distances in the transition state have led to divergent results. Although the theoretical basis for the values of - 17 to - 20 cc/mole of $\Delta V_p^{\neq}$ has not been defined yet, the difference between them and the experimental values listed in table IV, suggests that the nature of the transition state is appreciably affected by the surrounding monomers. Ogo found an empirical relation connecting $\Delta V_p^{\neq}$ to ΔV (volume contraction during the polymerization)

$$\Delta V_p^{\neq} = \frac{1}{2} \Delta V - 11$$

On the contrary, the values of $\Delta V_t^{\neq}$ and $\Delta V_i^{\neq}$ are very different according to the monomer. The rates of initiator dissociation and of bimolecular chain- termination (which is diffusion - controlled) are moderately reduced by an increase in pressure, so that $\Delta V_i^{\neq}$ and $\Delta V_t^{\neq}$ are positive. There are conventional concepts that these two terms partially cancel in the expression of $\Delta V^{\neq}$, so that $\Delta V^{\neq}$ is largely determined by $\Delta V_p^{\neq}$.

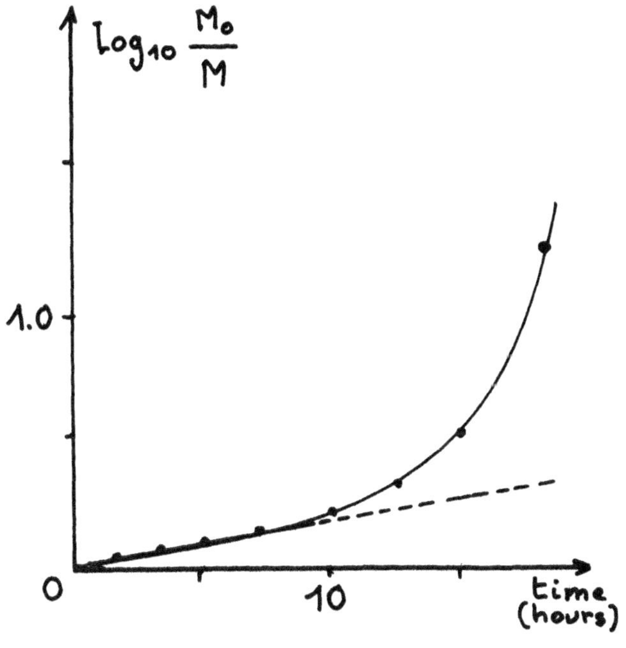

Fig. 2. First-order plot for the high pressure polymerization of isoprene (14 kbar) with peroxide as initiator (The gel effect becomes significant after ten hours)

It is apparent that the differences according to the monomer may be attributed to the difference in size of the molecule and especially to the viscosity of the medium. In fact viscosity plays an important role in termination processes and it is assumed that the mechanism of such diffusion - controlled termination actually changes with pressure.

Another effect related also to the increase in viscosity which often appears in free radical polymerizations at high pressures, is the Trommsdorff effect or gel effect (Fig. 2.)

In the first stage of the polymerization, the chemical species (molecules, radicals) still possess a good mobility, even under pressure, because the chains are short. The molecular weight increases slowly, although pressure accelerates greatly the propagation rate, but transfer reactions are also accelerated. However, as soon as the chains reach a sufficient length, they loose gradually their mobility, that is due to the considerable increase in viscosity so that diffusion acts as factor of mutual recombination. Pressure traps the polymer-radicals, increases their mean life time and hence the concentration of radicals. As a consequence, the overall polymerization rate is increased, as well as the molecular weight.

The pressure dependence of the polymer molecular weight can be estimated semi-quantitatively in the case of styrene [9]. In radical ideal polymerizations, the average degree of polymerization δ is given by :

$$\frac{1}{\delta} = \left[\frac{k_d\, k_t}{2}\right]^{1/2} \frac{1}{k_p} \frac{[I]^{1/2}}{[M]} + C_M + C_I \frac{[I]}{[M]}$$

C_M, C_I .: transfer constants to monomer and initiator respectively

The molecular weight of polystyrene (as an example) should increase approximatively about 4 times between 1 and 3000 bar. At higher pressures it should increase much more slowly , as C_M gradually falls with further rise of pressure.

II. 1.3 Non-ideal polymerizations. Some particular monomers cannot be polymerized in the usual way. A famous example is the homopolymerization of maleic anhydride which is worthwhile to be treated in some detail.

II. 1.3.1 Homopolymerization with degradative transfer - Maleic anhydride.

The reluctance of maleic anhydride to homopolymerize was principally attributed to two factors : low reactivity of the monomer towards its radical and steric repulsion in the hypothetical polymer chain. It was found that maleic anhydride can be polymerized by radical way in presence of high concentrations of initiator.

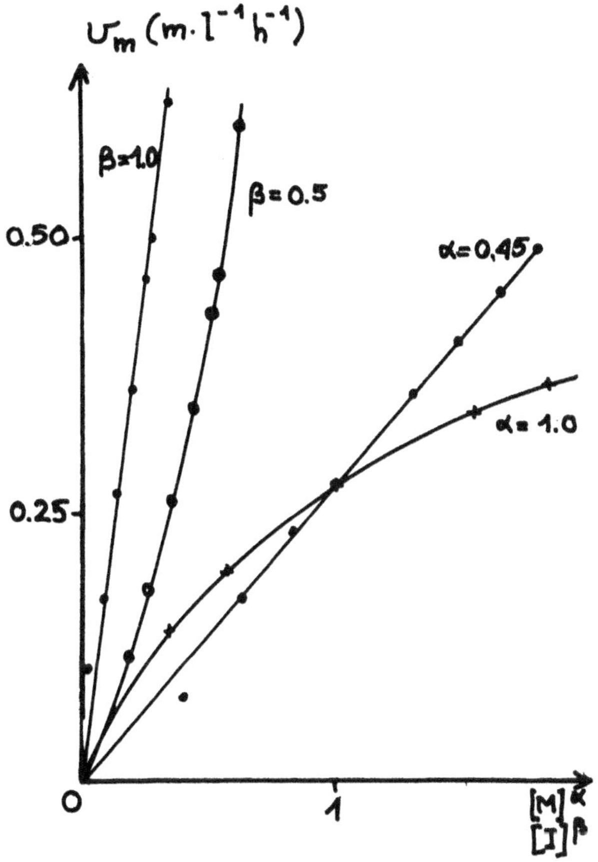

Fig. 3. Homopolymerization of maleic anhydride. Kinetic curves obtained with different exponents
(α = 1.0 and β = 0.5 : ideal polymerization)
(α = 0.45 and β = 1.0 : true exponents)

In fact recent studies [10] show that this polymerization is actually a complex reaction and does not follow the usual mechanism. This is supported by the reaction orders $\alpha = 0.45$ and $\beta = 1$, with following kinetic equation :

$$v = K [M]^{0.45} [I]$$

Fig. 3 shows the kinetic curves calculated first with the above values, then with the exponents characterizing a normal free-radical polymerization.

One can see immediately that the polymerization of maleic anhydride does not follow the usual kinetic equation, except when the monomer polymerizes for low concentration either in monomer or in initiator.

According to Scott and Senogles (Table II) the above equation would be similar in nature to the kinetic equation of a polymerization with degradative chain transfer, mainly to the monomer and little to the initiator and solvent.

The investigation of the pressure effect on this reaction confirms the degradative transfer. Pressure greatly increases the maximal yield, by lowering the concentration of radicals and probably the rate of termination reactions with a less rapid consumption of the initiator and a chain extension as consequences (Table VI)

Table VI. Pressure effect on the polymerization rate of maleic anhydride

$[M] = 5.00$ m/1 $[I] = 0.25$ m/1 T = 70.0°C
Lauroyl peroxide is the initiator and benzene is used as solvent.

Pressure (bar)	V_m (maximal rate) (mole.l^{-1} h^{-1})	Intrinsic viscosity (dl/g)
1	0.485	0.0485
500	0.515	0.0489
1000	0.560	0.0494
1500	0.615	0.0499
2000	0.700	0.0502
3000	solidification of reaction mixture	

From these data, the activation volume is easily calculated :

$$\Delta V^{\neq} = -4.6 \text{ cc/mole}$$

This is a very low negative value and is far away from those obtained in usual radical polymerizations (- 17 to - 25 cc/mole).

Now if the chain transfer to monomer is assumed as total :

$$v = \frac{k_d \, k_p}{k_{tr}} [I]$$

k_{tr} : rate constant for transfer

we have for $\Delta V^{\neq}$

$$\Delta V^{\neq} = \Delta V_p^{\neq} + \Delta V_d^{\neq} - \Delta V_{tr}^{\neq}$$

$\Delta V_{tr}^{\neq}$ is the activation volume relative to degradative transfer

$\Delta V_{tr}^{\neq}$ is always negative as well as $\Delta V_p^{\neq}$, but the quantity $\Delta V_d^{\neq} - \Delta V_{tr}^{\neq}$ is positive (since $\Delta V_d^{\neq} > 0$) so that $\Delta V^{\neq}$ is slowly negative and even can be positive, if the transfer becomes important.

Thus the pressure study reveals the abnormal character of the polymerization and confirms the hypothesis of desactivating transfer to the monomer which has as main effect to produce polymers with low molecular weight. Moreover pressure has very little effect on the degree of polymerization (see table VI).

From the scheme given in table III, the overall kinetic equation becomes [11] :

$$\frac{[I]\,[M]^2}{v^2} = \frac{1}{K} + \frac{C_M}{K} \, \frac{k^2 p}{k_t} \; \frac{2 \dfrac{k_{tr}}{k_{i'} . k_p}}{1 + \dfrac{v \, k_{tr}}{k_{i'} \, k_p \, [M]^2}}$$

C_M : transfer constant relative to monomer

$$K = 2 f \frac{k_d . k^2 p}{k_t}$$

$$\text{or} \quad y = A + \frac{B}{1 + Cx}$$

$$\text{with} \quad x = \frac{v}{[M]^2} \quad \text{and} \quad y = \frac{[I]\,[M]^2}{v^2}$$

The experimental points verify this equation with same small scattering along the y - axis (Fig. 4)

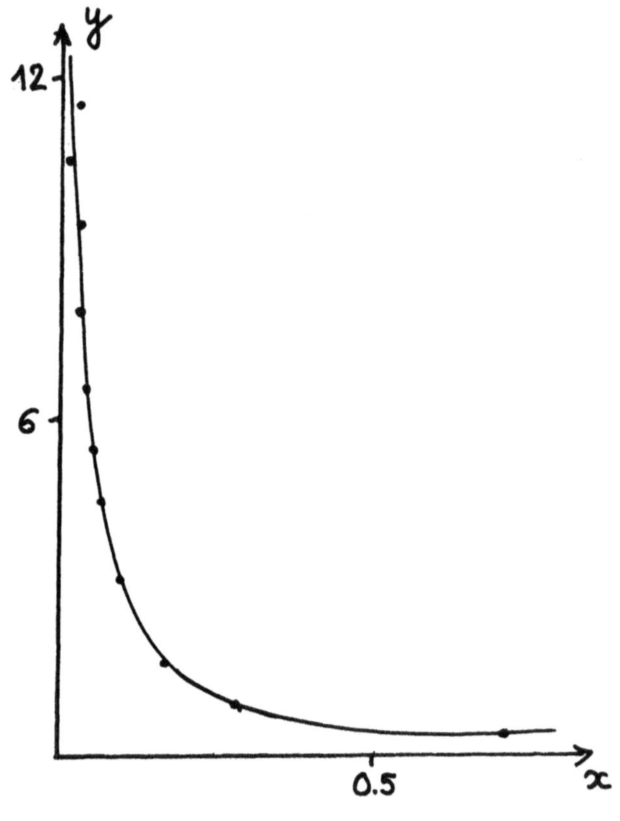

Fig. 4. Plot of Deb's equation for the polymerization of maleic anhydride (see mathematical expressions for x and y in text)

- Other monomers

Another reaction which may be included at least partially in this category is the polymerization of acenaphtylene [12], but in this reaction, the orders versus monomer and initiators seem to be pressure dependent. The activation volume (- 6 cm^3/mole) should confirm partial degradative kinetics.

The free-radical polymerization of propylene exhibits also degradative transfer to monomer [13] ($\Delta V^{\neq}$ = - 9 cc/mole at 80°C).

The case of allyl acetate is ambiguous [14]. At atmospheric pressure, molecular chains are chiefly terminated by monomer transfer in the polymerization of allyl acetate. Transfer is believed to involve the abstraction of an H atom from the α - CH_2 group, with the formation of a resonance-stabilized allyl radical. When pressure is increased, chain growth and transfer with allylic hydrogene are both strongly pressure - accelerated, however competition between both reactions is little affected. In addition, the resulting allylic radicals readily start new chains under pressure, so that the degradative nature of chain transfer becomes unimportant.

II. 1.3.2 Homopolymerization with depropagation effect. There are some systems for which the scheme given in table III is inadequate and in which the observed expressions for the overall polymerization rate and degree of polymerization can only be accounted for by postulating the significant participation of the reverse of the propagation process (called depropagation)

$$P_n^{\circ} \longrightarrow P_{n-1}^{\circ} + M$$

The rate for the propagation step becomes

$$v_p = \left(k_p [M] - k_{-p}\right) [P^{\circ}]$$

k_{-p} : depropagation constant

Dainton and Ivin introduced the concept of ceiling temperature T_c in following equation :

$$T_c = \frac{\Delta H}{\Delta S + R \ln [M]_e}$$

where ΔH is the heat of polymerization, ΔS the corresponding entropy and $[M]_e$ the monomer concentration when the equilibrium is reached.

The dependence of T_c on pressure follows the Clapeyron-Clausius equation :

$$\frac{d T_c}{d P} = T_c \frac{\Delta V}{\Delta H}$$

Both ΔV and ΔH may vary with pressure, but their ratio is constant or nearly constant.

According to this equation, T_c increases with increasing pressure. This effect is potentially important in the polymerization of substances which do not react at ordinary pressures.

The free-radical polymerization of α-methylstyrene is a good example of this type of polymerization. In the usual conditions (1 bar, T > 60°C), the polymerization is inexistent or occurs scarcely [15]. Application of the Clapeyron-Clausius equation leads to a value for ΔV = - 14.7 cc/mole. This is of the approximate magnitude to be expected for the algebric difference between the estimated values of the activation volume for chain growth and depropagation. A sharp maximum in the rate of polymerization as a function of pressure was also observed. This is probably due to solidification of the monomer. The region in which polymerization of α-methylstyrene is possible is therefore limited by the ceiling temperature and the freezing points which both increase markedly with pressure.

Some years ago, Busfield and Whalley [16] studied the equilibrium between chloral and its polymer up to 1500 bar. Gonikberg and Zhulin as well as Whalling and Osugi [17] investigated the equilibrium between butyraldehyde and polybutyraldehyde and found that, to overcome depolymerization, it is necessary to use pressures as high as 8000 bar and more and temperatures less than 20°C.

In conclusion, generally the activation volume for depropagation is assumed to be + 10 cc/mole, so that depropagation will be retarded at high pressures. Pressure thus is a potential parameter to induce polymerizations of monomers which fails to form polymers at ordinary pressure.

II. 2. Radical homopolymerization in the gaseous phase

II. 2.1 Kinetics of homopolymerization. The free-radical polymerization of ethylene displays most of the typical characteristics of a vinyl polymerization. Omitting chain initiation which is probably identical to that in other vinyl polymerizations with typical organic initiators and poorly understood with oxygene, the main reaction step which a chain radical can undergo and which should be considered, appear to be [18] (table VII). Not all these reactions take place under a given set of conditions. The polymerization kinetics can be interpreted in terms of fugacity or activity coefficients as in the liquid phase. The last has been developped in the preceeding section, while the first approach is normally made in analyzing gaseous equilibria to moderately high pressures, where the variation in activity coefficients with pressure can be expressed in terms of the variation of fugacity coefficients with pressure. The method is equally valid in principle, but requires a method of calculating the ratio of the fugacity coefficients.

Table VII. Kinetic scheme of polymerization of ethylene

Propagation	$C_2H_4 + P_n^{\circ} \xrightarrow{k_p} P_{n+1}^{\circ}$
Termination by coupling	$P - CH_2^{\circ} + P' - CH_2^{\circ} \xrightarrow{k_t} P - (CH_2)_2 - P'$
Termination by disproportionation	$P - CH_2^{\circ} + P'' - CH_2 - CH_2^{\circ} \xrightarrow{k_{t'}} P - CH_3 + P'' - CH = CH_2$
Chain transfer with a transfer agent X°	$P - CH_2 - CH_2 + X \xrightarrow{k_m} X^{\circ} + P - CH_2 - CH_3$
Intramolecular H - transfer	$P - CH_2-CH_2-CH_2-CH_2-CH_2^{\circ} \xrightarrow{k_b} P - \overset{\circ}{CH} - C_4H_9$
Intermolecular H - transfer	$P^{\circ} + P' - CH_2 - R'' \xrightarrow{k_1} P - H + P' - \overset{\circ}{CH} - R''$
β - scission	$P - CH_2 - \underset{R''}{\overset{\circ}{C}} - CH_2 - P' \xrightarrow{k} P^{\circ} + CH_2 = \underset{R''}{C} - CH_2 - P'$
	(R'' : H or alkyl)

The results show that under carefully controlled conditions and with initiators other than oxygene, the effect of pressure on the rate is chiefly due to an increase in the ratio k_p/k_t which can at sufficiently high pressures be interpreted in terms of the volumes of activation. When oxygene - initiation is used and possibly when there is phase separation, it is doubtful if steady - state kinetics are followed, and the conventional rate equations are inapplicable.

II. 2.2 Internal transfer in ethylene polymerization. As the pressure is increased, the molecular weight of the polymer increases ; but the structure of the polymer molecules is also very dependent on the pressure at which they are formed. The amount of branching decreases as the pressure is increased until the molecules have few, if any, side chains. The short side - chains of low density polyethylene are thought to arise from an internal chain transfer reaction (rate constant k_b). The transfer occurs via a 5- or 6- membered cyclic transition state and further chain growth occurs from the secondary radical. The decrease in the number of side - chains at high pressures shows that this reaction becomes relatively less important. According to transition state theory, the methyl/methylene ratio depends on the difference in volume of activation for the branching and propagation steps ($\Delta V_b^{\neq}$ - $\Delta V_p^{\neq}$) and if all the short - chain branches were to form via following mechanism :

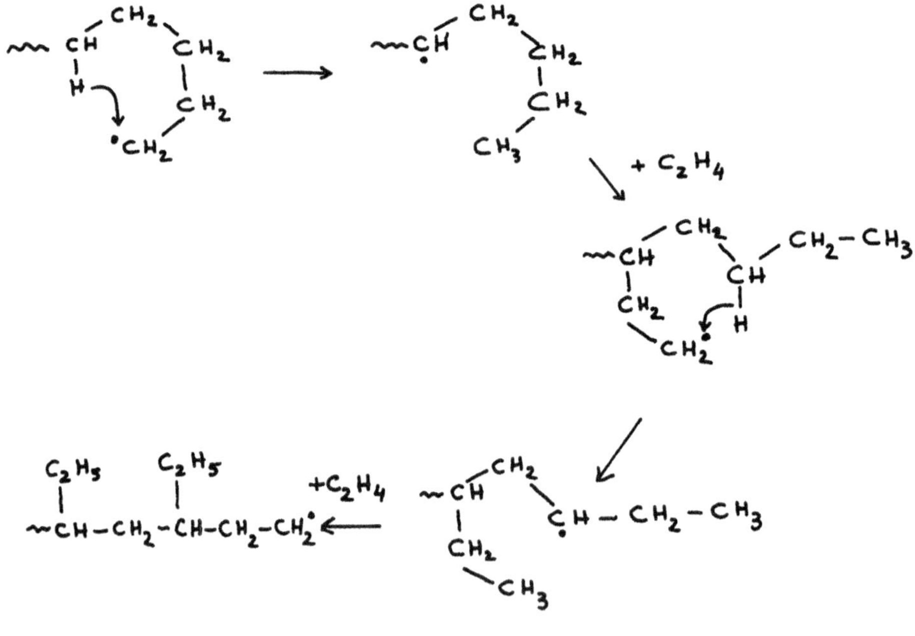

the pressure coefficient of the CH_3/CH_2 ratio would represent $n(\Delta V_b^{\neq} - \Delta V_p^{\neq})$ with n probably in the neighborhood of 2, whereas the factor n would be unity, if all branches were to form via a single type of transfer step. In this case, $\Delta V_p^{\neq} - \Delta V_b^{\neq}$ is negative and very similar in magnitude to $\Delta V_p^{\neq}$ itself. This would make $\Delta V_b^{\neq}$ rather close to 0. According to the former interpretation, $\Delta V_b^{\neq}$ would be approximately - 10 cc/mole. This value would also be in accordance with a variety of data on polymerization rates and polymer structure which indicate that abstraction reactions are accelerated substantially by pressure, though normally less so than the propagation step in vinyl polymerization.

It is probable that the formation of vinyl and vinylidene type unsaturation and the processes leading to molecular termination are closely associated with the short - chain branching mechanism.

The reaction of molecular termination by β-scission of a tertiary radical is also pressure dependent. Low - density polyethylene usually contains 0.1 to 1 vinylidene groups per molecule and so β-scission of a tertiary radical might be expected to be the dominant mechanism for molecular termination under some experimental conditions. This was shown to be so at low polymerization rates and at pressures of 1600 to 1800 atm at temperatures as low as 140°C by measuring number average molecular weights as a function of the polymerization rate v. At low v- values, molecular weight control by β-scission should become increasingly dominant over mutual termination as the temperature is increased and as the pressure is reduced, because both of these factors are associated with an increase in vinylidene content of the polymer.

II. 3 Ionic and pseudo-ionic polymerizations at high pressures

Ionic polymerizations at high pressure are not easy to study. There are several reasons for it : first ionic polymerizations involve species whose life - time is limited, the shorter as they are very sensitive to oxygene, moisture, alcohols, many impurities, even in slight amount, a second difficulty arises from the kinetics, because they are usually short - lived reactions, it is therefore necessary to make a choice amont them and to study those whose complete transformation requires several hours under pressure. Below one hour, kinetic measurements are to be avoided, because they run the risk of being disturbed by the increase in temperature of the medium accompanying the compression of the reaction mixture.

II. 3.1 Anionic polymerization. In an anionic polymerization, the growing end of the macromolecule carries a negative charge or is negatively polarized. We take as examples the anionic polymerization of two conjugated dienes at high pressures : 2-methylbutadiene (MB) and 2,3 dimethylbutadiene (DMB) carried out with organolithium compounds.

The reactional environment acts basically upon the reaction kinetics and the structure of anionic polydienes produced in presence of n-butyllithium. The actual mechanism has not yet been well clarified in spite of the lot of work which has been devoted to it. One of the main difficulties proceeds from the fact that each monomer has a different behaviour and thus must be considered separately so that no generalization can be made from one diene to another. However the anionic polymerization of MB is nowadays better known than the others.

If we adopt the simplified reaction scheme of Winstein [19], we write the possible propagation forms as :

$$M^-Li^+ \rightleftharpoons M^- // Li^+ \rightleftharpoons M^- + Li^+$$

tight ion pairs   loose ion pairs   free ions

The different configurations of the growing end depend on the experimental conditions. The ion pairs, for example, are intimately surrounded by a fluctuating environment of solvent molecules which may be able to superimpose their force field upon the electrostatic interaction between the ions of the pair. Thus tight ion pairs and loose ion pairs as well as the equilibrium established between the pairs are very sensitive to changes in the force fields.

As reaction parameter, pressure perturbs the energy levels of the molecules in solution and these perturbations depend on the character of the inter- and intramolecular forces. Therefore pressure affects drastically the kinetics of anionic polymerizations and also the structure of the resulting polymers.

II.3.1.1. Kinetic effect. In all cases the polymer yield in anionic polymerization is considerably increased by pressure. The mechanism is chiefly determined by the degree of ionic interactions between the initiating anionic species and the gegenion. In hydrocarbons, organic lithium compounds are easily associated forming aggregates in which the C-Li bond is highly polarized :

$$m\ BuLi \rightleftharpoons (BuLi)_m$$

$$BuLi + M \longrightarrow BuM^-\ Li^+ \qquad \text{initiation}$$

$$m'\ BuM_n^-\ Li^+ \rightleftharpoons (Bu\ M_n^-\ Li^+)_{m'}$$

$$Bu\ M_n^-\ Li^+ \longrightarrow Bu\ M_{n+1}^-\ Li^+ \qquad \text{propagation}$$

It is reasonable to consider that pressure accelerates both initiation and propagation steps and leads to higher yields in polymer. Moreover it has been shown that pressure shifts slightly the association equilibria towards the associated species. The addition of an oxygenated solvent to the reaction mixture destroys the lithium associations and produces selective solvatations. The solvated cation Li^+ is an electron -acceptor for the monomer which polymerizes therefore faster in ethers than in hydrocarbons.

For the anionic polymerization of DMB at room temperature, two activation volumes have been reported [20] :
- in n-heptane $\Delta V^{\neq}$ = - 36 cc/mole
- in diethylether $\Delta V^{\neq}$ = - 48 cc/mole

The apparent activation volume is very negative, indicating an important contribution of electrostatic effects in the transition state. These values should be very probably connected with a transformation of Winstein's equilibria towards more polarized or more dissociated species. In fact it is known that, in solution, ionization is accompagnied by a volume contraction (ΔV) resulting principally from the polarization and the attraction of solvent molecules by ions or ionic species (electrostriction). ΔV can be evaluated in first approximation by the Drude-Nernst's equation :

$$\Delta V = (- q^2/2\, r\varepsilon^2)\ \partial\varepsilon / \partial P$$

q : charge
r : ionic radius
ε : dielectric constant of the medium

Le Noble [21] observed that a pressure as low as 1000 bar is sufficient to convert fluorenyl-lithium from tight into loose ion pairs in tetrahydrofurane ($\Delta V \simeq$ - 15 cc/mole). It is thus demonstrated that the conversion of tight ion pairs into loose ion pairs is thermodynamically favoured by an increase in pressure. Therefore pressure shifts Winstein's equilibria to the right side, favouring those which contribute to create ion pairs occupying the smallest volume, e.g. even in a solvent as weakly polar and non-solvating as heptane, contact ion pairs can appear which lead to accelerate markedly the reaction rate.

II.3.1.2 Structural effect. MB can add to itself through four different ways (1,4 cis - 1,4 trans - 3,4 - 1,2 additions) whereas for DMB there are only three addition modes, 3,4 units being identical with 1,2 units due to symmetry in the molecule. It has been found that pressure is able to alter sometimes very significantly the structure of the two investigated dienes.
- Anionic poly (MB)

Fig. V shows the variation of the content in 1,4 cis units which diminishes with increasing pressure in solvents such as heptane as well as anisol, whereas the 1,4 trans addition is slightly favoured by pressure [19, 22].

The pressure - induced changes of 3,4 and 1,2 additions are represented in the next figure (Fig. VI). Concerning the 3,4 addition, we observe that the concentration of 3,4 units is clearly increased by pressure, when the polymerization is carried out in hydrocarbons or anisol, however no change occurs when dioxane is used as solvent. The situation looks like for the 1,2 addition, except that this addition does not appear when anisol is the solvent. The absence or the low proportion of 1,2 units may be easely explained by the steric hindrance of the methyl group in the MB molecule. It seems that pressure is able in the case of heptane to overcome this hindrance, at least to a small extent (10 to 15%) since on the other hand no pressure effect on the 1,2 addition is observed in the case of dioxane.

Thus pressure appears to act like solvents such as dioxane in which tight ion pairs are produced, probably by influencing the association equilibria of the growing chain. The structural changes are reflected by the values of the activation volumes (Table VIII).

Table VIII. Activation volumes for the various additions in the anionic polymerization of MB

Activation volumes (cc/mole)	Solvent n-heptane	anisol
$\Delta V^{\neq}_{1,4\ cis} - \Delta V^{\neq}_{1,4\ trans}$	+ 6.5	+ 0.5
$\Delta V^{\neq}_{1,4\ cis} - \Delta V^{\neq}_{1,2}$	+ 9.0	-
$\Delta V^{\neq}_{1,4\ cis} - \Delta V^{\neq}_{3,4}$	+ 17.0	+ 2.0

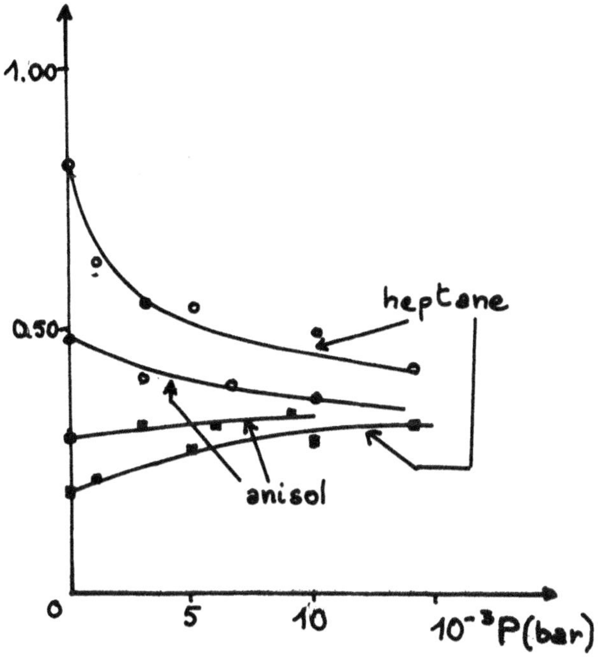

Fig. 5. Pressure effect upon the 1,4 cis and 1,4 trans additions in the anionic polymerization of isoprene
—o— : 1,4 cis
—▪— : 1,4 trans

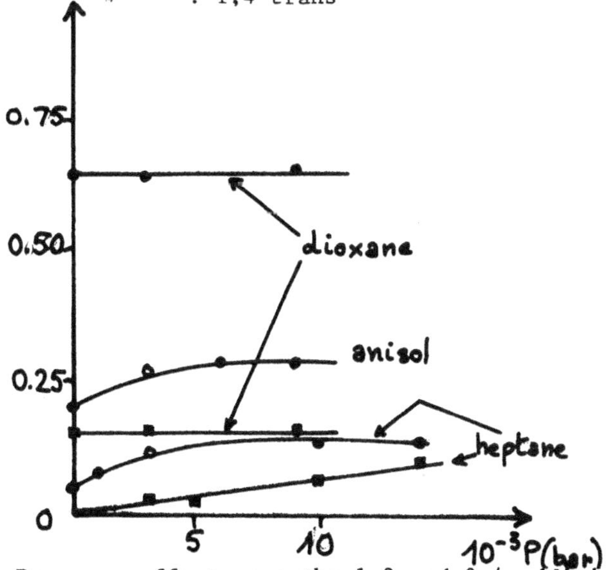

Fig. 6. Pressure effect upon the 1,2 and 3,4 additions in the anionic polymerization of isoprene
—o— : 3,4 —▪— : 1,2

Obviously the 3,4 addition is the most influenced by pressure especially when hydrocarbons are used as solvent.

The propagation reactions of the four possible additions are characterized by overall propagation rate constants which are composite quantities, i.e. the sum of the rate constants for the different propagation steps corresponding to each type of active centers and by an overall activation volume :

$$\Delta V^{\neq} = \Delta V_s^{\neq} + \Delta V_e^{\neq}$$

It is evident that in presence of species more or less ionized as they exist in anionic polymerizations, the solvent influences the value of $\Delta V_e^{\neq}$ and hence that of $\Delta V^{\neq}$. The apparent role of pressure is to act like an intensifier of the electric field in the same way as solvents leading to the formation of an electric dipole between the gegenion and the end of the growing chain :

$$MLi \rightleftharpoons \overset{-}{M} — \overset{+}{Li}$$

This can explain why the 1,2 and 3,4 additions are positively influenced by pressure. Nevertheless the decrease of the content of 1,4 cis units seems to be also related with steric effects. In fact the 1,4 cis poly (MB) has a lower density than the trans isomer, thus the simple application of Le Châtelier's principle enables to predict that the trans poly (MB) is favoured by an increase in pressure more than the cis isomer. This is effectively the case, when MB is polymerized in a medium in which the electrostatic effects are small (see next section).

Therefore the pressure effect upon the structure of anionic poly (MB) must be explained simultaneously by electrostatic as well as by steric considerations. The same explanation can be put forward for the pressure effect on the structure of anionic poly (DMB) [23] .

II. 3.2. Cationic and pseudo-cationic polymerizations. The few cationic polymerizations which have been investigated hitherto were shown as strongly pressure accelerated. The catalysts usually used ($SnCl_4$, $TiCl_4$...) do not necessary suit for high pressure studies, especially in the polymerizations of styrene and conjugated dienes, because the kinetics is too fast and the reactions become explosive under pressure.

As a Lewis acid, hydrogene iodide can be a priori considered as a cationoid polymerization catalyst. However the presence or the absence of water leads to quite different results in the high pressure polymerization of 2-methybutadiene (MB) with HI [24].

II.3.2.1. Activation volume. When water is carefully excluded from the medium, no polymerization occurs and only a typical Diels-Alder cyclodimerization is observed. With the progressive addition of water, the polymer appears and the overall yield increases, but it is necessary to raise greatly the concentration of water to observe a significant change of the yield. This is not the case in pure cationic polymerizations, where the presence of very small amounts of water (acting then as cocatalyst) increases considerably the polymerization rate. Whereas dimerization is too slow at low pressures, with water the reaction rate of polymerization can be appreciated even at atmospheric pressure. The following activation volumes were found (Table IX).

Table IX. Activation volumes in the HI - catalyzed polymerization of 2-methylbutadiene (*)

Solvent	$\Delta V^{\neq}$ apparent (cc/mole)
toluene	- 7.9
ethyl acetate	- 9.8
dichlormethane	- 10.8

(*) room temperature

Thus the influence of HI in the high pressure polymerization of MB appears to be very complex. The presence of water is necessary to initiate the polymerization which gives support for a cationic mechanism. However it is unlikely that carbonium ions still can live in a medium with so high water concentrations. Moreover, the polymerization being once initiated, the reaction rate is actually little sensitive to an increase of the water concentration, since the yield only doubles, when the concentration increases ten-fold. Besides the activation volume is low in any solvent. Even in the free - radical polymerization of MB, $\Delta V^{\neq}$ is more negative (- 18 cc/mole) (see section II.1.2). Because all these reasons the mechanism seems to be neither cationic nor radical.

The results lead to consider another active species which would be a coordinated organic iodide. One could imagine that a little part of HI would give with MB an organic iodide activated by coordination of one or more molecules of iodine. All features encountered in the high pressure polymerization of MB catalyzed by HI agree with a pseudo-cationic mechanism.

II. 3.2.2. Dependence of polymer structure on pressure . The analysis of the structure of the polymer derived from this MB polymerization shows that at atmospheric pressure, the 1,4 cis units represent about 20% of the total units number, this point also disagrees with the hypothesis of a cationic as well as of a free - radical mechanism.

However the pressure effect upon the structure is important, since the 1,4 cis addition disappears at sufficiently high pressures (table X).

Table X. Pressure effect on the structure of poly (MB) obtained by pseudo-cationic polymerization

Pressure (bar)	1,4 cis (*)	1,4 trans (*)
1	0.20	0.64
5000	0.17	0.68
9000	0.09	0.73
12000	0.05	0.76
14000	0.00	0.78

(*) The 3,4 and 1,2 additions which are present too, are only very slightly affected by pressure.

The polymerization being qualified as pseudo-cationic, the transition state should be not more polar than the initial state, at least for a not too great extent. Therefore the diminution of the 1,4 cis units as well as the increase of the concentration of 1,4 trans units due to an increase in pressure would result mainly from steric effects occuring during the crossing from initial to transition state. These effects may be responsible in part of the pressure - induced structural changes observed in the anionic polymerization of MB.

II.4. High pressure polymerizations in the solid state

II. 4.1 Introduction. When passing from liquid state to solid state the mechanism of polymerization is profoundly affected by changes in both the orientation of the monomeric units in the crystal and by the presence and distribution of impurities in the lattice. The method of initiating solid - state polymerization by radiation presents a means for investigating the mobility of a growing polymer chain within a solid monomer matrix in the absence of chemical impurity which must otherwise be introduced as a catalyst for the reaction. Since initiating species are distributed at random throughout the system, the course of the polymerization under given conditions is influenced by the presence of crystallographic irregularities (stresses, dislocations) either initially present or subsequently formed by the propagating chain. Otherwise propagation reactions are found to be independent of the radiation process and in general, terms appear identical with those occuring in chemically initiated polymerizations. In order to produce a crystallisable polymer, it is assumed that it is necessary for each monomer unit, at the moment of addition, to retain its crystalline orientation. This in turn will only be feasible when the unit cell of the monomer crystal has similar dimensions to that of the polymer.

There are various factors affecting solid - state polymerization. Particularly the effects of both the crystal structure and crystalline state of monomers on the polymerization are most important. Crystallographic structure of monomers, formation of modifications, formation of defects and holes and migration of these dislocations are all relevant here.

By studying the pressure effect on the solid - state polymerization, one may obtain useful information about the polymerization mechanism in the solid state from the point of crystalline structure, and also about the mobility of monomer molecules in the solid state in connection with the polymerization mechanism. Additional information can be gained about the nucleation process, the formation of defects, the change of lattice constants and the annealing of the system.

In solids, it has been found that polymerizations may be either retarded or accelerated by raising the pressure, although these directional effects of pressure are strongly dependent on temperature. A likely explanation is that the rates of these polymerizations are governed by diffusional processes. Nevertheless the kind of molecular motion needed to cause the reaction of organic molecules in the solid state is different from that in diffusion, so that the measurement of diffusion coefficients may add little to the understanding of reactions occuring in molecular crystals. Moreover, by compression, the molecular motion will be inhibited. Therefore, to achieve the reactions in the solid state, as pointed above, the initial geometrical arrangement of reacting sites is considered to be necessary.

In other terms, the application of an external applied stress would be expected to affect the course of a solid state polymerization, favouring the propagation reaction for monomers in which the polymer is formed along a preferred crystallographic axis but reducing the rate where it tends to oppose the tensile stress exerted by the growing polymer chain.

Additionally, combination of high pressures and temperatures in the solid state will favour the synthesis of new polymers.

II. 4.2. Examples of solid - state polymerizations. There are many examples of such reactions exhibiting different features.

II. 4.2.1 Polymerization through the C≡N bond. Malonitrile and succinonitrile can only be polymerized under very high pressures ($\geqslant$ 30 kbar) and temperatures ($\geqslant$ 200°C) in the solid state [25].

The polymerizability of these monomers is essentially attributed to an activation of the C≡N bond due to compression, which decreases the distance of reacting sites and makes it short enough to cause the reaction. The possible steric effect will be removed or lessened.

In these extreme conditions, the polymerization area is limited by the decomposition of the monomer. The decomposition occurs as a rather fast and sometimes explosive reaction. An explosive reaction may occur if the energy produced in the reaction will raise the temperature to a sufficient level and permits the neighboring molecules to react before the local energy has been transferred to a bigger portion of the substance. The compression of solid substances will aid the storage of local energy which is also associated with the effect of shear stress. Shearing stresses may be produced in "squeezer" apparatus during compression and could cause the rapid generation of high concentrations of dislocations and free radicals. Even it has been found that deliberate shearing under pressure causes the polymerization of some monomers at room temperature abready.

In general it seems that the reactions require a suitable combination of hydrostatic pressure, temperature and sometimes shearing stress. Their mechanisms are probably very complex.

Polymers having the backbone $(-\underset{R}{C}=N-)_n$ are obtained from saturated dinitriles, although tight packing of the molecules under high pressure allows the reaction to proceed intermolecularly with formation of crosslinked polymers.

II. 4.2.2 Polymerization of acrylonitrile. In the solid - state polymerization of acrylonitrile [26] (table XI), it has been observed that the highest reaction rate lies in the vicinity of - 100°C for the pressure range 0 - 5000 bar. It is quite interesting that the effect of pressure is very different above and below the transition temperature of the monomer and this suggests that the mechanism is different too. The pressure effect is not due to a change in activation entropy by pressure, but to a change in the activation enthalpy.

Table XI. Activation volumes in the solid - state polymerization of acrylonitrile

Type of crystal	T (°C)	Pressure region (bar)	$\Delta V^{\neq}$ (cc/mole)
β-form	- 98	1 - 1000	2.67
		1000 - 2000	2.01
		2000 - 3000	1.67
metastable form	- 196	1 - 2500	1.37
of α-type crystal		2500 - 6000	0

These results strongly indicate that the polymerisation at - 196°C proceeds without any change of volume in the system above 2500 bar. On that basis Tabata proposed a mechanism called "collective excitation process".

II. 4.2.3. Polymerization of acrylamide. The pressure effect on the solid state polymerization of acrylamide [26] has been attributed to the pressure dependence of $\Delta H^{\neq}$ and $\Delta S^{\neq}$ of the propagating step. Unlike the former case (II.4.2.2) the polymerization of acrylamide proceeds very slowly in the initial stages of polymerization. It is accelerated autocatalytically in a later stage. Pressure effects are also very different between the two stages. In other words, the rate of polymerization is suppressed by pressure in the crystalline phase of monomer, while it is accelerated in the heterogeneous phase where a mixture of monomer crystallites and amorphous polymer is produced. Activation volume changes are depending on the conversion of monomers. A positive value was obtained for the initial stage ($\Delta V^{\neq}$ = + 2.77 cc/mole at 28°C) while a negative value for the latter stage.

Of course, almost each monomer must be considered a part, because each one may crystallize in a different form. In solid state polymerization, the activation volume is a function not only of pressure and temperature, but also of monomer conversion.

II.5 High pressure synthesis of new polymers.

Polymers which cannot normally be obtained in usual conditions may be produced when the monomers are subjected to high pressures. In the preceeding chapter, we have seen that polymalononitrile and polysuccinonitrile are produced under high pressures.

A few years ago it has been found [27] that cyclic olefines including cyclohexene and internal olefines such as tetramethylethylene give saturated polymers by 1,2 addition polymerization at 300°C and 65 kbar. No polymerization was observed below 200°C.

Under lower pressures (23.5 kbar) between 100° and 250°C, Bartlett et al. [28] have polymerized acetylenic monomers including 2-butyne, 1-butyne. Korte [29] obtained a polymer (molecular weight 1000 - 1200) from phenylacetylene (9 kbar, 220°C). Earlier studies dealed with ring-opening polymerizations (60°, 12 kbar), polymerization of dinitrile and so on [9] .

It must be noticed that, with only a few exceptions, very high pressures are needed to achieve the synthesis of new polymers. The cost and the complexity of the experimental equipment explain the relative low number of works devoted to research on new polymers.

III. HIGH PRESSURE MULTICOMPONENT POLYMERIZATIONS

The term includes of course binary systems (copolymerizations) and systems containing n monomers ($n \geq 2$). We will limit in the present course n to 3, because with increasing number of components in the system, there is also an increase of experimental and calculation difficulties.

The pressure effect has been widely studied for free radical copolymerizations, while it has been very little considered hitherto on telomerizations and never on ternary systems (terpolymerizations).

III. 1. Free - radical copolymerizations.

III. 1.1 Copolymerization equations and their pressure dependence. Considering two monomers A and B and their corresponding radicals and assuming that the reactivity of the new radical formed from the addition of a radical to the monomer A or B depends only on the last fixed monomer unit (terminal model), we have the well known scheme for the propagation step :

$$\sim A^\circ + A \longrightarrow \sim A\,A^\circ \quad (k_{AA})$$
$$\sim B^\circ + B \longrightarrow \sim B\,B^\circ \quad (k_{BB})$$
$$\sim A^\circ + B \longrightarrow \sim A\,B^\circ \quad (k_{AB})$$
$$\sim B^\circ + A \longrightarrow \sim B\,A^\circ \quad (k_{BA})$$

with following reactivity ratios

$$r_A = k_{AA} / k_{AB}$$

$$r_B = k_{BB} / k_{BA}$$

If the copolymerization equation derived from the terminal model fails to represent the experimental results, it is necessary to take into account the penultimate unit of the growing chain, the number of distinct propagation reactions is increased to eight (penultimate model)

A generally accepted concept is that the monomer reactivity is governed by the following factors :

- steric factors
- conjugation of the double bond with unsaturated side groups
- polarity of the double bond.

Alfrey and Price have set up a scheme of correlation (the Q - e scheme), where the monomer reactivity is described in a general and at least semi-quantitative way and arrived at following relationship :

$$r_A = \frac{Q_A}{Q_B} \exp\left[- e_A (e_A - e_B)\right]$$

Q_A, e_A : resonance and polarity parameters respectively for the monomer A

Actually the Q - e scheme must rigorously be written :

$$r_A = \frac{Q_A}{Q_B} \exp\left[- e_A^{\circ} (e_A - e_B)\right]$$

e_A° : polarity parameter of the radical issued from A.

The pressure dependence of the monomer reactivity ratios is governed by the difference between the activation volumes for the homopropagation and the crosspropagation relations :

$$\frac{\partial \operatorname{Ln} r_A}{\partial P} = - \frac{\Delta V_{AA}^{\neq} - \Delta V_{AB}^{\neq}}{RT}$$

$$\text{or } (r_A)_{P_2} = (r_A)_{P_1} \exp\left\{ - \left[\frac{\Delta V_{AA}^{\neq} - \Delta V_{AB}^{\neq}}{RT}\right](P_2 - P_1)\right\}$$

Since $\Delta V_{AA}^{\neq} - \Delta V_{AB}^{\neq}$ has always a low value ($<$ 10 cc/mole) it can be assumed that this quantity varies linearly with pressure. Thus integration gives :

$$\operatorname{Ln} \frac{r_A^{\circ}}{r_A^{P}} = \frac{P - 1}{RT} (\Delta V_{AA}^{\neq} - \Delta V_{AB}^{\neq})$$

r_A°, r_A^{P} : reactivity ratios of monomer A at 1 bar and P

III. 1.2 Radical copolymerization based on the terminal model. Prediction of the directional effect of pressure.

III. 1.2.1 Investigation of some systems. The effect of pressure on the reactivity ratios has been summarized by Weale [9] for many copolymerization reactions. His tabulations show that the product of the reactivity ratios tends to approach unity as pressure increases, whereas no clear patterns appear for the influence of pressure on the individual r - values. Apparently the copolymerization moves to the ideal behaviour and the free radicals lose their individual selectivity with increasing pressure.

However it has been found that this concept is verified only in some cases. Even for many systems [30], the product $r_A \cdot r_B$ decreases with increasing pressure. Therefore it is important to be able to predict the directional effects of pressure upon the reactivity ratios.

The earlier interpretation of the pressure variations of r_A and r_B was based on polarity and resonance considerations. On that hypothesis, binary systems have been selected according to the e - values (in the Q - e scheme) of both monomers [31].

Table XII. Pressure effect on reactivity ratios

A	B	e_A	e_B	r_A (*)	r_B (*)
Acrylonitrile	diethyl fumarate	++	++	5.50	0.15
				9.20	0.10
Acrylonitrile	methyl acrylate	++	+	0.65	0.46
				0.76	0.43
Acrylonitrile	vinyl acetate	++	−	3.10	0.10
				3.50	0.05
Acrylonitrile	styrene	++	−	0.02	0.65
				0.05	0.53
Acrylonitrile	α-methylstyrene	++	−−	0.05	0.15
				0.04	0.10
Methyl acrylate	crotonaldehyde	+	+	20	0.02
				20	0.02
Methyl acrylate	methylvinylketone	+	+	0.50	1.80
				0.60	1.20
Methyl acrylate	2-vinyl pyridine	+	−	0.20	1.10
				0.30	1.05
n-Butyl acrylate	Styrene	+	−	0.20	0.62
				0.17	0.50
Methyl acrylate	α-methylstyrene	+	−−	0.10	0.27
				0.08	0.23
Styrene	2-vinyl pyridine	−	−	0.55	0.80
				0.70	0.80
Vinyl acetate	Isopropenyl acetate	−	−	0.90	0.30
				0.60	0.35

(*) For each couple, the first line refers to the value at 1 bar, while the second line gives the values for P = 3000 or 6000 bar.

Representing the polarity of the monomers, in following way :

++ for monomers ($e > 1$)
\+ d° ($0 < e < 1$)
\- d° ($-1 < e < 0$)
-- d° ($e < -1$)

Table XII sums up the results for some couples of monomers A and B different polarity.

From this results, following activation volumes are derived . (Table XIII).

Table XIII. Activation volumes in copolymerization (cc/mole).

A	B	$\Delta V^{\neq}_{AA} - \Delta V^{\neq}_{AB}$	$\Delta V^{\neq}_{BB} - \Delta V^{\neq}_{BA}$
acrylonitrile	diethyl fumarate	- 5.0	+ 3.7
acrylonitrile	methyl acrylate	- 1.5	≃ 0
acrylonitrile	vinyl acetate	- 1.1	+ 6.3
acrylonitrile	styrene	≃ 0	+ 1.9
acrylonitrile	α-methylstyrene	≃ 0	+ 3.7
methyl acrylate	crotonaldehyde	≃ 0	≃ 0
methyl acrylate	2-vinyl pyridine	- 3.7	≃ 0
n-butyl acrylate	styrene	≃ 0	+ 1.0
methyl acrylate	α-methylstyrene	+ 1.0	+ 0.7
styrene	2-vinyl pyridine	- 2.2	≃ 0
vinyl acetate	isopropenyl acetate	+ 1.8	- 1.2

Turning first to table XIII, we observe that the values of the activation folumes are effectively low. Pressure modifies the reactivity of some radicals in a more or less fashion according to the sign and the magnitude of $\Delta V^{\neq}$. The most negative values are found in systems involving acrylonitrile as one of the comonomers.

Table XII shows that the reactivity ratios may increase, decrease or, in some cases, remain constant when pressure is raised.

III. 1.2.2 Prediction of the directional effect of pressure. It has been recently proposed that the Q and e-values of the Alfrey-Price scheme may be influenced by pressure [31] .

$$\frac{\partial}{\partial P}(\mathrm{Ln}\, r_A) = \frac{\partial}{\partial P}\left[\mathrm{Ln}\,\frac{Q_A}{Q_B} - e_A(e_A - e_B)\right]$$

Methyl acrylate was chosen as reference, since for this monomer :

$$\frac{\partial Q}{\partial P} = \frac{\partial e}{\partial P} \simeq 0$$

Compilation of the pressure results for many binary systems leads to new values for Q and e at high pressures (standard pressure was 3000 bar). Replacing these values in the Alfrey-Price equation the directional effect of pressure on reactivity ratios can be determined and satisfactory agreement was obtained between predicted and effectively observed variations.

From these studies, it has been found that the Q - value is generally little or not at all affected by pressure, except for ethylene for which the Q - value is increased three times in the 0 - 3000 bar region.

Evans et al. [32] discussed the relation between the resonance stabilization of the radical and the reactivity of monomer and have suggested that the Q - value of a monomer can be identified with the quantity :

$$Q_A = \exp (R_A - R_A^{\circ}) / RT$$

where R_A and R_A° are the resonance energies of monomer and its adduct radical respectively.

No effective deformation of electron configurations is believed to exist at low pressures (up to 1000 - 2000 bar) and hence the conjugation of the double bond with unsaturated substituents will scarcely be influenced by pressure.

Therefore the pressure dependence upon the reactivity ratios is mainly related to the pressure dependence of the quantity $e_A - e_B$ as shown in :

$$\frac{\partial}{\partial P} (\ln r_A \cdot r_B) \simeq \frac{\partial}{\partial P} \left[- (e_A - e_B)^2 \right]$$

Despite the oversimplification of the Q - e scheme where the e-value is described in terms of electrostatic attraction, the e-value is probably related with the tendency to form the transition state. Since the r-values are pressure dependent the e-value will be so too :

$$RT \frac{\partial}{\partial P} \left[(e_A - e_B)^2 \right] \simeq - \left[(\Delta V_{AA}^{\neq} - \Delta V_{AB}^{\neq}) + (\Delta V_{BB}^{\neq} - \Delta V_{BA}^{\neq}) \right]$$

Unfortunately the e-value has not been yet correlated hitherto with a well defined physical parameter, so that question remains open.

Another interpretation about the directional effect of pressure on reactivity ratios has been recently proposed [33] . An activation volume can always be expressed as an addition of two volumetric contributions, viz. the partial activation volumes of a radical and a monomer. Then, by simplifying, it is assumed that these partial activation volumes are characteristic of the monomers and the radicals, regardless of the combination involved.

Hence

$$\Delta V_{AA}^{\neq} = \Delta V_{A^{\circ}}^{\neq} + \Delta V_{A}^{\neq}$$

$$\Delta V_{AB}^{\neq} = \Delta V_{A^{\circ}}^{\neq} + \Delta V_{B}^{\neq}$$

Combination with following equation

$$\frac{\partial \mathrm{Ln}\, r_A}{\partial P} = \frac{\Delta V_{AA}^{\neq} - \Delta V_{AB}^{\neq}}{RT}$$

gives

$$\frac{\partial \mathrm{Ln} r_A}{\partial P} = \frac{\Delta V_A^{\neq} - \Delta V_B^{\neq}}{RT}$$

Since the radical contributions cancel out, the effect of pressure of both r - values may be described by the difference of partial monomer activation volumes and consequently

$$\frac{\partial \mathrm{Ln}\, r_A r_B}{\partial P} = 0.$$

So if the concept is valid, the r - values will change with increasing pressure in opposite direction with $r_A . r_B$ = constant. The authors pointed further out that : "In case the activation volume cannot be described as the sum of independent partial volumes, it may at least be expected that $\frac{\mathrm{Ln}\, r_A}{P}$ and $\frac{\mathrm{Ln}\, r_B}{P}$ are opposite in sign.

The validity of this concept has been checked only on two copolymerization systems involving ethylene as one of the monomers [34].

However there is a doubt concerning the additivity of independent partial volumes to give an activation volume. In many systems investigated at high pressures, as indicated above (Section III.1.2.1), the product $r_A . r_B$ is not constant, when pressure is increased. Perhaps this seems the case, when the product is nearly equal to unity at atmospheric pressure (ideal copolymerizations), then it does not change (or only a little) with pressure [35] (table XIV).

Table XIV. Pressure effect on calculated reactivity product for quasi-ideal copolymerizations.

System	$r_A . r_B$ 1 bar	$r_A . r_B$ 3000 bar
ethylene - vinyl chloride	0.91	0.96
ethylene - vinyl acetate	1.00	0.91
ethylene - vinyl pivalate	0.96	1.00
propylene - isobutylene	0.95	0.96
styrene - indene	1.00	0.97
acrylonitrile - methylvinylketone	1.00	0.96

III. 1.3 Radical copolymerizations involving penultimate effects or charge transfer complex. Compositional deviations from those predicted by terminal polymerization mechanism have been explained by many authors using penultimate (and even antepenultimate) model or charge-transfer model. However it is rather difficult to attribute on absolute manner the deviational effects to one or other model.

III. 1.3.1 Pressure effect on the penultimate model. Imoto et al. [36] investigated several copolymerizations showing penultimate effect. This effect is related with steric or electronic interactions between substituents of polymer radical and monomer. Accordingly an unusual pressure effect has been expected on these copolymerization systems because the volume of transition state would be abnormally large in order to interact between the adding monomer and the remote unit of polymer radical.

III. 1.3.2 Charge - transfer model. Electropositive and electronegative monomers can form charge - transfer complexes (CTC) each one with the other. When this complex is attacked, both monomers in the complex add as a unit. Strong CTC tend to produce polymers of constant composition, while weaker CTC compete with normal copolymerization. Here composition would not be constant as feed composition changes. For a pure charge - transfer copolymerization, we can write :

$A + B \rightleftharpoons (AB)$ formation of the CTC

$R^{\circ} + (AB) \longrightarrow R - (AB)^{\circ}$ initiation

$R - (AB)^{\circ}_{n} + AB \longrightarrow R - (AB)^{\circ}_{n+1}$ propagation

However copolymerizations exhibiting pure CTC behaviour do not exist. The general copolymerization equation must take into account terminal, penultimate and CTC effects [37] . Of course it is a rather complex formula [38], when we consider the four additional propagation reactions :

$\sim A^{\circ} + AB \longrightarrow \sim AAB^{\circ}$

$\sim A^{\circ} + BA \longrightarrow \sim ABA^{\circ}$

$\sim B^{\circ} + AB \longrightarrow \sim BAB^{\circ}$

$\sim B^{\circ} + BA \longrightarrow \sim BBA^{\circ}$

If we design the CTC equilibrium constant by K_{CT} :

$$K_{CT} = \frac{[AB]}{[A][B]}$$

Pressure obviously is able to shift the CT bond and to change K_{CT} leading to a stronger CTC.

III. 1.3.3 Examples of copolymerizations involving penultimate or charge - transfer model. Maleic anhydride (MA) possesses strong electrophilic properties which make the monomer apt to copolymerize. Moreover the rigidity of the molecule can induce penultimate effects in its copolymerizations. Therefore it is expected that, in the copolymerizations involving maleic anhydride, both effects are present.

When MA is polymerized in presence of a non-homopolymerisable monomer, their copolymerization can be considered as a cross - reaction and the two reactivity ratios are very low or equal to zero. This occurs for example in the systems : MA - cyclopentene [39] , MA - Stilbene [40] . Both systems are characterized by the formation of a CTC and are strongly accelerated by pressure. For equimolar composition in the feed, the reported overall activation volumes have following values (table XV).

Table XV. Activation volumes (cc/mole) for some MA copolymerizations (equimolar composition in the feed)

Comonomer	Solvent	T°C	$\Delta V^{\neq}$
trans stilbene [40]	$CHCl_3$	60	- 24
trans stilbene [40]	THF	70	- 28.5
α -methylstyrene [41]	$CHCl_3$	70	- 30
cyclopentene [39]	benzene	60	- 36
cyclohexene [41]	$CHCl_3$	70	- 45

It seems - and this has been recently verified [41] that such copolymerization systems show a large volume change in the propagating transition state as the result of the steric interaction between monomer substituents and the propagating polymer end.

In the two systems MA - cyclopentene and MA - stilbene, the copolymer composition differs from the calculated composition on the basis of the terminal model. By using either the penultimate model or the CTC model, good agreement has been obtained in the MA - ST system between calculated and observed composition values.

In the system MA - CP, as well as in other copolymerizations showing penultimate effect, it was found that high pressure is generally presumed to be effective for eliminating the penultimate effect [36] .

III.2 Terpolymerizations

The problem of approaching the polymerization of more than two monomers is quite similar to that of approaching binary copolymerizations. Similar steady-state assumptions are employed, so that the composition relation of a terpolymerization is written :

Fig. 7 Pressure effect on the contour maps for styrene in the ternary system : styrene (ST) - acrylonitrile (ACN) - diethyl fumarate (FE)
(the composition lines are drawn from different mole fraction of styrene in the feed)

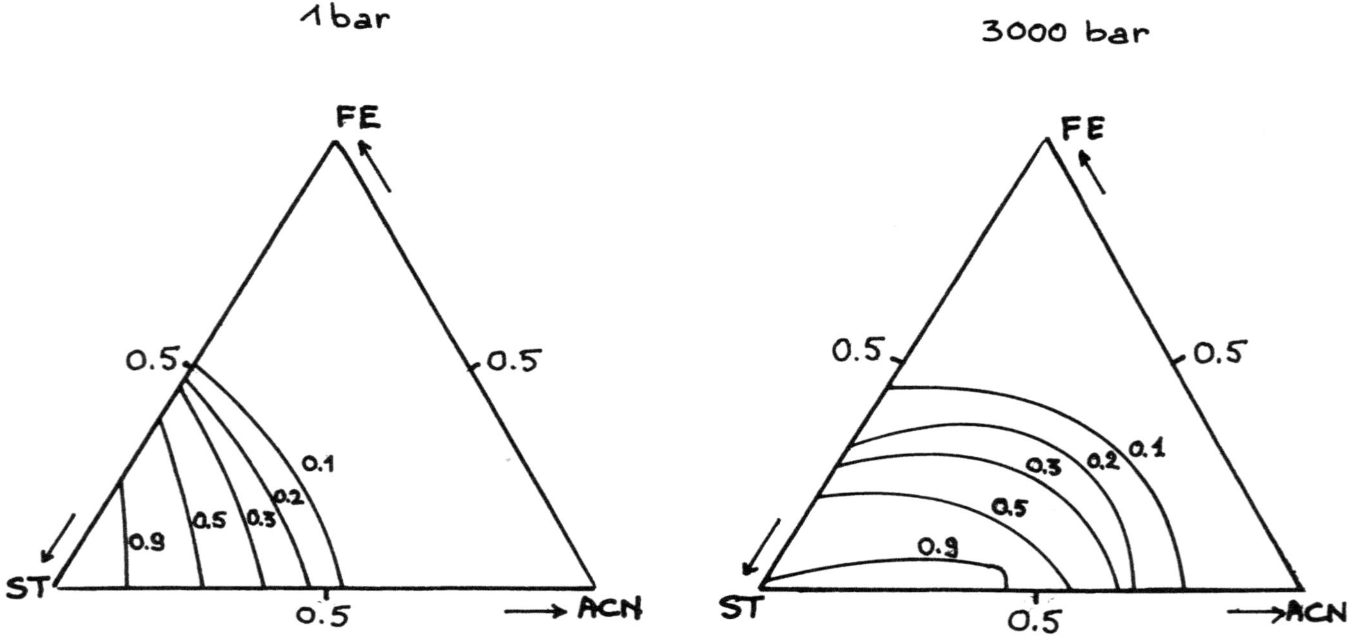

$F_A / F_B / F_C = \alpha / \beta / \gamma$

F_A : molar fraction of monomer A in the terpolymer
f_A : molar fraction of monomer A in the feed

$$F_A = \frac{\alpha}{\alpha + \beta + \gamma}$$

$$\alpha = f_A \left(\frac{f_A}{r_{CA}\, r_{BA}} + \frac{f_B}{r_{BA} r_{CB}} + \frac{f_C}{r_{CA} r_{BC}} \right) \left(f_A + \frac{f_B}{r_{AB}} + \frac{f_C}{r_{AC}} \right)$$

β and γ are obtained by permutation.

As pressure can affect the reactivity ratios, the terpolymer composition can be changed by pressure too. For example, in the ternary system : acrylonitrile (ACN) - diethyl fumarate (FE) - styrene (ST), there are drastic changes of the contour maps for the composition of the terpolymer (Fig. VII).

Superposing the contour maps for the three monomers with a triangular grid of equally spaced lines, the obtained points of intersection between lines on the grid and contour lines of equal value in mole fraction represents points of azeotropy. Connection of these points gives the locus of partial azeotropy . Partial azeotropy can occur when one monomer has the same composition in the polymer and feed. A true azeotrope in terpolymerization would occur when the mole fraction of all three monomers in the polymer initially produced was the same as that one in the feed.

Two ternary systems showing a true azeotrope have been investigated under pressure [42] (table XVI).

Table XVI. Pressure effect upon ternary azeotropes

System	P (bar)	f_A	f_B	f_C
ACN - FE - ST	1	0.36	0.14	0.52
	3000	0.38	0.13	0.49
ACN - MAM - αMS (*)	1	0.39	0.17	0.44
	3000	0.40	0.17	0.43

(*) MAM : methyl methacrylate
αMS : α-methyl styrene

It can be observed that for these two systems no change in the coordinates of the azeotrope point can be detected in the examined pressure range. However for both systems, the lines of partial azeotropy are strongly influenced by pressure. (see Fig. VII).

The fact that pressure affects partial azeotropy only, but not the ternary azeotrope is very surprising, since the reactivity ratios in both systems change with pressure. This needs a throughly explanation as soon as more experimental work will be done.

III.3 Telomerizations

A telomerization reaction is the polymerization of an unsaturated compound in the presence of a substance (called telogene), the molecules of which terminate the growth of the polymeric radical chain while taking up the terminal positions in the polymer.

$$nM + ZX \longrightarrow Z(M)_nX$$

There are only very few high pressure studies in this research area and they have been centered on ethylene telomerizations. Interesting compounds were obtained in the polymerization of ethylene (in the 1000 bar range) in presence of silicon tetrachloride or tetrachlorethylene [43].

Very recently the telomerization of various monomers in presence of hexachlorobutadiene (HCB) has been investigated [44] (table XVII).

Table XVII. Telomerization of indene and isoprene with HCB.

Pressure (bar)	N (*)	
	Indene	Isoprene
1	42	114
2500	38	90
5000	25	85
7500	18	64
10000	9	48

(*) N : number of monomer units for one unit of HCB

For other monomers (vinyl acetate, crotonaldehyde), the telomerization reaction with HCB affords telomers with high chlorine content increased by pressure.

As it can be seen from table XVII, pressure influences greatly the degree of incorporation of HCB in the growing chain. Comparision of HCB with saturated chlorinated hydrocarbons shows that the latters do not react as telogene. On the contrary if unsaturated chlorinated hydrocarbons (tri- and tetrachloroethylene for instance) are used, telomers are produced in the same manner as with HCB [44]. This is very interesting, because such telomers are well resistant against inflammability.

The activation volume of the above telomerizations are a little more negative than it would be calculated in the corresponding homopolymerization (table XVIII).

Table XVIII. Activation volumes in telomerizations with HCB

Monomer	$-\Delta V^{\neq}$ (cc/mole)	
	telomerization	homopolymerization
isoprene	24.0	18.0
indene	23.5	21.0

But these values are probably not a general feature of high pressure telomerizations, they are partially related with steric interactions in the transition state due to the bulky size of the HCB molecule.

IV. PRESSURE EFFECT UPON THE PROPERTIES OF POLYMERS

IV. 1 Structure of polymers

As for inorganic and organic materials, pressure changes sometimes to a great extent the properties of polymers. In the first part of this course, we talked already about the pressure effect on the structure of the polymers (see sections II.2.2, II.3.1.2, II.3.2.2.).

The effect of pressure upon the tacticity of polymers was investigated a long time ago by Walling and Asai [45] . The polymers of many substituted ethylenes can exist in stereoregular forms on which the configurations of successive asymmetric carbon atoms are either all identical (isotactic structure) or alternate in some regular sequence (syndiotactic structure). If $\Delta V^{\neq}$ is different for the various modes of the chain growth, pressure will alter the degree or type of stereoregularity exhibited by the polymer. The occurrence of this effect has been shown in investigations of polymethylmethacrylate carried out up to 8000 bar (table XIX).

Table XIX. Pressure effect on the tacticity of polymethylmethacrylate

Pressure (bar)	isotactic (i)	atactic (a)	syndiotactic (s)
1	4.2 ± 0.4	38 ± 1.3	57 ± 1.0
7900	11.4 ± 0.1	49 ± 3.0	40 ± 3.4

The decrease in syndiotacticity and increase in isotacticity are related with the difference in activation volumes $\Delta V_i^{\neq} - \Delta V_s^{\neq}$ and can therefore be analyzed in terms of copolymerization : the tendency of a growing chain to add the next unit isotactically or syndiotactically depends only on the stereochemistry of the preceeding linkage, and not on the more remote chain structure, at all pressures. It was concluded that the transition state for syndiotactic addition is slightly larger than that for isotactic addition to the growing chain.

Additionaly pressure may induce cross - linking in many polymers [19, 46, 47] and thus create three - dimensional lattice macromolecules. With high shearing stress, cross - linking is the predominant reaction and usually insoluble polymers with complicated network are produced [48].

IV. 2. Thermodynamic properties

IV. 2.1 Glass transition. The concept of internal ordering parameters which assume fixed values at the glass transition temperature Tg to describe the glass transition process as well as the glassy state leads to the following identity :

$$\Delta\beta \cdot \Delta C_p = VT\,(\Delta\alpha)^2$$

V, T : volume and temperature at the transition point

β, α, C_p : compressibility, thermal expansivity and heat capacity at constant pressure.

The pressure effect on Tg can be described by the Ehrenfest's relation (second - order transition)

$$\frac{\partial Tg}{\partial P} = TV\,\frac{\Delta\alpha}{\Delta C_p}$$

$$\frac{\partial Tg}{\partial P} = \frac{\Delta\beta}{\Delta\alpha}$$

therefore the glass transition is not only temperature, but also pressure dependent. It is assumed that the free volume concept of the glass transition holds and that the initial effect of pressure is to squeeze out the free volume and thus compress the polymer. At temperature above Tg, the effect of pressure is to cause the polymer to vitrify, when a certain critical free volume is attained during the application of pressure (Pg). Experimentally this critical free volume is evidenced by the change in slope over a narrow range, in the pressure versus change in volume plot. Once Pg occurs, the compressibility change with pressure is that of the glassy polymer and, in general, is much less than that for the rubbery solid.

IV.2.2 Pressure - volume - temperature studies. For the purpose of thermodynamic studies of polymers, it is indispensable to determine PVT relations. Many studies have been reported on PVT determinations of polymers. The basic measurements are constituted by compressibility measurements at constant temperature. From those, one obtains isobares and isochores (curves at constant volume) and hence an, at least empirical, equation of state can be derived.

Numerous studies have shown that the compressibility depends largely on the physical state in which is the polymer : crystal, amorphous, polymer liquids. For the polymeric equations of state, I can recommend following articles referenced [49, 50, 51] .

The density as function of pressure may be calculated via the Tait equation :

$$\frac{V_{1,T} - V_{P,T}}{V_{1,T}} = C \; Ln \left(\frac{B + P}{B} \right)$$

C and B being constants idependent on pressure

$V_{1,T}$ and $V_{P,T}$: volumes at constant temperature T at pressures 1 and P respectively

In a general way, the linear compressibility β is strongly decreased by increasing pressure in the low pressure region. With raising pressure, β decreases more and more slower. An interesting point is the compressibility behaviour of a polymer possessing amorphous and crystalline parts simultaneously. Assuming such a polymer (polyhexamethylene adipinamide, for example, [52]), its compressibility is largely pressure depending, in the cristalline parts of polyamide the molecules are denser than in the amorphous parts, when pressure is increased, the molecular segments become closer in the amorphous parts. At higher pressures, compression is only possible through a change of the valence angles or a shortening of atomic bonds. In the cristalline parts, pressure has to act against the repulsive forces between the atoms, this supposes high pressures, so that the compressibility of the cristalline parts has only low values.

I do not want to comment more extensively the P-V-T studies, since they are treated elsewhere in this course. I only will remark that, at the Tg point, the plot of compressibility versus temperature become discontinuous at all pressures for amorphous polymers.

IV. 3. Transport properties

Transport phenomena include viscosity, thermal conductivity, self-diffusion. Only the two first will be shortly treated here.

IV. 3.1 Pressure effect on the viscosity of polymeric materials. Visco-elastic losses, relaxation time and phase changes in thermoplastics and elastomers are generally associated with a free volume providing space for appropriate movement of chain segments. It is clear that hydrostatic pressure must play a very direct part in these phenomena and provide some insight into their mechanism.

Complex equations have been put forward to describe the pressure dependence of the viscosity for polymeric liquids [53]. Surprisingly for some polysiloxanes, the viscosity is relatively insensitive to pressure and temperature, while for all other polymers, viscosities are very sensitive to pressure.

IV. 3.2 Thermal conductivity. Little experimental work has been done to examine the variation of thermal conductivity of polymers with pressure. From measurements of λ, the specific heat can be calculated. Empirically, it is written :

$$\lambda = \left(\frac{B}{A}\right)\rho^{4/3}$$

ρ : density
B : constant depending on molecular weight
A : function of temperature.

The thermal conductivity of polyethylene was recently investigated up to 25 kbar [54] and increases strongly with pressure :

$\lambda_{25000}/\lambda_1$ = 2,75 for high density polyethylene

$\lambda_{25000}/\lambda_1$ = 2,19 for low density polyethylene

The differences observed for each type of polyethylene has been analyzed in terms of separate contributions from the amorphous and the crystalline parts.

On the basis of Wada's theory, the thermal conduction is mainly governed by interchain lattice waves. When pressure is applied to crystalline polyethylene, the compression is not isotropic, but occurs chiefly in the plane perpendicular to the fiber axis, thus changing the average Van der Waals potential energy and its anharmonicity. The strong increase observed for λ with pressure may therefore be interpreted as a consequence of the increase in the interchain thermal conduction.

IV. 4 Phase changes

It is understood that in this section only first - order transition will be treated. It is known that at the transition point there is a discontinuity in entropy, corresponding to a latent heat, and the various modes of transition are all governed by the relation :

$$\frac{dP}{dT} = \frac{\Delta H}{T\Delta V}$$

ΔH, ΔV : increase in enthalpy and volume at the transition point for one mole of substance

IV. 4. 1 Pressure effect on the melting curve of polymers. As ΔV is a positive quantity for polymers, the melting point will increase with pressure. Usually the pressure effect on the melting point is described by the semi - empirical equation of Simon and Glatzel [2] :

$$\frac{P}{a} = \left(\frac{T}{T_o}\right)^c - 1$$

a,c : constants
T_o : melting point at 1 bar.

Unfortunately for polymers, it seems inappropriate. Knowledge of the melting transition of crystalline polymers at high pressures is important, because both thermodynamic and structural information can be gained which experiments at atmospheric pressure do not yield directly. Moreover in order to crystallize a polymer under pressure at a predetermined degree of supercooling, it is necessary to know accurately the effect of pressure on the melting temperature.

Many crystalline polymers have been melted at various pressures [55, 56] (table XX).

Table XX. Effect of pressure on the melting curve for some polymers.

Polymer	Pressure (bar)	T_m (°C)
Polyethylenoxide	1	60
	1500	80
	3000	94
Polypropylene	1	174
	1000	226
	2000	243
Polyethylene	1	136
	1500	177
	3000	204

In general, the melting curve of polymers behaves in the same manner as for other compounds. No maximum in the melting temperature - pressure curve appears possible, unless pressure induces chain rupture or solid - solid transition due to a more compact structure.

Since the volume entropy change decreases with pressure while the configurational entropy change increases, the relative amounts of these opposing pressure effects determine whether the total entropy change during melting will increase or decrease with pressure.

However the overall melting process is far more complex than expected, when the melting transition is treated as a simple first - order transition.

IV. 4.2 Solid - solid transition. The solid - solid transition is the polymorphic change governed by the Clausius - Clapeyron's equation too. While polymorphism is frequently encountered in high pressure chemistry of solids, only recently studies have been devoted to polymorphism in polymers.

One of them however has been thoroughly investigated (teflon) [57, 58]. Four phases have been found from which one is a high pressure phase (teflon III) : the unit cell is orthorhombic and in this phase, the molecules have a planar zigzag arrangement different from that of a helix at atmospheric pressure.

In cis-polyisoprene [59] as well as in polyethylene, there exists a high pressure phase found as being a disordered hexagonal packing of polymer chains which have an "in-chain" repeat unit identical to that of the normal monoclinic crystals.

IV. 4.3 High pressure crystallization processes. An increase in pressure allows crystallization at elevated temperatures and leads to the formation of extended - chain crystals. The crystallization process has been found as a complicated one [60]. To clarify the phenomenon, it was necessary to develop many precise measurements under high pressure, especially microscopic observations.

The morphology of polyethylene crystallized at various pressures up to 6000 bar is very different according to the crystallization pressure [61]. Up to 2500 bar, folded - chain crystals are observed. Above 4000 bar, the character of the polyethylene resulted from crystallization is radically different since the high pressure process produces extended chain crystals. In the intermediate region, polyethylene crystallizes in a polymer with unknown structure (lens - like) which may be a nematic liquid crystal having the C - axis parallel to the lens - axis. Recently a better understanding was gained from Bassett's investigations [62]. From 3 kbar, polyethylene tends to crystallize as a disordered hexagonal phase for which the all - trans conformation is lost. Because this phase has a low entropy of fusion, it gives much thicker crystals than does crystallization of the usual orthorhombic form.

IV. 5 Electrical properties

Highly conjugated organic semi - conducting polymers are been shown to be very pressure sensitive [63] . The conductivity pressure coefficient is known to be related to the conductivity, its activation energy and to the chemical structure. In solid polymers, hyperelectronic polarization has been observed, which is due to the induced moments of mobile charges on long molecules. The charge is confined to the molecule and moves essentially without resistance along the molecule. When a small external field is applied, the charges are displaced and a net polarization is observed. From such a model, it would be expected that the pressure effect would be similar to the conductivity pressure effect.

Pressure acts upon the dielectric constant ε according to a Tait - like expression (equation of Owen and Brinkley) :

$$1 - \frac{\varepsilon_1}{\varepsilon_p} = A \ Ln \left(\frac{B + P}{B + 1} \right)$$

A, B : constants
changes of ε with pressure in non-polar polymers were observed by Japanese authors [64].

IV. 6 Mechanical properties

It can be concluded from several papers [65, 66] that major changes do occur in many polymers in such important mechanical characteristics as elastic and shear moduli, yield and ultimate strengths and deformation and fracture response. Some of the observed effects of pressure on these properties appear to be common features of all polymers, while others vary markedly from polymer to polymer and sometimes from sample to sample of the same material.

A common manifestation reported by most investigators is that the imposition of a high pressure environment causes significant increases to occur in both modulus and strength. These increases, though differing in degree from one polymer to another, may amount to as much as 50 to 100% increase per kilobar pressure.

However the influence of pressure on the ductility of polymers varies markedly from polymer to polymer. For example, it has been noted that in some materials, such as polytetrafluoroethylene, no local necking occured at any pressure and that the material progressively becomes brittle with increasing pressure.

But in another crystalline polymer (high - density polyethylene), localized necking has been intensified although cold - drawing was completely eliminated at high pressures.

Along with the new data on effects of hydrostatic pressure on mechanical behavior of polymers, various advances have also been made in the understanding of the phenomena involved. Several theories of yielding have been put forward and finite strain elasticity theory has been invoked to explain the dependence of elastic modulus on pressure. Description of the phenomena has also been sought in terms of molecular concepts such as reduction of free volume by pressure.

V. EXPERIMENTAL PROCEDURES IN HIGH PRESSURE POLYMERIZATIONS

In other parts of the present course, some lecturers have developped, (or will so do) various high pressure techniques. Therefore I have not as aim to recall the experimental methods for producing and measuring high pressures. However I think interesting and useful to show some particular techniques suitable for the investigation of high pressure polymerizations.

V. 1 Reaction tubes

The two fundamental properties of a high pressure reaction tube are as follows :

- the tube material must transmit fully and hydrostatically the applied pressure
- the tube must remain tight at any pressure in order to preserve the reaction mixture to be analyzed from any contamination of the transmitting fluid.

Various kinds of tubes have been constructed. Three basic designs exist up to day :

+ deformable tubes (metal or preferently teflon) which according to temperature may be used up to 150°C. Above, the thickness of the walls is too important so that their deformation becomes problematic
+ mercury - piston tubes

In this system the reaction mixture is isolated from the transmitting medium by mercury which has the advantage to suppress the problem of tightness, but the tubes are not easy to manipulate for several reasons [67] and nowadays are pratically given up.

+ piston - cylinder assemblies

All - glass hypodermic syringe is the simplest, but cannot

withstand to high pressures. More elaborated tubes made of stainless steel are usually used even to very high pressures and temperatures. Such tubes can be re-utilized indifinitely and the filling and extraction of the reaction mixture is easy.

V. 2 Measuring rates at high pressures

Overall reaction rates can be measured in two ways, either by a discontinuous method : series of samples which will be withdrawn each after the other from the apparatus after the desired time or by following continuously the reaction yield. In the latter case, various techniques should be used : measurement of the pressure drop during the polymerization by dilatometry or piezoresistivity, high pressure sampling at different intervals [68] etc...

Individual reaction rates (of initiation, propagation and termination) are more difficult to measure. An adaption of the technique on the photochemically initiated polymerization at high pressure under conditions of intermittent illumination (rotating - sector method) enables the effect of pressure on the individual reactions to be separated.

However, the technique on the optical experiment under high pressure has some special difficulties. For instance, the optical window must be transparent to the radiation used, and an additional problem is the selection of the metal surface and packing with which it will be in contact. Ogo [69] used a high pressure reactor with a window installed on the bottom of the reactor, made it possible to irradiate the medium. The window was a cylindrical block of fused quartz with optically polished ends mounted directly against an optically flat steel surface of the outlet plug and some leather packing surrounding the quartz prevented leakage of the pressure - transmitting fluid. The reactor was a hypodermic syringe with an optically flat end, which was placed in contact directly with the window. The rotating sector was made of a black disk having a 90° sector. Radical lifetimes are determined by this method whose theory (given by Matheson) is not described here. Photopolymerizations must be carried out at a constant concentration of photosensitizer. After the experiments of photopolymerization under steady illumination, the rates of polymerization under intermittent illumination are determined for a wide range of sector speeds. Finally the rate under steady illumination is determined again to ensure the results to be correct.

REFERENCES

1. G. Guarise, G. Palma, E. Siviero, G. Talamini, Polymer, 11, 613 (1970).
2. from S. Babb, Rev. Modern Phys., 35, 400 (1963).
3. R. Steiner, Chem. Ing. Techn., 48, 533 (1976) and references therein.
4. G. Scott, E. Senogles, J. Macromol. Sci., Rev., C 9, 49 (1973).
5. T. Yala, G. Jenner, unpublished results (1975).
6. H. Asai, T. Imoto, J. Chem. Soc. Japan, 84, 871 (1963).
7. M. Yokawa, Y. Ogo, Makromol. Chem., 177, 429 (1976)
8. M. Yokawa, J. Yoshida, Y. Ogo, Makromol. Chem., 178, 443 (1977)
9. K. Weale, "Chemical reactions at high pressures", Spon Ed., London (1967).
10. M. Kellou, G. Jenner, Europ. Polymer J., 12, 883 (1976)
11. P. Deb, Europ. Polymer J., 11, 31 (1975).
12. M. Romani, K. Weale, Trans. Faraday Soc., 62, 2264 (1966).
13. K. Hamanoue, Rev. Phys. Chem. Japan, 38, 103 (1969).
14. C. Walling, J. Pellon, J. Amer. Chem. Soc., 79, 4782 (1957).
15. T. Mitani, Y. Ogo, T. Imoto, Rev. Phys. Chem. Japan, 42, 25 (1972).
16. W. Busfield, E. Whalley, Trans. Faraday Soc., 59, 679 (1963).
17. V. Zhulin, M. Gonikberg, Vysokomol. Soedin., 3, 262 (1961).
Y. Ohtsuka, C. Walling, J. Amer. Chem. Soc., 88, 4167 (1966).
J. Osugi, M. Okamoto, Proceeding 4th Intern. Conference High Press. (Kyoto), 689 (1974).
18. P. Ehrlich, G. Mortimer, Adv. Polymer Sci., 7, 386 (1970).
19. G. Jenner, J. Macromol. Sci., Chem., A 9, 83 (1975).
20. A. Khalilpour, G. Jenner, A. Deluzarche, Bull. Soc. Chim. France, 583 (1976).
21. W. Le Noble, A. Das, J. Phys. Chem., 74, 3429 (1970).
22. G. Jenner, J. Hitzke, M. Millet, Bull. Soc. Chim. France, 1183 (1970).
23. G. Jenner, A. Khalilpour, Europ. Polym. J., 12, 105 (1976).
24. H. Abdi, G. Jenner, C. Brun, Makromol. Chem., 164, 149 (1973).
25. K. Hara, Rev. Phys. Chem. Japan, 40, 73 (1970)
26. Y. Tabata, S. Shu, K. Oshima, Makromol. Chem., 104, 236 (1967).
27. B. Anderson, C. Hoover, O. Vogl, Macromolecules, 2, 686 (1969).
28. K. Bartlett, G. O'Neill, H. Turner, W. Wall. Brit. Polymer J., 4, 503 (1972).
29. W. Jane, D. Bienik, F. Korte, Naturwiss., 62, 391 (1975).
30. T. Yala, G. Jenner, A. Deluzarche, Bull. Soc. Chim. France, 609 (1973).

31. G. Jenner, S. Aieche, J. Polymer Sci., Pt A-1, in press (1977).
32. T. Alfrey, L. Young, "Copolymerization" Ed. G. Ham, Interscience, New York (1964).
33. F. de Kok, D. Heikens, Proceedings 4th Intern. Conf. High Press. , Kyoto, 677 (1974).
34. R. van der Meer, A. German, Proceedings, 15th Meeting EHPRG, Darmstadt, 17 (1977).
35. G. Jenner, J. Polymer Sci., Pt A1, in press (1978).
36. Y. Ogo, T. Imoto, Rev. Phys. Chem. Japan, 42, 34 (1972).
37. J. Seiner, M. Litt, Macromolecules, 4, 308 (1971).
38. C. Pittman, T. Rounsefell, Macromolecules, 8 , 46 (1975).
39. M. Yokawa, Y. Ogo, T. Imoto, Makromol. Chem., 163, 135 (1973).
40. M. Kellou, G. Jenner, European Polymer J., 13, 9 (1977).
41. M. Kellou, G. Jenner, work in progress.
42. S. Aieche, Dissertation, Université Louis Pasteur, Strasbourg (1976).
43. H. Hopff, N. Balint, Personal communication (1974).
44. G. Jenner, Angew. Makromol. Chem., 64 , in press (1977).
45. C. Walling, D. Tanner, J. Polymer Sci., A1, 2271 (1963).
H. Asai, T. Imoto, J. Chem. Soc. Japan, 84, 868 (1963).
46. T. Sasuga, M. Takehisa, J. Macromol. Sci. Phys., B 11, 389 (1975).
47. F. Rackley, H. Turner, W. Wall, R. Haward, J. Polymer Sci., Pt B-1, 12, 1355 (1974).
48. V. Zhulin, Proceedings, 10th Meeting EHPRG, Strasbourg (1972).
49. R. Smith, J. Polymer Sci., Pt A-2, 8, 1337 (1970).
50. J. Curro, J. Macromol. Sci., Rev. Macromol. Chem., C 11, 321 (1974).
51. R. Simha, P. Wilson, O. Olabisi, Kolloid Z. Polym., 251 , 402 (1973).
52. H. Tautz, L. Strobel, Kolloid Z. Polym., 202, 33 (1965).
53. F. Cogswell, J. Mc Govan, Brit. Polymer J., 4, 183 (1972).
54. P. Andersson, G. Bäckström, J. Appl. Phys. , 44, 2601 (1973).
55. E. Baer, J. Kardos, J. Polymer Sci., Pt A, 3, 2827 (1965).
56. S. Bhateja, K. Pae, J. Macromol. Sci., Rev. Macromol. Chem., C 3 , 77 (1975).
57. H. Flack, J. Polymer Sci., Pt A-2, 10, 1799 (1972).
58. C. Nakafuku, T. Takemura, Jap. J. Appl. Phys., 14, 599 (1975).
59. P. Phillips, B. Edwards, J. Polymer Sci., Polymer Phys. Ed., 14, 377 (1976).
60. M. Yasuniwa, T. Takemura, Polymer, 15, 661 (1974).
61. M. Yasuniwa, C. Nakafuku, T.Takemura, Polymer J., 4 , 526 (1973).

62. D. Bassett, Proceedings, 15th Meeting EHPRG, Darmstadt, 36 (1977).
63. H. Pohl, J. Wyhof, J. Polymer Sci., Pt A-1, 10, 387 (1972).
64. H. Sasabe, S. Saito, Polymer J., 3, 749 (1972).
65. E. Jones, D. Tabor, Polymer, 14 , 617 and 623 (1973).
66. K. Pae, S. Bhateja, J. Macromol. Sci, Rev. Macromol. Chem., C 3, 1 (1975).
67. G. Jenner, A. Deluzarche, High Temper. High Press.,2, 345 (1970).
68. R. Grieger, C. Eckert, AICHE Journal, 16, 766 (1970).
69. Y. Ogo, M. Yokawa, T. Imoto, Makromol. Chem., 171, 123 (1973).

PRESSURE EFFECTS ON RAPID EQUILIBRIA IN SOLUTION

Donald A. Palmer and Hartwig Kelm

Institute for Physical Chemistry, University of Frankfurt, Robert Mayer Strasse 11, D6000 Frankfurt am Main, West Germany.

I. ABSTRACT

The effects of pressure on such equilibria as are found in acid-base behaviour, ion-pairing and the formation of labile complexes, are described. Emphasis is placed on examples taken from inorganic coordination chemistry.

For the acid-base equilibria, the proposed empirical correlations are discussed, as well as the utility of volumes of ion-pairing for differentiating between inner and outer sphere ion-pairs. Finally, pressure effects on complex formation are rationalized using a simple electrostriction theory.

II. DEFINITIONS AND METHODS

Analogous to the mathematical description of pressure effects on rate constants (1), the pressure derivatives of equilibrium constants are formulated as follows:

$$d \ln K_x/dP = - \Delta\bar{V}/RT \qquad (1)$$

$$d \ln K_c/dP = - \Delta\bar{V}/RT + \Delta\nu\kappa RT \qquad (1a)$$

where K_x and K_c are the equilibrium constants for a system in which the concentrations are expressed in mole fractions and

Kelm, H. (Ed.) High Pressure Chemistry. 421–434. *Copyright © 1978. D. Reidel Publishing Company. All Rights Reserved.*

molarities, respectively. As the units of molarity are pressure dependent, Eq. (1a) includes a correction term in which $\Delta\nu$ is the difference in the stoichiometric indices between the products and educts, while κ represents the compressibility coefficient of the solvent. The volume change during the reaction is symbolized by $\Delta\bar{V}$. The volume of reaction may be defined as the difference in the partial molar volumes, $\bar{V}_i$, of the products and educts (Eq. (2)):

$$\Delta\bar{V} = \Sigma_i\ (\bar{V}_{products}) - \Sigma_i\ (\bar{V}_{educts}) \tag{2}$$

Thus, besides directly studying the pressure dependence of the equilibrium constant, $\Delta\bar{V}$ can also be determined by measuring the individual partial molar volumes, which are obtained most conveniently from precision density data.

Certainly $\Delta\bar{V}$ can be defined and measured at different pressures, P. The corresponding values of $\Delta\bar{V}_p$ may be derived either from the slopes of (ln K) versus P plots at the particular pressures, or from density measurements at these pressures. However the latter are technically more difficult and for many systems fail to meet the necessary accuracy requirements.

The pressure dependence of the volume of reaction is referred to as the compressibility coefficient $\Delta\beta$.

$$d\ \Delta\bar{V}/dP = \Delta\beta \tag{3}$$

It may be of interest to review and discuss the effects of pressure on a number of fast equilibria involving charged species in solution. Examples of these are to be found in the dissociation of weak acids and bases, ion-pair formation, and complex formation. Such data can also be useful in the discussion of more complex kinetic systems. It is, however, by no means intended to present a complete review of this field here. The choice of examples is definitely subjective and determined by pedagogical criteria. Supplementary information is available in the literature (2).

III. PRESSURE EFFECTS ON WEAK ACID-BASE EQUILIBRIA

Within the class of weak electrolytes, weak acids and bases - although a special group - play an important role. Their comparable and almost predictable behaviour is a consequence of the exceptional volume effects caused by the release of a H^+ ion

(- 4.5 cm^3 mol^{-1}) and, even more so, a OH^- ion (- 17 cm^3 mol^{-1}) in aqueous solution.

The fact that the structures of aliphatic and aromatic carboxylic acids can be systematically varied has made them the objects of detailed investigations. In Table 1 are shown a selection of volumes of dissociation for mono- and dicarboxylic aliphatic acids, together with their dissociation constants. From an inspection of Table 1 a nearly constant value of (- 13.5 ± 1) cm^3 mol^{-1} emerges for the volume change resulting from the dissociation of the first proton, $\Delta\bar{V}_o(1)$. Only the first examples in each series (i.e. n = 0) show smaller negative values than expected on the basis of a common effect. Hamann (2b)

Table 1. Volumes of reaction for the dissociation of mono- and dicarboxylic acids in water at 25°C.

Acid	K_1	K_2	$\Delta\bar{V}_o(1)$	$\Delta\bar{V}_o(2)$	Ref.
$H-(CH_2)_n-COOH$	kg mol^{-1}		cm^3 mol^{-1}		
n = 0	1.74×10^{-4}		-8.8		2a,3,4
n = 1	1.75×10^{-5}		-12.2		2a,3,4
n = 2	1.34×10^{-5}		-12.9		2a,4
n = 3	1.52×10^{-5}		-13.7		4
$(CH_2)_n-(COOH)_2$					
n = 0	5.90×10^{-2}	6.40×10^{-5}	-6.7	-11.9	5,6
n = 1	1.49×10^{-3}	2.03×10^{-6}	-10.1	-18.6	5,6
n = 2	6.89×10^{-5}	2.47×10^{-6}	-12.9	-13.6	5,6
n = 3	4.58×10^{-5}	3.89×10^{-6}	-13.2	-13.6	5,6
n = 4	3.71×10^{-5}	3.87×10^{-5}	-13.5	-13.5	5,6
n = 5	3.09×10^{-5}	4.88×10^{-6}	-14.1	-13.5	5,6

explains this with the hypothesis of a more strongly "hydrated", initial, unionized acid molecule via hydrogen bonding. The initial contraction which results from this increased "hydration" is thought to reduce the negative volume effect upon dissociation. The exceptionally strong negative value of $\Delta\bar{V}_o(2)$ for the second dissociation step of malonic acid was explained by Høiland (6) to be the result of the "cis"-configuration of this substrate. The close neighbourhood of two negative charges should lead to an increase in electrostriction as the latter is proportional to the square of the charges. However, an acceptance of this explanation raises the following question. Why does this effect not apply to oxalic acid (n = 0), thereby offsetting the "hydration" effect used to rationalize the abnormally small negative $\Delta\bar{V}_o(1)$ value?

Benzoic acid derivatives (7,8), as well as phenols (9),

Table 2. Volumes of reaction for the dissociation of weak "inorganic" acids in water at 25°C.

Acid	K_1	K_2	$\Delta\bar{V}_o(1)$	$\Delta\bar{V}_o(2)$	Ref.
	mol kg^{-1}		cm^3 mol^{-1}		
H_2SO_3	1.4×10^{-2}		- 21.		10
H_3PO_4	7.1×10^{-3}		- 16.		10
H_2CO_3	4.2×10^{-7}		- 29.		2c,11
H_2S	1.5×10^{-7}		- 15.		14
H_2O	$2. \times 10^{-16}$		- 21.		2b,15-17
HSO_4^-		1.2×10^{-2}		- 20.	12
$H_2PO_4^-$		6.2×10^{-8}		- 28.	13
HCO_3^-		4.7×10^{-11}		- 28.	2c
$Fe(OH_2)_6^{3+}$	2.2×10^{-3}		+ 3.0		18
$Cr(OH_2)_6^{3+}$	7.4×10^{-5}		- 3.8		19
$Co(NH_3)_5OH_2^{3+}$	2.5×10^{-7}		- 0.7		20

exhibit volumes of dissociation of (- 9 ± 1) and (-12 ± 2) cm^3 mol^{-1}, respectively, depending on the nature of the substituent on the benzene ring. No correlation could be detected between $\Delta\bar{V}_o$ and the acid dissociation constants.

Dissociation constants of weak inorganic acids are definitely more pressure dependent as can be seen from Table 2, in which a number of typical examples are listed. The range of $\Delta\bar{V}_o(1)$ values for the first dissociation step, K_1, is relatively large compared to the carboxylic acids, indicating that "individual" or specific solvation occurs in water for the anionic product species. For the second dissociation step, K_2, larger values of $\Delta\bar{V}_o(2)$ are expected since electrostriction increases substantially.

The self-ionization of water was investigated over a large range of temperature and pressure, revealing that water behaves like a strong electrolyte at high temperatures and pressures.

A substantial increase in the acidity of a water molecule occurs when it is bound to a metal ion, especially to a cationic species. The deprotonation of this aquo ligand is accompanied by surprisingly small volume changes. Here the release of electrostricted solvent water from around the complex ion following the charge reduction obviously compensates the negative partial molar volume of the proton, which is normally taken to be - 4.5 cm^3 mol^{-1} (21). The examples of complex aquo ions included in Table 2 underline the fact that an estimation of the volumes of reaction for protonation-deprotonation equilibria cannot be made on the basis of a simple electrostatic theory. More experimental data are needed to establish a useful rule.

A rather constant value for $\Delta\bar{V}_o$ is obtained for the base dissociations, as illustrated in Table 3. Regardless of the structure of the substrate, experimental values of $\Delta\bar{V}$ concentrate around (28 ± 1) cm^3 mol^{-1}. These data reflect the dominant role played by the OH^- ion compared to the cationic counter ions.

The pressure dependence of the dissociation constants of weak acids and bases could be shown to follow relatively simple empirical correlations. One of these (Eq. (4)) was proposed by El'yanov and Gonikberg (23) and later modified by El'yanov and Hamann (24):

$$K_p/K_o = - \Delta\bar{V}_o\Phi/T \tag{4}$$

Here K_o and K_p are the dissociation constants at 1 bar and pressure P, respectively, while $\Delta\bar{V}_o$ is the volume of reaction at 1 bar and Φ is an empirical function tabulated for different pressures. Indeed, Eq. (4) describes weak acids very well over

Table 3. Volumes of reaction for the ionization of weak bases in water at 25°C.

Base	K mol kg^{-1}	$\Delta\bar{V}_o$ cm^3 mol^{-1}	Ref.
Ammonia	1.75×10^{-5}	- 29.4	22
Monomethylamine	4.19×10^{-4}	- 27.5	4,22
Dimethylamine	5.89×10^{-4}	- 27.4	3,4
Trimethylamine	6.28×10^{-5}	- 28.1	3,4
Aniline	3.82×10^{-10}	- 28.6	4
Pyridine	1.71×10^{-9} *	- 28.1	4

* 20°C

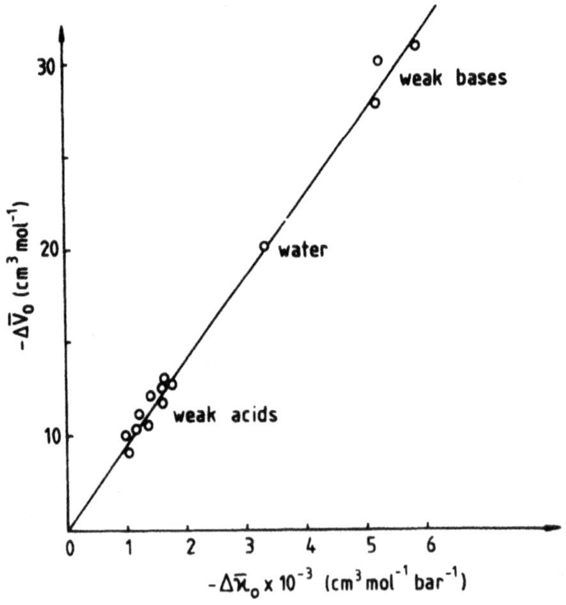

Fig. 1. Correlation between the partial molar volume change upon ionization, $\Delta\bar{V}_o$, and the change in the partial molal isothermal compressibility of ionization, $\Delta\bar{\kappa}_o$.

a pressure range from 1 to 3000 bar and for temperatures between 25° and 250°C. It is a typical linear free energy relationship with $\Delta\bar{V}_o$ and Φ as variable parameters. As in the case of the analogous kinetic equation, the parameter Φ can be approximated using the Tait equation.

Wynne-Jones and co-workers (25) proposed Eq. (5) involving the parameters $\Delta\bar{V}_o$ and $\Delta\bar{\kappa}_o$, where $\Delta\bar{\kappa}_o$ represents the partial molal isothermal compressibility of ionization, $(- d\, \Delta\bar{V}_o/dP)$.

$$RT[\ln(K_p/K_o)] = \Delta\bar{V}_o(P-1) + 0.5[\Delta\bar{\kappa}_o(P-1)^2] \tag{5}$$

In addition, an empirical relation between $\Delta\bar{V}_o$ and $\Delta\bar{\kappa}_o$ was found by these authors and is depicted in Fig. 1. This is similar to a treatment developed by Swaddle et al. (26) for the pressure dependence of $\Delta V^{\neq}_{exp}$.

North (27) postulated that the pressure dependence of $\Delta\bar{V}_o$ is a function of the number of water molecules released from the solvation sphere during the course of the reaction.

Using either the correlation shown in Fig. 1, or the tabulated values of Φ, weak acid and base dissociation constants at higher pressures and temperatures can be calculated from $\Delta\bar{V}_o$ data, with reasonable accuracy, using Eqs. (4) and (5). In case $\Delta\bar{V}_o$ is determined from density measurements, <u>no high pressure experiments</u> are needed.

Thus the situation appears to be satisfactory for acid-base equilibria as long as extreme accuracy is not required. Especially for complex kinetic systems, which include such equilibria, the empirical relationships mentioned here are sufficient when combined with high pressure kinetic data of normal accuracy.

IV. PRESSURE EFFECTS ON ION-PAIR FORMATION

General ion-pair formation is, in principle, not very much different from the dissociation of weak acids and bases, except that the specific ions, H^+ and OH^-, are replaced by a variety of cations and anions. This replacement introduces additional variations, namely intrinsic and solvation contributions, into the partial molar volume quantities and therefore into $\Delta\bar{V}_o$. This complication and the fact that only a limited amount of data is available, are probably responsible for the absence of any noticeable trends or correlations in the volumes of ion-pairing.

Furthermore, the situation is complicated by the fact that different types of ion-pairs exist. There are those in which the ions approach each other with their solvation shells intact, and others in which their solvation spheres at least partially overlap. Thus one must differentiate between ion-pairs in which the counterions are separated by one or several solvent molecules and those in which they are in direct contact. Terms such as "loose" and "tight", or "outer" and "inner sphere" ion-pairs are used to qualitatively describe these situations.

The values for the volumes of ion-pair formation are therefore expected to be rather different in magnitude depending on the kind of ion-pair formed. For outer sphere ion-pair formation, where charge separation is maintained, only moderately positive contributions to $\Delta\bar{V}_0$ are observed, due to the partial release of electrostricted solvent molecules. No intrinsic contribution to $\Delta\bar{V}_0$ is expected in these cases since the solvated ions do not interpenetrate. The situation is different for inner sphere ion-pair formation. Here an essential part of the individual charges is neutralized. An additional intrinsic contribution has to be considered due to the interpenetration of the solvent sheaths, giving rise to a simultaneous release of solvent molecules into the bulk medium. As a rule this contribution, $\Delta\bar{V}_{intr}$, is considered to be a positive quantity, although little is known about its quantitative contribution, due to the release of formerly oriented solvent molecules into the bulk medium.

While discussing the different types of ion-pairs another "floating" borderline should be mentioned here. Kineticists who deal with ionic species, especially coordination compounds, differentiate between ion-pair formation - sometimes called precurser-complex formation - involving a well defined coordination compound and a solvated, free counterion on the one hand and the final product complex after substitution, on the other hand. The latter involves the formation of a new coordination compound with the former counterion as ligand. Volume effects associated with the overall substitution reaction are not considered to be volumes of ion-pair formation. The difference between such a substituted coordination compound and an ion-pair in close contact (inner sphere ion-pair) is mainly determined by kinetic stability (i.e. the life time) of the species. For instance, the possibility that one could, in principle, isolate the former may be used as a criterion. Such differentiations are certainly subject to some discretion and frequently complicate the interpretation of volumes of ion-pairing.

Nevertheless, attempts have been made to experimently prove the existence of different types of ion-pairs. A fortunate situation is provided by the sodium fluorenyl/THF system. Spectral changes were detected resulting from the transition from tight to loose ion-pairing (28). Application of pressure shifts the equilibrium in favour of loose ion-pairs with a rather large $\Delta\bar{V}_o$ value of - 24.4 cm^3 mol^{-1}. However one must remember at this point that nonpolar solvents like THF give rise to rather large electrostriction effects when charge separation occurs.

For the majority of ion-paired systems, spectral changes cannot be detected so that electrical conductivity measurements are often conducted instead, normally in conjunction with a concentration-pressure variation study. Ion-pair formation constants, K_{IP}, and volumes of ion-pair formation at normal pressure, $\Delta\bar{V}_{o,IP}$, are calculated on the basis of the well known equations of Kohlrausch, Onsager and Debye-Hückel (29). Furthermore, application of the various relationships of Bjerrum (30), Denison-Ramsey (31) and Fuoss (32) yields the distances of closest approach of the ions within the ion-pair from the K_{IP} values. These, in turn, indicate whether an inner or outer sphere ion-pair is present in a particular system, when combined with the conclusions drawn from the absolute values of $\Delta\bar{V}_{o,IP}$. Distances larger than 3 Å normally suggest an outer sphere ion-

Table 4. Volume effects associated with ion-pair formation in water at 25°C: $A^m + B^n \rightleftharpoons A.B^{m+n}$.

A	B	K_{IP} kg mol^{-1}	a_o Å	$\Delta\bar{V}_{o,IP}$ cm^3 mol^{-1}	Ref.
Rb^+	NO_3^-			+5.	33
Tl^+	NO_3^-			+12.	33
Ni^{2+}	SO_4^{2-}	$3.3x10^{-3}$	3.9	+8.6	34
Zn^{2+}	SO_4^{2-}	$3.7x10^{-3}$	3.6	+8.0	35
Co^{2+}	SO_4^{2-}	$1.5x10^{-3}$	3.0	+7.0	36
Mn^{2+}	SO_4^{2-}	$4.4x10^{-3}$	3.8	+7.4	37
Mg^{2+}	SO_4^{2-}	$5.8x10^{-3}$	4.2	+7.3	38
Ca^{2+}	SO_4^{2-}	$4.9x10^{-3}$	4.0	+10.2	38
$Co(NH_3)_5NO_2^{2+}$	SO_4^{2-}	$2.1x10^{-3}$	3.3	+9.5	39
La^{3+}	$Fe(CN)_6^{3-}$	$1.6x10^{-4}$	7.2	+8.0	40

pair, whereas those smaller than 3 Å suggest inner sphere association.

It must be emphasized, however, that the simplifications made in the derivation of the above-mentioned equations call for critical evaluations in each case. Whenever possible, the distances derived by the various procedures should be compared with the sum of the ionic radii obtained from determinations of the partial molar volumes of individual ions.

The ion-pair formation constants, together with the distances of closest approach, a_o, and the volumes of ion-pairing are compiled in Table 4. Only examples of electrolytes with the same charge on both ions are presented here. For the first two entries (i.e. for singly charged ions), the volumes of ion-pair formation were determined from partial molar volume measurements. Therefore no direct values for the association constants were reported. Although the $\Delta\bar{V}_{o,IP}$ values are expected to be positive, the large difference between these two quantities is difficult to rationalize, unless one assumes that thallium nitrate forms inner sphere ion-pairs, while rubidium nitrate prefers the outer sphere type.

Within the group of doubly charged ions the association constants scatter over a fairly narrow range of 1.5 to 5.8 x 10^{-3} mol kg^{-1}. Consequently, the distances of closest approach reported in the literature, supplemented by values calculated from the Denison-Ramsey equation, fall into the range of 3 to 4 Å. These values suggest that both the cations and anions are hydrated in the ion-pair, except for the $Co(NH_3)_5NO_2^{2+}$ cation with its stable and defined , first coordination sphere incorporating the ammonia and nitro ligands.

The similarity in the volumes of association (+ 8. ± 1.5 cm^3 mol^{-1}) within this group is in agreement with the formation of solvent-separated ion-pairs, especially since the observed positive volume change is substantially smaller than expected for a total charge neutralization. This conclusion is readily substantiated if one introduces the listed a_o parameters into simple electrostatic calculations.

The last entry in Table 4 concerns triply charges ions. Here the $\Delta\bar{V}_{o,IP}$ value of + 8.0 cm^3 mol^{-1} is much smaller than may be anticipated. However the relatively small value of K_{IP} and, consequently the large a_o parameter, indicate the weakness of the association between two large, hydrated species. Considering the magnitude of the distance between the ionic centers, the observed moderate volume effect is understandable.

Although it is possible to qualitatively rationalize the data in Table 4, more data are needed to fully understand ion-pairing processes and their volume effects. This is especially true when one turns to ion-pairing between differently charged ions.

Volume changes incurred in the association between trivalent cations and univalent anions, for instance, are reported as 0.8; +3.4; +4.2; +4.6; +17.5; and +23.6 $cm^3\ mol^{-1}$ for $CeCl^{2+}$ (20); $CeNO_3^{2+}$ (20); $EuNO_3^{2+}$ (20); $FeCl^{2+}$ (41); $FeNCS^{2+}$ (42); and $CeOOCCH_2CH_3^{2+}$ (20), respectively. Similarly, values of $\Delta\bar{V}_{o,IP}$ for the association of trivalent cations with divalent anions are +7; +15; +24; and +25 $cm^3\ mol^{-1}$ for $Co(NH_3)_6SO_4^+$ (43); $CeSO_4^+$ (20); $LaSO_4^+$ (44); and $EuSO_4^+$ (45), respectively.

In order to explain the large variation in $\Delta\bar{V}_{o,IP}$ in these series, one must consider each ion-pair individually in terms of either inner or outer sphere ion-pairing. Preferably independent experiments, such as spectral measurements, are required to specifically characterize the nature of the ion-pair. However, in view of the preceding discussion, this might be especially difficult for kinetically labile systems.

Moreover, for certain systems (33,34,36,43), concentration, ionic strength and temperature are found to strongly influence $\Delta\bar{V}_{o,IP}$.

VI. PRESSURE EFFECTS ON RAPID COMPLEX FORMATIONS

As already indicated in the preceding section, for practical purposes there is no difference between inner sphere ion-pairing and the formation of labile complexes. When neutral or anionic complexes react with neutral or anionic free ligands, ion-pairing becomes virtually negligible and the processes can be considered to involve straightforward complex formation.

Indeed, examples of these reactions exist in which changes in coordination number also occur, depending on the concentration of the free ligand, temperature and pressure. Perturbation of the absorption spectra resulting from such changes in configuration, makes these systems most suitable for detailed studies. Thus the pressure induced chromism of solutions of cobalt(II) halides have attracted considerable attention (46-48). Here the colour changes are attributed to transitions between octahedral and tetrahedral configurations according to the overall equation:

$$CoS_6^{2+} + nX^- \rightleftharpoons CoS_{4-n}X_n^{2-n} + (2+n)S \qquad (6)$$

with $0 \leqq n \leqq 4$, S = solvent and X = halide ion.

From recent studies by Osugi and coworkers (49) on the cobalt(II) chloride and bromide systems in acetone, a general scheme involving essentially two coupled equilibria emerged:

$$CoX_2S_2 + 4S \rightleftharpoons CoS_6^{2+} + 2X^- \quad (7)$$

$$CoX_2S_2 + X^- \rightleftharpoons CoX_3S^- + S \quad (8)$$

Species with less than two halogeno ligands may be ignored in nonaqueous solvents, particularly in the presence of free halide ions. On the other hand, formation of tetra(halogeno) complex ions occurs only in extremely concentrated halide solutions, if not in molten halides.

From a simple estimation of electrostriction effects, one expects a negative volume of reaction for the dissociation equilibrium in Eq. (7), which should also be solvent dependent. This effect on $\Delta\bar{V}_o$ will be enhanced by a negative intrinsic contribution originating from the formation of two additional metal - ligand bonds. In fact, the $\Delta\bar{V}_o$ values for this equilibrium range from -16 $cm^3\ mol^{-1}$ in water, through -34 $cm^3\ mol^{-1}$ in acetone, to -45 $cm^3\ mol^{-1}$ in ethanol for the chloride system.

The formation of the tetrahedral tri(halogeno) complex, which does not involve any change of coordination number (Eq. (8)), is almost pressure independent. This indicates that processes in which both the overall charge and number of ligands remain constant are subject to volume effects that are small and, more or less, compensate each other. Only the accumulation of negative charge during the formation of a tetra(halogeno) species is expected to be again favoured by pressure.

The qualitative picture drawn here for the formation of the tetrahedral cobalt(halogeno) complex is based on simple electrostriction considerations and should also be appropriate for labile complexes of other configurations and coordination numbers.

VII. ACKNOWLEDGMENTS

The authors wish to thank various colleagues and publishers for permission to reproduce Figure 1 and the data in Tables 1,2 and 3.

VIII. REFERENCES

1. H. Kelm and D.A. Palmer, "Determination and Interpretation of Volumes of Activation", this volume.
2. a) S.D. Hamann, "Physico-Chemical Effects of Pressure", Butterworths, London, 1957.
b) S.D. Hamann, "Chemical Equilibria in Condensed Systems" in "High Pressure Physics and Chemistry", Vol. 2, R.S. Bradley, Ed., Academic Press, London, 1963.
c) B.B. Owen and S.R. Brinkley, Chem. Rev., 29, 461 (1941).
3. S.D. Hamann and W. Strauss, Trans. Faraday Soc., 51, 1684 (1955).
4. S.D. Hamann and S.C. Lim, Aust. J. Chem., 7, 329 (1954).
5. G. Körtüm, K. Vogel and L. Andrussow, "Dissociation Constants of Organic Acids in Aqueous Solutions", Butterworths, London, 1967.
6. H. Høiland, J. Chem. Soc., Faraday Trans., I, 71, 797 (1975).
7. R.J.H. Clark and A.J. Ellis, J. Chem. Soc., 247 (1960).
8. A. Fischer, B.R. Mann and J. Vaughan, J. Chem. Soc., 1093 (1961).
9. S.D. Hamann and M. Linton, J. Chem. Soc., Faraday Trans., I, 70, 2239 (1974).
10. A.J. Ellis and D.W. Anderson, J. Chem. Soc., 1765 (1961).
11. A.J. Ellis, J. Chem. Soc., 3689 (1961).
12. J.M. Klotz and C.F. Eckert, J. Am. Chem. Soc., 64, 1878 (1942).
13. J.S. Smith, in "Electrochemical Constants", by B.B. Owens, Nat. Bur. Stand. Circular, 524, 193 (1943).
14. A.J. Ellis, J. Chem. Soc., 4678 (1961).
15. A. Bodanszky and W. Kauzmann, J. Phys. Chem., 66, 177 (1962).
16. H.G. Daud and S.D. Hamann, Trans. Faraday Soc., 55, 72 (1959).
17. E.U. Franck, Angew. Chem., 73, 309 (1961).
18. A Jost, Ber. Bunsenges. Phys. Chem., 80, 316 (1976).
19. T.W. Swaddle and P.C. Kong, Can. J. Chem., 48, 3223 (1970).
20. T.C. Spiro, A. Revesz and J. Lee, J. Am. Chem. Soc., 90, 4000 (1968).
21. F.J. Millero, "The Partial Molar Volumes of Electrolytes in Aqueous Solutions", R.A. Horne, Ed., in "Water and Aqueous Solutions", Wiley-Interscience, New York (1972).
22. J. Buchanan and S.D. Hamann, Trans. Faraday Soc., 49, 1425 (1953).
23. B.S. El'yanov and M.G. Gonikberg, Russ. J. Phys. Chem., 46, 856 (1972).
24. B.S. El'yanov and S.D. Hamann, Aust. J. Chem., 28, 945 (1975).
25. D.A. Lawn, H.R. Thirsk and Lord Wynne-Jones, J. Chem. Soc., 2073 (1968).
26. W.E. Jones, L.R. Carey and T.W. Swaddle, Can. J. Chem., 50, 2739 (1972).
27. N.A. North, J. Phys. Chem., 77, 931 (1973).

28. S. Claesson, B. Lundgreen and M. Szwarc, J. Chem. Soc., 3053 (1970).
29. R.A. Robinson and R.H. Stokes, "Electrolyte Solutions", Butterworths, London, 1965.
30. N. Bjerrum, Kgl. Danske Videnshab. Selskab., 7 No 9 (1926).
31. J.T. Denison and J.B. Ramsey, J. Am. Chem. Soc., 77, 2615 (1955).
32. R.M. Fuoss, J. Am. Chem. Soc., 80, 5059 (1958).
33. W.L. Masterton, H. Welles, J.H. Knox and F.J. Millero, J. Sol. Chem., 3, 91 (1974).
34. K. Shimizu, N. Tsuchihashi and Y. Furumi, Rev. Phys. Chem., Japan, 46, 30 (1976).
35. Y. Taniguchi, T. Watanabe and K. Suzuki, Bull. Chem. Soc., Japan, 48, 3023 (1975).
36. K. Shimizu and T. Okamoto, Science and Engineering Review of Doshisha University, 16, 58 (1975).
37. F.H. Fisher and D.F. Davies, J. Phys. Chem., 69, 2595 (1965).
38. E. Inada, K. Shimizu and J. Osugi, Rev. Phys. Chem., Japan, 42, 1 (1972).
39. M. Ueno, K. Shimizu and J. Osugi, Rev. Phys. Chem., Japan, 43, 33 (1973).
40. S.D. Hamann, P.J. Pearce and W. Strauss, J. Phys. Chem., 68, 375 (1964).
41. B.B. Hasinoff, Can. J. Chem., 54, 1820 (1976).
42. K.R. Brauer, J. Am. Chem. Soc., 90, 5401 (1968).
43. M. Nakahara, K. Shimizu and J. Osugi, Rev. Phys. Chem., Japan, 40, 12 (1970).
44. F.H. Fisher and D.F. Davies, J. Phys. Chem., 71, 819 (1967).
45. C.F. Hale and F.H. Spedding, J. Phys. Chem., 76, 2925 (1972).
46. F.G. Wick, Proc. Am. Acad. Arts. Sci., 58, 555 (1923).
47. Y. Kitamura, Rev. Phys. Chem., Japan, 39, 1 (1969).
48. H.D. Lüdemann and E.U. Franck, Ber. Bunsenges. Phys. Chem., 71, 455 (1967); 72, 514 (1968).
49. J. Ishihara, K. Hara and J. Osugi, Rev. Phys. Chem., Japan, 44, 11 (1974).

THE EFFECTS OF PRESSURE ON THE RATES OF REACTIONS OF COORDINATION COMPOUNDS IN SOLUTION

Donald A. Palmer and Hartwig Kelm

Institute for Physical Chemistry, University of Frankfurt, Robert Mayer Strasse 11, D6000 Frankfurt am Main, West Germany.

I. INTRODUCTION

The use of high pressure techniques in studies of the kinetics of inorganic substrates is still relatively new with the majority being conducted over the last ten years. A short review of these data was written in 1974 by Stranks (1). The method has been used almost exclusively to aid in the elucidation of reaction mechanisms of transition metal complexes. Thus the volume of activation, $\Delta V^{\neq}$, derived from the pressure dependence of the rate coefficient (Eq. (1)), supplements the existing, and more commonly used enthalpy and entropy of activation.

$$(d \ln k/dP)_T = - \Delta V^{\neq}/RT \qquad (1)$$

We shall confine ourselves exclusively to a discussion of mechanisms of such coordination compounds and will attempt to demonstrate that $\Delta V^{\neq}$ is indeed more descriptive and, in some cases, more readily rationalized than, for instance, $\Delta S^{\neq}$. It should be mentioned that the relative significance of activation parameters in this field was recently discussed in detail by Swaddle (2).

The majority of reactions of coordination compounds studied under high pressure deal with ligand substitution, which is the subject of this review, whereas only a limited number of rearrangement and redox processes have been investigated by this technique.

Kelm, H. (Ed.) High Pressure Chemistry. 435–466. *Copyright © 1978. D. Reidel Publishing Company. All Rights Reserved.*

The volume of activation is defined as the difference in partial molar volumes between the transition state and reactants:

$$\Delta V^{\neq} = \bar{V}_{\neq} - \Sigma_i \bar{V}_i \tag{2}$$

Since it is theoretically possible to determine the partial molar volumes of reactants and products from precision density measurements, a combination of these data with $\Delta V^{\neq}$ facilitates the assignment of an absolute volume to the transition state, thereby enabling a total volume profile of the reaction to be constructed. Valuable conclusions concerning the reaction mechanism can be drawn from such profiles.

In order to extract information from the volume of activation alone, the experimentally accessible quantity, $\Delta V^{\neq}_{exp}$, can be considered as comprising two components: an intrinsic part, $\Delta V^{\neq}_{intr}$, originating from changes in the nuclear positions of the reactants during the activation process; and a solvation part, $\Delta V^{\neq}_{solv}$, which describes the change in solvation during the formation of the transition state.

$$\Delta V^{\neq}_{exp} = \Delta V^{\neq}_{intr} + \Delta V^{\neq}_{solv} \tag{3}$$

$\Delta V^{\neq}_{intr}$ is indeed diagnostic of the mechanism because bond making should result in a negative value and bond breaking in a positive value, as illustrated in the following scheme.

$$A + B \rightleftharpoons (A\text{----}B)^{\neq} \rightleftharpoons A\text{–}B$$

Forward reaction - bond formation: $-\Delta V^{\neq}_{intr}$.

Reverse reaction - bond stretching: $+\Delta V^{\neq}_{intr}$.

$$A^{+} + B^{-} \rightleftharpoons (A^{\delta+}\text{-----}B^{\delta-})^{\neq} \rightleftharpoons A\text{—}B$$

Forward reaction - charge quenching: $+\Delta V^{\neq}_{solv}$.

Reverse reaction - charge separation: $-\Delta V^{\neq}_{solv}$.
(formation)

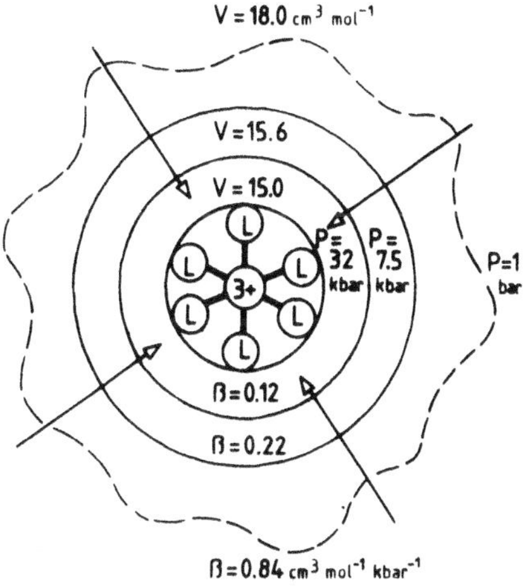

Fig. 1. Solvation of a trivalent metal complex cation in water, where V represents the partial molar volume of water; β is the compressibility coefficient of water; and P is the internal pressure exerted on the solvent.

reflect in a pressure dependence of $\Delta V^{\neq}_{exp}$, as will be exemplified later.

It should be noted that other views have been recently put forward (3) asserting that there exists a solvation sphere between the ordered, compressed inner solvation sphere/s and the bulk solvent in which the solvent molecules are unstructured with a higher entropy and compressibility than the bulk solvent. This model predicts (4) that the alignment of solvent molecules within the first and second solvation spheres will lead to an overall increase in volume, in contrast to the above mentioned model. Furthermore, although a similar picture may be drawn for anionic complexes, the situation is more complicated (3) so that a formal treatment of the type proposed by Stranks (1) has not yet been attempted. Fortunately the vast majority of charged metal complexes are cationic.

II. CLASSIFICATION OF MECHANISMS

The mechanisms of substitution reactions of coordination compounds can be classified into two groups involving either associative or dissociative rate-determining steps, closely related to Ingold's differentiation (5) between S_N2 and S_N1

Changes in solvation during the rate determining step often obscure such a clear distinction. In fact, in cases where charges are created, changed, or quenched, $\Delta V^{\neq}_{solv}$ may substantially outweigh $\Delta V^{\neq}_{intr}$. As one generally deals with charged species in coordination chemistry, an essential part of high pressure studies is concerned with the attempt to separate the two terms in Eq. (3). The methods used for this purpose will be introduced later during the discussion of specific systems.

The pressure derivative of the volume of activation can be defined by Eq. (4)

$$\Delta\beta^{\neq} = -(d\Delta V^{\neq}/dP)_T = -\{(d\bar{V}_{\neq}/dP)_T - (d\bar{V}_i/dP)_T\} \tag{4}$$

where $\Delta\beta^{\neq}$ is also known as the compressibility coefficient of activation. It has already been shown from a number of different approaches that reacting molecules and transition states may be treated as incompressible within a moderate pressure range (i.e. up to 3 kbar). Thus, should plots of (ln k) versus pressure exhibit curvature, then this effect may be solely attributed to changes within the solvent structure involving the transition of solvent molecules from and to the bulk solvent to and from the solvation spheres surrounding the reacting species. Notwithstanding the difficulties incurred in obtaining accurate data-fitting for such (ln k) versus pressure plots, meaningful $\Delta\beta^{\neq}$ values can be calculated. In order to minimize the errors in $\Delta\beta^{\neq}$ it is desirable to conduct a large number of kinetic measurements over small pressure increments up to 2 to 3 kbar. To rationalize the sign and magnitude of $\Delta\beta^{\neq}$ some understanding of the composition of the solvation sheaths is essential. Stranks (1) has proposed a simplified model for cationic complexes in which the complex ion exerts a centro-symmetrical electrostatic pressure on the solvent, as shown in Fig. 1. The magnitude of this pressure may be approximated as a function of the charge on the ion, the bulk dielectric constant of the medium, and the distance from the center of the ion. In water this amounts to 32 kbar at a distance of 3 - 3.5 Å from the center of a 3+ ion (this corresponds to the first solvation sphere) so that the partial molar volume of water in this layer is reduced to 15 cm^3 mol^{-1} compared to 18 cm^3 mol^{-1} in the bulk solvent and its compressibility is drastically reduced to 8.0×10^{-6} bar^{-1} compared to 46.4×10^{-6} bar^{-1} at 1 bar. Although the electrostatic pressure decreases rapidly with distance, for a 3+ ion a second solvation shell should still exist under a pressure of ca. 7.5 kbar which is sufficient to reduce the molar volume of water to 15.6 cm^3 mol^{-1} and the compressibility to 14×10^{-6} bar^{-1}. Thus it is argued that the displacement of solvent molecules from these various regions, including the coordination sphere itself, should processes, often used in organic chemistry. These definitions were further refined by Langford and Gray (6) to include the limiting cases of a D-mechanism at one extreme in which the leaving group X is completely separated from the metal center in the transition state, to the other extreme of an A-mechanism where the incoming group Y forms an additional bond with the central atom. Between these limiting cases a continuum of possibilities exist in which the leaving and incoming groups are both present in the first and second coordination spheres. During the rate-determining step they exchange positions following a so-called interchange-(I)-mechanism. Depending on whether bond stretching or bond making is predominant during the exchange, the mechanisms are characterized by the symbols I_d and I_a, resp.. These definitions permit the establishment of a number of boundary conditions by which one can theoretically distinguish between each mechanism.

II.1 Characterization of the D-Mechanism

For a ligand substitution process which follows a D-mechanism, as symbolized by Eq. (5), a number of characteristics can be predicted:

$$ML_nX + Y \longrightarrow \{ML_n + X + Y\}^{\neq} \longrightarrow ML_nY + X \qquad (5)$$

i) An intermediate species of higher energy must exist with a lower coordination number than the substrate.
ii) The rate parameters must be independent of the nature of the incoming nucleophile Y, but depend strongly on the leaving group X.
iii) The maximum rate of ligand exchange, k_{ex}, should be greater than, or equal to, the rate of substitution of X by Y, k_{sub}.
iv) Entropies of activation are generally positive due to the release of the previously bound ligand X. However $\Delta S^{\neq}$ is also very sensitive to the properties of the solvent, the nature of X, and the degree of structural rearrangement that occurs during the formation of the meta-stable ML_n species.
v) For uncharged leaving groups, where solvation changes are expected to be generally small or negligible, $\Delta V^{\neq}_{exp}$ must be positive resulting from the volume expelled as the M-X bond is elongated, with its absolute value not exceeding the partial molar volume of X.
vi) The compressibility coefficient of activation should be relatively large and positive for X = solvent, due to the diffusion of the solvent molecule from the virtually incompressible first coordination sphere to the compressible bulk solvent.

II.2 Characterization of the I_d-Mechanism

The I_d-mechanism has been schematically represented by Tobe (7) and is shown here, Eq. (6), in a vastly simplified form:

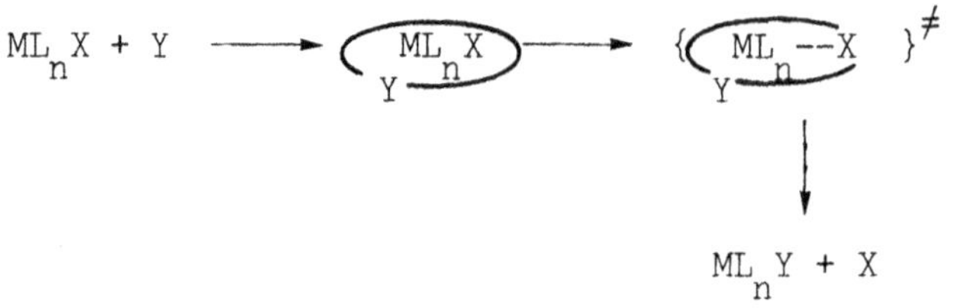

(6)

This mechanism is mainly characterized by the following observations:

i) M-X bond stretching is considered to be synchronous with bond formation so that no intermediate exists in the reaction profile, with the interchange of X and Y mainly involving the first and second coordination spheres. In other words, a precursor entity is formed (often called an "encounter complex" or an ion-pair when ions are involved), within which the exchange occurs. Bond stretching is dominant over M-Y bond formation, although the extent to which the former takes place in the transition state can vary within a continuous spectrum of values limited by the extreme case of bond fission, i.e. the D-mechanism.
ii) The rate parameters remain strongly dependent on the nature of X, however the properties of Y also become important in dictating the magnitude of the formation constant of the precursor complex.
iii) The maximum rate of ligand exchange should still be greater or, in the limiting case at high [Y], equal to the rate of substitution by Y, although allowances must be made for the possible effects on k_{ex} of encounter complex formation.
iv) $\Delta S^{\neq}$ values become less predictable because of encounter complex formation.
v) $\Delta V^{\neq}_{intr}$ remains positive, however smaller than for a D-mechanism.
vi) $\Delta V^{\neq}_{exp}$ is generally not so significantly pressure dependent as in the case of a D-mechanism. In fact, for solvent exchange reactions, $\Delta\beta^{\neq}$ is expected to be virtually zero as the interchange occurs between two solvent molecules in the first solvation sphere thereby not involving the movement of other solvent molecules to or from the bulk solvent within the duration of the activation process.

II.3 Characterization of the A-Mechanism

The A-mechanism is depicted in Eq. (7):

$$ML_nX + Y \longrightarrow \{ML_nXY\}^{\neq} \longrightarrow ML_nY + X \qquad (7)$$

i) An intermediate or transition state of higher coordination number as the substrate exists in the energy profile diagram.
ii) The rate parameters must be independent of X, but dependent on Y, giving straight forward second-order rate dependencies with respect to the concentration of Y.
iii) The second-order rate constant for substitution by Y may be greater or equal to the corresponding rate constant for exchange. For an anation reaction: $k_{sub} > k_{ex}$, where k_{ex} is the rate constant for solvent exchange.
iv) $\Delta S^{\neq}$ is generally large and negative.
v) $\Delta V^{\neq}_{intr}$ must be negative due to bond formation.
vi) As in the case of a D-mechanism, $\Delta\beta^{\neq}$ should be large but negative, so that substantial curvature of the (ln k) versus pressure plot is observed.

II.4 Characterization of the I_a-Mechanism

The reaction sequence presented in Eq. (8) is analogous to that shown for the I_d-mechanism.

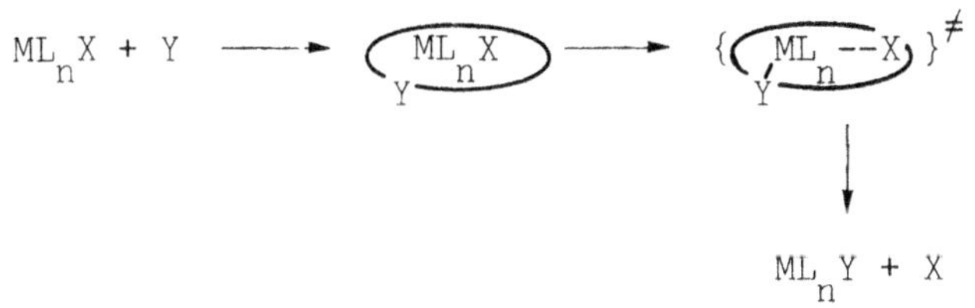

(8)

i) A similar situation as in the I_d-mechanism prevails, however with bond making more developed in the transition state.
ii) The rate parameters should be dependent on both X and Y.
iii) The rate of substitution should exceed, or be equal to, the rate of solvent exchange, analogous to II.3.iii.
iv) $\Delta S^{\neq}$ would normally be expected to be small and negative, but is also highly sensitive to the charges on X and Y.
v) $\Delta V^{\neq}_{intr}$ can be expected to be somewhat negative.
vi) For solvent exchange reactions $\Delta\beta^{\neq}$ should be approximately zero for the same reasons already stated for the I_d-mechanism.

These criteria are summarized in Table 1.

Table 1. An approximate classification of reaction mechanisms.

$$ML_nX + Y \longrightarrow ML_nY + X$$

Properties	Dissociative		Associative	
	D	I_d	A	I_a
M-X bond	broken	stretched	intact	partially stretched
M-Y bond	not formed	partially formed	formed	partially formed
$k \propto X$ or Y	X	X and Y	Y	X and Y
k^*_{ex}/k_{sub}	$\geqq 1$	$\geqq 1$	$\leqq 1$	$\leqq 1$
$\Delta S^{\neq}$	large positive	small	large negative	small
$\Delta V^{\neq}_{intr}$	positive	small positive	negative	small negative
$\Delta\beta^{\neq}$ *	positive	small	negative	small

* refers to solvent exchange reactions

It must be emphasized that we have not included all possible means by which different mechanisms can be distinguished, i.e. linear free energy relationships; an investigation of the steric course of the reaction; solvent and solvent-mixture variations, etc. However the major rate parameters have been introduced so that we are now in a position to look at individual kinetic systems within this general framework. The majority of kinetic data exists for octahedral complexes so that it is convenient to begin with these.

III. HIGH PRESSURE STUDIES OF OCTAHEDRAL COMPLEXES

Octahedral coordination is a common phenomenon in transition metal chemistry. These complexes are generally kinetically labile in solution, with reaction rates which are sufficiently fast so as to necessitate the application of special techniques to investigate their dynamic behaviour. Although these techniques are usually suitable for application in high pressure studies, very little has been done in this direction.

On the other hand, complexes of trivalent cobalt, chromium,

rhodium, iridium and ruthenium, as well as those of tetravalent platinum, are kinetically relatively stable. Consequently their substitution reactions are sufficiently slow to be studied by conventional methods. High pressure studies have mainly been performed within this group of substrates, with the bulk of the data being accumulated for cobalt and chromium complexes.

III.1 Substitution reactions involving uncharged ligands

As mentioned in the introduction, the interpretation of $\Delta V^{\neq}_{exp}$ is most conclusive when it can be assumed that the contribution from solvation is small or negligible. Such a situation is presented when the leaving and entering groups are uncharged. The most interesting and informative group of reactions of this type are solvent exchange processes, characterized by the rate constant k_{ex}. The stereochemistry of these reactions is also important.

Stranks and Vanderhoek (8) recently reported the effects of pressure on the isomerization of trans- and cis-diaquobis(ethylenediamine)cobalt(III) ion in acidic aqueous solution, shown in Fig. 2. It had been previously shown (9) that the rate of stereoretentive exchange of the water ligand in the trans complex is only 1.7 times faster than the rate of isomerization, k_{isom}, while the rate of water exchange with the cis complex is 250 times faster than k_{isom}. As the equilibrium in fact lies to the right

Fig. 2. Isomerization of trans-$Co(en)_2(OH_2)_2^{3+}$ in acidic aqueous solution. (en ≡ ethylenediamine).

with the cis isomer representing 98% of the total complex at equilibrium, either two different paths or one path consisting of consecutive steps with more than one transition state, must exist for the exchange and isomerization reactions. In 0.05M $HClO_4$ the volume of activation for isomerization is relatively large and positive, i.e. + 14.2 ± 0.2 $cm^3\ mol^{-1}$, and strongly pressure dependent. As it may be assumed that $\Delta V^{\neq}_{exp} \sim \Delta V^{\neq}_{intr}$, the large positive value is strong evidence for a dissociative mechanism. It should be kept in mind that for the extreme limit of a D-mechanism, a maximum value of + 18 $cm^3\ mol^{-1}$ would be anticipated. The compressibility coefficient of activation, $\Delta\beta^{\neq}$, at 0.05M $[HClO_4]$ is equal to $(1 \pm 0.2) \times 10^{-3}\ cm^3\ mol^{-1}\ bar^{-1}$ (8). The estimations of this quantity based on the concept illustrated in Fig. 1 are:

for an I_d-mechanism, $$\Delta\beta^{\neq} = (\beta_{\text{first solvation sphere}} - \beta_{\text{coordination sphere}}) \ll 6 \times 10^{-5}\ cm^3\ mol^{-1}\ bar^{-1}$$

for a D-mechanism, $$\Delta\beta^{\neq} = (\beta_{\text{bulk water}} - \beta_{\text{coordination sphere}}) = 8 \times 10^{-4}\ cm^3\ mol^{-1}\ bar^{-1}$$

Both calculations correspond to the transfer of one water molecule. It is obvious that the later is in very good agreement with the observed value for $\Delta\beta^{\neq}$. Finally, the entropy of activation is also large and positive, namely (+ 103 ± 5) $J\ K^{-1}\ mol^{-1}$. Keeping in mind the dictates of the above mechanism classification, this value also better fits a D-mechanism. Fortunately Swaddle et al. (10) have nicely complemented this study by investigating the water exchange reaction of trans-$Co(en)_2(OH_2)_2^{3+}$ under pressure. They reported a $\Delta V^{\neq}_{exp}$ of (+ 5.9 ± 0.2) $cm^3\ mol^{-1}$ which is independent of pressure up to ca. 3 kbar at 35°C. The $\Delta S^{\neq}$ value is slightly less positive (+ 84 $J\ K^{-1}\ mol^{-1}$) (10) than for the isomerization. Therefore applying the same reasoning used for the isomerization reaction the mechanism appears to be I_d.

A reaction scheme for the overall kinetic behaviour of the system has been proposed (8,10) and is shown in Fig. 3. Here it is proposed that the exchange processes take place through a square-pyramidal transition state, while the isomerization, which involves a D-mechanism, necessitates the rearrangement of the transition state into a trigonal-bipyramidal intermediate.

Further aquo ligand exchange reactions were studied for a number of systems and these data are summarized in Table 2.

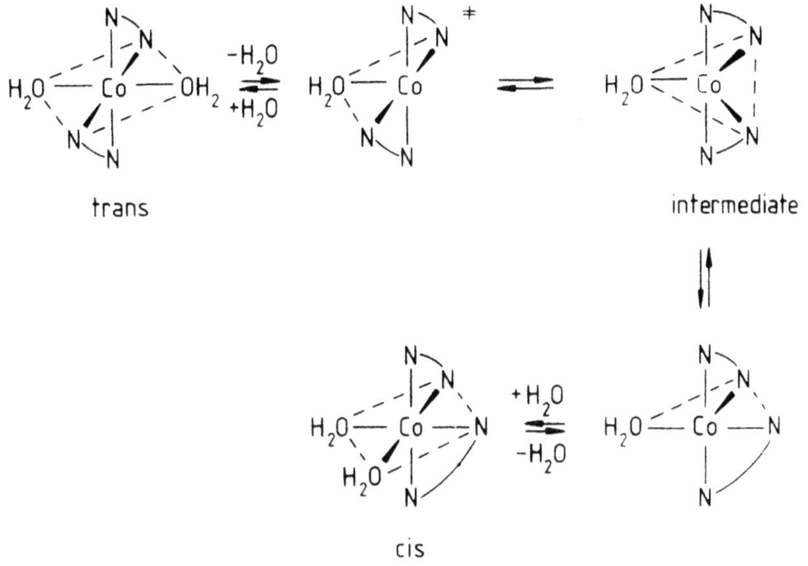

Fig. 3. The complete reaction scheme for the isomerization of trans-$Co(en)_2(OH_2)_2^{3+}$, incorporating the water exchange reactions of both isomers.

Table 2. Activation parameters for a number of water exchange reactions.

Complex Ion	$\Delta V^{\neq}_{exp}$ $cm^3\ mol^{-1}$	$\Delta S^{\neq}$ $J\ K^{-1}\ mol^{-1}$	Temp. ^oC	Ref.
$Co(NH_3)_5OH_2^{3+}$	+ 1.2 ± 0.2	+ 28 ± 4	25/35	11
$Rh(NH_3)_5OH_2^{3+}$	- 4.1 ± 0.4	+ 3 ± 5	25	12
$Ir(NH_3)_5OH_2^{3+}$	- 3.2 ± 0.1	+ 11 ± 4	70	13
$Cr(NH_3)_5OH_2^{3+}$	- 5.8 ± 0.1	0 ± 7	25	12
$Cr(OH_2)_6^{3+}$	- 9.3 ± 0.3	+ 1 ± 4	45	14

It can be seen that the $\Delta S^{\neq}$ values are rather small and inconclusive, in view of the magnitude of the experimental error limits. At best they point towards an interchange mechanism.

On the other hand the $\Delta V^{\neq}_{exp}$ values are of much higher accuracy and with the assumption of a negligible solvation contribution, they present a strong mechanistic criterion. Although the $\Delta V^{\neq}_{exp}$ values are not extremely large , their signs alone indicate that all the reactions, except for the one involving the cobalt complex, have a basically associative character. The pressure independence of $\Delta V^{\neq}_{exp}$ for all the reactions listed, i.e. $\Delta\beta^{\neq} \sim 0$, also provides evidence of a prevailing interchange mechanism.

Similar conclusions can also be drawn for nonaqueous systems, such as the solvent ligand exchange processes involving $Cr(DMSO)_6^{3+}$ (15) and $Cr(DMF)_6^{3+}$ (16) for which $\Delta V^{\neq}_{exp}$ values of - 11.3 and - 6.3 $cm^3\ mol^{-1}$, respectively, were found.

A number of rapid equilibrium reactions have been investigated which dealt with more general substitution processes involving uncharged ligands. Caldin et al. (17) studied the kinetics of complex formation between ammonia and pyridine-2-azodimethylaniline (PADA) and the metal ions of Co^{2+} and Ni^{2+} by a laser-induced temperature jump method. They reported $\Delta V^{\neq}_{exp}$ values in the region of + 5 to + 8 $cm^3\ mol^{-1}$ indicating a significant contribution from bond stretching during the rate-determining step, in agreement with the proposed Eigen-Wilkins mechanism (18) for a dissociative interchange mechanism.

Interesting results were reported by Merbach et al. (19,20) for the ligand exchange reactions of octahedral tantalum and niobium halides employing high resolution magnetic resonance techniques at elevated pressures.

$$MX_5L \ + \ L' \rightleftharpoons MX_5L' \ + \ L \qquad (9)$$

where M = Nb or Ta; X = Cl or Br; L = Me_2O, Me_2S, Me_2Se, or Me_2Te. These uncharged ligand exchange processes, for which no major $\Delta V^{\neq}_{solv}$ contributions need be considered, follow first-order kinetics when L = Me_2O, but are second-order for the remaining ethers. The volumes of activation impressively underline the dissociative - associative crossover, as can be seen for one example of each presented in Table 3.

III.2 Substitution reactions involving charged ligands

III.2.1 Hydrolysis reactions

One of the more common types of substitution reactions, in which at least one of the exchanging ligands is a charged species, are the hydrolysis reactions:

Table 3. Activation parameters for ligand exchange processes of tantalum pentabromides.

$TaBr_5L$ L	$\Delta H^{\neq}$ (19) $kJ\ mol^{-1}$	$\Delta S^{\neq}$ (19) $J\ K^{-1}\ mol^{-1}$	$\Delta V^{\neq}_{exp}$ (20) $cm^3\ mol^{-1}$
Me_2O	74.8 ± 3.3	+ 62 ± 12	+ 10 ± 3
Me_2S	29.0 ± 1.6	- 102 ± 6	- 12 ± 2

$$ML_5X^{m+n} + H_2O \longrightarrow ML_5OH_2^{\ m} + X^n \qquad (10)$$

In Table 4 are presented a series of volume data for hydrolysis reactions of pentaammine complexes of cobalt(III) and chromium(III). At first glance, the overall negative values of $\Delta V^{\neq}_{exp}$ - except for the water exchange of $Co(NH_3)_5OH_2^{3+}$ - suggest that an associative mechanism applies throughout. However it must be recognized that electrostriction, caused by the effect of charge separation on the solvent resulting from possible stretching of the M-X bond, may give rise to a large negative volume change ($\Delta V^{\neq}_{solv}$), when X carries a charge.

Values of $\Delta S^{\neq}$, like those of $\Delta V^{\neq}_{exp}$, are composite quantities and are very sensitive to solvation changes. Although a relationship exists between these two parameters (25), only limited conclusions can be drawn from the former.

Table 4 reveals that in the case of the cobalt(III) complexes, the overall change in volume, $\Delta\bar{V}$, is more negative than $\Delta V^{\neq}_{exp}$, whereas for the corresponding chromium(III) complexes, the values of $\Delta V^{\neq}_{exp}$ are more negative than $\Delta\bar{V}$. Thus in the chromium case the transition state must be considered to be more compact than the products. This can at least be taken as an indication that bond making makes a substantial contribution to $\Delta V^{\neq}_{exp}$ in the reactions of the chromium complex ions.

To illustrate this point further, just as linear free energy relationships are often used as a criterion in the discussion of reaction mechanisms, the relationship between the volume of activation and the volume of reaction may also be used for these pentaammine complexes. This relation is shown in Figure 4.

A good linear correlation exists for the cobalt series with a slope of approximately unity, indicating a strong similarity between the transition state and products, which in turn suggests

Table 4. Volume changes for the aquation reactions of pentaammine complexes of cobalt(III) and chromium(III) at 25°C and μ = 0.1 M.

X^n	$\Delta V^{\neq}_{exp}$ $cm^3\ mol^{-1}$	$\Delta\beta^{\neq}$ $cm^3\ mol^{-1}\ bar^{-1}$	$\Delta\bar{V}$ $cm^3\ mol^{-1}$	Ref.
	$Co(NH_3)_5X^{3+n}$ + H_2O ⟶ $Co(NH_3)_5OH_2^{3+}$ + X^n			
Cl^-	- 10.6 ± 0.4	- 0.0021	- 11.6	21
Br^-	- 9.2 ± 0.2	- 0.0020	- 10.8	21
SO_4^{2-}	- 18.5 ± 0.7	- 0.004	- 19.2	21
NO_3^-	- 6.3 ± 0.4	- 0.0010	- 7.2	21
OH_2	+ 1.2 ± 0.1	0	0	11
DMSO	- 1.7 ± 0.9	ca. 0	-	22
NCS^- *	- 4. ± 1.	-	-	21
	$Cr(NH_3)_5X^{3+n}$ + H_2O ⟶ $Cr(NH_3)_5OH_2^{3+}$ + X^n			
Cl^-	- 10.8 ± 0.3	- 0.0010	- 8.4	23
Br^-	- 10.2 ± 0.3	- 0.0010	- 7.2	23
I^-	- 9.4 ± 0.2	- 0.0010	- 6.0	23
OH_2	- 5.8 ± 0.2	0	0	12
NCS^-	- 8.6 ± 1.0	-	-	24

* measured at 88°C.

a dissociative mechanism for this group. Although less well established, the chromium pentaammine series show a substantially smaller slope of approximately 0.5 which was considered (23) as evidence for an associative rate-determining step with about 50% bond stretching in the transition state.

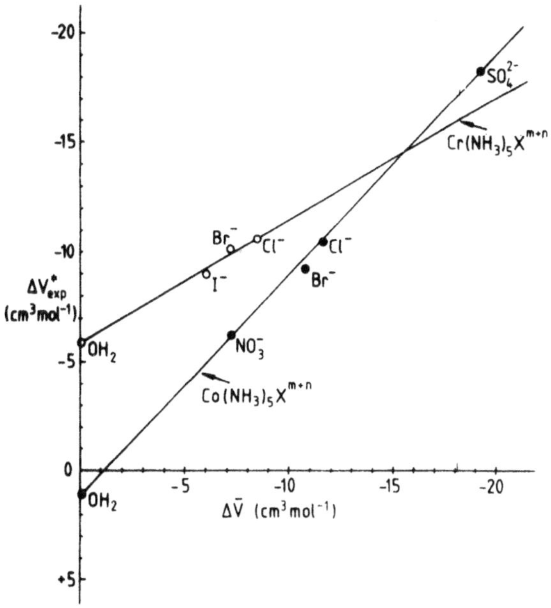

Fig. 4. A plot of $\Delta V^{\neq}_{exp}$ against $\Delta\bar{V}$ for the hydrolysis reactions of pentaammine complexes of cobalt(III) and chromium(III) at 25°C (taken from Ref. 21 and 23).

The values of $\Delta\beta^{\neq}$, which are also listed in Table 4, are markedly dependent on the charge on the leaving group. Their interpretation is obviously more complicated than for the simple water exchange processes treated in the preceeding section. The substitution of a charged species normally involves a large change in solvation incorporating several water molecules.

Accordingly, Stranks (1) and Swaddle (21) have individually tried to interpret these $\Delta\beta^{\neq}$ values in slightly different ways, using the solvation shell model in Fig. 2 and a modified Tait equation, respectively. For the extreme case of the hydrolysis of $Co(NH_3)_5SO_4^+$, Swaddle (21) concluded that about 8 additional water molecules participate in the formation of the transition state.

It should be mentioned at this stage that some reactions of square-planar complexes exhibit $\Delta\beta^{\neq}$ values which are so large that the same treatment results in an unreasonably large number of participating water molecules (26). Only the accumulation of further reliable data will prove the validity and meaningfulness of the proposed concepts.

A new approach to the rationalization of the volume of activation was recently developed in our laboratory employing the so-called "volume equation" (22). From precision density measurements the apparent molar volume of a solute can be determined. From its concentration dependence, the partial molar volume may then be evaluated. With the assumption that the partial molar volume of the potassium ion is 4.5 $cm^3\ mol^{-1}$, volumes can be assigned to the individual complex ions and ligands.

On the basis of a D-mechanism for the hydrolysis of the pentaammine complex ions, the following equation can be written:

$$M(NH_3)_5X^{3+n} \xrightarrow{\Delta V^{\neq}_{exp}} \{M(NH_3)_5{}^{3+} + X^n\}^{\neq} \longrightarrow M(NH_3)_5OH_2{}^{3+} + X^n$$

$$\bar{V}_{RX} + \Delta V^{\neq}_{exp} = \bar{V}_R + \bar{V}_X \qquad (11)$$

Here $\bar{V}_{RX}$, $\bar{V}_R$ and $\bar{V}_X$ represent the partial molar volumes of the substrate, $M(NH_3)_5X^{3+n}$, the five-coordinate intermediate, $M(NH_3)_5{}^{3+}$, and the free ligand X^n, respectively. The individual volume data, determined either by experiment or calculation, for the pentaamminecobalt(III) series are listed in Table 5. Using the partial molar volume of $Co(NH_3)_5OH_2{}^{3+}$, which was found to be 54.0 ± 0.7 $cm^3\ mol^{-1}$, $\Delta\bar{V}_{cal}$ (the total volume change for the reaction) can be calculated according to Eq. (12) and compared

Table 5. Volumes and volume changes associated with the aquation of $Co(NH_3)_5X^{3+n}$. (Volumes are given in $cm^3\ mol^{-1}$).

X^n	$\bar{V}_{RX}$	$\Delta V^{\neq}_{exp}$	$\bar{V}_X$	$\bar{V}_R$	$\Delta\bar{V}$	$\Delta\bar{V}_{cal}$
Cl^-	85.5	-10.6	21.7_5	53.8	-11.6	-10.1
Br^-	95.0	-9.2	29.4	56.4	-10.8	-11.6
$SO_4{}^{2-}$	96.5	-18.5	23.4	54.6	-19.2	-19.1
$NO_3{}^-$	94.5	-6.3	33.3	54.9	-7.2	-7.2
DMSO	125.2	-1.7	68.8	54.7	-	-2.4
OH_2	54.0	+1.2	0	55.2	0	0
				54.9 ± 0.9		

with $\Delta\bar{V}$ obtained experimentally from dilatometry. Both quantities

$$\Delta\bar{V}_{cal} = \bar{V}_{ROH2} + \bar{V}_{X} - \bar{V}_{RX} - \bar{V}_{H2O} \tag{12}$$

are included in Table 5 and agree very well, thereby establishing the consistency of the data. The values tabulated for $\bar{V}_R$ are indeed quite similar, amounting to 54.9 ± 0.9 $cm^3\ mol^{-1}$.

Stranks (1) concluded from compressibility calculations that a five-coordinate intermediate (ML_5^m) should occupy the same space as a six-coordinate complex ion (ML_6^m) containing the same type of ligands and the same charge. On this basis one would expect the value of $\bar{V}_R$ to be comparable to the partial molar volume of hexaamminecobalt(III), which was in fact determined to be (55.1 ± 1.1) $cm^3\ mol^{-1}$. This is in perfect agreement with the above concept.

It should be noted that for X = H_2O - the solvent exchange reaction - $\bar{V}_X$ must be taken as zero in order to fit the "constancy" of $\bar{V}_R$ and to fulfill Eq. (12). Although this may be due to specific solvation of the aquo ligand in $Co(NH_3)_5OH_2^{3+}$, this phenomenon is not fully understood at this time. Nevertheless the volume equation treatment appears to be rather consistent and leads to the conclusion that either a D-mechanism prevails for the cobalt(III) series rather than an I_d-mechanism as generally proposed (7,21,27), or that diffusion of the leaving group out of the first solvation sphere of $Co(NH_3)_5^{3+}$ involves only a slight volume change. However the latter possibility is not consistent with the results of pressure studies (28) made on ion-pair equilibria.

It may be of interest to point out that the linkage isomerization of nitritopentaamminecobalt(III) to give the nitro isomer does not fit the volume equation treatment outlined above (29). In this case a different mechanism must be assumed, namely an intramolecular mechanism, as was proposed earlier (30) based on the absence of significant exchange of coordinated nitrite with isotopically labeled, free nitrite ions.

The corresponding data for the hydrolysis of the $Cr(NH_3)_5X^{3+n}$ series are summarized in Table 6. Application of the volume equation treatment results in $\bar{V}_R$ values which vary considerably. In addition, their mean value of 50 ± 3 $cm^3\ mol^{-1}$ deviates substantially from the partial molar volume of the hexaammine-chromium(III) ion of 68.6 ± 0.7 $cm^3\ mol^{-1}$. A D-mechanism can therefore be excluded for the chromium series. The evidence presented in the previous section for the associative character of the pentaamminechromium(III) reactions, together with the

Table 6. Volumes and volume changes associated with the aquation of $Cr(NH_3)_5X^{3+n}$. (Volumes are given in $cm^3\ mol^{-1}$).

X^n	$\bar{V}_{RX}$	$\Delta V^{\neq}_{exp}$	$\bar{V}_X$	$\bar{V}_R$	$\Delta\bar{V}$	$\Delta\bar{V}_{cal}$
Cl^-	83.2	-10.6	21.7_5	50.7	-8.4	-8.9
Br^-	91.9	-10.2	29.4	52.3	-7.2	-9.9
I^-	97.1	-9.4	41.0	46.7	-6.0	-3.5
NCS^-	102.6	-8.6	40.2	53.8	-	-9.8
OH_2	70.6	-5.8	18.0	46.8	0	0
				50 ± 3		

obvious variation of $\Delta V^{\neq}_{exp}$ with X and a semiquantitative treatment of the total volume data (26), affirm the assignment of an I_a-mechanism to this chromium series.

III.2.2 Formation reactions

In our discussion of hydrolysis reactions the leaving ligand was generally a charged species, which lead to difficulties in the interpretation of the volume of activation. The entering group in these reactions was an uncharged solvent molecule. For formation reactions (Eq. (13)), this situation is reversed.

$$ML_5OH_2^{\ m} + X^n \longrightarrow ML_5X^{m+n} + H_2O \qquad (13)$$

It is generally accepted that complex ions and entering ligands initially form an encounter complex, or ion-pair, before the rate-determining substitution within the ligand sphere occurs. With the knowledge that ion-pairs are rapidly formed (diffusion controlled) in a preequilibrium, the observed rate constants, k_{obs}, are then products of the ion-pair equilibrium constant, K_{IP} and the real rate constant, k. Similarly, the experimentally obtained volumes of activation consist of the sum of the volume change associated with precursor complex formation and the actual volume of activation, $\Delta V^{\neq}$, of the rate-determining step (Eq. (14)). Both volume quantities can again be interpreted in terms of intrinsic and solvation contributions, with the bulk of solvation changes due to electrostriction, originating from the $\Delta\bar{V}_{IP}$ term.

$$k_{obs} = K_{IP}k \quad \text{or} \quad \Delta V^{\neq}_{exp} = \Delta\bar{V}_{IP} + \Delta V^{\neq} \tag{14}$$

Volume changes for ion-pair formation are often theoretically assessed using a simple electrostatic approach. Actual measurements of the pressure dependence of K_{IP} are to be prefered, although they are not always possible.

Even without the desired separation into $\Delta\bar{V}_{IP}$ and $\Delta V^{\neq}$, useful information can be obtained from a qualitative discussion of $\Delta V^{\neq}_{exp}$, whereby values expected for a certain type of mechanism are estimated and compared with the experimental results. This approach was chosen for the successive formation reactions of the oxalato-aquo-chromium(III) system (Eq. (15) to (17)).

$$Cr(OH_2)_6^{3+} + C_2O_4H^- \xrightarrow{k_{12}} Cr(C_2O_4)(OH_2)_4^+ + H_3O^+ + H_2O \tag{15}$$

$$Cr(C_2O_4)(OH_2)_4^+ + C_2O_4H^- \xrightarrow{k_{23}} cis\text{-}Cr(C_2O_4)_2(OH_2)_2^- + H_3O^+ + H_2O \tag{16}$$

$$cis\text{-}Cr(C_2O_4)_2(OH_2)_2^- + C_2O_4H^- \xrightarrow{k_{34}} Cr(C_2O_4)_3^{3-} + H_3O^+ + H_2O \tag{17}$$

At pH = 2.7 the free oxalate ions are almost completely converted to bioxalate and essentially none of the aquo ligands are present in its deprotonated form. Furthermore, at this acidity all three reactions proceed with comparable rates. The volumes of activation (31) amount to $\Delta V^{\neq}_{12,exp} = -2.2 \pm 1$; $\Delta V^{\neq}_{23,exp} = -8.2 \pm 0.5$; and $\Delta V^{\neq}_{34,exp} = -10.0 \pm 0.5$ cm^3 mol^{-1}, respectively.

Ion-pair formation, although experimentally not detectable, should be strongest for the first step and negligible for the last, considering that in the latter case two anionic species are involved. These preequilibria are associated with volume changes which should clearly be positive for the first step, slightly positive for the second, and - if present at all - negative for the third step. These qualitative estimations can then be combined with possible $\Delta V^{\neq}$ values associated with the rate-determining steps, for which either a dissociative or an associative mechanism can be assumed.

A dissociative mechanism would conceivably involve the same elementary process for all the formation reactions in this system, i.e. the stretching of the chromium-(water)-oxygen bond, with its small positive effect on $\Delta V^{\neq}_{intr}$. Further solvation changes, besides those which occur during ion-pair formation, are not to be expected in this picture. Therefore the overall $\Delta V^{\neq}_{exp}$ should be distinctly positive for the first reaction, slightly or moderately positive for the second, and presumably close to zero for the final formation reaction.

The possibility that these reactions are associative, i.e. bond making between (oxalate)-oxygen and chromium, would necessitate negative values for $\Delta V^{\neq}_{intr}$. Combining this with the volume changes arising from ion-pair formation results in a $\Delta V^{\neq}_{exp}$ of about zero for the first step, becoming increasingly negative for the following two reactions.

A comparison of these estimations with the experimental data clearly verifies the associative character of these substitutions, underlying the general behaviour of chromium(III) complex ions discussed so far.

A particular formation reaction which can be treated more quantitatively was reported by Stranks and Vanderhoek (32). It involves the anation of cis-diaquobis(ethylenediamine)cobalt(III) by oxalate at various acidities (Eq. (18)).

$$\text{cis-Co(en)}_2(\text{OH}_2)_2^{3+} + \text{C}_2\text{O}_2\text{H}_n^{n-2} \longrightarrow \{\text{ion-pair}\}$$

$$\longrightarrow \text{cis-Co(en)}_2(\text{C}_2\text{O}_4)^+ + n\text{H}^+ \qquad (18)$$

Table 7. Volume changes in the anation reactions between cis-diaquobis(ethylenediamine)cobalt(III) ion and various protonated forms of oxalate ion. (units are $cm^3\ mol^{-1}$)

Reactants	$\Delta\bar{V}_{IP}$	$\Delta V^{\neq}$
$\text{cis-Co(en)}_2(\text{OH}_2)_2^{3+} + \text{H}_2\text{C}_2\text{O}_4$		+ 4.8 ± 0.2
$\text{cis-Co(en)}_2(\text{OH}_2)_2^{3+} + \text{HC}_2\text{O}_4^-$	+ 10	+ 4.8 ± 0.2
$\text{cis-Co(en)}_2(\text{OH}_2)(\text{OH})^{2+} + \text{C}_2\text{O}_4^{2-}$	- 1	+ 4.6 ± 0.4

A study of the kinetic effect of pressure on this reaction enabled the authors to assign volumes to the ion-pair formation equilibria, $\Delta\bar{V}_{IP}$, as well as to the kinetic substitution processes, $\Delta V^{\neq}$. The results for the reactions with oxalate, bioxalate and oxalic acid are presented in Table 7. The values of $\Delta V^{\neq}$ are identical within experimental error and are similar in magnitude and sign to $\Delta V^{\neq}_{exp}$ for the water exchange reaction with the diaquo-substrate, i.e. + 5.9 cm^3 mol^{-1} (10). As $\Delta V^{\neq}$ is also pressure independent, it was concluded that, as in the water exchange reaction, the mechanism is I_d-dissociative. Thus the rate-determining step is the dissociation of a water ligand within the ion-pair forming a square-pyramidal transition state for which both water and oxalate compete for entry.

III.2.3 Combined hydrolysis-formation reaction studies

Only a few systems have been studied with respect to both their hydrolysis and formation reactions. Among these, the acido-pentacyanocobaltate(III) and chloro-aquo-rhodate(III) systems were investigated in our laboratory.

The reactions of the pentacyanocobaltate system, presented in Eq. (19), have long been considered (27) as classic examples of a D-type mechanism.

$$Co(CN)_5X^{3-} \xrightarrow{\Delta V^{\neq}_{exp,aq}} \{Co(CN)_5^{2-} + X^-\}^{\neq} \xleftarrow{\Delta V^{\neq}_{exp,an}} Co(CN)_5OH_2^{2-} + X^- \qquad (19)$$

Therefore this is an ideal system for high pressure studies, especially as little is known about the solvation of anionic complexes in aqueous solution. For a threefold variation of X values of $\Delta V^{\neq}_{exp}$ were obtained for the aquation and formation reactions (33) and are shown in Table 8.

The volumes of activation are all positive. Those for anation are constant within experimental error, consistent with a common rate-determining step in which the cobalt-(water)-oxygen bond is stretched. However the absolute value of $\Delta V^{\neq}_{exp,an}$ is perhaps less than one would expect for a D-mechanism, i.e. ca. + 15 to + 18 cm^3 mol^{-1}. A negative volume change associated with ion-pair formation has been discussed earlier, although ion-pairing between anionic species is likely to be minimal. The anticipated large values of $\Delta\beta^{\neq}$ were not observed. Plots of (ln k) versus pressure showed no curvature up to 1.5 kbar. This may however be due to the higher ionic strength employed and/or to the different nature of the solvation sheaths surrounding the

Table 8. Volumes of activation for the aquation and anation reactions of $Co(CN)_5X^{3-}$ and $Co(CN)_5OH_2^{2-}$ ions at 40°C and μ = 1 M. (Volumes given in $cm^3\ mol^{-1}$)

Complex Ion	$\Delta V^{\neq}_{exp,aq}$	$\Delta V^{\neq}_{exp,an}$
$Co(CN)_5Cl^{3-}$	+ 7.8 ± 0.5	
$Co(CN)_5Br^{3-}$	+ 7.6 ± 0.6	+ 8.4 ± 1.0
$Co(CN)_5I^{3-}$	+ 14.0 ± 0.7	+ 9.4 ± 1.6
$Co(CN)_5NCS^{3-}$		+ 8.2 ± 0.9

anionic substrate.

Charge separation occurs during the aquation step giving rise to a small positive contribution from $\Delta V^{\neq}_{solv}$ (note that the overall charge is not changed). In addition, an intrinsic contribution origins from stretching the Co-X bond and should dominate the positive values of $\Delta V^{\neq}_{exp,aq}$. Therefore, although these composite high pressure kinetic results do establish that the reaction mechanism is dissociative, they do not distinguish between D and I_d.

The first two successive aquation and anation reactions of the hexachlororhodate(III) system in acidic aqueous solution can be described by Eq. (20).

$$RhCl_6^{3-} \rightleftharpoons RhCl_5OH_2^{2-} + Cl^- \rightleftharpoons cis\text{-}RhCl_4(OH_2)_2^- + Cl^- \qquad \text{........(20)}$$

Both aquation reactions (i.e. that of $RhCl_6^{3-}$ and $RhCl_5OH_2^{2-}$) showed positive volumes of activation amounting to + 21.5 ± 0.6 and + 14.3 ± 0.5 $cm^3\ mol^{-1}$ at μ = 2 M and 15°C. If one accepts a dissociative mechanism for this system, then the large positive $\Delta V^{\neq}_{exp,aq}$ values are not surprising in view of the complementary effects of a positive $\Delta V^{\neq}_{intr}$ and a positive $\Delta V^{\neq}_{solv}$ due to charge separation. Both anation reactions involving $RhCl_5OH_2^{2-}$ and cis-$RhCl_4(OH_2)_2^-$ exhibit identical, positive volumes of activation of + 15.7 ± 6.5 and + 14.7 ± 1.6 $cm^3\ mol^{-1}$, respectively, within the rather large error limits. Here the dissociation of a water ligand determines the development of the transition state. The absolute values of $\Delta V^{\neq}_{exp,an}$, which approach the molar volume of water, provide strong evidence for a D-type process, although, here again, $\Delta\beta^{\neq} \sim 0$ for each reaction step.

IV. HIGH PRESSURE KINETICS OF FOUR-COORDINATE PLANAR COMPLEXES

Coordination compounds of metal ions with d^8 electron configurations have a square-planar geometry. Such examples are known for univalent rhodium and iridium; divalent nickel, palladium, platinum; and trivalent gold. Their substitution reactions (i.e. replacement of ligand X by nucleophile Y) are subject to a two term rate law (7,27) of the form:

$$-d[\text{complex}]/dt = k_1[\text{complex}] + k_2[Y][\text{complex}] \qquad (21)$$

with $$k_{obs} = k_1 + k_2[Y] \qquad (22)$$

The second-order rate constant k_2 is basically dependent on the nucleophilicity of the entering group Y, while k_1 is normally small and independent of Y. Fig. 5 illustrates this behaviour for the reaction of trans-$Pt(py)_2Cl_2$ with various nucleophiles in methanol.

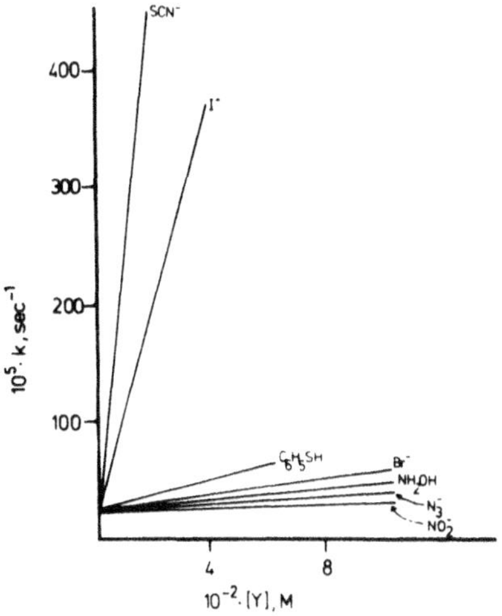

Fig. 5. Plots of the pseudo-first-order rate constant k_{obs} versus nucleophile concentration for reactions of trans-$Pt(py)_2Cl_2$ in methanol at 30°C (taken from Ref. 7 and 27).

Both the k_1 and k_2 paths are believed to be associative in character with the following schemes being proposed:

$$L_3M\text{-}X + S \xrightarrow[-X]{k_1} L_3M\text{-}S \xrightarrow[-S]{+Y} L_3M\text{-}Y \qquad (23)$$

During the rate-determining step, k_1, the ligand X is substituted by a solvent molecule S forming a solvent-containing complex, which in turn rapidly reacts with the nucleophile Y to give the final product. The reaction characterized by k_2 appears to be a straight forward associative displacement of X by Y via a five-coordinate transition state:

$$L_3M\text{-}X + Y \longrightarrow \{L_3M{<}^{X}_{Y}\}^{\neq} \longrightarrow L_3M\text{-}Y + X \qquad (24)$$

A combined variation of pressure and nucleophile concentration leads to the evaluation of the volumes of activation for both the k_1 and k_2 pathways. An example of this is given in Fig. 4 of the chapter entitled "Determination and Interpretation of Volumes of Activation". The k_1 values have generally larger errors as they are determined from the intercept of the k_{obs} versus [nucleophile] plot. Some typical results (26) are summarized in Table 9.

Table 9. Volume changes involved in the substitution reaction of $Pt(dien)X^+$ with Y^- in aqueous solution at 25°C.

X	Y	$\Delta V^{\neq}_{exp,1}$ $cm^3\ mol^{-1}$	$\Delta V^{\neq}_{exp,2}$ $cm^3\ mol^{-1}$	$\Delta\bar{V}$ $cm^3\ mol^{-1}$
Cl^-	N_3^-	- 17 ± 2	- 8.2 ± 1.3	- 2.7 ± 1.0
Br^-	N_3^-	- 15 ± 1	- 8.5 ± 0.2	- 1.2 ± 1.0
Br^-	NO_2^-	- 18 ± 1	- 6.4 ± 0.7	+ 0.9 ± 2.2
Br^-	OH^-	- 18 ± 2		
I^-	N_3^-	- 18 ± 2	- 8.2 ± 0.7	+ 0.8 ± 0.6
N_3^-	I^-	< 0 *	- 12.2 ± 0.1	- 0.8 ± 0.6
N_3^-	SCN^-	< 0 *	- 7.3 ± 0.1	+ 11.8 ± 1.0

* values of k_1 were too inaccurate to calculate $\Delta V^{\neq}_{exp,1}$.

The values for k_1, represented by $\Delta V^{\neq}_{exp,1}$, are all relatively constant and approximate the partial molar volume of water. The latter may be fortuitous, however these characteristics definitely point towards an A-mechanism. In addition, the plots of ($\ln k_1$) versus pressure are markedly curved which again fulfills the requirements outlined by Stranks (1) for an A-mechanism. Looking at the k_2 path with its volume of activation, $\Delta V^{\neq}_{exp,2}$, one sees that these are also negative. More significant are the $\Delta V^{\neq}_{exp,2}$ values for which azide ion is the attacking nucleophile and where the potential leaving group varies: Cl^-, Br^- and I^-. It is clear that $\Delta V^{\neq}_{exp,2}$ is completely independent of the leaving group. The volume of activation is, however, dependent on the nature of the incoming nucleophile. Again these characteristics reflect an A-mechanism.

A volume profile diagram can be constructed, as shown in Fig. 6, for the reaction between $Pt(dien)I^+$ and N_3^-, for which the reverse substitution was also investigated. The partial molar volumes of the reacting species are also known (26). When calculated under the assumption that an A-mechanism prevails, the same volume of the transition state was found, within experimental error (151.9 ± 2.6 $cm^3\ mol^{-1}$), for both the forward and reverse reactions. Clearly the results are compatible with the extreme A-type mechanism.

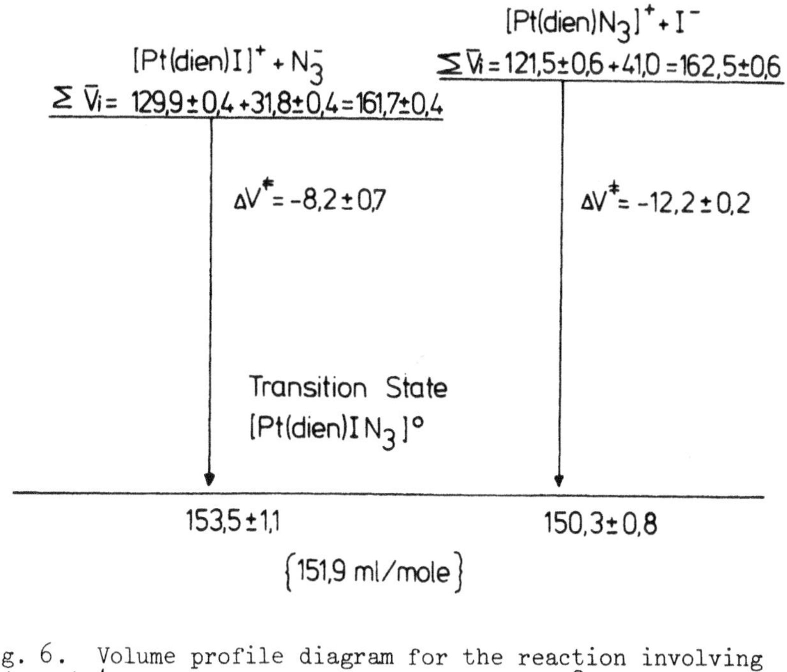

Fig. 6. Volume profile diagram for the reaction involving $Pt(dien)I^+$ and N_3^- in aqueous solution at 25°C.

Reactions of the corresponding palladium(II) complexes are substantially faster, as can be seen from the comparisons given in Table 10.

Table 10. Rate constants for the substitution reactions of $M(dien)X^{+}$ with the nucleophiles Y^{-} in aqueous solution.

M	X	Y	k_1	k_2	μ	Temp.	Ref.
			sec^{-1}	$l\ mol^{-1}\ sec^{-1}$	M	^{o}C	
Pd	Cl^{-}	I^{-}	27.	2900.	0.05	15	35
Pd	Br^{-}	I^{-}	19.9	5500.	0.05	15	35
Pt	Cl^{-}	N_3^{-}	0.00011	0.0025	0.20	25	26
Pt	Br^{-}	N_3^{-}	0.00009	0.0064	0.20	25	26

It is quite apparent from these data that an enormous difference in reactivity exists between palladium(II) and platinum(II) complexes. The rates of substitution of the former may be reduced by introducing steric hinderance into the molecule, thereby impeding the approach of the incoming nucleophile or solvent molecule. It has been argued (27) that when steric crowding is sufficiently accute so that the incoming species can no longer readily bind to the palladium center, the mechanism will tend to become more dissociative. Steric crowding may be subtly introduced by attaching alkyl groups (R = methyl or ethyl) to the amine functional groups of the coordinated dien ligand.

$$Pd(R_n\text{-dien})X^{+} + Y^{-} \longrightarrow Pd(R_n\text{-dien})Y^{+} + X^{-} \qquad (25)$$

The values of k_1 listed in Table 11 clearly show this drastic reduction with increasing steric hinderance. A detailed inspection of the k_1 values reveals that they fall into three groups. Firstly, the "dien" ligands, dien, Me_3dien and Et_3dien where the rate constant is virtually uneffected by the presence of the alkyl groups. Secondly, the intermediate case of Me_5dien where the rate is reduced by a factor of 10^2. Finally, the extreme case where the rate is reduced a further factor of 10^3. Within the last group there is a further small decrease in rate when a methyl group is substituted onto the secondary amine of the 1,1,7,7-tetraethyltriamine ligand. The rate constant k_2 experiences a

Table 11. Effect of (dien) alkylation on the rates of ligand substitution reactions of $Pd(R_n\text{-dien})X^+$ in aqueous solution at 15°C.

Substrate	Nucleophile	k_1 sec^{-1}	k_2 $l\ mol^{-1}\ sec^{-1}$	Ref.
$Pd(dien)Cl^+$	OH^-	19.1	-	35
	I^-	27.	2900.	35
$Pd(Me_3dien)Br^+$	OH^-	19.9	-	35
	I^-	29.	2900.	35
$Pd(Et_3dien)Cl^+$	OH^-	24.	-	35
	I^-	21.	1100.	35
$Pd(Me_5dien)Br^+$	I^-, OH^-	0.132	<0.1(I^-)	35
$Pd(Et_4dien)Cl^{+*}$	OH^-	0.0020	0.055	36
	I^-	0.0019	0.0009	37
	$S_2O_3^{2-}$	0.0018	5.6	37
$Pd(MeEt_4dien)Cl^{+*}$	I^-, OH^-	0.00060	-	35
	$S_2O_3^{2-}$	0.00058	0.029	35
$Pd(Et_5dien)Cl^{+*}$	OH^-	0.00070	-	35
	$S_2O_3^{2-}$	0.00059	0.019	35

* 25°C

similar dramatic decrease in rate down the series. It must be remembered that $S_2O_3^{2-}$ is a much stronger nucleophile than I^-, which accounts for the apparent increase in rate at the end of the series.

The exact nature of the steric crowding can best be shown with the aid of the proposed structure of the $Pd(Et_4dien)X^+$ ion, illustrated in Fig. 7. It can be seen that the axial positions above and below the plane of the complex are partially blocked.

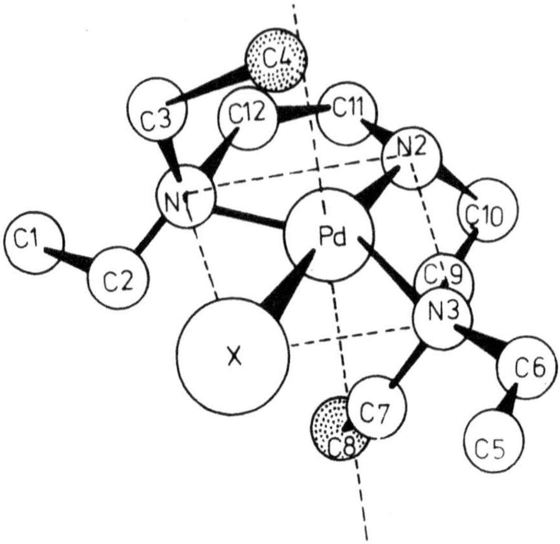

Fig. 7. Proposed structure of $Pd(Et_4dien)X^+$, taken from the known structure (38) of $[Pd(Et_4dien)NO_2]NO_3.H_2O$

Table 12. Rate constants and activation parameters for the nucleophile independent substitution reaction of $Pd(Et_4dien)X^+$ in aqueous solution at 25°C.

X	10^4k_1 * sec^{-1}	$\Delta H^{\neq}$ $kJ\ mol^{-1}$	$\Delta S^{\neq}$ $J\ K^{-1}\ mol^{-1}$	$\Delta V^{\neq}_{exp}$ $cm^3\ mol^{-1}$
Cl^-	21.6	65.5	- 74	- 14.9
N_3^-	4.59	71.8	- 71	- 13.9
Br^-	17.4	65.5	- 75	- 13.3
I^-	2.65	71.4	- 69	- 11.5
NCS^-	1.52	75.2	- 67	- 10.3
NH_3	0.130	93.2	- 28	- 3.0

* rate constants were measured or extrapolated to μ = 0.05 M.

It is precisely in these "5th and 6th coordination sites" that the solvent or nucleophile is thought to attack. Therefore it is conceivable that such steric crowding may enhance a dissociative component of the mechanism and thus reflect in the volumes of activation. The results of these pressure studies (39) are summarized in Table 12.

The complexes are listed in the anticipated order of increasing Pd-X bond strength. It is quite apparent that there is a reasonable correlation between bond strength and the rate of "aquation" (i.e. the k_1 path). The latter varies by a factor of ca. 170. This trend cannot be ascribed to an increase in the overall steric crowding caused by the increasing size of the leaving group because iodide, which is undoubtedly the largest group in terms of crosssectional area, lies in the middle of the series. Thus the variation of k_1 with X already indicates that these reactions do possess some dissociative character. Very significantly there is a parallel trend in the volumes of activation. Although the values are negative they are complicated by bond stretching which must result in a marked increase in electrostriction of the solvent. Of course, if these reactions were purely associative , then $\Delta V^{\neq}_{exp}$ would be independent of X. Therefore we conclude that there is some bond breaking, together with Pd-OH_2 bond formation, in the transition state.

If one assumes that the partial molar volume of the

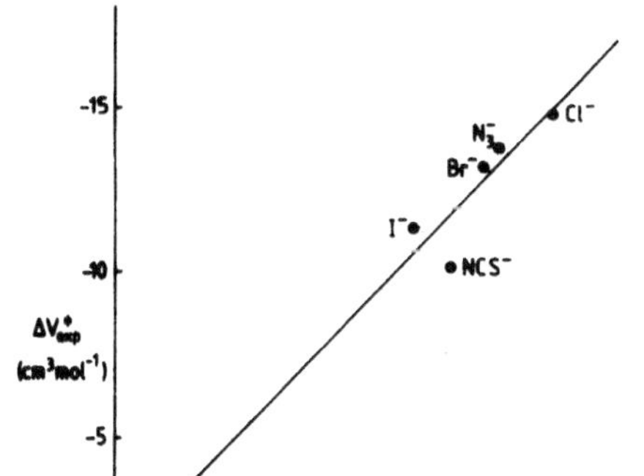

Fig. 8. A plot of $\Delta V^{\neq}_{exp}$ versus $\Delta\bar{V}$ for the reaction (k_1 path) of Pd(Et_4dien)X^+ at 25°C.

$Pd(Et_4dien)NH_3^{2+}$ ion, which can be readily isolated as a perchlorate salt, is the same as that of the reactive $Pd(Et_4dien)OH_2^{2+}$ species, which has as yet not been isolated in a pure form, then one can estimate the $\Delta\bar{V}$ values for these reactions using the appropriate partial molar volumes of the reactants and the corresponding X^- ions. As for the pentaammine complexes a plot of $\Delta V^{\neq}_{exp}$ versus $\Delta\bar{V}$ can be made (see Fig. 8) resulting in a reasonable straight line with a slope of 0.5. Thus, as assumed for the pentaamminechromium(III) complexes, this corresponds to an I_a-mechanism which also fits the above results and conclusions exactly.

V. CONCLUSIONS

This review was written to demonstrate the usefulness of high pressure kinetic investigations and thus the volume of activation in the discussion of the reaction mechanisms of coordination compounds in solution. It is not meant to be comprehensive so that only specific examples have been taken from the various classes of reactions.

It is our belief that in simple systems, such as solvent exchange reactions, $\Delta V^{\neq}_{exp}$ provides direct mechanistic information since it can be equated directly to $\Delta V^{\neq}_{intr}$. Where the situation is complicated by additional contributions from $\Delta V^{\neq}_{solv}$, either qualitative comparisons can be made between the expected and obtained values of $\Delta V^{\neq}_{exp}$, or exact volume profiles can be drawn. It is hoped that with the increasing availability of volume data and a deeper understanding of solvation phenomena, $\Delta\beta^{\neq}$ values will in future be precisely interpretable in terms of intimate molecular processes.

VI. ACKNOWLEDGMENTS

The authors wish to thank several colleagues and publishers for their permission to reproduce Figures 2 and 5 and the data in Tables 2, 3, 4 and 7. The results from our laboratory were supported by various grants of the Deutsche Forschungsgemeinschaft.

VII. REFERENCES

1. D.R. Stranks, Pure Appl. Chem., 38, 303 (1974).
2. T.W. Swaddle, Coord. Chem. Rev., 14, 217 (1974).
3. B.G. Cox, G.R. Hedwig, A.J. Parker and D.W. Watts, Aust. J. Chem., 27, 477 (1974).
4. R.A. Horne, "Water and Aqueous Solutions, Structure, Thermodynamics, and Transport Processes", Wiley-Interscience, p 259-260 (1971).
5. C.K. Ingold, "Structure and Mechanism in Organic Chemistry", Cornell University Press, Ithaca (1953).
6. C.H. Langford and H.B. Gray, "Ligand Substitution Processes", Benjamin, New York (1965).
7. M.L. Tobe, "Inorganic Reaction Mechanisms", Nelson, London (1972).
8. D.R. Stranks and N. Vanderhoek, Inorg. Chem., 15, 2639 (1976).
9. W. Kruse and H. Taube, J. Am. Chem. Soc., 83, 1280 (1961).
10. S.B. Tong, H.R. Krouse and T.W. Swaddle, Inorg. Chem., 15, 2643 (1976).
11. H.R. Hunt and H. Taube, J. Am. Chem. Soc., 80, 2642 (1958).
12. T.W. Swaddle and D.R. Stranks, J. Am. Chem. Soc., 94, 8357 (1972).
13. S.B. Tong and T.W. Swaddle, Inorg. Chem., 13, 1538 (1974).
14. D.R. Stranks and T.W. Swaddle, J. Am. Chem. Soc., 93, 2783 (1971).
15. D.L. Carle and T.W. Swaddle, Can. J. Chem., 51, 3795 (1973).
16. S.T.D. Lo and T.W. Swaddle, Inorg. Chem., 14, 1878 (1975).
17. E.F. Caldin, M.W. Grant and B.B. Hasinoff, J. Chem. Soc. Faraday Trans. I, 3, 560 (1973).
18. M. Eigen and R.G. Wilkins, Adv. Chem. Ser., No 49, 55 (1965).
19. R. Good and A.E. Merbach, Inorg. Chem., 14, 1030 (1975).
20. A.E. Merbach, I.C.C.C., 17th (1976), Hamburg.
21. W.E. Jones, L.R. Carey and T.W. Swaddle, Can. J. Chem., 50, 2739 (1972).
22. D.A. Palmer and H. Kelm, Inorg. Chem., 16, 3119 (1977).
23. G. Guastalla and T.W. Swaddle, Can. J. Chem., 51, 821 (1973).
24. D.L Gay and R. Nalepa, Can. J. Chem., 48, 910 (1970).
25. M.V. Twigg, Inorg. Chim. Acta, 24, L84 (1977).
26. D.A. Palmer and H. Kelm, Inorg. Chim. Acta, 19, 117 (1976).
27. F. Basolo and R.G. Pearson, "Mechanisms of Inorganic Reactions", 2nd Ed., John Wiley & Sons (1967).
28. K. Shimizu, N. Tsuchihashi and Y. Furumi, Rev. Phys. Chem., Japan, 46, 30 (1976).
29. M. Mares, D.A. Palmer and H. Kelm, Inorg. Chim. Acta, 27, 153 (1978).
30. R.K. Murmann and H. Taube, J. Am. Chem. Soc., 78, 4886 (1956).
31. C. Schenk and H. Kelm, J. Coord. Chem., 2, 71 (1972).
32. D.R. Stranks and N. Vanderhoek, Inorg. Chem., 15, 2645 (1976).
33. D.A. Palmer and H. Kelm, to be published.

34. K. Hyde, H. Kelm and D.A. Palmer, Inorg. Chem., accepted.
35. M. Kotowski, PhD Thesis, Frankfurt (1977).
36. W.H. Braddley and F. Basolo, J. Am. Chem. Soc., 88, 2944 (1966).
37. D.A. Palmer and H. Kelm, Inorg. Chim. Acta, 14, L27 (1975).
38. N. Bresciani, M. Calligaris, L. Randaccio, V. Ricevuto and U. Belluco, Inorg. Chim. Acta, 14, L17 (1975).
39. D.A. Palmer and H. Kelm, submitted for publication.

PRESSURE EFFECTS ON BIOCHEMICAL SYSTEMS

K. Heremans

Department of Chemistry,
Katholieke Universiteit te Leuven, Belgium.

ABSTRACT. A general review is given on pressure effects on : (1) conformational transitions in polypeptides, proteins, nucleic acids and membranes. Applications to anesthesia, lipid-protein interactions and the isolation of subcellular particles with the ultracentrifuge are discussed; (2) relaxation and transient kinetics of enzyme reactions and the interaction of dyes with nucleic acids; (3) molecular assembly is treated with ribosome subunit interaction, glutamate dehydrogenase and microtubuli as examples.

1. INTRODUCTION

There are good reasons for doing high pressure studies on biological systems. First of all a large part of the biomass on our planet is embedded in the high pressure medium of the ocean. A crucial question is the mechanism of adaptation to this environment. We shall not be concerned about this question here but refer to the excellent introduction into that field by Macdonald (1).

For the physical chemist pressure is an unique tool to obtain thermodynamic and kinetic information which characterize chemical equilibria. The physical biochemist uses these principles to study reactions of biological interest. This approach is used in this study to discuss pressure effects on proteins, nucleic acids and membranes and their interactions with small molecules as well as their mutual interaction in molecular assembly processes. A special feature of pressure effects in biochemical systems is that the effect is predominantly on noncovalent interactions such as hydrogen bonding, electrostatic interactions and nonpolar interactions in aqueous solutions.

The choice of topics reflect the authors personal interest and

Kelm, H. (Ed.) High Pressure Chemistry. 467–487. *Copyright © 1978. D. Reidel Publishing Company. All Rights Reserved.*

activities in the field. Besides the book mentioned on deep sea biology, several books and reviews treat the behaviour of various biological systems under pressure (2-7).

2. PHASE TRANSITIONS IN PROTEINS; NUCLEIC ACIDS AND MEMBRANES

The first systematic observation on the effect of pressure on proteins was made by Bridgman in 1914. He observed that after treatment at 7500 atm egg albumen looked as if it was boiled at 100°C. The effect was irreversible. The picture that high pressure invariably denatures proteins was refined and it was found that lower pressures have a protective effect against temperature denaturation (8).

The observation that temperature induces reversible transitions in macromolecular structures has provoked pressure studies on helix-coil transitions in polypeptides (9), folding-unfolding processes in proteins (10) melting of the double helix into single stranded DNA (11) and recently phase transitions in membranes (12).

2.1 Proteins and nucleic acids

Following Zimm and Bragg (13) these transitions have been treated in terms of a mathematical model used by Ising to describe ferromagnetism. We consider the cis-trans equilibrium in the polypeptide Poly-L-proline studied by Rifkind and Applequist (9). Both forms of the polymer are helical but the trans form is more extended than the cis form. If f_c is the fraction of the residues in the cis form then the following equation describes the pressure transition of cis to trans

$$f_c = \frac{1}{2} + \frac{s - 1}{2[(1-s)^2 + 4\sigma s]^{1/2}}$$

The stability constant s is the equilibrium constant for the conversion of a residue from trans to cis at the end of a cis sequence; the interruption constant σ is the equilibrium constant for a process in which a polymer containing νcis sequences is converted to a polymer containing $(\nu+1)$cis sequences, keeping the number of cis peptide units fixed.

It is clear that

$$\frac{d \ln s}{dP} = - \frac{\Delta V}{RT}$$

where ΔV is the change in volume in going from trans to cis for one unit. From both equations we can calculate the slope of the pressure transition curve at the midpoint (i.e., s = 1)

$$\frac{d\,f_c}{dP} = -\frac{\Delta V}{4RT\sigma^{1/2}}$$

A similar expression can be derived for the temperature dependence

$$\frac{d\,f_c}{d\,T} = +\frac{\Delta H}{4RT^2\sigma^{1/2}}$$

These equations may be combined to obtain the familiar Clausius-Clapeyron equation

$$\frac{d\,T}{dP} = \frac{T\Delta V}{\Delta H}$$

If one wants volume changes or enthalpy changes for the transition of one residue, one has to make assumptions of the value of σ. For a discussion of this problem we refer to the original paper.

The above equations have been used to describe the thermal denaturation of DNA (14) and the melting of the hydrocarbons of phospholipids (15).

Brandts (16) and Hawley (17) have made the interesting observation that proteins show a quite different pressure-temperature stability behaviour in contrast to nucleic acids. In detailed studies Hawley showed that the P-T coexistence lines for proteins show considerable curvature while nucleic acids (18) show no curvature. Several proteins exhibit this phenomena of maximum stability at certain T and P. The situation becomes even more complex when the pH is included in the analysis as shown by Zipp and Kauzmann (19). The phenomenon has been attributed to the exposure of hydrophobic groups to the solvent upon unfolding (20).

2.2 Hydrophobic interactions

An interesting aspect of these studies is the magnitude of the volume changes observed for protein folding. They are small compared to the total volume of the protein. It is difficult to explain this if one accepts the picture that native proteins are chiefly stabilized by hydrophobic interactions (21). On the basis of model systems one expects much larger volume changes. Li et al. (22) have pointed out the possible significance of the methods used to study the denaturation. These are uv, visible and fluorescence spectroscopy and could thus reflect local changes on the protein. It would therefore be desirable to use techniques which look to the protein as a whole and not just to one spot on the surface or the interior.

Another way to explain the discrepancies is to have a closer look to the model systems. In this respect it seems justified to make a distinction between aliphatic and aromatic compounds. Several literature reports support this statement (22-23). A simplified picture would be that aliphatic hydrocarbons show positive volume changes on association (24) while aromatic show negative volume changes (25). Since proteins contain both residues, compensation effects may explain the small volume changes observed upon denaturation. The different composition of proteins and nucleic acids might also explain the peculiar temperature-pressure behaviour of proteins compared with nucleic acids.

2.3 The two state hypothesis of protein denaturation

Brandts (20)has pointed out that in theory any transition can conveniently be defined as a two state transition by any arbitrary division of the microscopic assembly into two parts. This approach has been used in the above discussed results.

Hawley and Mitchell (26) in their kinetic studies of chymotrypsinogen denaturation at pH 2, found evidence for a two state model. But the authors point out that this does not mean that the existence of low levels of intermediate states are excluded.

Li et al. (27) on the other hand "disprove the two state hypothesis" in their thermodynamic study of chymotrypsinogen and lysozyme. Observing the protein fluorescence they find a first domain of the protein which denatures below 8 kbar. With ANS binding in the case of chymotrypsinogen and protein fluorescence in the case of lysozyme they find a second independant domain between 8 and 11 kbar. Thus these studies reveal a plurality of pressure-denaturated forms in both proteins. The same authors (22) find only one domain however up to 10 kbar for the riboflavin binding protein of egg white.

For a fuller discussion of protein folding we refer to the reviews by Baldwin (28) and Pohl (29).

2.4 Phase transitions in phospholipids

The study of biological membranes has increased rapidly in recent years. Especially the model proposed by Singer and Nicholson made a tremendous contribution in this respect. This model is that of a lipid bialyer with some proteins partially embedded in the lipid matrix and some completely penetrating. The phospholipid molecules form a bilayer due to their amphipatic nature, i.e. the hydrocarbon tails avoid the water, while the polar headgroups interact with the solvent. Prolonged ultrasonic irradiation produces vesicles which are closed spheres enclosing solvent.

It has been shown by Chapman and coworkers (30) that pure phospholipids may undergo thermal phase transitions. This transition (α crystalline → liquid crystal) occurs at a temperature (T_t) which is characteristic for the molecule. Trauble and Haynes (31) inter-

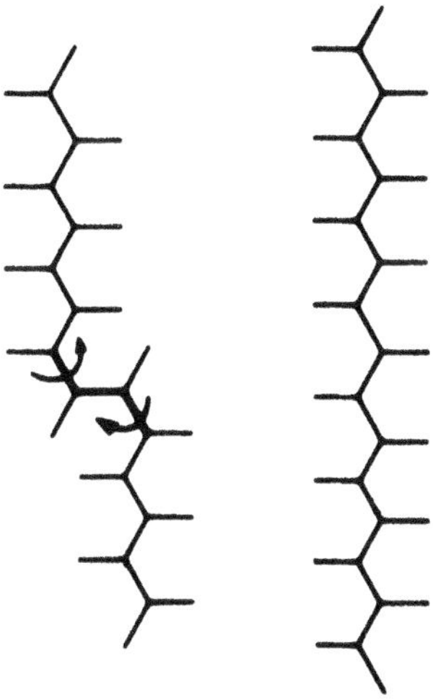

Fig. 1. Rotational isomers of lipid hydrocarbon chains (a) 2gl-kink and (b) all-trans configuration.

preted these transitions in terms of the formation of rotational isomers in the hydrocarbon chains and loosening of the polar head group arrangement. The simplest rotational isomer is the so-called 2gl kink where two trans configurations are transformed into a gauche configuration as shown in fig. 1.

This has important consequences for the dimensions of the lipid bilayer : (i) : the bilayer decreases in thickness by 5 Å; (ii) lateral expansion occurs so that the area per molecule increases from 48 to 58 $Å^2$; (iii) an increase in volume is observed of 1.4 %.

For vesicles, no change in size and shape is observed as measured by the average diffusion coefficient (32).

In close collaboration with Dr. H. Desmedt, we have measured the pressure dependence of T_t by observing changes in light scattering under pressure (12). We found that dT/dP is pressure independent up to 2500 atm as shown in fig. 2 for dipalmitoyllecithin. This was predicted by Nagle (33) and also confirmed by Liu and Kay (34). These authors found that ΔV is pressure independent up to 250 atm. In contrast to dT/dP values for proteins and nucleic acids, the value for lipids is quite high : 20K/1000 atm (35). This is largely due to smaller ΔH changes for lipid transitions(36)

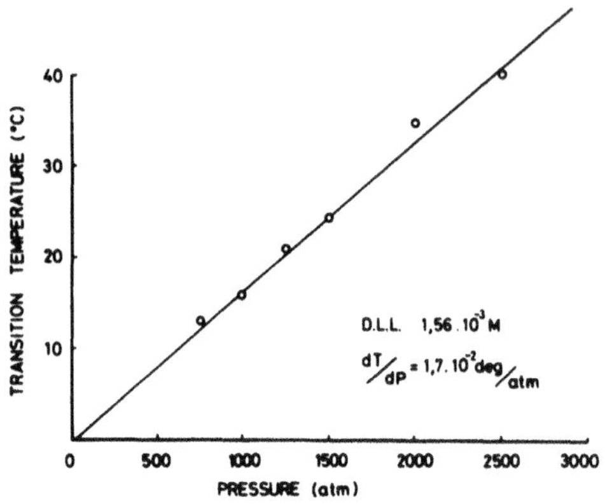

Fig. 2. Pressure dependence of the transition temperature of dipalmitoyllecithin as determined with high pressure light scattering.

compared to proteins (37). dT/dP is somewhat dependent on the length of the hydrocarbon and the nature of the headgroup. We also made the interesting observation that the presence of small molecules which have a large effect on the value of T_t, do not affect dT/dP (12). This is easily understandable in terms of the colligative properties of the bilayer. It has important consequences for the interpretation of combined effects of pressure and drugs in biological systems. Similar observations are made when the transition is shifted by proteins.

Biological membranes contain mixtures of lipids and one thus obtains phase diagrams. Trudell (39) has studied pressure effects up to 136 atm on these diagrams.

2.5 Pressure effects on biological membranes

In this section we briefly describe three examples of pressure effects on biological systems where the results can be explained on the basis of pressure effect on phospholipid transitions.

Wattiaux and coworkers (40) studied the effect of experimental procedure on the isolation of mitochondria in a sucrose gradient in the ultracentrifuge. They observed deterioration of the particles at high rotor velocity (i.e. high pressure) the effect being counteracted by an increase in temperature. Protection could also be obtained with drugs which lower the transition temperature of lipids. All the observations can be explained assuming that the particles deteriorate when the phospholipids are below a critical temperature.

Pressure is still an important tool in the study of the mechanism of anesthesia. In 1950 Johnson and Flagler (41) reported that the application of 100-120 atm restored swimming activity of anesthesized tadpoles. Lever et al. (42) used mice and helium pressure to obtain the same results. It was concluded that the primary site of the anesthetics is the membrane bilayer. Trudell et al. (39) showed that the same antagonistic effect can be studied with mixed phospholipid bilayers.

The conclusion that anesthetics act primarily on the membrane has not found general acceptance and several authors conclude from their experiments that proteins embedded in the membrane are the primary site of action. For a recent discussion see (43). In recent years, however, it has become evident that the physical state of the lipids in the membrane can have strong effects on the membrane proteins (44). It then becomes easier to reconcile both views on the mechanism of anesthesia.

The results which we have obtained on the effect of pressure on the breaks in Arrhenius plots in an enzyme shed some light on the interaction of lipids with proteins.

Nitrogenase is a multisubunit enzyme which catelyzes the reduction of nitrogen to the ammonium ion (45). The activity of the enzyme can conveniently be measured with the gaschromatograph to follow the reduction of acetylene or HCN to C_2H_4 or CH_4. When the activity is measured as a function of temperature, a sharp break is observed in the Arrhenius plot (fig. 3b) which Thornely et al. (46) attributed to a change in the conformation of the subunits. Ceuterick et al. (47) determined the pressure dependence of the temperature at which the break occurs, as shown in fig. 3a, and find $dT/dP = 20$ K/1000 atm. This closely corresponds to the change of transition temperature with pressure we have described in section (2.4) for phospholipids. A change in conformation in one of the protein subunits would show a much smaller pressure dependence (35).

We have also independent biochemical evidence that the break in the Arrhenius plot in nitrogenase is due to a transition in the phospholipids.

There have been a few other reports in the literature describing pressure effects on systems which involve lipid transitions (48, 49). None of them however describes the combined effects of temperature and pressure on these systems.

Macdonald and Miller (50) have recently reviewed pressure effects on biomembranes and tissues.

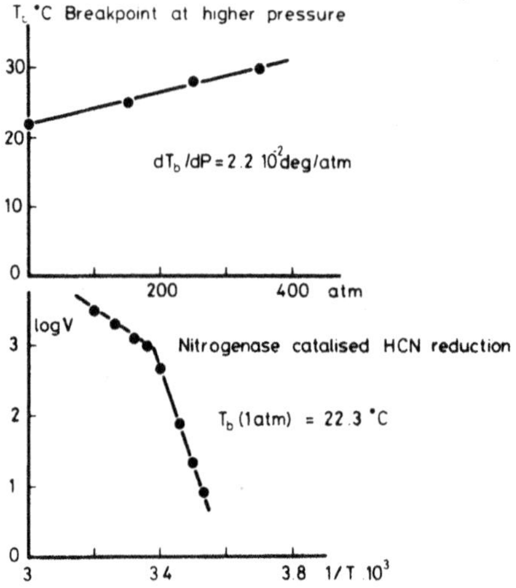

Fig. 3. (a) Pressure dependence of the temperature at which a break occurs in the Arrhenius plot of nitrogenase. (b) Break at 1 atm.

3. INTERACTIONS OF SMALL MOLECULES WITH MACROMOLECULES

In this section we describe pressure effects on the interaction of substrates and substrate like molecules with enzymes including enzyme catalysis. We also discuss the interaction of drugs with nucleic acids.

3.1 Enzyme catalysis

Michaelis and Menten proposed the simplest mechanistic picture for enzyme catalysis.

$$E + S \underset{k_{-1}}{\overset{k_1}{\rightleftharpoons}} ES \overset{k_2}{\rightarrow} E + P$$

The substrate (S) is bound to the enzyme (E) to form a complex (ES) which slowly decomposes to enzyme and product (P). The velocity of the reaction as a function of substrate concentration is given by

$$v = \frac{V_{max}\,(S)}{K_m + (S)}$$

where $v = V_{max}$ when $(S) \gg K_m$. The hyperbolic curve is analogous to a Langmuir adsorption curve. K_m is the Michaelis-Menten constant defined as

$$K_m = \frac{k_2 + k_{-1}}{k_1}$$

When $k_2 \ll k_{-1}$, K_m is a dissociation constant for the (ES) complex.

Laidler (51) has discussed the experimental conditions to obtain the activation parameters for this reaction scheme. It can however easily be shown that a more complicated mechanism can be described by the same formalism (52). The expression for V_{max} and K_m becomes then very complicated so that a simple interpretation of the activation parameters becomes very difficult if not meaningless. An example of such a situation is found in the paper of Morild (53).

Two other possibilities must be considered. The enzyme may be destroyed irreversibly by the action of pressure or the enzyme may dissociate into subunits. Both situations can sometimes be exploited to obtain information as shown by the work of Ludwig et al. (54) and Schmid et al. (55).

In a few cases breaks have been observed when plotting ln (activity) versus pressure. In one case this has been interpreted as evidence that at high pressure both the activity as well as the binding of the substrate are being affected by pressure (56). In the case of nitrogenase however the break is due to a transition in the lipids which affect the activity (47). These effects can be discerned by observing the breaks at different temperatures (see section 2.5).

3.2 High pressure transient and relaxation kinetics of enzyme reactions

From the discussion in the preceding section it is evident that before any reasonable interpretation of enzyme activation parameters is attempted, information about the mechanism is needed from other experimental sources.

In this respect transient and relaxation methods are powerful techniques to observe the formation and decomposition of intermediates in enzyme reactions (52, 57).

Chymotrypsin and Trypsin are in this respect the most extensively studied examples. The following scheme represents the minimum number of reaction steps for chymotrypsin and trypsin.

$$ECH_2OH + RCOX \rightleftharpoons ECH_2OH{:}RCOX \xrightarrow{k_2} ECH_2OCOR + HX$$

$$ECH_2OCOR + H_2O \xrightarrow{k_3} ECH_2OH + RCOOH$$

In this scheme ECH_2OH represents the Ser-195 side chain of the enzyme and RCOX is an amide or ester substrate. k_2 is the rate constant for the enzyme acyl intermediate formation and k_3 gives the rate of decomposition.

The reversible binding of the substrate to the enzyme to form the Michaelis complex can be studied with molecules that bind to the active site but are not transformed by the enzyme. Proflavin is such a molecule. The binding can be followed by observing the changes in the spectrum of the dye. We have studied the pressure dependence of the binding with chemical relaxation spectrometry (58). The results are given in the Table I.

The results reflect differences in the substrate binding site of both enzymes. While trypsin is specific for charged amino acid residues, chymotrypsin is specific for nonpolar side chains. The binding of proflavin, a dye with a positive charge, to trypsin reveals a charge neutralisation while in chymotrypsin no charge is present in the binding pocket.

The pressure dependence of k_3 has been studied with steady state kinetics while k_2 can only be studied with stopped-flow.

Under steady state conditions (i.e. when $(S) \gg K_m$) k_3 is rate limiting. Neumann and coworkers (59) have studied the pressure dependence of k_3 for various substrates. The activation volumes are negative but smaller than the corresponding values for the

TABLE I

Kinetic and Thermodynamic parameters for the interaction of Proflavin with Trypsin and Chymotrypsin obtained with High Pressure Relaxation Spectrometry. pH 7.7 (Tris-buffer), 16°C.

	Chymotrypsin	Trypsin
k_{12} (1 atm)M^{-1} sec^{-1}	4.6×10^7	4.6×10^7
k_{21} (1 atm) sec^{-1}	8.9×10^3	1.0×10^4
$\Delta V^{\#}_{12}$ ml/mole	8.7	17.5
$\Delta V^{\#}_{21}$ ml/mole	9.5	1
ΔV ml/mole	0	17.8

base catalysed hydrolysis in the absence of enzyme. The results were interpreted as a repositioning of the CO-group of the acyl-enzyme towards the NH-group which are thought to be involved in the stability of the hydrogenbond formed for the charge stabilization.

The study of the pressure dependence of k_2 was recently done with a newly developed high pressure stopped-flow (60). Preliminary results with N-acetyl-L-tryptophaneethylester revealed that the activation volume is also negative. This is consistent with activation volumes obtained for base catalysed hydrolysis of the formation of the acyl-intermediate (61). A more detailed study is in progress to see whether these activation volumes are affected by the nature of the substrate.

Another example where the combined use of stopped-flow and relaxation kinetics under pressure has given information on a biological redox reaction, is given in the accompanying paper in this volume.

3.3 Cooperative interactions in multisubunit systems

When substrates interact with enzymes, changes in conformation of the protein may be observed. It is then to be expected that when the enzyme is composed of several subunits, these changes in conformation are transferred to the other subunits. This in turn can affect the binding of substrate molecules. In most cases this enhances the binding which gives rise to sigmoidal instead of hyperbolic substrate saturation curves (62).

Formally these curves can be represented by the following equation

$$v = \frac{V_{max}\ (S)^n}{K_m + (S)^n}$$

where n is a parameter reflecting the cooperativity of the system. n has been called the Hill Coefficient. When n = 1 there is no cooperativity. Even from this simplified picture it is clear that we have a device to regulate the activity of the enzyme. The significance of cooperative interactions in enzymes is therefore great. The effect of pressure on the cooperativity of some systems has been studied, but the results are not easily interpreted in molecular interaction terms.

Hemoglobin, the oxygen carrier of the blood, is one of the most intensely studied systems which show cooperative interactions. A recent study (63) has revealed that pressure does not affect the R-T transition in terms of the Monod-Wyman-Changeux scheme, but affects the ligand binding process of oxygen and n-butylisocyanide.

A study has also been made on the cooperative interactions in nitrogenase (64). The Hill coefficient decreases with increasing pressure.

The fact that cooperative interactions can be pressure sensitive has some important consequences for the adaptation of live to the high pressure conditions in the deep sea. One example has been given where the cooperative interactions of a regulatory enzyme isolated from a deep sea fish shows no pressure dependence (65).

3.4 The interaction of dyes with nucleic acids

The study of the binding of acridine and phenanthridine dyes to DNA has attracted considerable interest because of their mutagenic and antimalaria activity respectively. The planar dyes intercalate into the bases pairs of DNA and the structure of some dye complexes with dinucleosides monophosphates has been resolved to atomic resolution. Li and Crothers (66) proposed a two step binding mechanism from their temperature-jump studies

$$\text{Dye} + \text{DNA} \rightleftharpoons C_{out} \underset{k_4}{\overset{k_3}{\rightleftharpoons}} C_{in}$$

C_{out} stands for an outside bound dye and C_{in} for an intercalated dye molecule.

Yves Van Nuland has studied the effect of pressure on this equilibrium up to 2 kbar (58, 67). For Ethidium bromide and Ditertiarybutylproflavin, an intercalating and nonintercalating dye respectively, the volume change on binding to DNA is small. This means that outside binding and intercalation are pressure insensitive. From kinetic studies with relaxation techniques the outside binding of proflavin is also found to be pressure independent. This indicates that the DNA-bound counterions are being replaced by dye molecules producing no net volume change. The activation volume for the intercalation is + 9 ml and is similar to that found for the interaction of proflavin with chymotrypsin (see section 3.2).

Somewhat different results have been obtained with tRNA (58). It is not clear whether an intercalation process occurs as observed with DNA (68). Negative volume changes are also observed when dyes are bound to uncoiled DNA, i.e. when the bases pair in their unstacked conformation. This explains the pressure dependence of the melting behaviour of DNA as observed with proflavin as a probe (67).

At present it is not clear how to reconcile the small ΔV observed for helix-coil transition of DNA and intercalation with ΔV observed for dye stacking (69) and the stacking of purine bases (23, 25).

A possible explanation for the small ΔV observed with DNA might be compensating changes in volume due to changes in counterion binding.

4. MOLECULAR ASSEMBLY

A feature of biological macromolecules is their tendency to form supramolecular structures. These structures have properties which are not present in the constituing subunits. A typical example is hemoglobin. The sygmoidal shape of the oxygen saturation curve is not present in the subunits. Pressure effects on allosteric interactions have been considered in section 3.3.

The formation of membranes starting from phospholipids is another example where a functional cooperation is established. Proteins and nucleic acids interact to form chromatin or ribosomes, structures which play an essential role in the self-reproduction mechanism of the cell.

In this section we consider pressure effects on ribosome subunit interaction and two typical examples of protein assembly : glutamate dehydrogenase and microtubuli.

Two important points should be mentioned in relation to these studies. The first is that pressure effects on molecular assembly can be dramatically affected by temperature and/or solvent conditions. Glutamate dehydrogenase and microtubuli are examples to be discussed in detail. Striking examples of solvent conditions are the presence of double charged anions such as sulfate or phosphate which have rather specific effects on the behaviour of serum albumin under pressure (70). While no change in light scattering is observed in the absence of salts, association numbers of 80 have been observed at 3000 atm. Similar observations have been made by Suzuki et al. on egg albumin (71). The presence of small amounts of toluene in the case of glutamate dehydrogenase, reverses the sign of the pressure effect (72).

The second point concerns the experimental conditions. While protein associations under pressure have been studied with optical techniques already in 1963 by Suzuki et al. (71), it seems that the importance of pressure as a regulating factor has only been realised as an irritating complexity in the ultracentrifuge (73). Examples are ribosome association and myosin polymerization. The reason is quite obvious. These systems show very strong pressure effects and can therefore easily be detected in the ultracentrifuge. Glutamate dehydrogenase on the other hand shows no appreciable effects in the ultracentrifuge while light scattering observations can easily be made (72).

4.1 Ribosome subunit assembly

A number of studies on protein synthesis in vivo as well as in vitro have revealed that the ribosome is the primary site of action of pressure (4). Recent studies have therefore concentrated on the subunit association in ribosomes. Formally the process can be written as

30 S subunit + 50 S subunit ⇌ 80 S ribosome

Each subunit is composed of several proteins together with RNA. Most studies have been performed on E. Coli ribosomes but ribosomes from eukariotyc systems are equally pressure sensitive (74). Nieuwenhuysen et al. (75) have recently studied pressure effects on polysomes from Artemia Salina. From these studies it is clear that mRNA on which the ribosomes are bound, stabilizes the interaction of the ribosome subunits against pressure in the sence that dissociation is much slower.

An interesting question is the source of the large ΔV associated with subunit assembly. It is known that Mg ions are involved in the reaction. A simple scheme to explain pressure effects would be electrostriction around the Mg ion.

$$30\ S + 50\ S + Mg^{2+} \rightleftharpoons \text{Ribosome}$$

Two possible mechanisms have been proposed for the stabilization of ribosomes by Mg : Mg can be considered as a true reactant in the assembly with specific interactions sites or Mg is a parameter like pressure or temperature. In the latter case Mg would reduce the repulsion between the subunits. The specificity of the interaction would then results from hydrogen bonding and/or hydrophobic interactions (76).

Recently pressure-jump techniques have been used to study the kinetics of subunit association (78). Wishnia et al. (79) announced pressure-jump work performed under pressure with E. Coli ribosomes.

4.2 Protein assembly

In the previous section an example was discussed where two subunits form a dimer. In hemoglobin only tetramers are formed. These so-called closed type associations are in contrast to open types of association where very long polymers can be formed by the subsequent addition of monomers to already existing polymers. Two possibilities may then arrise : the addition of the next monomer may be cooperative or not, i.e. either a monomer binds to a dimer without interaction to the other monomer or it interacts to form a nucleus from which the polymer grows. A schematic diagramme can be depicted as follows

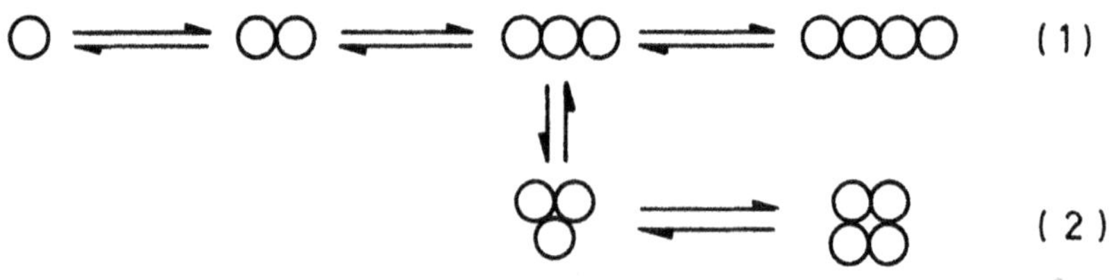

Both systems behave quite different from an experimental point of view. In scheme (1) polymers are present at all concentrations of protein, i.e. the transition from monomer to polymer is not sharp. Polymerization according to scheme (2) is quite different. Polymer is only formed above a critical concentration of protein. The behaviour of both models is depicted in fig. 4.

Sutherland (80) has recently discussed the discrepancy between calorimetric and van 't Hoff enthalpies observed with some systems.

Fig. 4. Simulated behaviour of protein polymerization according to scheme (1) or (2) of the text. In (1) the ratio between nucleation and propagation is one; in (2) the ratio is 10^{-3}. The concentration of polymer (c_p) is plotted against the total concentration of protein (c_t).

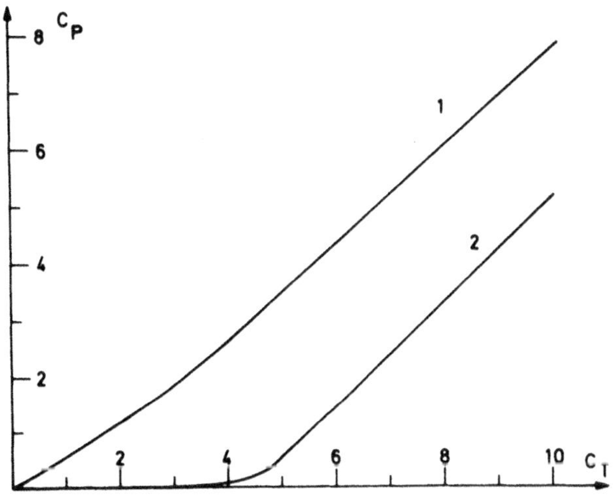

4.3 Glutamate Dehydrogenase

A typical example of a linear polymerization (scheme 1) is observed with glutamate dehydrogenase. The kinetic studies of Thusius (81) have revealed that a random association model applies to this system, i.e. reaction not only occurs between monomer and polymer but between dimers and tetramers, etc. Thermodynamics can not make a distinction between the sequential and random association model.

In a detailed study of the thermodynamics of this system, Reisler and Eisenberg (82) showed that both ΔH and ΔS vary with temperature. We showed with high pressure light scattering studies (72) that ΔV changes from 33 ml at 10°C to 10 ml at 40°C. These volume changes are attributed to hydrophobic interactions at low temperature while at higher temperatures electrostatic interaction may become important. Low concentrations of toluene which at normal pressure enhance the polymerization, inhibit the dissociation at low pressure while at higher pressures (500-1000 atm) the association is enhanced. This effect is most pronounced at 25°C (70).

A recent report by Gauper et al. (83) shows that most of the temperature effects must be considered as caused by specific solvent-solute interactions. In the presence of dioxane or ethylene glycol the reaction becomes enthalpy driven while in the absence of these chemicals the reaction is entropy driven.

4.4 Microtubuli

Microtubuli are molecular fibres, present in all eukariotic cells, formed by the reversible polymerization of the protein tubulin. They have not only a structural role but their assembly and disassembly controls several cell functions.

Reversible pressure effects on important cell functions have been observed long before the existence of microtubuli was known and their involvement in cellular processes was recognized. The effect of pressure on mitotic spindle, cell motility, integrity of heliozoan axopoda, etc. have been interpreted as a disintegration of microtubuli (7).

O'Connor et al. (84) have found no effect of pressure on the stability of microtubuli of neuronal origin, in vivo as well as in vitro. These experiments were performed at 37°C. In collaboration with Dr. Engelborghs (85) we showed that at lower temperatures, where microtubuli of nonneuronal origin are usually studied, neuronal microtubuli also showed a strong pressure dependence (fig. 5).

Light scattered by microtubuli solutions is nearly independent of the length of the polymer so that it is a good mean to measure c_p, the number concentration of protomers present in the polymer. If the light scattered is plotted as a function of the

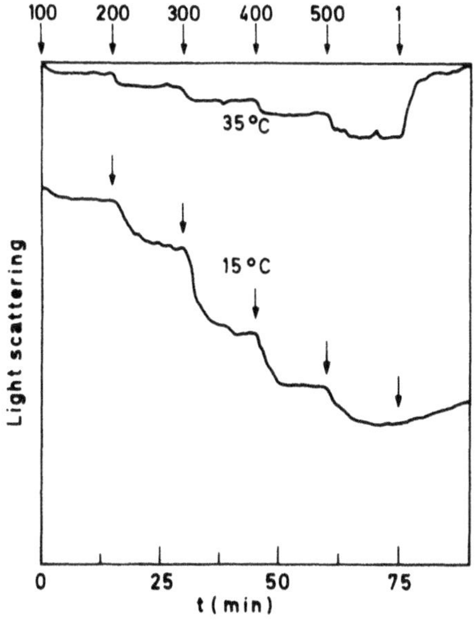

Fig. 5. The changes in scattering of microtubules solutions as a function of pressure and temperature.

total concentration of protein, a plot as in fig. 6 is obtained. It is seen that the system behaves like a helical polymerization, i.e. according to scheme (2) of section 4.2.

Engelborghs (85) showed that with the model of Oosawa (86) it is possible to analyze the system in more detail, i.e. in terms of the equilibrium constants for propagation and nucleation. The results show that the pressure effect is essentially linked to the nucleation at low temperature. The pressure dependence of the propagation equilibrium is small at all temperatures.

The results give some indication on the types of bonds involved in the assembly of microtubuli. The small pressure dependence at temperatures between 25 and 35°C is probably the consequence of the dominance of hydrophobic interactions which are usually associated with small volume changes; large volume changes are generally found in ionic interactions. It is thus possible that lowering the temperature decreases the stability of hydrophobic interactions and changes the conformation which leads to the exposure of charged groups.

A link between the pressure and temperature effect on protein interactions has been observed with other systems such as Flagellin (87) and Casein (88). The case of glutamate dehydrogenase has

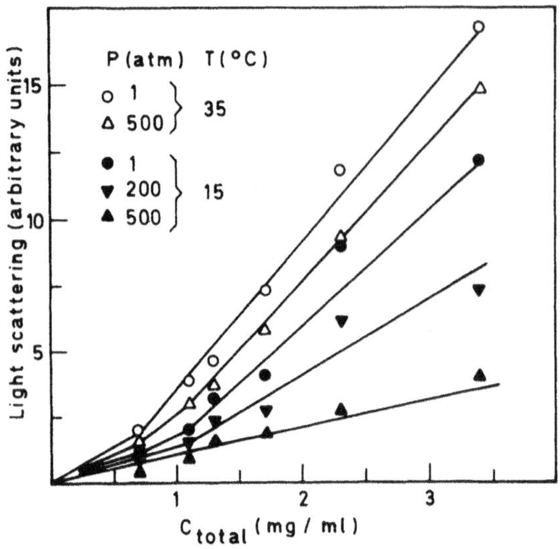

Fig. 6. Changes in scattering of microtubuli solutions at different total protein concentration as a function of temperature and pressure.

been discussed in section 4.3.

ACKNOWLEDGEMENT

The author acknowledges the encouragement and support from Professor Leo De Maeyer.

REFERENCES

1. A.G. Macdonald, Physiological aspects of deep sea biology, C.U.P., Cambridge, 1975.
2. M.A. Sleigh and A.G. Macdonald, eds, The effects of pressure on organisms, C.U.P., Cambridge, 1972.
3. P.W. Hochachka, Biochemistry at depth, Pergamon Press, Oxford, 1976.
4. R.E. Marquis, in : Advances in Microbial Physiology, 14, 159 (1976).
5. G.N. Somero and P.W. Hochachka, in : Adaptations to Environ-

ment, R.C. Newell, ed., Butterworths, London, 1976.
6. F.H. Johnson, H. Eyring and B.J. Stover, The theory of rate processes in biology and medicine, Wiley-Interscience, New York, 1974.
7. A.M. Zimmerman, ed. High Pressure Effects on Cellular Processes, Academic Press, New York, 1970.
8. K. Suzuki and Y. Taniguchi, Effect of pressure on Biopolymers and model systems, in ref. 2.
9. J.M. Rifkind and J. Applequist, J. Am. Chem. Soc., 90, 3650 1968.
10. S.J. Gill and R.L. Glogovsky, J. Phys. Chem., 69, 1515 1965.
11. K. Suzuki, Y. Miyosawa and Y. Taniguchi, J. Biochem., 69, 595, 1971.
12. H. De Smedt, H. Olbrechts and K. Heremans, Europhysics Conf. Abstr., 1A, 98, 1975.
13. B.H. Zimm and J.K. Bragg, J. Chem. Phys., 31, 526, 1959.
14. J. Applequist and V. Damle, J. Am. Chem. Soc., 88, 3895, 1966.
15. D. Marsh, A. Watts and P.F. Knowles, Biochim. Biophys. Acta, 465, 500, 1977.
16. J.F. Brandts, R.J. Oliveira and C. Westort, Biochemistry, 9, 1038, 1970.
17. S.A. Hawley, Biochemistry, 10, 2436, 1971.
18. S.A. Hawley and R.M. Macleod, Biopolymers, 16, 1833, 1977.
19. A. Zipp and W. Kauzmann, Biochemistry, 12, 4217, 1973.
20. J.F. Brandts, in Biological Macromolecules, eds., G. Fasman and S. Timasheff, Vol. 2., Marcel Dekker, 1969.
21. W. Kauzman, Proc. 4th Int. Conf. High Pressure, Kyoto, 1974.
22. T.M. Li, J.W. Hook III, H.G. Drickamer and G. Weber, Biochemistry, 15, 3205, 1976.
23. D. Pörschke and F. Eggers, Eur. J. Biochem., 26, 490, 1972.
24. K. Suzuki, Y. Taniguchi and T. Watanabe, J. Phys. Chem., 77, 1918, 1973.
25. U. Gaarz and H.D. Lüdemann, Ber. Bunsen. Phys. Chem., 80, 607, 1976.
26. S.A. Hawley and R.M. Mitchell, Biochemistry, 14, 3257, 1975.
27. T.M. Li, J.W. Hook III, H.G. Drickamer and G. Weber, Biochemistry, 15, 5571, 1976.
28. R.L. Baldwin, Ann. Rev. Biochem., 44, 453, 1975.
29. F.M. Pohl, Protein folding and unfolding, in : Chemical Relaxation in Molecular Biology, eds., I. Pecht and R. Rigler, Springer Verlag, Berlin, 1977.
30. D. Chapman, Quart. Rev. : Biophysics, 8, 185, 1975.
31. H. Träuble and D.H. Haynes, Chem. Phys. Lipids, 7, 324, 1971.
32. F. Ceuterick, K. Heremans, H. De Smedt, P. Nieuwenhuysen and J. Clauwaert, Meeting Belgian, Dutch and German Biophysics Societies, Liege, 1977.
33. J.F. Nagle, J. Chem. Phys., 58, 252, 1973.
34. N. Liu and R.L. Kay, Biochemistry, 16, 3484, 1977.

35. K. Heremans, Proc. Int. Conf. High Pressure, Boulder, 1977.
36. S. Mabrey and J.M. Sturtevant, Proc. Natl. Acad. Sci., 73, 3862, 1976.
37. P.L. Privalov and N.N. Khechinashvili, J. Mol. Biol., 86, 665, 1974.
38. E.J. Shimshick and H.M. McConnell, Biochemistry, 12, 2351, 1973.
39. J.R. Trudell, D.G. Payan and J.H. Chin, Proc. Natl. Acad. Sci., 72, 210, 1975.
40. R. Wattiaux, Mol. & Cel. Biochem., 4, 21, 1974.
41. F.H. Johnson and E.A. Flagler, Science, 112, 91, 1950.
42. M.J. Lever, K.W. Miller, W.D.M. Paton, et al., Nature, 231, 368, 1971.
43. R.D. Kaufman, Anesthesiology, 46, 49, 1977.
44. R.B. Gennis and A. Jonas, Ann. Rev. Biophys. Bioeng., 6, 195, 1977.
45. R.R. Eady and J.R. Postgate, Nature, 249, 805, 1974.
46. R.N.F. Thorneley, R.R. Eady and M.G. Yates, Biochim. Biophys. Acta, 403, 269, 1975.
47. F. Ceuterick, J. Peeters, K. Heremans, H. De Smedt and H. Olbrechts, submitted to Eur. J. Biochem.
48. R.B. Murphy and W.F. Libby, Proc. Natl. Acad. Sci., 73, 2767, 1976.
49. A. Péqueux, Proc. Int. Conf. High Pressure, Boulder, 1977.
50. A.G. Macdonald and K.W. Miller, Biochem. Biophysical Perspectives Mar. Biol., 3, 117, 1976.
51. K.J. Laidler, Arch. Biochem. Biophys., 30, 226, 1951.
52. H. Gutfreund, Prog. Biophys. Molec. Biol., 29, 161, 1975.
53. E. Morild, Biophys. Chem., 6, 351, 1977.
54. K.O. Greulich and H. Ludwig, Biophys. Chem., 6, 87, 1977.
55. G. Schmid, H.D. Lüdemann and R. Jaenicke, Biophys. Chem., 3, 90, 1975.
56. T. Ikkai and T. Ooi, Biochim. Biophys. Acta, 234, 190, 1971.
57. H. Gutfreund, Ann. Rev. Biochem., 40, 315, 1971.
58. K. Heremans, J. Snauwaert, H. Vandersypen and Y. Van Nuland, Proc. Int. Conf. High Pressure, Kyoto, 1974.
59. R.C. Neuman, Jr., D. Owen and G.D. Lockyer, Jr., J. Am. Chem. Soc., 98, 2982, 1976.
60. K. Heremans, J. Snauwaert and J. Rijkenberg, Proc. Int. Conf. High Pressure, Boulder, 1977.
61. R.C. Neuman, Jr., G.D. Lockyer, Jr. and J. Marin, J. Am. Chem. Soc., 98, 6975, 1976.
62. G.G. Hammes and C. Wu, Ann. Rev. Biophys. Bioeng., 3, 1, 1974.
63. F.G. Carey, F. Knowles and Q.H. Gibson, J. Biol. Chem., 252, 4102, 1977.
64. J. Peeters, A. Van Rossen and K. Heremans, in preparation.
65. T. Mustafa, T.W. Moon and P.W. Hochachka, Am. Zool., 11, 451, 1971.
66. H.J. Li and D.M. Crothers, J. Mol. Biol., 39, 461, 1969.

67. Y. Van Nuland and K. Heremans, in preparation.
68. D.R. Kearns, Ann. Rev. Biophys. Bioeng., 6, 477, 1977.
69. K. Suzuki and M. Tsuchyia, Bul. Chem. Soc. Japan, 44, 967, 1971.
70. K. Heremans, Verh. Kon. Acad. Wet. België, 1977.
71. K. Suzuki, Y. Miyosawa and C. Suzuki, Arch. Biochem. Biophys., 101, 225, 1963.
72. K. Heremans, Proc. Int. Conf. High Pressure, Kyoto, 1974.
73. W.F. Harrington and G. Kegeles, Methods in Enzymology, 27, 306, 1973.
74. P. Nieuwenhuysen, J. Clauwaert and K. Heremans, Arch. Int. Physiol. Biochim., 83, 983, 1975.
75. P. Nieuwenhuysen, K. Heremans and J. Clauwaert, in preparation.
76. A. Wishnia and A.S. Boussert, J. Mol. Biol., 116, 577, 1977.
77. G.H.B. Hoa, M. Graffe and M. Grunberg-Manago, Biochemistry, 16, 2800, 1977.
78. E. Schulz, R. Jaenicke and W. Knoche, Biophys. Chem., 11, 253, 1976.
79. A. Wishnia, B. Flaig and F.L. Lin, Proc. Int. Conf. High Pressure, Boulder, 1977.
80. J.W.H. Sutherland, Proc. Natl. Acad. Sci., 74, 2002, 1977.
81. D. Thusius, Glutamate Dehydrogenase Self-Assembly, in : Chemical Relaxation in Molecular Biology, Eds., I. Pecht and R. Rigler, Springer Verlag, Berlin, 1977.
82. E. Reisler and H. Eisenberg, Biochemistry, 10, 2659, 1971.
83. F.P. Gauper, K. Markau and H. Sund, Eur. J. Biochem., 49, 555, 1974.
84. T.M. O'Connor, L.L. Houston and F. Samson, Proc. Natl. Acad. Sci., 71, 4198, 1974.
85. Y. Engelborghs, K. Heremans, L. De Maeyer and J. Hoebeke, Nature, 259, 686, 1976.
86. F. Oosawa, J. Theor. Biol., 27, 69, 1970.
87. B.R. Gerber and H.J. Nochuch, J. Mol. Biol., 26, 197, 1967.
88. T.A.J. Payens and K. Heremans, Biopolymers, 8, 335, 1969.

KINETIC PHENOMENA IN GASES AT HIGH PRESSURE

Jürgen Troe

Institut für Physikalische Chemie, Universität Göttingen
Tammannstrasse 6, D-3400 Göttingen / Germany

ABSTRACT. The influence of density variations on elementary kinetic phenomena is considered. Starting from low pressure gases, some fundamental processes are followed up into highly compressed gases and liquids. Examples for photolytic cage effect, atom recombination, dissociation, isomerization, association, and hot atom reaction are discussed. Theoretical models such as many-particle trajectory calculations, diffusion equations for reactive systems, and modified versions of low pressure gas phase theories are presented.

1. INTRODUCTION

The influence of high pressures on kinetic phenomena in condensed phases has been studied in detail with conventional methods of chemical kinetics and high pressure technology (see e. g. the excellent reviews in refs. [1] - [4]). However, most of these earlier investigations provide only a rough and global picture of the elementary processes involved. The advent of new powerful experimental techniques promises a much more direct insight in the very details of elementary processes. The picosecond laser spectroscopy in particular will allow for time resolved studies on the scale of very fast intra- and intermolecular motions. Nanosecond laser spectroscopy provides a direct access to many fast elementary reaction steps. These laser techniques are particularly useful for high pressure experiments because of their geometry: the intense light beams can enter a reaction vessel easily through a small window.

Although only few experimental studies with these techniques have been accomplished as yet, they will lead soon to attempts to describe kinetic phenomena in compressed gases and liquids on a much more

Kelm, H. (Ed.) High Pressure Chemistry. 489–520. *Copyright © 1978. D. Reidel Publishing Company. All Rights Reserved.*

detailed level than this was done before. The classification of pressure effects on rate constants in terms of activation volumes $\Delta v^{\neq}$,

$$\left(\frac{\partial \ln k}{\partial P}\right)_T \equiv - \frac{\Delta v^{\neq}}{RT} \tag{1.1}$$

is very useful in practice. However, it gives only a macroscopic picture of the intermediate stages of a reaction. It is not clear to what extent the underlying concept of transition state theory holds for reactions in compressed gases and liquids. It also provides only little insight into the details of the elementary steps and motions involved. In this situation, new and more detailed experiments call for new models. Very promising are computer calculations of inter- and intramolecular trajectories in many-particle systems. Such calculations can be done for time scales which now also become accessible in experiments with pico- and nanosecond time resolution. One can also try to apply the classical theory of Brownian motion to reactive processes in dense media. In doing this, the "dynamical friction" of the medium has to be known. Trajectory calculations may provide a microscopic picture of this important quantity. With the diffusion-type equations from the theory of Brownian motion, the transition state theory can be controlled and its limitations identified. In this way, the meaning of activation volumes can be understood much more rigorously than on the basis of transition state theory.

In general, the understanding of low pressure gas phase reactions, as well on the experimental as on the theoretical side, is much further advanced than that of condensed phase reactions. Many reaction systems have been analyzed to their ultimate details. One might try to profit from these results in an attempt to understand compressed gas and liquid phase reactions. For instance, one could study a reaction in the gas phase, from low up to very high pressures, as well as in a liquid solvent at various pressures. In this way, the density of the medium can be varied over a large range and the typical high density effects are continuously introduced. This concept has been followed by the present author for a variety of elementary chemical reactions. It appears that paying more attention to such density variations, can lead to a better understanding of various kinetic phenomena in liquids. This will be demonstrated in the following by a number of representative examples of particularly simple elementary reactions.

2. PHOTOLYTIC CAGE EFFECT

Primary quantum yields of photolytic processes can be decreased by increasing density of the surrounding medium via several elementary mechanisms which cannot always be separated rigorously. One of these processes may be termed "photolytic cage effect". We under-

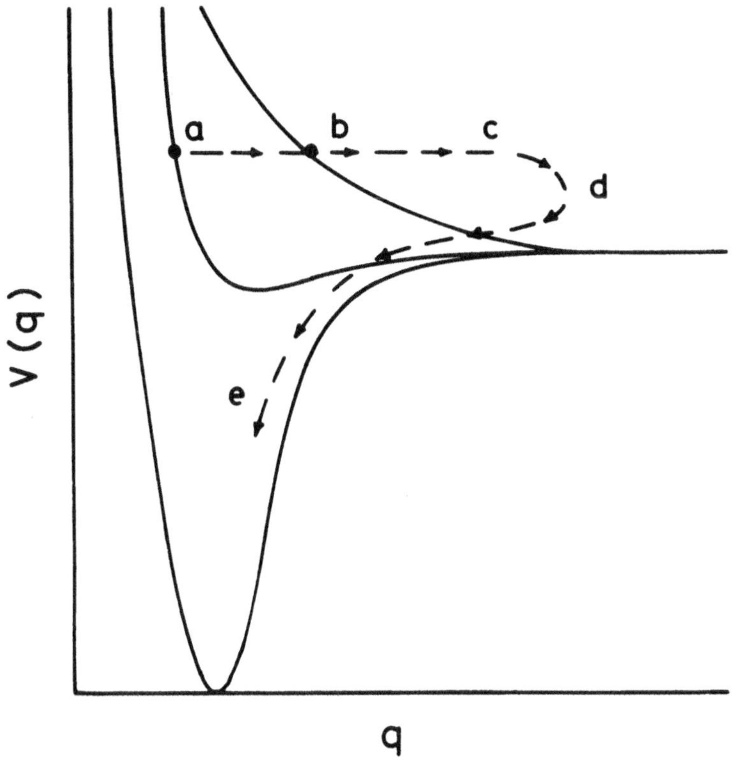

Fig. 1. Potential curves and schematic pathway for photolytic cage effect

stand by this a situation illustrated in fig. 1. After absorption of light, molecules can dissociate out of bonding (a) or repulsive (b) electronic states. The fragments start to separate (c); however, in a dense medium this motion can be reversed by collisions (d) with the surrounding molecules and the two initial fragments recombine soon into the bonding ground state (e). This process occurs on a picosecond time scale. It is to be distinguished from later recombinations of fragments from different initial molecules, generally involving longer migrations of the fragments and much longer time scales. It is also to be distinguished from collisional deactivation of excited molecules in the bonding range of the chemical forces. Obviously there is no well defined separation from the latter process in the case of very short lifetimes of the excited species.

One of the cleanest examples of a photolytic cage effect is probably observed in the photolysis of iodine molecules from the weakly bonding $A(^3\Pi_{1u})$ state. Light absorption in the 600 - 800 nm absorption quasi-continuum directly excites this state. Experiments

[5] have been performed by the use of radiation from a ruby laser at 694.3 nm (= 41.18 kcal mol^{-1} = 14403 cm^{-1}). The iodine dissociation energy E_o (= ΔH_o^o) is equal to 35.570 kcal mol^{-1} (= 12440.9 cm^{-1}). After light absorption, the separating iodine atoms therefore carry 5.6 kcal mol^{-1} and some additional energy from the thermal rovibrational preexcitation (about 1 kcal mol^{-1}) in the form of relative translational energy. The quantum yield Φ of this photolysis in low pressure gases is certainly equal to 1, although it was never measured directly. Flash experiments in steel reaction cells with sapphire windows allow the measurement of the change of this quantum yield at high pressures of added inert gases. To do this, the amount of iodine decomposed is observed directly ($\simeq$ 100 ns) after the flashes from the ruby laser, before the later recombination of separated atoms sets in. Whereas no pressure influence of the quantum yield $\Phi(P)$ could be detected in He, Ne, Ar, and H_2 at pressures up to 1000 atm, a decrease was observed in Kr, Xe, O_2, CO_2, CH_4, C_2H_6, and C_3H_8 [5].

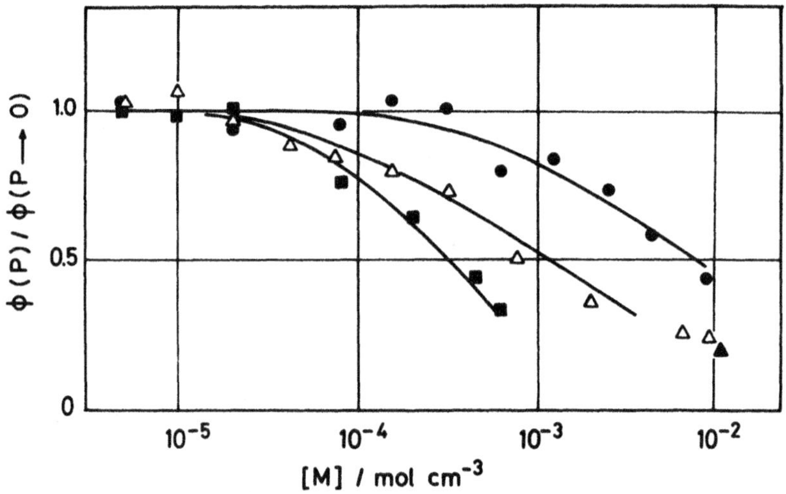

Fig. 2 Photolytic cage effect of iodine at 694.3 nm in high pressure CH_4 (●), C_2H_6 (△), C_3H_8 (■) at 41 °C, and in liquid C_2H_6 (▲) at 25 °C; from ref. [5].

Fig. 2 shows the onset of this photolytic cage effect of iodine near one atmosphere in CH_4, C_2H_6, and C_3H_8 and its continuation down to quantum yields near 0.2 at higher gas pressures, extending up to liquid densities for C_2H_6. Interesting enough, experiments in gaseous C_2H_6 at 41 °C and $[C_2H_6] = 10^{-2}$ mol cm^{-3}, and in liquid C_2H_6 at 25 °C and the same density, give the same quantum yield of 0.2. Earlier steady-state experiments in the liquid phase, with added scavengers, led to analogous values of the quantum yield, e.g. $\phi = 0.13$ for liquid hexane at 643 nm [6]. In compressed gases, the photolytic cage effect was shown to have only small temperature coefficients between 30 and 100 °C [7]. It would of course be interesting to study the wavelength dependence of the photolytic cage effect in compressed gases. Such experiments are now becoming feasable by the use of tunable dye lasers which again allow for direct quantum yield measurement after rapid ($\simeq$ 1 μs) flash photolysis [8]. These experiments can be extended to wavelengths near the iodine photolysis threshold at 803.8 nm. The onset of the photolytic cage effect, because of the small translational energy of the separating iodine atoms, is then expected to occur at lower inert gas pressures.

New and very detailed insight into the photolytic cage effect stems from laser picosecond spectroscopy. Although the only available experiments have been performed for iodine in liquid CCl_4 and hexadecane, they are relevant to the phenomenon in highly compressed gases as well, and they will certainly soon be extended to high pressure gas phase studies. In these experiments [9], iodine is excited by picosecond pulses (pulse width $\simeq$ 5 ps) from a frequency doubled Nd^{+3}-glass laser at 530 nm. The major part of the light absorption leads into the bonding B $(^3\Pi_{o^+u})$ state, a smaller part into the repulsive 1 u$(^1\Pi)$ state. The transmission of the reaction cell at 530 nm is followed by an optically delayed portion of the light pulse. With variable optical delays of the probe pulses, the transmission and hence the concentration of absorbing I_2 molecules is monitored from 0 to 800 ps after the excitation flash, see fig. 3. During the first 20 ps after the short flash, a transmission rise is observed which is attributed to collision induced predissociation from the bonding B $(^3\Pi_{o+u})$ state (which probably also absorbs at 530 nm) possibly into the repulsive 1 u $(^1\Pi)$ state. The subsequent decrease of the transmission is then attributed to the reappearance of the ground state I_2 molecules, after the iodine atoms have separated, migrated in the cage and recombined. Quite clearly, the cage "geminate" recombination at the 100 ps time scale can be distinguished from the ordinary "non-geminate" recombination which would occur at the μs time scale. The remaining change of the transmission signals after about 1 ns corresponds to those iodine atoms which have escaped from the cage. This determines the resulting quantum yield of the photolytic cage effect which can also be measured with nanosecond laser flashes as described earlier.

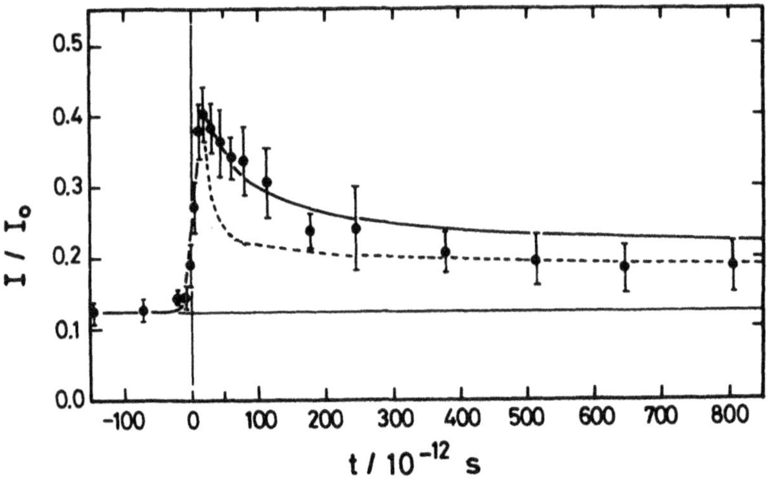

Fig. 3 Photolytic cage effect of iodine at 530 nm in liquid CCl_4; from ref. [9]. Solid and dashed curves from "random flight in a continuum" model of ref. [11].

As a first approximation, one can try to understand the density influence on the photolytic cage effect, its quantum yield and its time dependence, by the "random flight in a continuum" model of Noyes [10]-[12]. The probability of recombination at a given time t after the flash from the theory of random flights is given [11],[13] by a function

$$h(t) \simeq \frac{a \exp(-b/t)}{t^{3/2}} \tag{2.1}$$

where the parameters a and b are related to the encounter diameter, and the frequency and distance of diffusive displacements. The solid and dashed curves in fig. 3 have been fitted to the experiments [9]; they correspond to jump sizes of 0.5 - 1 Å and jump frequencies of (1 - 5) x $10^{12} s^{-1}$ in the solvent CCl_4. In spite of the simplicity of the model, qualitative agreement is produced with not unreasonable values of the input parameters.

The quantum yield for the cage effect in the "random flight in a continuum" model, in its simplest form, is given by [12]

$$\phi = 1 - \left(1 + \frac{\sqrt{m(E-E_o)}}{6\pi\eta a^2}\right)^{-1}\left(1 + \frac{\sqrt{mkT/24}}{\pi\eta a^2}\right)^{-1} \qquad (2.2)$$

Here $E - E_o$ denotes the initial excess kinetic energy of the separating atoms, η is the viscosity of the surrounding medium, m is the mass of the atoms and a their radius. Eq. (2.2) has been shown to predict experimental values reasonably well for a number of liquid solvents [12]; however, at small $E - E_o$, it was always underestimating the extent of cage effect. As an example, in liquid hexane ($\eta = 29 \cdot 10^{-4}$ poise, T = 298 K) a quantum yield of 0.13 was measured at λ = 643 nm, whereas eq. (2.2) predicts a quantum yield of 0.4. This discrepancy was attributed to "proximity effects"; at small excess kinetic energies $E - E_o$, the atoms do not separate very far; the surrounding molecules reflect the atoms back toward each other, and the continuum model looses its significance. The same observation is made, if eq. (2.2) is applied to the high pressure gas phase and liquid phase experiments of fig. 2. For high pressure gas phase ethane at a density of 10^{-2} mol/cm^3 (at 41 °C) and for liquid ethane at the same density (at 25 °C) one estimates [14] a viscosity of $3.6 \cdot 10^{-4}$ poise. For λ = 694.3 nm, eq. (2.2) with this viscosity predicts ϕ = 0.8, substantially larger than the value of 0.25 from fig. 2. Nevertheless, eq. (2.2) qualitatively gives the observed decrease of ϕ with increasing density and hence increasing viscosity.

The quantitative failure of the simple continuum model stimulated molecular dynamics type calculations of the cage effect in dense gases and liquids [15].[16]. In ref. [15], two iodine atoms and 26 solvent molecules were placed into a cubic box with a 14 Å edge, corresponding to liquid CCl_4 densities. Lennard-Jones potentials for the I - CCl_4, and CCl_4 - CCl_4 interactions and a I - I Morse potential were used. Then, trajectories of the iodine atoms and the solvent molecules were calculated using specularly reflecting walls of the box. Periodic boundary conditions were believed to require at least 100 solvent molecules and were not used in this work. This treatment indeed indicated smaller quantum yields than predicted by eq. (2.2), although most attention was directed to the distinction between primary and secondary cage effects and the time dependence of the iodine displacements. In ref. [16], a direct attempt was made to compare the quantum yields with the results from highly compressed gases [5]. Here, two iodine atoms and 22 Kr solvent atoms were placed in spherical boxes, with reflecting walls, and with diameters chosen to give the considered densities. The trajectories have been calculated up to 10^{-11}s after excitation. with light of wavelength of the ruby laser. The quantum yields obtained are shown in fig. 4 and compared with the experiments of ref. [5].

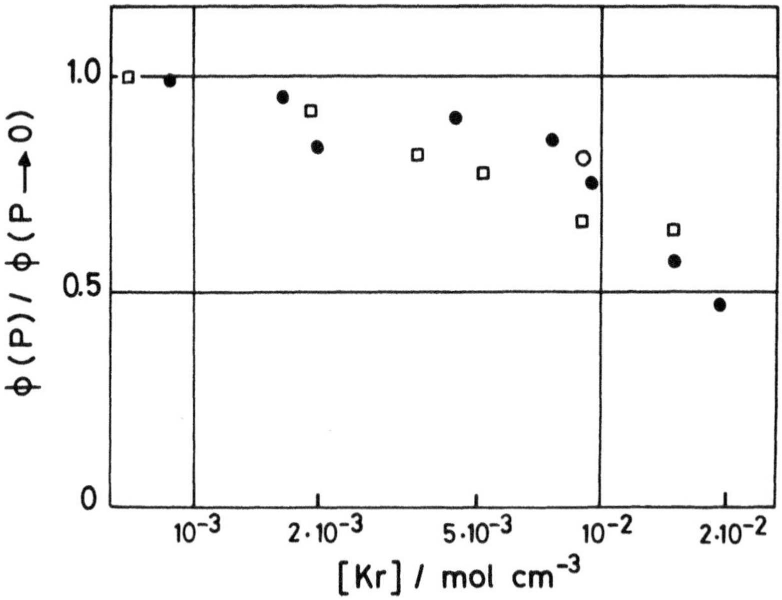

Fig. 4 Photolytic cage effect of iodine in Kr at 694.3 nm.
● = experiments, 41 °C, from [5]; □ = theory, 41 °C, from [16]
○ = theory, 71 °C, from [16].

The agreement is remarkly good, although the possibility of secondary cage effects at later times cannot be ruled out. Common to the two trajectory studies of refs. [15], [16] is the observation that the main part of the cage recombination occurs within a few ps, whereas fig. 3 showed slower cage recombination at the 100 ps scale. This discrepancy may be due to the mentioned delayed collisional predissociation observed in the experiments at 530 nm.

Experiments at longer wavelengths with ps time resolution are therefore required. Furthermore, since highly excited ground state molecules of I_2 may have a different spectrum than normal I_2, the time profiles in fig. 3 may correspond to cage recombination followed by vibrational deactivation of the highly excited molecules which can take a rather long time.

It is quite evident that the photolytic cage effect, because of its simplicity and easy accessibility, will also in the futur play a central role in the understanding of cage phenomena in compressed gases and liquids. The need to single out an independent variable, which can be varied over broad ranges, has been emphasized earlier (see e. g. ref. [15]). Obviously, the density is the variable of choice. Only little information on cage phenomena in a medium of variable density is available for other reaction systems. One of the few existing examples is the study of cage recombination of methyl radicals in propane 1.2 degrees above the critical temperature 96.8°C with $[C_3H_8] = 2 \cdot 10^{-3} - 6 \cdot 10^{-3}$ mol cm^{-3} [17].

3. ATOM RECOMBINATION AND DISSOCIATION OF DIATOMICS

Atoms, which have been produced by flash photolysis and which have escaped from the solvent cages as described in section 2, can recombine by "volume recombination". In the liquid phase, this process is considered a diffusion controlled reaction which obeys a second order rate law. At low pressures in gases the reaction order is three; this is interpreted in terms of the energy transfer and the intermediate complex mechanisms again with the iodine system: the rate determining step of the energy transfer mechanism at low pressures is the collisional stabilization of unstable I_2^* pairs in collisions with surrounding molecules M; the rate determining step of the intermediate complex mechanism at low pressures is the exchange reaction of iodine atoms with weakly bound IM complexes; in the former mechanism, a $I + I \rightleftharpoons I_2^*$ preequilibrium is established, in the latter a $I + M \rightleftharpoons IM$ preequilibrium. It is, of course, interesting to follow the changes of the elementary mechanism from low pressure gases via high pressure gases to the liquid phase. Such experiments have been performed in ref. [18] with ruby laser flash photolysis, similarly as described in section 2. The volume recombination occurs at the μs time scale, i. e. much slower than cage recombination. Fig. 5 shows a representative example of the second order recombination rate constant of iodine atoms in argon. In the log-log representation, the 45° degree slope at low pressures corresponds to a third order rate law; the reaction approaches a second order rate law at high pressures. This change of reaction order at densities of the order of liquid densities may be understood, from the viewpoint of the liquid phase reaction, as a control of the reaction by diffusion of iodine atoms toward each other. From the viewpoint of the gas phase reaction, it may be

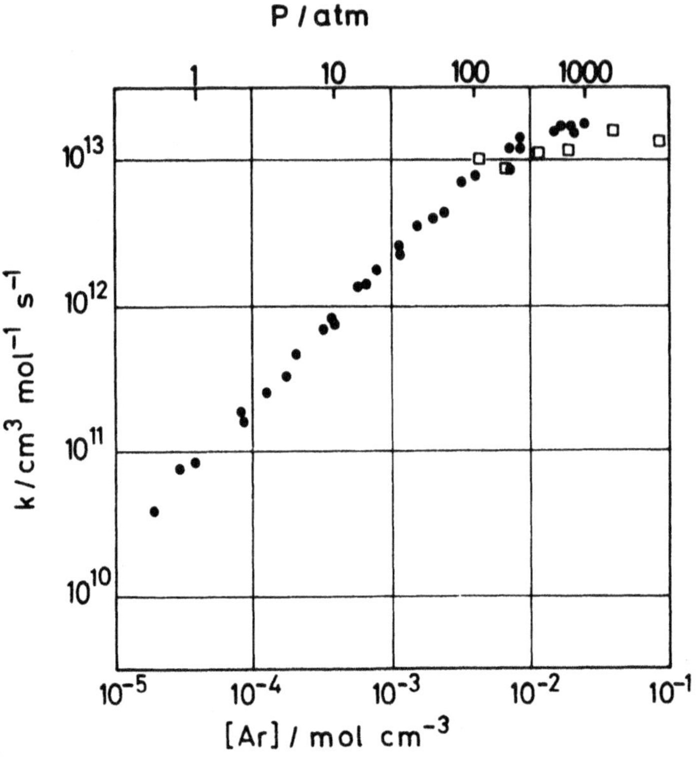

Fig. 5 Recombination of iodine atoms in compressed Ar at 41 °C. ● = flash photolysis experiments from [18], □ = many-body trajectory calculations from [20].

understood from the assumption that all free iodine atoms are attached by weak bonds to single particles or "many-body collision complexes" of the surrounding species M, such that collisions or encounters of the "I-M_n" complexes become rate determining. From the density, where the change of the reaction order occurs, one obtains some information on the equilibrium constant for some average "I-M_n complex" or "I-M_n many-body collision pair". For Ar one has $[IAr_n]/[I][Ar_n] = K \simeq 130\ cm^3\ mol^{-1}$. This value is considerably smaller than it was derived initially for I-Ar complexes from the low pressure rate constant. However, molecular beam scattering experiments between iodine and rare gas atoms have confirmed the only weak van der Waals type attraction. Nevertheless, it should be emphasized that the pressure dependence of the recombination rate constant reflects only some average interaction changing from two-body toward many-body collision pairs. In the experiments of ref. [18] besides Ar, the inert gases He, Ne, Kr, Xe, H_2, N_2, O_2, CO_2, CH_4, C_2H_6 and C_3H_8 have been studied. We do not want to go into further details, but repeat that one can obtain from the pressure dependence of all of these experiments information on the weak binding forces between the iodine atoms and the "solvent" molecules.

The pressure dependence of the recombination rate constant in high pressure gases can be interpreted on two levels of sophistication. Starting from low pressure gases a sequence of formal reaction steps of the type $I + M_m \rightleftharpoons IM_m$, $I + M_n \rightleftharpoons IM_n$, $IM_n + IM_m \rightleftharpoons I_2 + M_{n+m}$ can be written up and solved. Of course, these steps have only a symbolical meaning, reflecting some average transition from binary to many-body collisions [18]. Considering for the moment only the simplest steps

$$I + Ar \longrightarrow IAr \tag{3.1}$$

$$IAr \longrightarrow I + Ar \tag{3.2}$$

$$IAr + I \longrightarrow I_2 + Ar \tag{3.3}$$

$$IAr + IAr \rightarrow I_2 + 2\,Ar \tag{3.4}$$

one obtains a rate law

$$\frac{d[I_2]}{dt} = \frac{K_1[Ar]\left(k_3 + k_4 K_1[Ar]\right)}{\left(1 + K_1[Ar]\right)^2}\left([I] + [IAr]\right)^2 \tag{3.5}$$

where $K_1 = k_1/k_2$. This rate law shows a change of the reaction order from three to two, typically at $[Ar] \simeq K_1^{-1}$. Depending on the values of k_3 and k_4, also higher reaction orders than three can occur. Including higher collision pairs leads to more complicated expresious where the various rate constants are difficult to estimate. If one assumes, however, that k_3, k_4, and the corresponding rate constants of higher complexes are all equal, and, on the other hand, that K_1 and the analogous equilibrium constants of higher complexes are equal, the problem can be handled. In this way, the value of $K_1 \simeq 8 \cdot 10^{-3}$ mol cm^3, mentioned above for M = Ar, was determined.

On the other hand, the limiting diffusion controlled reaction can be described [12] in terms of effective diffusion coefficients,

$$k \simeq 4\pi \rho D N_L \tag{3.6}$$

where D is the diffusion coefficient of the combining species in the solvent, ρ their collision diameter, and N_L is Loschmidt's number. For iodine recombination in liquid CCl_4 at 25 °C, $\rho \simeq 4.8 \times 10^{-8}$ cm, $D \simeq 6 \times 10^{-5}$ $cm^2\,s^{-1}$ and hence $k \simeq 1.9 \times 10^{13}$ $cm^3\,mol^{-1}s^{-1}$.

This value agrees well with the limiting high pressure rate constants in compressed gases such as demonstrated in fig. 5. However, neither the effective diffusion coefficient D nor the effective collision diameter ρ can be interpreted rigorously on a molecular, microscopic basis. Therefore, both of these simple treatments have only a "symbolic" meaning and it is not clear to what extent their language corresponds to reality. Again, as for cage recombination, many-particle trajectory calculations can give an answer. It should be emphasized that the typical low density trajectory calculations, involving two-, three- and four-particle systems only [19], are hardly more useful than the simple model described above. However, trajectories have now been calculated [20] with more particles in the test volume and with realistic I - I, I - Ar, and Ar - Ar interaction potentials. These calculations cannot be extended down to low densities, because of the increasing length of the trajectories to be calculated. However, at high densities they give remarkably good agreement with the experiments of fig. 5.

An analysis of the many-particle trajectory calculations with respect to the number of atoms present in the recombing collision pair is of particular interest. This analysis has been made in ref. [20] and is illustrated in fig. 6. For the conditions of fig. 5, the percentage contribution of IAr_n - IAr_m collision pairs as intermediate complexes in the recombination processes is indicated; the IC (Ar_3) curve, e. g. corresponds to n + m = 3. The contribution from the energy transfer mechanism

$$I + I \longrightarrow I_2^* \tag{3.7}$$

$$I_2^* \longrightarrow I + I \tag{3.8}$$

$$I_2^+ + Ar \rightarrow I_2 + Ar \tag{3.9}$$

is also indicated (ET). Quite obviously, many-particle collision events become increasingly important in the intermediate complex mechanism at high densities. The decreasing importance of the energy transfer mechanism is also interesting to note. This is probably due to the decreasing number of free iodine atoms; in other words, the energy transfer and the intermediate complex mechanism are coupled and cannot be considered separately.

With increasing temperature, the concentration of bound I - Ar_n pairs is expected to decrease. Unbound I - Ar_n collision pairs then take their place to a certain extent, but more free atoms, and a larger contribution from the energy transfer mechanism, can also be expected. Experiments at high temperatures and high pressures are more difficult to perform. In one series of experiments, the thermal dissociation of iodine in Ar has been studied in high pressure shock waves near 1100 K [18] . In these experiments, for the first time a deviation from a second order rate law in a

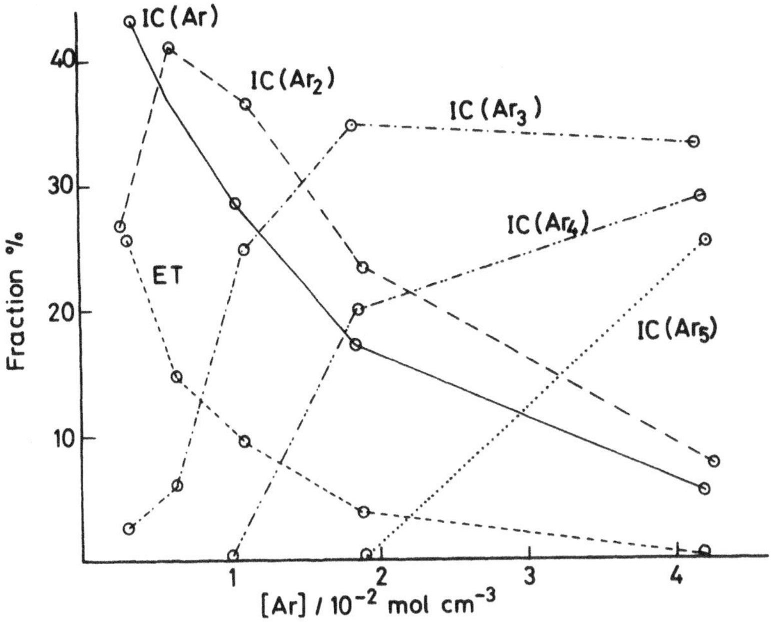

Fig. 6 Percentage contribution of energy transfer mechanism (ET) and higher complexes in the intermediate complex mechanism (IC) in iodine recombination at 300 K, from calculations in [20] (Ar_n indicates the number of Ar atoms in the recombination collision pair).

dissociation of a diatomic molecule at high pressures was observed. The results have been transformed via the equilibrium constants into recombination rate constants and represented in fig. 7. They are compared with many-body trajectory calculations from ref. [20]. The corresponding analysis with respect to I - Ar_n many-body collision pairs is given in fig. 8. It appears that at higher temperatures the increase in the importance of higher complexes is shifted to high densities. This observation can be attributed to the change in the ratio kT/ε, where ε denotes the depth of the I - M potential, since the results with M = He at 300 K and with M = Ar around 1200 K give quite similar results. In spite of the

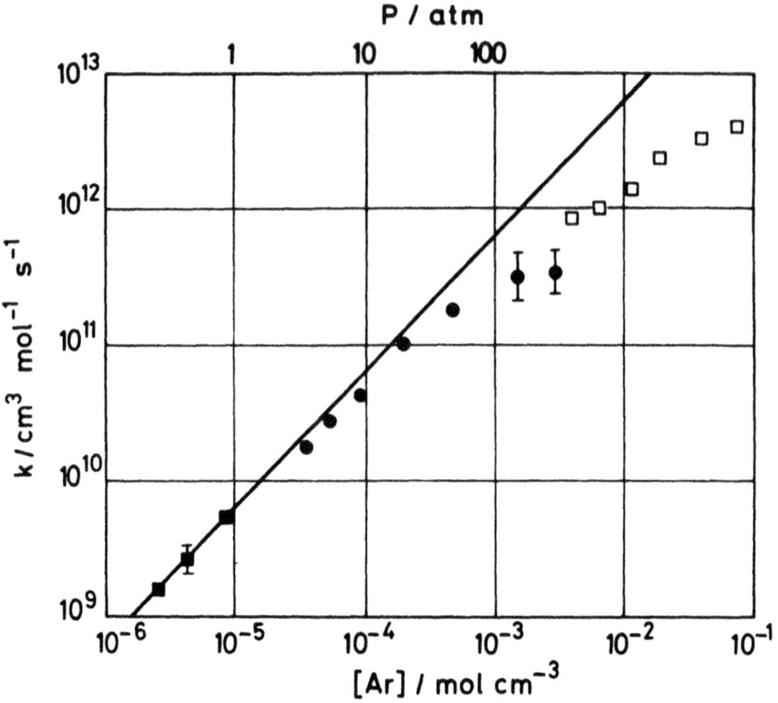

Fig. 7 Dissociation of iodine molecules in Ar at 1060 K. (■,● from [18], via equilibrium constants transformed into recombination rates; □ many-body trajectory calculations at 1173 K, from [20]).

good agreement between experiment and theory, one major and hitherto unsolved problem of the calculations has to be mentioned. In the calculations for 300 K a fraction of 5/16 of the collisions where assumed to contribute to the recombination in contrast to 1/16 around 1200 K. This is because of the presence of altogether 16 electronic states correlating with electronic ground state atoms, one of which is the strongly bonding electronic ground state, and four of which are presumably weakly bonding, contribute at low, but do not contribute at high temperatures. Clearly, the unknown role of these states can be of crucial importance.

It should be mentioned that more complex density dependences of the recombination rate constant can occur. The recombination of iodine atoms in propane [18] seems to show a reaction order of three below [M] = 10^{-4} mol cm^{-3}, and a reaction order of about 3.3 between 10^{-4} and 10^{-3} mol cm^{-3}. The high value of the rate constant of

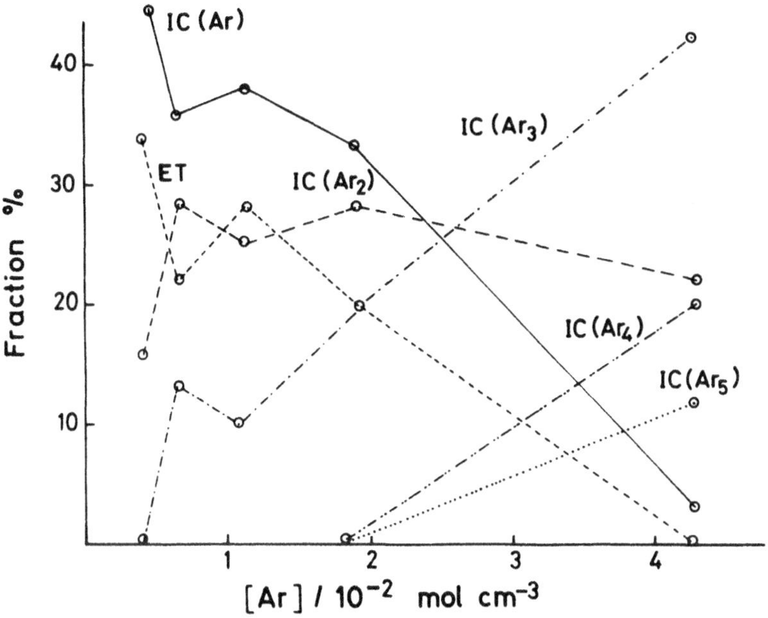

Fig. 8 As fig. 6 for 1173 K

$2 \cdot 10^{13}$ cm^3 mol^{-1} s^{-1} is reached already at [M] = 10^{-3} mol cm^{-3}. Below the critical temperature, the recombination value in the liquid phase at 10^{-2} mol cm^{-3} is also about $2 \cdot 10^{13}$ cm^3 mol^{-1} s^{-1}, which implies that at densities between 10^{-3} and 10^{-2} mol cm^{-3} the recombination rate constant must run through a more or less pronounced maximum. The recombination of chlorine atoms in high pressure CO_2 and N_2 also seems to show a maximum in the rate constants at pressures around 10 atm [21]. These surprising effects, which might be attributed to particular "chemical" interactions between the "solvent" molecules and the reacting atoms, have to be verified by further experiments. Finally, further many-body trajectory calculations of H atoms in high pressure gases should also be mentioned [22].

IV. QUENCHING OF DISSOCIATION AND ISOMERIZATION PROCESSES

In the dissociation of a diatomic molecules, the dissociative elementary step is the separation of the two atoms on a Morse type two-body potential. In polyatomic molecules, the presence of several coupled oscillators introduces a time lag into the reaction such that the final separation of the two fragments generally is proceeded by a much slower intramolecular rearrangement process. Intermolecular energy transfer in collisions can remove the excitation from the excited molecule during this rearrangement process, and, therefore, lead to quenching of the product formation. This phenomenon is found in systems with thermal, chemical, photochemical, and other types of activation; in the present section we shall consider optical and chemical activation only.

An efficient quenching of the dissociation or isomerization process occurs when the collision frequency of the excited molecules becomes larger than the rate of the intramolecular rearrangement process. The latter rate can well be characterized by the theory of unimolecular processes [23]-[25] which gives the specific rate constant for a unimolecular process by

$$k(E) = \frac{W^{\ddagger}(E-E_0)}{h\,\rho(E)} \qquad (4.1)$$

E is the vibrational energy of the molecule, E_0 is its threshold energy for dissociation or isomerization, h is Planck's constant, $\rho(E)$ is the density of states and $W^{\ddagger}(E - E_0)$ is the number of open vibrational reaction channels for product formation. To a good approximation (except at $E \simeq E_0$), eq. (4.1) can be represented by

$$k(E) \simeq \bar{\nu} \left(\frac{E-E_0+a(E-E_0)E_z^{\ddagger}}{E+a(E)E_z} \right)^{s-1} \qquad (4.2)$$

where $\bar{\nu}$ is a suitable average frequency, s is the total number of oscillators of the molecule, E_z is the zero point energy of the vibrations, a(E) is the Whitten-Rabinovitch correction factor, the symbol $\neq$ denotes "activated complex" properties. We have performed a representative set of calculations with $\bar{\nu} = 10^{14}\ s^{-1}$, $E_0 = 50$ kcal mol^{-1}, $E_z = s \times 1.924$ kcal mol^{-1}, $E_z^{\neq} = (s - 1) \times 1.924$ kcal mol^{-1}, and $a(E) = 1 - \beta w$, where $\beta = 1.419$ and $\log w = -1.0506 \times (E/E_z)^{0.25}$; (for $E = E_0$, $k(E) = 1/h\,\rho(E)$ has been evaluated separately). Fig 9 shows the resulting specific rate constants k(E).

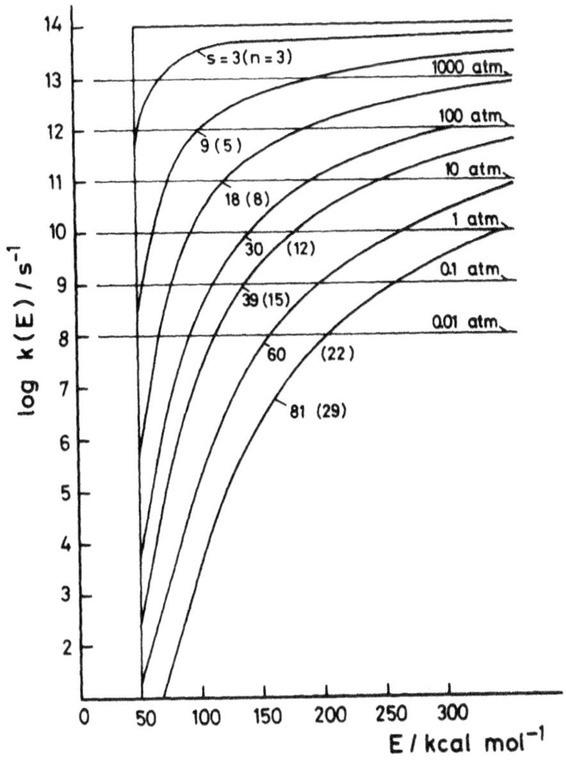

Fig. 9 Simplified statistical calculation of specific rate constants for unimolecular reactions, eq. (4.2); see text.

For simplicity, we further assume that the rate of collisional stabilization is equal to the collision frequency (in reality, a collision efficiency factor γ between about 0.01 and 1 accounts for stepwise energy removal [26]). Finally, we assume that the collision frequency concept can still be used in high pressure gases at pressures up to several hundred atm. We use a standard collision frequency Z of about 2.5 x 10^{14} cm^3 mol^{-1} s^{-1}. Then, the crossing of the horizontal lines in fig. 9 with the k(E) curves indicates, up to which energies collisional quenching of the unimolecular process is efficient at a given pressure. Although, the present arguments give only order of magnitude estimates, the general trends can nicely be recognized. In order to stabilize a large polyatomic molecule up to high energies, small pressures are sufficient. Small molecules, on the other hand, require high densities for efficient collisional stabilization.

At first we illustrate this general behaviour by two photochemical examples. The quantum yield for reaction, in the first approximation, here is given by

$$\phi \simeq \frac{k(E)}{\gamma Z[M] + k(E)} \tag{4.3}$$

Represented in the usual Stern-Volmer form of ϕ^{-1} as a function of [M], this gives a straight line with the slope of $\gamma Z/k(E)$. Fig. 10 shows Stern-Volmer plots for NO_2 photolysis at wavelengths below about 400 nm [27]. As indicated qualitatively in fig. 9, very high gas pressures are required to quench the photolysis. Near the threshold energy (λ near 400 nm), some ten atm

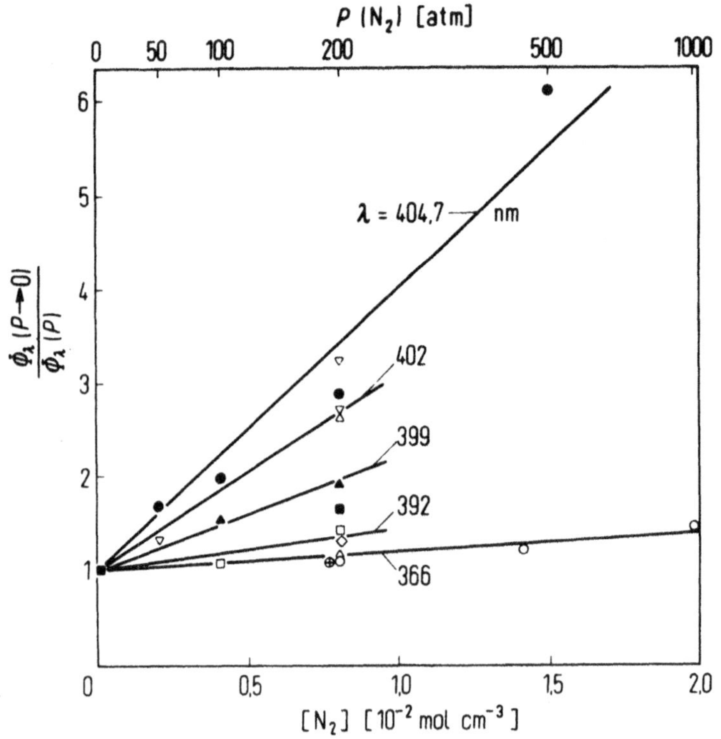

Fig. 10 Stern-Volmer plots of the primary quantum yield for photolysis of NO_2 in N_2; from [27] (symbols not connected by lines correspond to intermediate wavelengths).

are sufficient for quenching; near 360 nm several hundred atm are required; near 310 nm even one thousand atm do not suffice to quench the excited molecules because of their short lifetimes near 10^{-13}s. As long as no more direct method for measuring these short lifetimes is available, the "collision frequency clock" proves very useful indeed, although the limitations at high densities are quite clear. Nevertheless, this concept seems to operate up to very high densities. The full analysis of the data for fig. 10 gives specific rate constants as indicated in fig. 11. It would of course be interesting to continue these experiments to pressures in the 10 000 atm range in order to see where molecules excited at 310 nm start to be quenched. Possibly, collisional quenching and photolytic cage effect then overlap .

Another representative example, the photoisomerization of cyclo-heptatriene toward toluene, is illustrated in fig. 12. In the

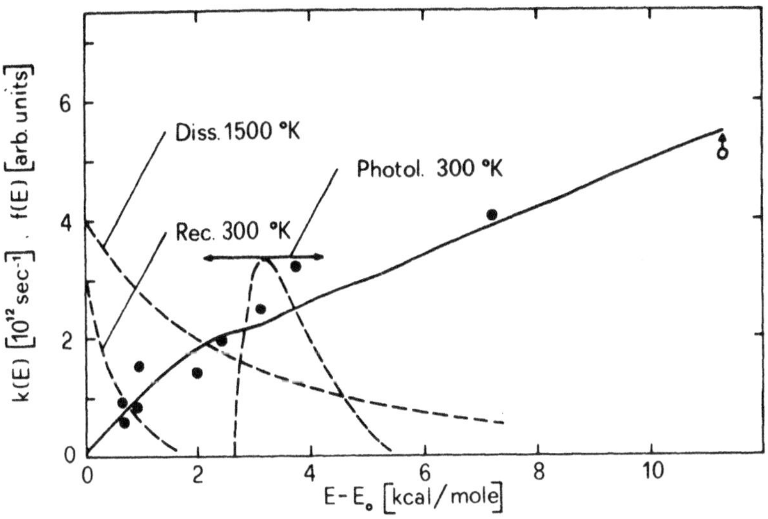

Fig. 11 Specific rate constants for dissociation of NO_2 near E_o = 71.86 kcal mol^{-1}; from [28]. Points = experiments, solid line = calculations from [35], dashed lines = populations of excited states in thermal dissociation, thermal recombination and photolysis.

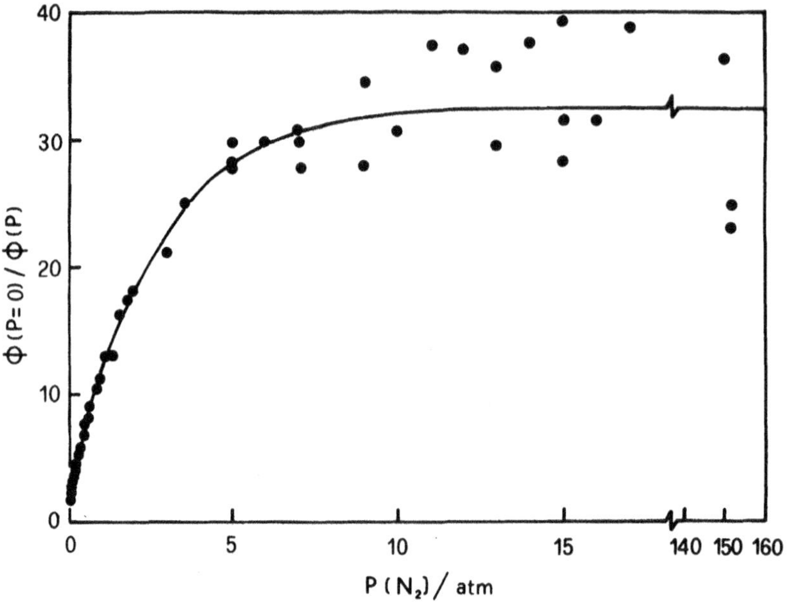

Fig. 12 Stern-Volmer plots of the quantum yield of photoisomerization of cycloheptatriene in N_2 at 265 nm, from [26].

center of the nearest UV absorption spectrum of this molecule near 260 nm, excited molecules with photoisomerization lifetimes near 10^{-8} s are produced. Accordingly, pressures below one atm suffice to quench the photoisomerization. The nearly linear Stern-Volmer relationship is followed down to quantum yields (for disappearance of the molecules) near 5 percent. Then, the curve levels off, and the quantum yield remains constant at about 3 percent. The same quantum yield is also observed in the liquid phase. Apparently, one has a second underlying reaction channel which cannot be influenced at ordinary densities. This is believed to be a photoisomerization leading to a bicyclic product. One has to conclude that this process involves a non-bonding higher electronic state with only very short lifetimes of the excited parent molecules.

In the following we discuss a chemical activation system. By addition of H atoms to butenes, chemically activated butyl radicals are formed which can undergo subsequent collisional stabilization or dissociate toward C_3H_6 and CH_3. Here, usually the ratio of dissociated (D) to stabilized (S) products is represented. In the simple model, used for photochemical activation, this would be given by

$$\frac{D}{S} = \frac{k(E)}{\gamma Z [M]} \tag{4.4}$$

Representing

$$k_a \equiv \langle k(E) \rangle \equiv Z [M] D/S \tag{4.5}$$

as a function of pressure, should therefore be a constant. The fact, that some pressure dependence is observed, is contributed to the pressure dependence of the formal collision efficiency and, therefore, to the stepwise collisional stabilization process. Fig. 13 shows a summary [29] of extensive experiments in the inert gas H_2 which have been performed from about 10^{-3} Torr up to about 200 atm. The mentioned pressure dependence of γ shows up as "turn-up" of the curves near 1 Torr. One may conclude, from the nearly

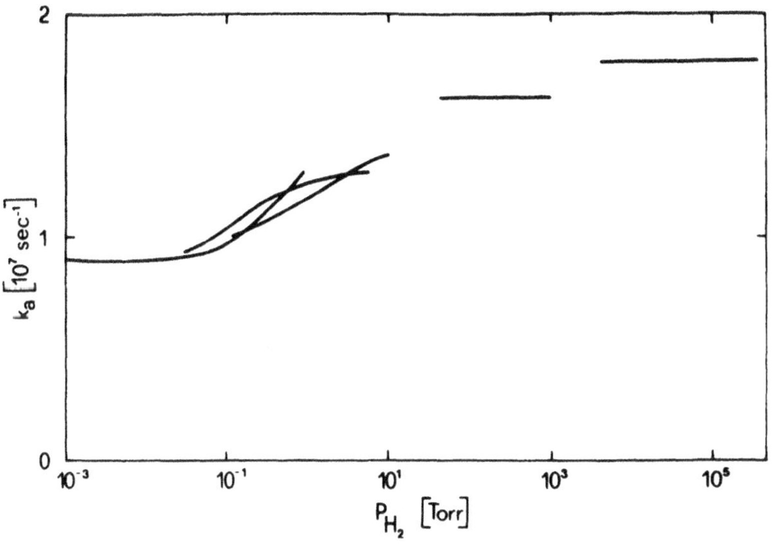

Fig. 13 Effective rate constant for decomposition of chemically activated sec-butyl radicals in H_2 (several sets of experiments, from [29]).

constant value of k_a from 1 to 200 atm, that the collision frequency concept essentially holds for this pressure range, in agreement with the conclusions on the NO_2 system. Furthermore, k(E) in this system must correspond to a nearly statistical distribution of excited molecules with randomized energies. This point will be explained in more detail in the following section.

Highly excited polyatomic molecules can also be produced, by another type of "chemical activation", in recoil hot atom substitution reactions. If, e. g. recoil tritium atoms collide with hydrocarbons, they can exchange with hydrogen atoms, such that highly excited tritium containing molecules are formed. These again can undergo collisional stabilization, or they break up and form other products. Different from photoactivation and ordinary chemical activation, broad distributions f(E) of excited states are produced in the tritium exchange process. Therefore, a broad spectrum of specific rate constants k(E) has to be considered. The pressure dependence of the product yields depends on both of these quantities. This is illustrated in fig. 14 for recoil tritium substitution with CH_4 and

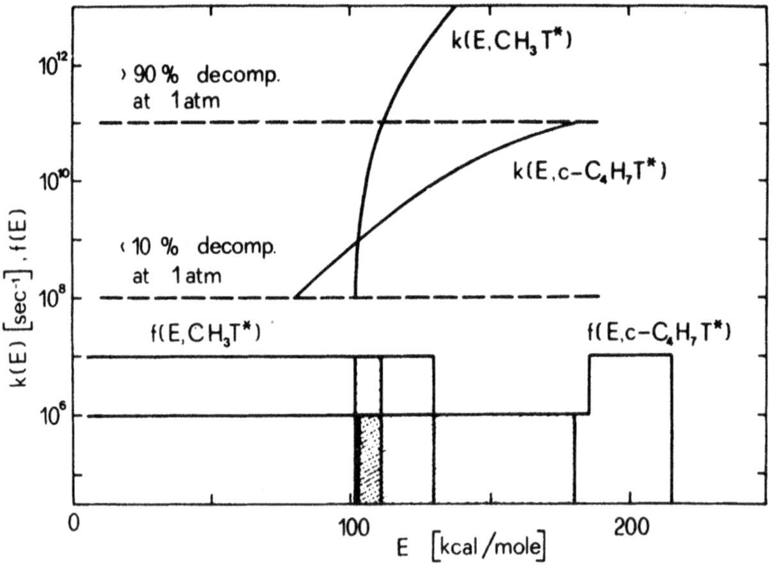

Fig. 14 Decomposition of activated CH_3T^* and cyclo-$C_4H_7T^*$ molecules from hot recoil tritium substitution; from [30].

with cyclo-C_4H_8 [30]. Whereas molecules from the indicated parts of the distribution functions f(E) are collisionally quenched below 1 atm, higher pressures are required for molecules at higher energies. Particularly highly excited molecules are still not quenched at liquid densities and break up because of their short lifetimes. Again, at the shortest lifetimes, collisional quenching of the intramolecular reagreement and cage effect will overlap.

V. NON-RANDOMIZATION TESTS

High pressure experiments can provide some insight into a central problem of unimolecular reactions, i. e. the randomization problem. After the excitation process, the energy either distributes fast over all of the oscillators of the molecule, and a statistical distribution of lifetimes results; or there is some bottleneck in this energy flow, resulting in nonrandomized energy distributions during the reaction and nonstatistical lifetime distributions appear. Such non-randomization effects, if present, should result in unusual pressure dependences of rate constants or of product distributions; they should set in, when the collisions compete with some slow intramolecular energy flow. No such effects became apparent in fig. 13; however, this test would be fairly insensitive. A more sensitive test is provided by experiments, in which excited molecules are prepared in two different ways, leading to different places of the initial excitation. If this initial excitation has the time to migrate to another place in the molecule, a chemical reaction there may set in. If it does not have this time, the reaction may occur only at the initial place of excitation. Characteristical times for energy flow between different quantum states in a molecule are believed to be of the order of 10^{-12} - 10^{-14}s (not considering radiationless transitions between different electronic states). Therefore, such non-randomization effects can only be expected for high pressure gases or liquids. Few examples have been observed so far; among them are the CF_2-elimination from chemically activated hexafluorobicyclopropyl [31], and the HCl-elimination from recoil tritium activated isopropyl chloride [30]. In all cases, only very minor non-randomization effects could be observed, with a slight reaction preference at the site of the initial excitation when the reaction was studied in a high density medium.

VI. THERMAL DISSOCIATION AND ADDITION REACTIONS

In this and the following section, thermal unimolecular dissociation and isomerization reactions, and their reverse processes, taking place in media of high density, are considered. Similar as described in section IV, at high pressures the collisional deactivation competes with the intramolecular rearrangement of excited molecules.

However, different from the situations considered in section IV, in thermal unimolecular reactions the activating process is a collisional process as well. Therefore, at high pressures, the fast collisional activations and deactivations are nearly balanced, and the intramolecular rearrangement becomes rate determining. For the reverse process of a unimolecular dissociation, i. e. the association of radicals, analogously the association process becomes rate determining. Formally, this situation is well represented by the simple Lindemann- Hinshelwood scheme of a unimolecular reaction,

$$A + M \longrightarrow A^{*} + M \qquad (6.1)$$

$$A^{*} + M \longrightarrow A + M \qquad (6.2)$$

$$A^{*} \longrightarrow \text{products} \qquad (6.3)$$

which gives a first order rate constant for unimolecular reaction of A

$$-\frac{d[A]}{[A]\,dt} = k = k_1[M]\frac{k_3}{k_2[M]+k_3} \qquad (6.4)$$

The change of reaction order, from second order with $k \propto [M]$ toward first order with $k = \text{const}$, occurs at densities of the medium $[M]$ where collisional deactivation competes with intramolecular rearrangement, i. e. $k_2[M] \simeq k_3$.

It can be shown [23] - [25] that k_3 corresponds to the specific rate constant k(E), discussed in section IV, at an energy $E \simeq E_o + s_{eff}kT$ where s_{eff} is a temperature dependent parameter accounting for the quantum nature of the s oscillators of the molecule (one has $s_{eff} < s$). From fig. 9 it is quite obvious that the change of reaction order occurs at low pressures for polyatomic molecules, because of their large number of oscillators and hence small k(E). On the other hand, very high pressures are required for this transition to occur in small molecules. Fig. 15 shows log k as a function of log [Ar] for the thermal dissociations of CH_3NO_2, $C_2H_5NO_2$, and $C_3H_7NO_2$. The shift of the transition of reaction order toward higher densities is nicely observed with smaller molecules. The change of reaction order occurs at several hundred atm for the spinallowed dissociation of triatomic molecules like NO_2 [33].

The corresponding transition between a low pressure third order reaction and a high pressure second order reaction in the recombination or association reaction of radicals is illustrated in fig. 16. In these experiments NO_2 is photolyzed in the presence of other reactants X and an inert gas (N_2) at varying pressures [34]. At low pressures, the quantum yield of the overall photolysis, which is determined by a fairly complex radical reaction mechanism,

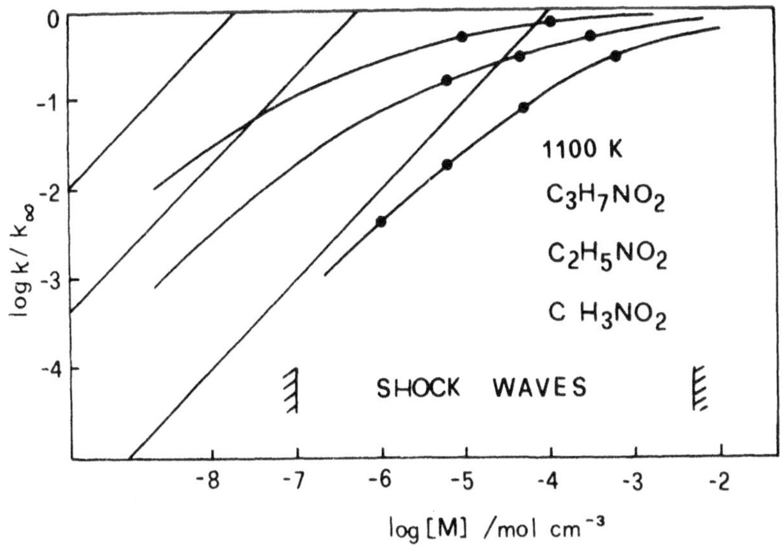

Fig . 15 Fall-off curves for thermal dissociation of nitroalkanes in shock waves near 1100 K; from [32].

is larger than 2. The rise of the inert pressure increases the rate of the addition of the primary photolysis product O to the added reactants X to form OX; this then leads to a decrease of the overall quantum yield. At high pressures, this addition becomes pressure independent and the quantum yield becomes constant. From the pressure dependence of the quantum yield curve, one therefore obtains the "half-quenching pressures" for stabilization of the OX complexes. If the species X is the NO_2 molecule itself, the OX complex is the NO_3 radical; this reaction has a half-quenching pressure near 5 atm. With X = C_3H_6 one has a half-quenching pressure near 1 atm, with X = C_2H_4 one has 20 atm, with X = C_2H_2 one has about 100 atm. Apart from NO_3 (for the reasons see ref. [34]), one again nicely observes the shift of the transition between a pressure dependent and pressure independent addition reaction toward higher pressures, when the number of oscillators of the addition complex decreases.

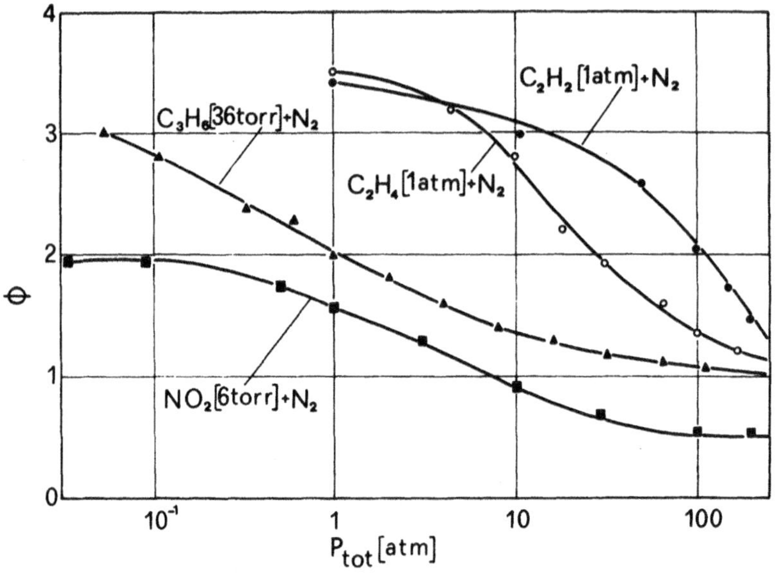

Fig. 16 Quantum yields for NO_2 photolysis at 366 nm in the presence of olefins and of high pressure N_2; from [34].

It would make no sense at all, to represent the pressure dependence indicated in this section in terms of an "activation volume". The pressure dependence of the rate constant here is entirely attributed to the competition between pressure dependent and pressure independent elementary steps of the mechanism. Only for the high pressure limiting rate constants the conditions to apply ordinary transition state theory become nearly fulfilled. On the other hand, dissociation and association reactions in very high density media become essentially diffusion controlled. Then, the conditions for applying transition state theory again are no more fulfilled, and an interpretation of activation volumes in terms of transition state theory becomes questionable. In the following section, we discuss this point in detail in considering thermal unimolecular reactions at ultra-high densities.

VII. "ACTIVATION VOLUMES" IN UNIMOLECULAR REACTIONS AT VERY HIGH DENSITIES

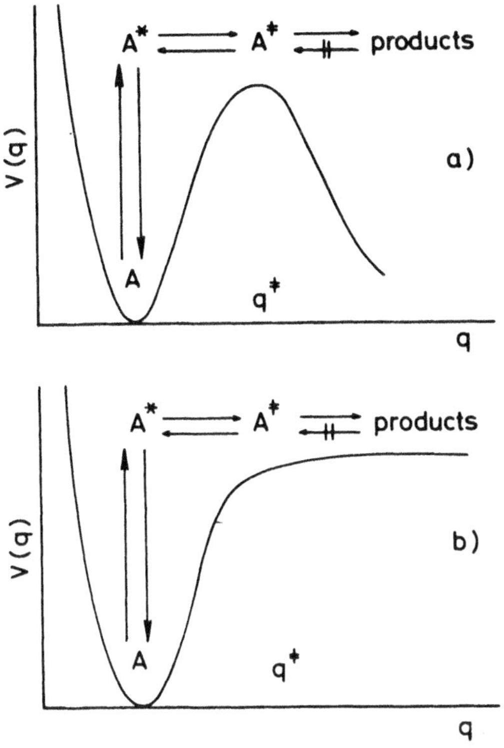

Fig. 17 Potential energy profiles along reaction coordinate illustrating transition state theory (see text)

In this section, we demonstrate that the concept of ordinary transition state theory (TST) in unimolecular reactions at high densities becomes misleading. We illustrate the basic concept of TST in fig. 17a und b. In TST the existence of a bottle neck is a crucial postulate. One assumes an intermediate structure $A^{\neq}$, whose coordinates except one (the "reaction coordinate q") are statistically populated relative to excited molecules A^*; on the other hand, in thermal systems at sufficiently high pressures, A^* has an equilibrium population relative to A. An essential assumption is then the postulate of a flow exclusively from systems A^* through $A^{\neq}$ toward products, and a totally vanishing flow in the backward direction from the product side through $A^{\neq}$ toward A^*. The localization of such a configuration $A^{\neq}$ may seem to pose no problems in the case of a pronounced barrier as shown in fig. 17a. If no such barrier exists, such as in the Morse potential of fig. 17b, the localization presents already more problems; however, here the choice of the maximum of the free energy may solve the problem [35]. Nevertheless, the question remains wether collisions can establish an equilibrium population of all configurations at the left of $A^{\neq}$ in fig. 17 under conditions where configurations at the right of $A^{\neq}$ are not populated by collisions at the same time. This question, which is of basic importance for high density media, cannot be answered by TST, but requires a more fundamental transport equation. Many-particle trajectory calculations, such as described in sections II and III, could provide a conclusive answer. We consider here a model equivalent to the simpler "random flight in a continuum model" also described earlier.

The motion of a particle of mass m in a force field $F(q) = -dV(q)/dq$, under the influence of a directed and a statistically fluctuating friction force, has been treated in the modern theory of Brownian motion. This starts from the Langevin equation [13]

$$\frac{d\dot{q}}{dt} = -\beta\dot{q} + A(t) + \frac{F(q)}{m} \tag{7.1}$$

and derives the Fokker-Planck equation from Liouville's theorem by assuming small average displacements of the particle under the influence of the fluctuating force A (t). Simplified to one coordinate, on has

$$\frac{\partial g(\dot{q},q,t)}{\partial t} + q\frac{\partial g}{\partial q} + \frac{F(q)}{m}\frac{\partial g}{\partial \dot{q}} = \beta\dot{q} + \beta g + C\frac{\partial^2 g}{\partial \dot{q}^2} \tag{7.2}$$

where $g(\dot{q}, q, t)$ is the population of states in phase space and

$$C = \frac{\beta kT}{m} \tag{7.3}$$

A steady-state solution for a potential of the form shown in fig. 17a has been derived by Kramers under the assumption that an equilibrium population of configurations $A^{\neq}$ is always established [36]. A combination of this solution with the low density Lindemann-Hinshelwood model has been given in ref. [23]. The solution of this model gives results which are in remarkable contrast to the postulates of transition state theory: the higher the dynamical friction ß (and hence the density), the more important becomes the back flow from the product toward the reactant side through $A^{\neq}$; the forward flow and the backward flow approach each other and only a small net flow remains; the configurations at the left but near to $A^{\neq}$ approach an equibrium population only at such densities where the back flow is already important. In other words, the usual "high pressure rate constant k_{∞}" of a thermal unimolecular reaction in a gas corresponds only to an intermediate situation; at higher densities, when the friction term ß increases with density, the rate constant drops again.

The transition state theory expresses k_∞ by

$$k_\infty = \frac{kT}{h}\,\frac{Q^{\neq}}{Q}\,\exp\left(-\frac{E_o}{kT}\right) \tag{7.4}$$

The Lindemann-Hinshelwood model describes the low density unimolecular reaction with eq. (6.4) by

$$k_{LH} = \frac{k_o\, k_\infty}{k_o + k_\infty} \tag{7.5}$$

where k_o is the pressure dependent low pressure rate constant. The Kramers model, in the generalization of ref. [23], gives the rate constant from very low up to very high densities by

$$k = k_{LH}\sqrt{\frac{\sqrt{\beta^2/4+\omega^2}-\beta/2}{\sqrt{\beta^2/4+\omega^2}+\beta/2}} \tag{7.6}$$

where ω is the "imaginary frequency" of the potenial of fig. 17a near E_o. The potential there is assumed to have the form

$$V(q) \simeq E_o - \frac{1}{2}\, m\, \omega^2 (q-q^{\neq})^2 \tag{7.7}$$

k/k_{LH} approaches unity at $\beta \ll \omega$, whereas one obtains

$$k \simeq k_\infty \frac{\omega}{\beta} \tag{7.8}$$

at $\beta \gg \omega$, i. e. for high densities. If the dynamical friction β of the particle in the medium can adequately be represented by Stokes law, one has

$$\beta \simeq \frac{6\pi\eta a}{m} \tag{7.9}$$

with the viscosity η and the radius of the particle a. Therefore one obtains

$$k \simeq k_\infty \frac{\omega m}{6\pi\eta a} \tag{7.10}$$

Eq, (7.6) and, at $\beta \gg \omega$, eq. (7.10) can be used to illustrate the meaning of the formal activation volume $\Delta v^{\neq}$ of eq. (1.1) in a unimolecular isomerization reaction at high densities or pressures. As seen from eq. (7.10), several factors contibute to the pressure dependence of k : (i) pressure influences on quantities related to potential parameters like the effective values of $Q^{\neq}/Q$ and ω and (ii) pressure influences on transport properties like the effective values of η and a. In TST, normally only the influence on $Q^{\neq}/Q$ is considered. In the following, we show that the viscosity part may well be of dominant importance. From eq. (7.10) we have

$$-RT\left(\frac{\partial \ln k}{\partial P}\right)_T \equiv \Delta v^{\neq} = RT\left(\frac{\partial \ln \eta}{\partial P}\right)_T + \dots \qquad (7.11)$$

where only the viscosity part of the pressure dependence is written explicitely at the right hand side. Although this expression has been derived for a potential of the type of fig. 17a, similar expressions should apply for unimolecular dissociations with potentials of the type of fig. 17b.

Evaluation of eq. (7.11) indicates the following: Viscosities of liquids, at room temperature and in the 1 - 10 kbar range, generally have pressure coefficients $\partial \ln \eta/\partial P$ between $0.8 \cdot 10^{-4}$ and $4 \cdot 10^{-4}$ atm^{-1} [14] (e. g. $0.8 \cdot 10^{-4}$ atm^{-1} for M = H_2O, $4 \cdot 10^{-4}$ atm^{-1} for M = isopentane). With eq. (7.11), this leads to activation volumes between + 2 and + 10 cm^3 mol^{-1}. Experimentally determined activation volumes fall well within this range [1] - [4]. Nevertheless, it has to be emphasized that the use of the solvent viscosity η for the friction of the reacting particles in the solvent can only be a rough approximation with deviations possible toward lower and higher values. Therefore, the pressure dependence of the transport properties in the rate constant contributes dominantly to the observed experimental activation volumes.

We conclude from the model of this section that the analysis of pressure dependences of rate constants of unimolecular reactions in high density media should not be done in the language of transition state theory alone. The "activation volume" should rather be considered a formal parameter. The models presented here, or more detailed many-particle trajectory calculations, give a more realistic understanding of this quantity.

REFERENCES

1. W. J. Le Noble, in Progr. Phys. Org. Chem. 5, 207 (1967).

2. K. E. Weale, Chemical Reactions at High Pressures, Spon, London 1967

3. G. Kohnstam, in Progr. React. Kinet. 5, 335 (1970)

4. C. A. Eckert, Ann. Rev. Phys. Chem. 23, 239 (1972).

5. K. Luther and J. Troe, Chem. Phys. Letters 24, 85 (1974).

6. L. F. Meadows and R. M. Noyes, J. Amer. Chem. Soc. 82, 1868 (1960)

7. H. van den Bergh and C. Dupuy, to be published.

8. K. Luther, M. Maier, J. Schröder and J. Troe, to be published.

9. T. J. Chuang, G. W. Hoffman and K. B. Eisenthal, Chem. Phys. Letters 25, 201 (1974).

10. R. M. Noyes, J. Chem. Phys. 22, 1349 (1954).

11. R. M. Noyes, J. Amer. Chem. Soc. 78, 5486 (1956).

12. R. M. Noyes, Z. Elektrochem 64, 153 (1960); Progr. in Reaction Kinetics, 1, 129 (1961).

13. S. Chandrasekhar, Rev. Mod. Phys. 15, 1 (1943).

14. R. C. Reid and T. K. Sherwood, The Properties of Gases and Liquids, 2nd ed. 1966, Mc Graw-Hill, New York.

15. D. L. Bunker and B. S. Jacobson, J. Amer. Chem. Soc. 94, 1843 (1972).

16. J. N. Murrell, A. J. Stace and R. Dammel. The author thanks Dr. Stace for sending these results prior to publication.

17. R. K. Lyon, J. Amer. Chem. Soc. 86, 1907 (1964).

18. H. Hippler, K. Luther and J. Troe, Chem. Phys. Letters 16, 174 (1972); Ber. Bunsenges. Physik. Chem. 77, 1104 (1973).

19. G. Burns and W. H. Wong, J. Amer. Chem. Soc. 97, 710 (1975).

20. A. J. Stace and J. N. Murrell, Mol. Physics 33, 1 (1977).

21. H. Hippler and J. Troe, Int. J. Chem. Kin. 8, 501 (1976).

22. A. J. Stace and J. N. Murrell, to be published.

23. J. Troe, in Pysical Chemistry. An Advanced Treatise. Vol. VIB. Kinetics of Gas Reactions, Academic Press, New York, 1975.

24. W. Forst, Theory of Unimolecular Reactions, Academic Press, New York, 1973.

25. P. J. Robinson and K. A. Holbrook, Unimolecular Reactions, Wiley, New York, 1972.

26. S. H. Luu and J. Troe, Ber. Bunsenges. Physik. Chem. 77, 325 (1973); 78 (1974).

27. H. Gaedtke and J. Troe, Ber. Bunsenges. Physik. Chem. 79, 184 (1975).

28. H. Gaedtke, H. Hippler and J. Troe, Chem. Phys. Letters 16 177. (1972).

29. I. Oref, D. Schuetzle and B. S. Rabinovitch, J. Phys. Chem. 75, 2164 (1971).

30. Y. N. Tang and F. S. Rowland, J. Phys. Chem. 72, 707 (1968).

31. J. D Rynbrandt and B. S. Rabinovitch, J. Phys. Chem.75, 216 (1971).

32. K. Glänzer and J. Troe, Helv. Chim. Acta 55, 2884 (1972); 56, 577 (1973); 56, 1691 (1973).

33. J. Troe, Ber. Bunsenges. Physik. Chem. 73, 144 (1969).

34. H. Gaedtke, K. Glänzer, H. Hippler, K. Luther and J. Troe, XIVth Intern. Symp. on Combustion, p. 295, The Combustion Institute, Pittsburgh (1973).

35. M. Quack and J. Troe, Ber. Bunsenges. Physik. Chem. 78, 240 (1974); 81, 329 (1977).

36. H. A. Kramers, Physica 7, 284 (1940).

HIGH PRESSURE EFFECTS ON PHOTOPHYSICAL AND PHOTOCHEMICAL PRIMARY PROCESSES

H.-D. Brauer, R. Schmidt and H. Kelm

Institute for Physical Chemistry, University of Frankfurt, Robert Mayer Strasse 11, 6000 Frankfurt am Main, West Germany

I. INTRODUCTION

Over the last two decades several authors [1-4] have published summarizing reports concerning studies of pressure effects on photophysical and photochemical primary processes. With the increasing availability of highly specialized apparatus these studies expanded into a wide variety of topics and involve large pressure ranges.

In principle, pressure exertion on a system can either change intramolecular bonds and angles or intermolecular distances and orientations. The first of these effects is reached only at ultrahigh pressures whereas the second effect is largest at low pressures and gradually decreases with rising pressure. Therefore the second effect with its influence on intermolecular interactions can be compared with a solvent effect, however with the advantage of causing gradual and controlled changes of the medium instead of stepwise alterations due to the use of different solvents.

Pressure applications present a powerful tool of studying spectroscopic phenomena, for instance, spectral bands (locations, widths, shapes and oscillator strengths), transition probabilities and energy transfers. At the same time proposed theories can be tested.

In this survey an attempt will be made to review the results of pressure studies on photophysical and photochemical primary processes of organic systems, especially in fluid and solid solutions.

Kelm, H. (Ed.) High Pressure Chemistry. 521-567. *Copyright* © 1978. *D. Reidel Publishing Company.*
All Rights Reserved.

In addition energy transfer involving inorganic compounds will also be discussed. No claim is made to give a complete review. Didactic reasons have influenced us to begin with a description of the mechanisms of electronic activation and deactivation within a molecule using a modified Jablonski diagram. Thereafter the effects of pressure on these primary processes are discussed. Experimental techniques are omitted, they are reported elsewhere [5-10].

II. DESCRIPTION OF PHOTOPHYSICAL PROCESSES

In this review we intend to report primarily on the influence of high pressure on the electronic behaviour of organic molecules in "inert solutions". The term "inert solutions" is hereby meant to indicate that the dissolved organic molecules can be considered as "isolated" absorbers. Furthermore, neither specific solute-solvent interactions nor bimolecular processes occur between the absorber and impurities in the solution. A further requirement is that the solvent molecules possess no low lying electronic states, i.e. that the solvent is completely transparent.

The absorption of light by a molecule in the condensed phase can lead to several subsequent processes. The total electronic exitation energy can be completely converted into thermal energy, although it can also be partially converted into radiation or used in a chemical reaction. Finally, electronic excitation energy can also be transferred to neighbouring molecules whereby part of the energy can dissipate into heat. For a description of those processes that occur within the molecule after absorption, a modified Jablonski diagram will be used [11,12].

Figure 1 shows such a modified Jablonski diagram for a typical organic molecule with an even number of electrons and a singlet ground state. The lower excited states are singlet and triplet states with the lowest being a triplet in accordance with Hund's Rule. Normally the energy difference between the ground state and the first excited singlet state is larger than the differences between excited states.

Each of the electronic states is associated with a number of vibrational and rotational states. For the sake of simplicity the diagram presents only the states of nuclear vibration. The arrows in the diagram indicate the most important changes of the energy states, especially processes of deactivation after absorption of radiation.

Electronic transitions of organic molecules, which do not contain any heavy atoms, are subject to strictly defined spin selection

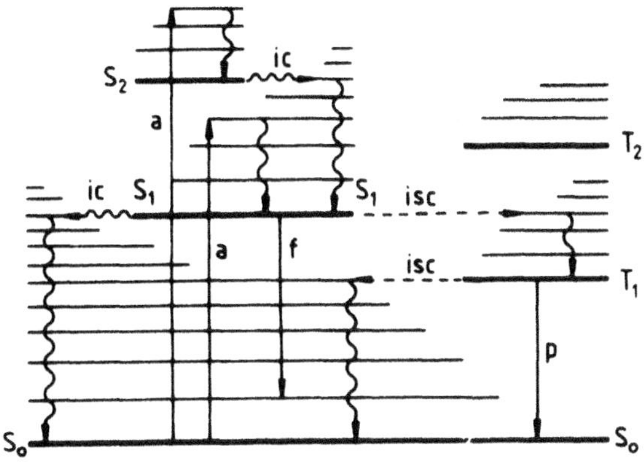

Fig. 1. Modified Jablonski diagram of a typical organic molecule ——→ radiative transition (a = absorption; f = fluorescence and p = phosphorescence), ∿∿∿→ thermal relaxation, ~~~~→ internal conversion (ic) and ----→ intersystem crossing (isc).

rule. Spin selection dictates that as a first approximation all electronic transitions during which the electron spin is reversed are forbidden.

In accordance with the spin selection rule the molecule is transferred from the singlet ground state - normally the lowest vibrational state thereof - to an excited singlet state upon absorption of a light quantum. Generally optical excitation can lead to a higher vibrational level of the excited electronic state depending on the wavelength of the absorbed light. Thus, an electronically excited molecule can also be a "hot" one. In principle, such a molecule can react directly, e.g. via spontaneous dissociation, as in the case of aliphatic compounds in the gas phase. Organic compounds with π-electron systems, however, behave differently. They possess excited states of low energy, from which a spontaneous dissociation is normally not possible. The electronic excitation of such a molecule survives multiple collisions with other molecules in the gas phase or in solution. The excited molecule only transfers its excess vibrational energy to the surroundings and attains a thermal equilibrium, which is normally reached at the lowest vibrational level of the corresponding excited electronic state. In the condensed phase this state is reached after about 1 picosecond. Such a thermal relaxation also occurs when the molecule is transferred primarily to a higher electronic state, e.g. the S_2 state. It then reaches the lowest vibrational level of this electronic state. Apart from special cases, the equilibrated higher electronic states are quickly deactivated via internal conversion to the lowest excited electronic state of equal multiplicity, e.g. S_2 converts into S_1. During internal conversion the electronic excitation energy of a molecule is spontaneously converted into vibrational excitation energy without the interference of another ground state molecule. Internal conversion is therefore an intramolecular process. As one can derive from Franck-Condon-Principle, such a process is the more probable, i.e. internal conversion proceeds faster, when the amount of electronic energy, which has to be converted into the vibrational form, is smaller. Internal conversion normally limits the lifetime of higher electronically excited states to ca. 10 picoseconds. In the example depicted in Figure 1, the molecule reaches a relatively high vibrational level of the S_1 state after leaving the lowest vibrational level of the S_2 state. The radiationless transitions are usually isoenergetic. Thereafter, the thermal relaxation proceeds again rapidly. Regardless of the state to which the molecule is excited primarily, it very soon reaches the thermally equilibrated S_1 state without radiation emission, provided no chemical reaction occurs in the meantime.

The S_1 state has an essentialy longer lifetime than any other excited singlet state because internal conversion to the S_o state is very slow. The reason for this is the large energy gap between the involved electronic states and the difference in nuclear configurations. This spin-allowed, but Franck-Condon-forbidden, conversion can practically be neglected in comparison to the other two deactivation processes: fluorescence emission and radiationless conversion to the T_1 state, known as intersystem crossing. Because of the spin restriction, such a process proceeds rather slowly, although still faster than the spin-allowed internal conversion since the energy difference between the S_1 and T_1 states is normally quite small. Fluorescence emission and radiationless intercombination limit the lifetime of the S_1 state to about 10 nanoseconds (ns).

Intersystem crossing of molecules from the S_1 to the T_1 state is exceptionally important in photochemistry, because for the majority of organic molecules it represents the essential process for populating the T_1 state. Excitation of the molecule from the S_o to the T_1 state by absorption of a photon is strongly forbidden by the spin selection rule and is only rarely observed.

The thermally equilibrated T_1 state has a longer life time than the S_1 state because radiationless deactivation to the S_o state is forbidden by the spin-selection rule, as well as the Franck-Condon-Principle. At the same time, radiative deactivation of T_1 state by phosphorescence emission is a spin-forbidden process, analogous to the corresponding excitation. In the absence of bimolecular processes, which can only be excluded in rigid solutions, radiationless deactivation and phosphorescence emission result in a life time of the T_1 state in the order of 1 second. In fluid solutions the lifetime can be shortened substantially by bimolecular processes of different kinds, for instance, energy transfer processes. Nevertheless, the lifetime of the T_1 state is the longest of all electronically excited states involved in the deactivation cascade.

Competing with the intramolecular deactivation processes described so far and conventionally called photophysical primary steps are two further processes:
a) intermolecular energy transfer processes
and
b) photochemical primary processes.

The electronic states, from which these processes originate, and their associated mechanisms will be discussed in later sections of this review.

III. PRESSURE EFFECTS ON PHOTOPHYSICAL AND PHOTOCHEMICAL PRIMARY PROCESSES

III.1. Absorption and Luminescence

III.1.1. Spectral Shifts and Band Broadening

The influence of high pressure on the position of absorption and emission bands of organic substances, especially aromatic hydrocarbons has been studied quite intensively over the last 20 years [1-4]. The investigations of aromatic hydrocarbons were conducted in the gase phase, in liquid and rigid solutions, and in the crystalline state. A typical example of the influence of high pressure on the singlet spectra of aromatic compounds is shown in Figure 2, where plastic polymethylmethacrylate (PMMA) is used as the solvent [13]. In general, increasing pressure results in a displacement of the electronic bands. In most cases a shift towards lower energies (red shift) is observed. The pressure induced shift is accompanied by an increase in the half width, $\Delta\tilde{\nu}_{1/2}$, of the band. This applies to absorption bands (A) as well as to fluorescence bands (F) of aromatics.

III.1.1.1. Absorption

Pressure effects on absorption spectra of aromatics in gases and in dilute solutions were investigated by Vodar, Oksengorn [2,14-18], Robertson [1,19-28] and recently by Okamoto and Drickamer [29]. The results obtained by these authors establish within experimental errors that in many cases the shift of the transition

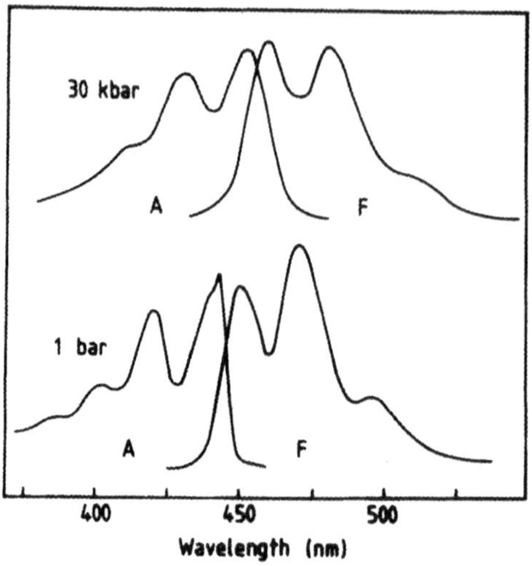

Fig. 2. Absorption and fluorescence spectra of perylene in PMMA at different pressures. The concentration is 10^{-4} M and the temperature is 296 K [3,13].

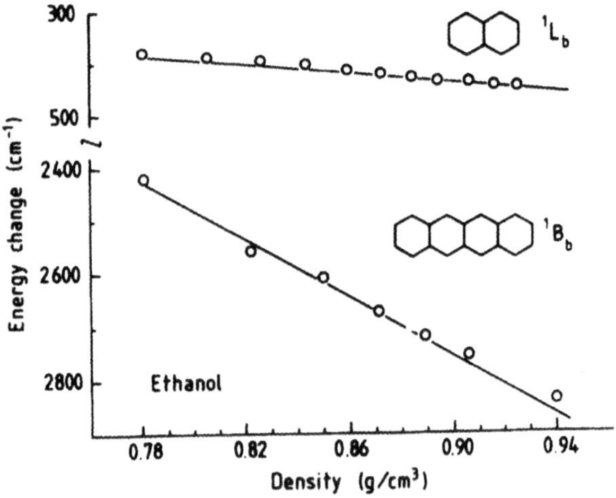

Fig. 3. The pressure shift of the ^1L_b absorption band of naphthalene and the ^1B_b absorption band of tetracene in ethanol [28].

frequency, ν, correlates linearily with the density of the diluent gas or liquid. This is illustrated in Figure 3 for the

^1L_b-absorption band of naphthalene and the ^1B_b-absorption band of tetracene in ethanol [28]. The range of densities indicated in Figure 3 corresponds approximately to a pressure range of 1 bar to 4 kbar.

At higher pressures measurements of band shifts of aromatics are preferably performed in plastic media [13,29-38] or in the crystalline state [4,29,39]. It was observed that the bands are shifted linearily with pressure up to about 30 kbar. At higher pressures, however, a deviation from linearity occurs as is demonstrated for example of the ^1L_a-absorption band of anthracene in PMMA (Figure 4). The pressure dependence of the half width, $\Delta\tilde{\nu}_{1/2}$, of the same band is also depicted in Figure 4. The half width increases linearily with pressure within the range investigated, i.e. up to about 75 kbar. This behavior was also found for other aromatics [29].

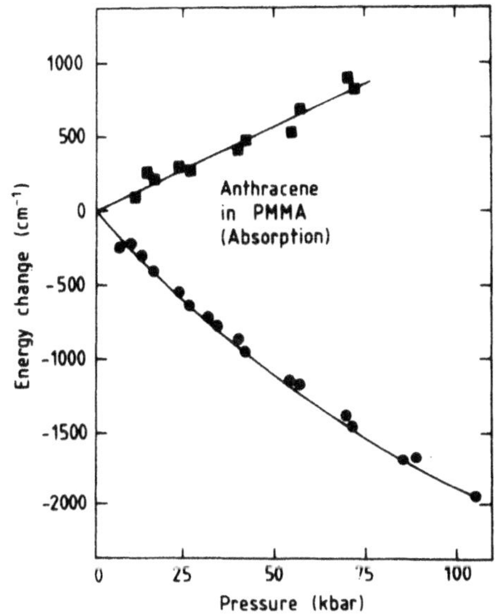

Fig. 4. Peak shift $\Delta\tilde{\nu}$ (●) and half-width change $\Delta\tilde{\nu}_{1/2}$ (■) of anthracene in PMMA (absorption) [29].

It is known from studies at normal pressure that the solvent influences the position of absorption bands of aromatic hydrocarbon compounds [40,41]. Investigations of the pressure effects on such bands revealed essentially the same results as can be seen from the data compiled in Table 1 for benzene, naphthalene, phenanthrene, anthracene and tetracene.

Table 1. Pressure-induced spectral shifts in different media

Aromatic Hydrocarbon	Absorpt. Band	Oscillator Strength	Pressure Range: 1 bar - 5 kbar, Liquids [27,28] $-d\tilde{\nu}/d\rho$ (cm^2/g)		1 bar - 30 kbar, Plastics [31,38] $-d\tilde{\nu}/dP$ ($cm^{-1}/kbar$)
			Pentane	Ethanol	PMMA
Benzene	^1L_b	0.002	650	-	-
Naphthalene	^1L_b	0.002	470	450	-
Naphthalene	^1L_a	0.18	2170	-	-
Phenanthrene	^1L_b	0.003	640	-	8
Phenanthrene	^1L_a	0.18	2070	-	22
Anthracene	^1L_a	0.10	1890	-	22
Tetracene	^1B_b	1.85	3100	2600	-

In column 2 of Table 1 some $\pi \rightarrow \pi^*$-absorption bands of these compounds are listed using the terminology of Klevens and Platt [43]. The ground states, ^1A, of all five compounds are non-polar. The lowest excited state of benzene, naphthalene and phenanthrene is a π^*-state where the nodes of the wave functions pass through the carbon atoms as depicted in Figure 5 for the example of naphthalene. This state labelled ^1L_b by Klevens and Platt has a moderate dipole moment. In the second excited state, ^1L_a, of these compounds the nodes are placed between the carbon atoms resulting in a somewhat larger dipole moment. The probability of a transition from the ground state to an excited state, as illustrated by the intensity of the arrows in Figure 5, depends on the change in polarity demonstrated. The oscillator strength, f, listed in column 3 of Table 1 is a measure for the transition probability and is related to the absorption coefficient ε as follows [11]:

$$f = 4.32 \times 10^{-9} \int \varepsilon(\tilde{\nu})d\tilde{\nu} \quad ,$$

where $\int\varepsilon(\tilde{\nu})d\tilde{\nu}$ is the integrated absorption band on a wavenumber scale.

As can be seen from Figure 5, the transition from the ^1A to the ^1B_b state is associated with the largest oscillator strength. The same is true for anthracene and tetracene. The ^1L_b and ^1L_a states of these latter two compounds are, however, reversed, i.e. the first excited state is a ^1L_a state whereas the second excited state is characterized by ^1L_b.

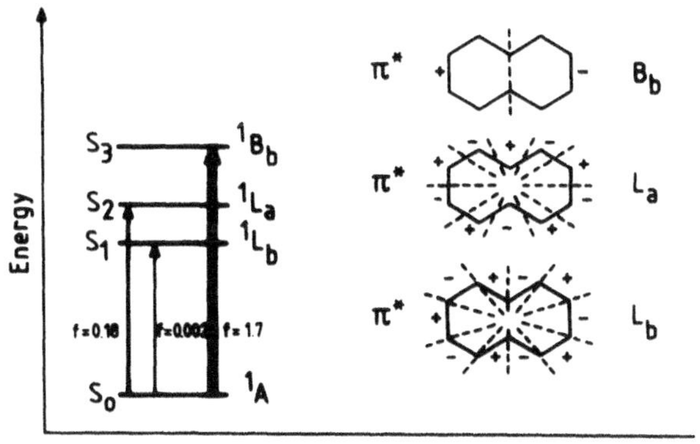

Fig. 5. Schematic energy-diagram of naphthalene

Column 4 and 5 of Table 1 show the density and pressure coefficients of the band maxima, respectively. In a first approximation one can assume that 1 cm^{-1}/kbar is equivalent to 30 cm^2/g. These data clearly show that the size of the pressure shift of an absorption band of aromatic compounds depends on the oscillator strength. This applies to liquid as well as to rigid solutions. Band shifts, $-d\tilde{\nu}/d\rho$, correlate linearily with the oscillator strength for $f > 0.1$, however, for $f < 0.1$ this linearity is not seen [1,27].

Variations in the extend to which the bands are shifted with different oscillator strengths results in a change in the energy gaps between excited states with increasing pressure. Thus the energy difference ^1L_a - ^1A, of a given aromatic hydrocarbon compound decreases more significantly with pressure than the energy difference, ^1L_b - ^1A. Because of these properties it should be possible to "press" the ^1L_a state of phenanthrene below the level of the ^1L_b state. Using an estimated energy difference $E = {}^1L_a - {}^1L_b = 5300$ cm^{-1} for phenanthrene in PMMA [44] and the pressure coefficient of Table 1, this cross-over is expected to occur at approximately 380 kbar.

For all hydrocarbon compounds, considered so far, excitation to the S_1 state was associated with an increase of the dipole moment. Consequently, the bands experienced a shift towards lower energies. However, should the dipole moment decrease during electron excitation, a shift towards higher energies is expected (blue shift). Indeed, this can be observed for compounds with heteroatoms containing double bonds, e.g. carbonyl compounds. The low energy $n \rightarrow \pi^*$-absorption bands of ketones in highly polar solvents are shifted towards shorter wavelengths upon pressure exertion [45]. In case of azulene, which has a permanent dipole moment in its ground state, contrary to the hydrocarbon compounds cited above, the excitation to the ^1L_b state is accompanied by a decrease (and reversal) in the dipole moment [46]. According to expectations this band shows a blue shift in ethanol as demonstrated in Figure 6. The excitation of azulene to the ^1B_b state is connected with an increase in the dipole moment [46] and consequently this band is shifted towards longer wave lengths (Figure 6).

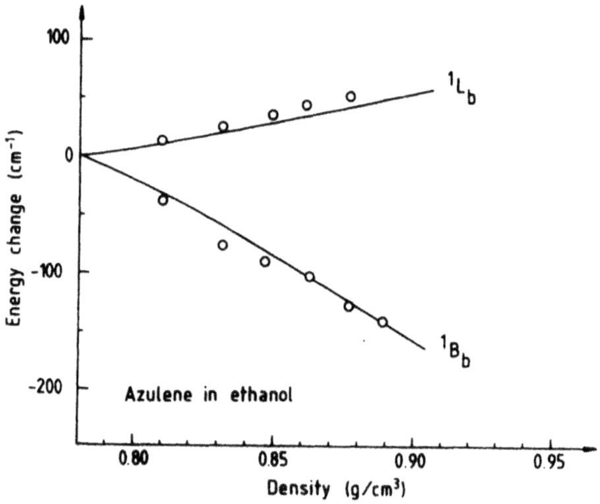

Fig. 6. The pressure shift of the ^1L_b and ^1B_b absorption bands of azulene in ethanol [28].

III.1.1.2. Emission

The fluorescence spectra of a series of aromatic hydrocarbon compounds were investigated in liquid and rigid solutions as a function of pressure [13,29,31,34,37-39,47-49]. For all the aromatics investigated, the fluorescence bands are shifted towards longer wavelengths with increasing pressure as has been already exemplified in Figure 2 for perylene. Some of the spectral shift data are compiled in Table 2.

Apart from a few exceptions, such as azulene, fluorescence emission occurs from the lowest excited singlet state (S_1-state). This state is for pyrene, coronene and chrysene, as well as naphthalene and phenanthrene, the ^1L_b-state, whereas for anthracene it is the ^1L_a-state. One can see from the results in Table 2 that the pressure induced band shift of fluorescence is larger than that for the corresponding absorption. This is very clearly shown for phenanthrene [38], anthracene [31] and pyrene [34].

Table 2. Pressure-shifts of spectra in PMMA at 296 K; $-d\tilde{\nu}/dP$ (cm^{-1}/kbar)

	Absorption		Fluorescence ($S_1 \rightarrow S_o$)		Phosphorescence ($T_1 \rightarrow S_o$)
	^1L_b-Bande	^1L_a-Bande	$^1L_b \rightarrow {}^1A$	$^1L_a \rightarrow {}^1A$	
Naphthalene [37,47]	-	-	7	-	7
Phenanthrene [38]	8	20	12	-	11
Chrysene [31,50]	10	28	10	-	7
Coronene [36,50]	8	22	8	-	7
Anthracene [31,51]	-	22	-	27	12**
Pyrene [34]	7	30	10	-	-

*Pressure: 1 bar - 30 kbar; **in single crystal [51].

The pressure coefficients, $-d\tilde{\nu}/dP$, listed in Table 2 apply only pressures up to ca. 30 kbar. At higher pressure $-d\tilde{\nu}/dP$ is no longer constant. Studies made by Okamoto and Drickamer [29] using anthracene in PMMA at 298 K underline this statement. As can be seen in Figure 7, the band shift for this system deviates from linearity beyond 40 kbar.

The pressure effects on the phosphorescence spectra of a number of aromatics have been investigated [38,39,47,50,51]. It is interesting to note from Table 2 that these pressure shifts are similar to those of the weak $^1L_b \rightarrow {}^1A$ fluorescence bands, e.g. of naphthalene or phenanthrene, despite the fact that the phosphorescence oscillator strengths are still smaller than the oscillator strengths of the weak fluorescence bands. One can assume that for emission bands with $f < 0.002$ the oscillator strength no longer corresponds to the pressure shift. Independent of the oscillator strengths one finds a minimum pressure shift of about 5 - 10 cm^{-1}/kbar for weak emission bands.

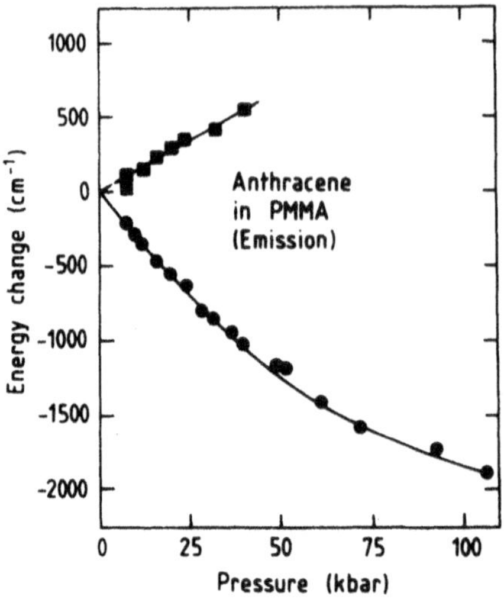

Fig. 7. Peak shift $\Delta\tilde{\nu}$ (●) and half-width change $\Delta\tilde{\nu}_{1/2}$ (■) of anthracene in PMMA (emission) [29].

III.1.1.3. Theoretical Considerations

As mentioned in the introduction of this survey, the effect of pressure on absorption and luminescence spectra of aromatics is frequently compared with solvent effects and therefore treated within the framework of solvent shift theories which were recently discussed by Okamoto and Drickamer [29].

In the past several attempts have been made to explain the density dependence of the pressure shift of absorption bands. From the second-order perturbation theory on dispersion forces in condensed media Equation (1) can be derived [52]

$$-h\Delta\tilde{\nu} = \rho(A\cdot f + B), \qquad (1)$$

where A and B are constants for a given non-polar solute - non-polar solvent system and f the oscillator strength of the transition. This description is, however, not always adequate since sometimes the shift of absorption band is not linear with density, e.g. anthracene in n-hexane [29].

An alternative approximation for the estimation of pressure shifts utilizes the electronic theory [40,41,52-56], which expresses environmental perturbations in terms of the measurable quantities n (refractive index) and ε (dielectric constant) of the medium. These bulk properties increase in the same manner as does the density. The solvent dielectric theory for non-polar molecules predicts the following correlation:

$$-h\Delta\tilde{\nu} = \frac{A'f + B'}{a^3} \cdot \frac{(n^2 - 1)}{(2n^2 + 1)} \tag{2.0}$$

in which a is the Onsager cavity radius. The applicability of this equation was tested by Bayliss [54] and McRae [52] using electron spectra at normal pressure. Recently Zipp and Kauzmann [57] also used this approach for optical absorption data of solutions at moderate pressures.

From investigations of the solvent dependence of the absorption at normal pressure the quotient $(A'f + B')/a^3$ can be obtained. Using the Lorentz-Lorentz equation* one can calculate the refractive index n of the solvent from its density at any pressure. In turn, the pressure dependence of the density can be determined from the Tait-equation**. Thus Equation (2.0) should predict frequency shifts of absorption bands with solvent density provided the effect of density upon the cavity radius a is known. (Assuming A' and B' to be pressure independent.)

For the application of Equation (2.0) two limiting conditions are to be considered:

a) In a first approximation the parameter a is assumed to be pressure dependent in as much as a^3 varies inversely with the solvent density. The Equation (2.0) can then be rewritten as:

$$-h\Delta\tilde{\nu} = K\rho \cdot \frac{(n^2 - 1)}{(2n^2 + 1)} \tag{2.1}$$

b) For the second case a is taken to be pressure independent with the consequence that

$$-h\Delta\tilde{\nu} = K\rho_o \cdot \frac{(n^2 - 1)}{(2n^2 + 1)} \tag{2.2}$$

Experience has shown that the shift of weak absorption bands of non-polar hydrocarbon compounds in non-polar solvents can be well

---

* Lorentz-Lorentz-equation: $R_M = \frac{n^2 - 1}{n^2 + 2} \cdot \frac{M}{\rho} = \frac{4\pi}{3} N_L \bar{\alpha}'$ with R_M = mole refraction, M = molecular weight and $\bar{\alpha}'$ = optical polarizability.

**Tait-equation: $\ln \rho_p/\rho_o = C \frac{(L + P)}{(L + 1)}$ with C and L being empirical constants.

described by Equation (2.2). This is illustrated in Figure 8, where the shift of the ^1L_b absorption band of naphthalene is shown according to Equation (2.2) (solid curve) and Equation (2.1) (dashed curve) [46]. A similar observation can be made for the ^1L_b absorption band of phenanthrene.

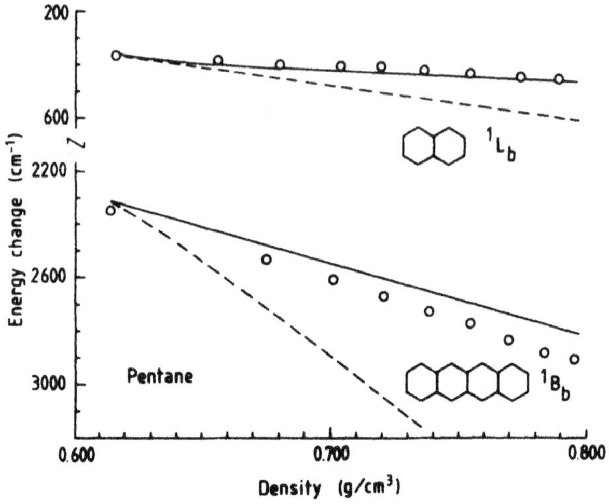

Fig. 8. The frequency shifts of the ^1L_b transition of naphthalene and the ^1B_b transition of tetracene are shown as functions of n-pentane solvent density. Using Equ. (2.1), the dashed curves assume a cavity volume for the absorber that varies inversely as the density ρ; Equ. (2.2), the solid curves assume a constant cavity volume. Circles represent the experimental pressure data. Frequency shifts are referred to the vapor absorption frequencies [28].

Strong absorption bands of non-polar aromatics in non-polar solvents are also better represented by Equation (2.2) than by Equation (2.1) as can be seen from Figure 8 in case of the ^1B_b band of tetracene in pentane.

Equation (2.1) is more suitable for the description of pressure shifts of strong absorption bands of non-polar aromatics in polar solvents. The observed pressure shifts of the ^1L_a band of anthracene in glycerol agree well with the values calculated according to Equation (2.1) as shown by the solid line drawn in Figure 9 [29]. Major deviations occur for anthracene in n-hexane.

Generally it is difficult to decide whether the deviations in the calculated band shifts from the observed experimental shifts are due to the pressure dependence of the solvent cavity radius or to the inadequacies of the dielectric theory to describe the interactions in the systems.

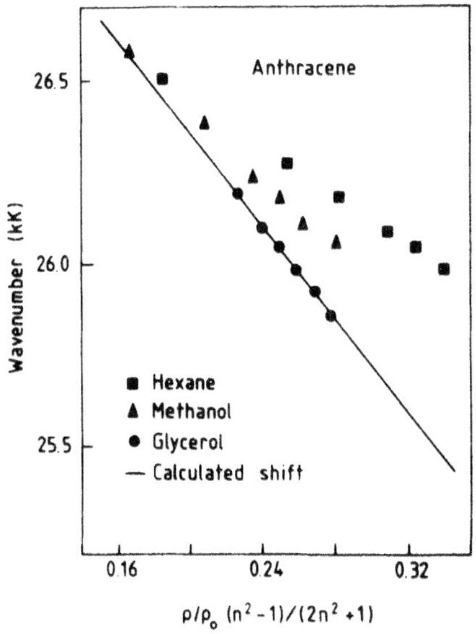

Fig. 9. Shift of absorption peak vs. $\rho/\rho_o(n^2 - 1)/(2n^2 + 1)$ for anthracene in various solvents [29].

Two summarizing statements can be made with respect to the applicability of the dielectric theory:

1) The dielectric theory predicts identical pressure shifts for absorption and emission of non-polar molecules in non-polar solvents. The experimental evidence, however, disproves this prediction [29].

2) The dielectric theory does not include the change of half widths of absorption and emission bands upon the exertion of pressure.

In this context it is of interest that Drickamer and coworkers [58] have developed a configuration coordinate model, which relates the shift in optical absorption and emission bands as well as changes in the half-width, with pressure to the change in volume of the system upon electron excitation, Δ, and the ratio R of force constants of the excited and ground states, and the coupling of the force constant to the bulk modulus of the medium.

A schematic configuration coordinate diagram is presented in

Figure 10. The authors assume the existence of harmonic potential wells with different force constants ω^2 for the ground and ω'^2 for the excited state. The shift in the configuration coordinate of the excited state relative to the ground state can be considered as a volume change of the system upon electron excitation. This is justified since the configuration coordinate represents the breathing mode of the lattice or solvent molecules. For the shift in the optical absorption or emission bands and the changes in half-widths, the following equations were derived [58]:

$$(\delta\tilde{\nu})_a = PR\Delta + P^2\frac{(R-1)}{2\omega^2} + \frac{R\Delta^2}{2}(\omega^2 - \omega_o^2) \tag{3}$$

$$(\delta\tilde{\nu})_e = \frac{P\Delta}{R} + P^2\frac{(R-1)}{2\omega^2R^2} - \frac{\Delta^2}{2}(\omega^2 - \omega_o^2) \tag{4}$$

$$(\delta\tilde{\nu}_{1/2})_a = N\left(\left|\omega R\Delta + P\frac{(R-1)}{\omega}\right| - |R\Delta\omega_o|\right) \tag{5}$$

$$(\delta\tilde{\nu}_{1/2})_e = N\left(\left|\frac{\omega\Delta}{R^{1/2}} + P\frac{(R-1)}{R^{3/2}} - \frac{\omega_o\Delta}{R^{1/2}}\right)\right. \tag{6}$$

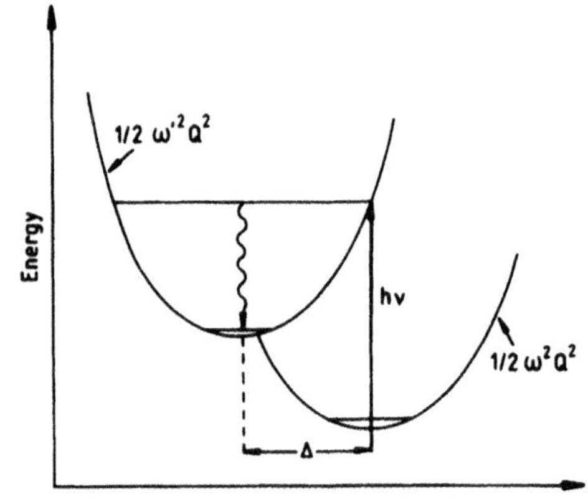

Fig. 10. Schematic configuration coordinate-diagram Δ = change in volume of the system upon electron excitation, ω^2 = force constant of the ground state and ω'^2 = force constant of the excited state.

The symbols used in the preceeding equations have the following meaning: $\tilde{\nu}$ = energy of band maximum, $\tilde{\nu}_{1\ 2}$ = half-width of the band, δ = the change in these parameters between pressure P and 1 bar, R = ratio of the force constants, ω_o = the square root of the force constant at 1 bar and $N = (8\ k\ T\ \ln 2)^{1/2}$ for Gaussian bands at ordinary temperatures. It is assumed that R and ω^2 are pressure independent and ω^2 is proportional to the bulk modulus of the crystal or the solvent. The authors define a dimensionless coupling coefficient ß by the relation $\omega^2 = ß(-dP/dV)$ where the quantity in parentheses is evaluated for the bulk crystal or solvent. Thus three parameters (Δ, R and ß) can be derived for the characterization of the electron excitation.

The configuration model has been applied to absorption data of benzene, phenanthrene and anthracene in various solid and liquid media [29] and to the absorption data of purine and pyrimidine bases and their nucleosides in different media [59]. It could be shown that the results are internally consistant.

One example should be discussed here in more detail. Okamoto and Drickamer [29] have studied the absorption and emission of anthracene in hexane, methanol and glycerol at room temperature over a pressure range of 1 bar to 10 kbar. The configuration coordinate parameters, as extracted from the Equations 3 through 6, are listed in Table 3:

Table 3. Configuration coordinate parameters for anthracene in solution.

Solvent	Δ ($cm^3\ mol^{-1}$)	R	ß
Hexane	- 9.2	1.00	1.7×10^{-3}
Methanol	- 7.4	1.00	2.9×10^{-3}
Glycerol	- 5.7	1.02	0.18

The negative values of Δ are consistent with the large dipole moment and stronger interaction of the ^1L_a state with the solvent molecules compared with the non-polar ^1A ground state of anthracene. The volume contractions upon excitation differ in the three media with the largest effect in the non-polar solvent hexane and progressively smaller effects in the polar solvents methanol and glycerol. According to Okamoto and Drickamer two factors are responsible for this variation. One is the difference in the interactions between the excited and ground states with the medium. The other is the deformability or compressibility of the medium. Indeed the compressibility of the solvent decreases in the series: κ (hexane) = 16.06×10^{-11}; κ (methanol) = 12.77×10^{-11}; κ (glycerol) = $3.72 \times 10^{-11}\ cm^2\ dyne^{-1}$. However, a quantitative correlation between κ and Δ cannot be found and the first factor might play an important role as well.

The values of R do not vary significantly from 1.00 for aromatics as could be shown by the study of anthracene.

The coupling to the surrounding medium is assumed to be strongly dependent on the rigidity of the medium. The ß values of Table 3 are consistent with this assumption, their ratio of 1 : 1.7 : 100 = hexane : methanol : glycerol may be compared with the ratio of the respective viscosities which are 1 : 1.7 : 1270.

One can summarize that the model of Drickamer has the advantage of good internal consistency and generality. The drawback of the latter, however, is that one gains only a limited amount of detailed knowledge about a particular system.

III.1.2. Intensities and Lifetimes

III.1.2.1. Absorption

The intensity of absorption, or the oscillator strength f, of aromatic hydrocarbons is approximately independent of pressure in inert matrices. The pressure induced increase in the band width is almost compensated by the decrease in the band maximum. However, the pressure independence is only a first-order approximation, as the intensity of the absorption bands can only be determined to within an uncertainity factor of ca. ± 50 % in an optical high pressure cell [3].

The oscillator strength f is directly proportional to the rate constant of fluorescence k_f, the reciprocal of which is equal to the inherent lifetime of the lowest excited singlet state τ_o. As f appears not to be a function of pressure, k_f and τ_o may also be considered to be independent of pressure.

III.1.2.2. Emission

Measurements of the fluorescence quantum yields ϕ_f and the fluorescence lifetimes τ_f allow one to draw conclusions involving the rate constants for the deactivation of the excited singlet states. By way of the Jablonski diagram one can derive Equations (7) and (8) for ϕ_f and τ_f:

$$\phi_f = k_f/(k_f + k_{isc} + k_{ic}) \qquad (7)$$

$$\tau_f = 1/(k_f + k_{isc} + k_{ic}) \qquad (8)$$

Herein k_{isc} and k_{ic} represent the rate constants for intersystem crossing and internal conversion, respectively. From these equations it follows that the fluorescence intensity I_f is proportional to the lifetime of fluorescence:

$$I_f = a\ \phi_f = a\ k_f\ \tau_f \qquad (9)$$

The proportionality constant a is not a function of pressure, as is k_f, at least as a first approximation. Thus,

$$I_f(P) \propto \tau_f(P) \qquad (10)$$

Pressure measurements of fluorescence intensity and lifetimes therefore lead to the same results. Both experimental methods enable conclusions to be drawn concerning the deactivation processes of excited singlet states.

The fluorescence lifetimes of a large number of aromatic hydrocarbons have been measured under pressure in PMMA matrices. The results are shown in Table 4. As can be seen, all lifetimes show

Table 4. Fluorescence and phosphorescence lifetimes of aromatic hydrocarbons in PMMA at room temperature

	Fluorescence*		Phosphorescence**	
	τ_f (ns)	$-d\ln\tau/dP$ ($Mbar^{-1}$)	τ_p (s)	$-d\ln\tau/dP$ ($Mbar^{-1}$)
Naphthalene	100	16	1.5	9.4
Naphthalene-d_8	95	14	14.2	16
Phenanthrene	56	6.9	2.5	9.1
Phenanthrene-d_{10}	57	6.9	11.2	12
Chrysene	42	4.4	1.2	2.5
Chrysene-d_{12}	43	4.9	7.9	12
Triphenylene	37	10	8.5	15
Triphenylene-d_{12}	42	11	13.2	25
Pyrene	307	24	0.35	3.9
Pyrene-d_{10}	277	21	2.5	12
Biphenyl			2.5	3.7
Biphenyl-d_{10}			6.4	11
Fluorene			4.5	11
Coronene	273	16	6.5	9.4
Acenaphthene	38	6.5	1.8	6.2
1,2-Benzanthracene	42	6.7	0.31	1.2
1,2,5,6-Dibenzanthracene	34	4.9	0.94	2.1
Ovalene	416	20		
9-Methylanthracene	15	6.7		
9,10-Diphenylanthracene	14	3.8		

* [60,61,63] ** [62]

a weak negative pressure coefficient, which is represented here by $d\ln\tau/dP$. The largest pressure effect was found for pyrene, whose lifetime is reduced by 50 % at 30 kbar. Obviously the pressure coefficient of the lifetime is proportional to the actual lifetime at normal pressure. Apparently the pressure induced perturbation of the environment of an excited molecule becomes stronger, when the electronic relaxation process is forbidden. The correlation, $d\ln\tau/dP \propto \tau$, agrees with the fact that deuterium substitution has no influence on the pressure coefficient.

As different compounds have different $d\ln\tau/dP$ values, it is clear that medium effects (e.g. the change in the refractive index with pressure) cannot completely explain the effect of pressure on lifetimes. Additional specific interactions with the environment must explain the variations in the pressure coefficients.

As illustrated in Table 4, not only do the fluorescence lifetimes τ_f decrease with increasing pressure, but also the phosphorescence lifetimes τ_p. Although there are seven orders of magnitude between τ_f and τ_p, the pressure coefficients of both emissions are similar. Nevertheless for the phosphorescence lifetimes the correlation, $d\ln\tau/dP \propto \tau$, still holds.

Exceptions

The pressure dependence of the fluorescence lifetimes and intensities of anthracene and a series of its derivatives deviates substantially from the previously discussed behaviour. Instead of the generally weak negative pressure coefficients of ϕ_f and τ_f, one finds for these compounds, which will be discussed later, not only strong negative but also positive $d\ln\phi/dP$ values and $d\ln\tau/dP$ values.

Johnson and Offen determined the pressure dependence of τ_f for anthracene, 9-methylanthracene (9-MA) and 9,10-diphenylanthracene (DPA) at 77 K dissolved in PMMA and ethanol [63]. These results are illustrated in Figure 11. 9-MA and DPA show the usual pressure dependence; in both solvents τ_f decreases slightly with a pressure coefficient of $d\ln\tau/dP = -5$ Mbar^{-1}.

However, anthracene shows a completely different behaviour. In the lower pressure region τ_f increases linearly. Above 25 kbar it approaches a limiting value, which is approximately equal to the inherent lifetime τ_0 of anthracene. For the interpretation of this unexpected behaviour it is useful to consider the energy of the lowest excited states of anthracene. As is demonstrated in Table 5, both the T_2- and T_1-states lie below the S_1-level, such that the energy difference between S_1 and T_2 is only ca. 500 cm^{-1}. However, the T_3-level of anthracene lies about 800 cm^{-1} above the S_1-state. At room temperature and normal pressure,

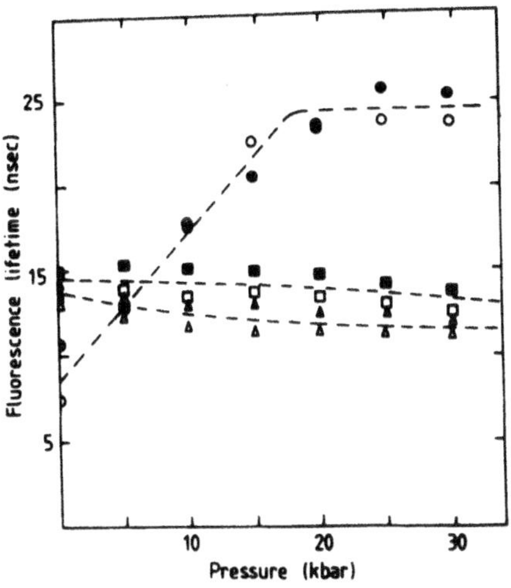

Figure 11. Fluorescence lifetimes of anthracene (o), 9,10-diphenylanthracene (□), and 9-methylanthracene (Δ) as a function of pressure in PMMA (solid points) and ethanol (open points) at 77 K [63].

Table 5. Energy of the lowest excited states of anthracene [64,66]

State	Energy (kK)
S_1	26.6
T_1	14.9
T_2	26.1
T_3	26.8

deactivation of the S_1-level occurs exclusively by fluorescence and intersystem crossing (isc) [65]. In this case isc consists of a temperature dependent and independent part [66]. Herein the temperature independent rate constant k_{isc} represents the exothermic transition to T_2 and the temperature dependent $k_{isc}(T)$ the endothermic transition to T_3. isc to the T_1-state cannot compete because of the large energy gap between S_1 and T_1. Thus:

$$\tau_f = 1/(k_f + k_{isc} + k_{isc}(T)) \qquad (11)$$

By cooling to 77 K the competing temperature dependent term $k_{isc}(T)$ disappears since the activation energy for the endothermic isc to T_3 can no longer be supplied. Therefore the lifetime increases to $\tau_f = 1/(k_f + k_{isc})$ and thereby reaches the starting point in Figure 11.

Johnson and Offen assumed that the S_1-state decreases in energy with pressure more than the T_2-level. It follows therefore that at elevated pressures the S_1- and T_2-states should cross, whereby isc to the T_2-state becomes endothermic. This should cause a decrease in k_{isc} with pressure, which continues until $k_f >> k_{isc}$ and therefore $\tau_f = \tau_o$ [63].

Although this interpretation seems at first to be reasonable, it must be remembered at this point that it defies Siebrand's energy gap law. This law demands that the propability of an exothermic transition between two levels increases continuously with a decreasing energy gap [67]. Thus τ_f should decrease at first with increasing pressure.

Recent investigations by Shaw and Nicol [68] of the pressure dependent red shift of absorption, phosphorescence, and T_1-T_2-absorption of anthracene in PMMA at room temperature seem to confirm Offen's hypothesis, in that the S_1- and T_2-states cross at ca. 20 kbar.

Tanaka and Osugi studied the fluorescence quantum yield of anthracene and a series of its derivatives as a function of pressure at room temperature in various solvents [69-71]. As shown in Figures 12 and 13, the authors found remarkable pressure effects in several cases, despite the relative small pressure range of 3 kbar. Figure 12 presents the pressure dependence of the fluorescence

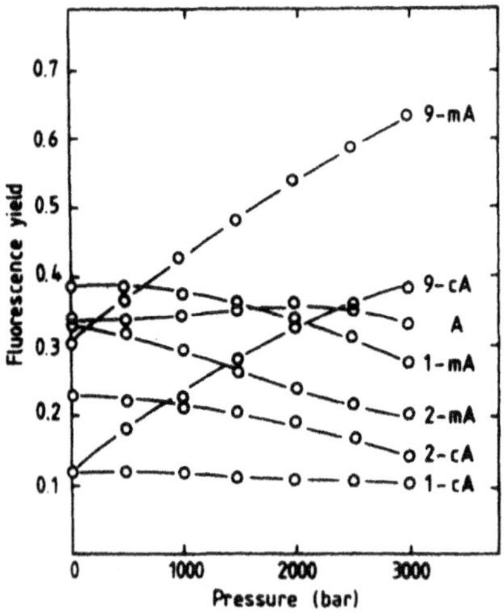

Figure 12. Pressure dependecies of fluorescence yields of anthracene, methylanthracene and chloroanthracene in hexane at room temperature[71].

quantum yield of 9-chloroanthracene (9-CA), 9-MA, 1-CA, 1-MA, 2-CA, 2-MA, and anthracene in hexane. Only the fluorescence of the 9-substituted derivatives increases, while for anthracene and 1-CA it remains relatively constant, whereas the fluorescence of the remaining compounds has a negative pressure coefficient. Furthermore, the pressure dependence of fluorescence is still clearly a function of solvent, as illustrated in Figure 13.

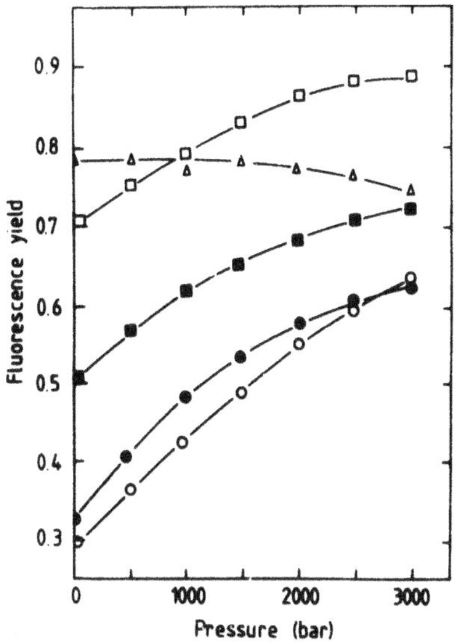

Figure 13. Pressure dependencies of fluorescence yields of 9-methylanthracene in various solvents at room temperature. Hexane (○), methylcyclohexane (●), mixture of hexane and toluene 1 : 1 (■), toluene (□), tetraline (Δ) [71].

For the interpretation of these unusual results the energy level scheme for anthracene must be kept in mind, as it can be taken as representative of the different substituted derivatives. It must only be remembered that their energy levels are shifted slightly relative to anthracene. It is known that the T_2-level of 9-MA in heptane lies approximately 720 cm^{-1} above the S_1-state, so that isc is endothermic [66]. For the 1- and 2-substituted derivatives, however, the T_2-state lies under the S_1-level, as in the case of anthracene [72-74]. The derivatives also resemble anthracene in that deactivation of the S_1-level occurs neither by internal conversion to the ground state nor through isc to T_1. Fluorescence emission and isc to the T_2-state - for some compounds also to the T_3-state - lead to the depopulation of the S_1-level [66].

In Tanaka's interpretation only fluorescence emission and isc to T_2 were taken into account. Thus:

$$\phi_f = k_f/(k_f + k_{isc}) \quad (12)$$

As k_f is pressure independent, the pressure dependence of the fluorescence yield must arise from the pressure dependence of k_{isc}.

In accordance with the energy gap law, Tanaka qualitatively proposed a bell shaped curve for the propability of transition from S_1 to T_2 as a function of the energy difference $\Delta E = E(T_2) - E(S_1)$, as illustrated in Figure 14. The transition propability, which is proportional to k_{isc}, is maximal when S_1 and T_2 are isoenergetic.

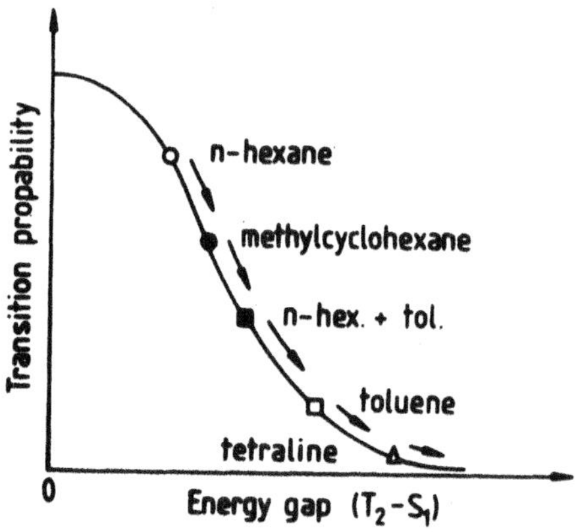

Figure 14. Propability of transition from S_1 to T_2 as a function of the corresponding energy gap [71].

By varying the solvent, the energy gap for each derivative should vary discontinously, as is demonstrated in Figure 14. Thereby k_{isc} also changes in a stepwise manner.

Tanaka concurs with Johnson and Offen in assuming that the S_1-level is energetically lowered with increasing pressure more than the T_2-level. By changing ΔE in this way, it is possible to supplement the stepwise variation of k_{isc} with solvent by a continuous variation of k_{isc} with pressure. For a given starting point on the curve, which is dictated by the ΔE of the derivative in the particular solvent, can a different pressure dependence of k_{isc} be found, as shown by the arrows in Figure 14.

These assumptions allow one to qualitatively describe Tanaka's results for 9-MA, for which ΔE in these solvents is positive. As shown in Figure 14, the energy difference should increase in the series from hexane to tetraline. Therefore k_{isc} should decrease and ϕ_f increase in this series. From Figure 13 it can be seen that this is the case.

Increasing pressure should give rise to an increase in the energy difference. The relationship between the transition propability and ΔE, given in Figure 14, indicates a continuous decrease in the transition propability and in k_{isc}, being stronger in hexane than in toluene or tetraline. This is demonstrated by the arrows in Figure 14. Therefore the pressure coefficient of ϕ_f decreases in this series, as can be seen in Figure 13.

This relatively good qualitative description fails for a number of derivatives for which ΔE is negative, e.g. for anthracene, 1-MA, or 2-MA. The weakness of this theory probably lies in the assumption that intersystem crossing for these compounds only occurs to the T_2-state. For example, from recent studies it is known that for anthracene at room temperature 30 % of the isc leads to the T_3-state and only 70 % to the T_2-state [64].

However, for a description of the pressure dependence of the fluorescence of 9-MA, where in fact only isc to T_2 is of significance, Tanaka's model is not yet established. A proof of the model's validity would be possible through an additional investigation of the temperature dependence of the fluorescence quantum yield in the above mentioned solvents, as then one could determine the transition propability W and ΔE by the equation:

$$k_{isc}(T) = W \exp(-\Delta E/RT) \qquad (13).$$

A plot of W against ΔE should then take the form of the right hand side of the bell shaped curve.

In a recent article Drickamer et al. reported an extraordinarily large, positive pressure coefficient of the fluorescence intensity for a series of 9-carbonyl-substituted anthracenes [75]. The pressure dependence of the emission spectrum of 9-acetylanthracene in PS is illustrated in Figure 15. This compound practically does not emit at normal pressure and room temperature. However, at low temperatures and high pressure emission is observed, increasing by two orders of magnitude with increasing pressure and decreasing temperature. In addition a strong red shift of the fluorescence is observed.

One can conclude from the temperature dependence of the fluorescence that isc from $S_1(\pi\pi^*)$ to $T_2(n\pi^*)$ proceeds over a thermal barrier. As the S_1-state is deactivated mainly by isc, it is suggested that the observed effects are due to the temperature

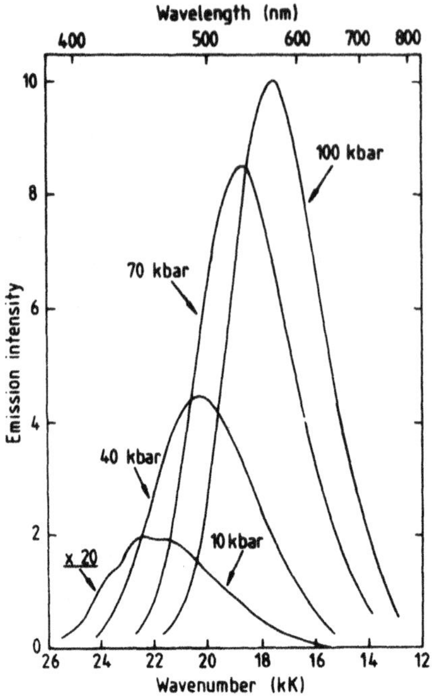

Figure 15. Pressure response of fluorescence emission for 9-acetyl-anthracene in polystyrene (PS) [75].

and pressure dependencies of k_{isc}. According to Drickamer at least two factors determine the pressure dependence of k_{isc}:

1. an increase in the energy gap between T_2 and S_1 with increasing pressure, which corresponds to a higher thermal barrier.

2. a decrease in the propability of transition from S_1 to T_2, which results from increased splitting of $S_1(\pi\pi^*)$ (red shifted) and $T_2(n\pi^*)$ (probably blue shifted) thereby leading to a reduced mixing of the $\pi\pi^*$ and $n\pi^*$ states.

In summarizing the interpretation of the so-called exceptions, the following remarks should be made.

A quantitative description of the pressure dependence of fluorescence lifetimes or intensities has not as yet been successful. In every case where a strong pressure dependence of the emission occurs, the energy difference between the S_1- and T_2-levels is slightly positive or negative. In each case it was assumed that with increasing pressure the S_1-level is lowered more than the

T_2-state, whereby isc is directly influenced.

For those compounds for which isc is thermally controlled, a pressure induced increase in the thermal barrier was assumed. However, it is uncertain how strong this increase is. In addition to these energetic restrictions Tanaka and Drickamer discussed a further influence on k_{isc} resulting from changes in the transition propability.

It is unclear which of these effects is dominant, although it should be possible to answer this question by combined pressure and temperature dependent measurements of fluorescence lifetimes or quantum yields.

III.2. Electronic Energy Transfer

We intend to discuss electronic energy transfer processes in terms of an intermolecular transfer, in which the electronic excitation energy is transferred from a primary excited donor D^* to the acceptor A in its electronic ground state. The donor

$$D^* + A \longrightarrow D + A^*$$

molecule is thereby deactivated to its ground state while the acceptor molecule is excited to a higher electronic state. Only in recent times has the pressure dependence of energy transfer processes been investigated. Here it was found that depending on the mechanism of energy transfer - resonance or exchange type - different pressure dependencies were observed.

III.2.1. Resonance Energy Transfer

According to the resonance mechanism the electronic excitation energy is transferred radiationlessly from a donor D^* to an acceptor A by dipole-dipole-interactions. Because of the nature of the

$$D^* + A \xrightarrow{k_{et}^r} D + A^*$$

interaction this process takes place over large distances - up to 100 Å - from D^* to A. Therefore it is not a collision process, but rather a long range one, which is almost independent of the viscosity of the surrounding medium and can occur in fluid as well as in frozen solutions [11,76].

As Förster and Dexter could show, the following Equation (14) holds for the rate constant k_{et}^r of energy transfer of the resonance type:

$$k_{et}^{r} = k\,(\tau_d^{o})^{-1}\,R_{da}^{-6} \int F_d(\tilde{\nu})\,\varepsilon(\tilde{\nu})\,\tilde{\nu}^{-4}\,d\tilde{\nu} \qquad (14)$$

k_{et}^{r} is proportional to both the reciprocal of the natural lifetime τ_d of the excited donor D*, and to the reciprocal of the sixth power of the donor-acceptor distance R_{da} in a solid matrix, and directly proportional to the integral of spectral overlap of normalized donor emission $F_d(\tilde{\nu})$ and acceptor absorption $\varepsilon(\tilde{\nu})$.

The overlap integral becomes larger as the acceptor absorption increases.

The proportionality constant k and τ_d^{o} in Equation (14) are pressure independent. R_{da} varies with pressure depending on the compressibility of the sample. The integral of spectral overlap can be a function of pressure depending on whether or not donor emission and acceptor absorption are shifted in the same manner. A knowledge of these shifts allows one to calculate the integral of spectral overlap as a function of pressure and so test Förster's equation by high pressure measurements. This was undertaken by Bieg and Drickamer using fcc KCl for example, which was doped with AgCl as donor and TlCl as acceptor [80].

The more important properties of this donor-acceptor pair are illustrated in Figure 16 at atmospheric pressure and 18 kbar. Absorption bands are represented by solid lines and emission bands by dashed lines. For clarity, bands associated with the Ag^+-donor are shown as hatched regions. The Ag^+emission - Tl^+absorption spectral overlap region, related to the degree of resonance between donor and acceptor, is shown cross hatched.

At normal pressure the donor Ag^+ absorbs the high energy (44000 cm^{-1}) incident photon. The electron then relaxes to the bottom of the Ag^+ excited state well. The remaining excitation energy may be passed to the acceptor Tl^+ nonradiatively through the resonance mechanism, followed by a photon emission at 32000 cm^{-1}. The probability of resonance transfer is proportional to the overlap of the donor emission and acceptor absorption bands, which is seen to be small in the system at normal pressure. Therefore upon donor excitation only a weak acceptor emission is observed. If energy transfer does not occur, the excited donor Ag^+ will simply emit a photon as the electron returns to the ground state.

The bottom diagram in Figure 16 shows the case at high pressure. The Ag^+ donor emission and Tl^+ acceptor absorption bands have shifted significantly resulting in a larger region of spectral overlap. In accordance with the transfer predicted by the Förster-Dexter theory, the efficiency of energy transfer, and thus the amount of Tl^+ emission, has increased substantially over the low pressure intensity.

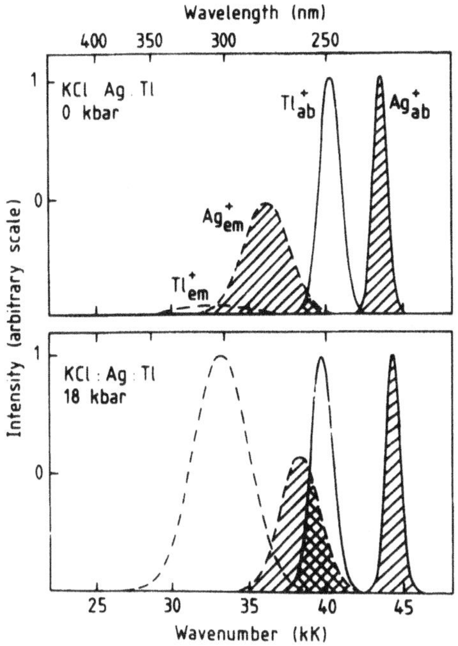

Figure 16. Schematic representation of spectra for KCl : Ag : Tl at low and high pressure [80].

As the donor and acceptor both emit and because energy transfer decreases the donor lifetime, the efficiency of energy transfer can be quantitatively determined by two methods:

1. by measurement of the intensity of donor emission (characteristic for nontransferred energy) and acceptor emission (characteristic of the amount of energy transferred).

2. by direct measurement of the Ag^+ donor lifetime.

The points shown in Figures 17 and 18 correspond to the values for the pressure dependence of the efficiency of energy transfer calculated according to the above-mentioned methods, while the drawn curves represent the efficiency-functions calculated after Förster and Dexter. The efficiency of energy transfer increases continuously with pressure due to an increase in the spectral overlap. The pressure coefficient amounts to ca. 5 to 10 % per kbar. Although the description of the intensity measurements is quite adequate, despite the obvious scatter, the life-

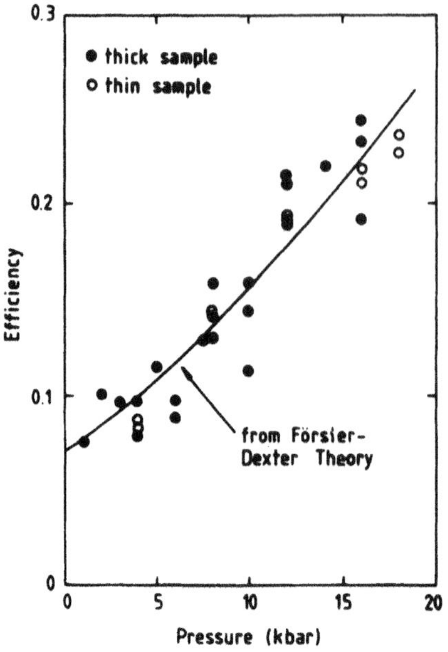

Figure 17. Comparison of transfer efficiency measured from relative emission intensities with theory for KCl : Ag : Tl [80].

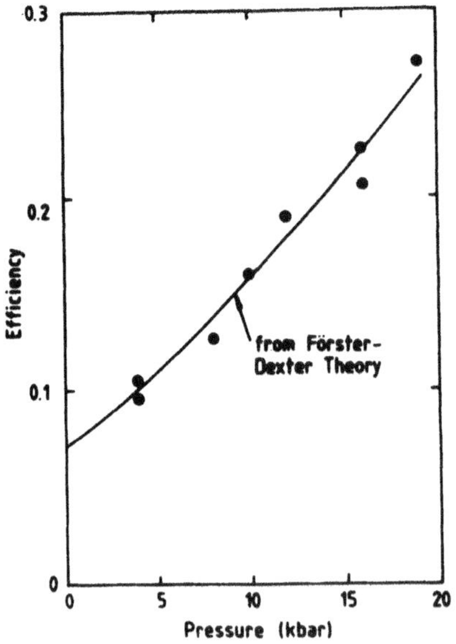

Figure 18. Comparison of transfer efficiency measured from sensitizer lifetimes with theory for KCl : Ag : Tl [80].

time measurements are in excellent agreement with the theory.

Thus Bieg and Drickamer were able to impressively demonstrate the validity of the theory developed by Förster and Dexter with the aid of spectroscopic measurements at high pressures.

III.2.2. Exchange Energy Transfer

A second type of energy transfer can take place only if the two molecules approach each other so closely that they may be considered to be in molecular contact - i.e. their centers are separated by the sum of their molecular radii. In this case D* and A are close enough to allow their electron clouds to overlap. In the region of overlap the electrons are indistinguishable so that an excited electron of D* may exchange with an unexcited electron of A - i.e. the exchange mechanism for energy transfer operates.

According to Dexter this mechanism operates for energy transfer processes in which the multiplicity is conserved. Such a process

$$^3D_1 + {}^1A_o \xrightarrow{k_{et}^{ex}} {}^1D_o + {}^3A_1$$

is a triplet-triplet- (TT-) energy transfer. In the case where this process is exothermic and back-transfer of energy is not possible, the rate of exchange energy transfer will only be limited by the rate of diffusion of donor and acceptor through the solvent. The rate constant k_{et}^{ex} is then basically determined by the Debye equation

$$k_{dif} = 8\,RT / 3000\,\eta \qquad (15)$$

in which η represents the viscosity of the solvent [11,76].

In that the viscosity of most organic solvents is strongly dependent on hydrostatic pressure (η approximately doubles per kbar), means that the pressure effect on the rate constant k_{et}^{ex} for diffusion controlled exchange energy transfer is significantly greater than the corresponding effect on the rate constant for resonance transfer k_{et}^{r}. Therefore the application of high pressure to energy transfer processes in solution can establish the nature of the exchange energy transfer. To illustrate this concept the following example was chosen [81].

The thermal decomposition of tetramethyl-1,2-dioxetane (TMD) quantitatively leads to the formation of acetone, whereby approximately 50 % of the acetone produced is in its excited triplet state and approximately 0.1 % in its excited singlet state

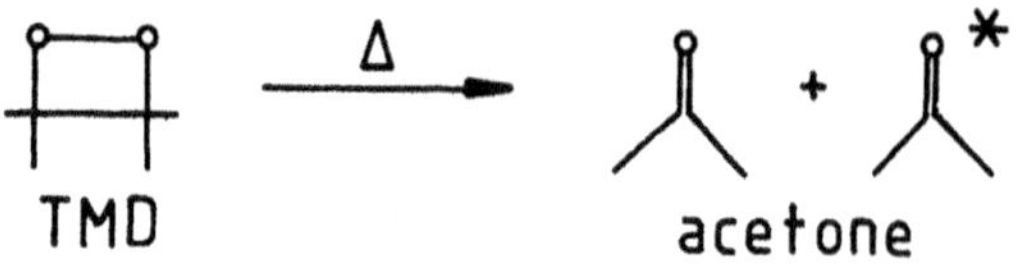

[82,83].

Due to the high yield of excited acetone the thermolysis of TMD is accompanied by chemiluminescence. The amount of light emitted by the reaction can be increased by a number of orders of magnitude when 9,10-dibromoanthracene (DBA) is added to the solution containing TMD. This is caused by an efficient energy transfer from triplet acetone to DBA as acceptor with the consequent fluorescence emission of DBA.

We were able to show that the transfer process in liquid solution occurs via an exchange mechanism [81]. Initially energy is transferred in a

$$^3A_1 + {}^1DBA_o \xrightarrow{k_{et}^{TT_2}} {}^1A_o + {}^3DBA_2 \qquad \text{TT-transfer}$$

$$^3DBA_2 \xrightarrow{\phi_{isc}^{TS}} {}^1DBA_1 \qquad \text{intersystem crossing}$$

spinallowed and exothermic manner to DBA, whereby both the T_1- and the T_2-states of DBA may be competively excited. However, the excited singlet state, from which the fluorescence emission occurs, can only be reached by isc from the T_2-state with a yield of ϕ_{isc}^{TS}. In competition with energy transfer, deactivation of triplet acetone to its ground state occurs with the rate constant k_{dea}^{TA}. By energy transfer, the acceptor DBA quenches the T_1-state of donor acetone by a diffusion controlled process with a rate constant k_{dif}^{DBA}.

Assuming that the T_1- and T_2-states are excited with the same probability one obtains for the efficiency of energy transfer ϕ_{et} from triplet acetone to S_1-DBA:

$$\phi_{et} = \frac{0.5\, k_{dif}^{DBA}\,[DBA]\,\phi_{isc}^{TS}}{k_{dif}^{DBA}\,[DBA] + k_{dea}^{TA}} \qquad (16)$$

From measurements of the intensity of indirect chemiluminescence as a function of DBA concentration it can be shown that $k_{dif}^{DBA} = 2.6 \times 10^{10}\ M^{-1}\ s^{-1}$, which compares favourably with the value calculated from Equation (16), i.e. $k_{dif} = 1.5 \times 10^{10}\ M^{-1}\ s^{-1}$. k_{dea}^{TA} was determined by quenching experiments with 2,4-hexadiene to be $5 \times 10^7\ s^{-1}$. Similar experiments conducted at higher pressures showed k_{dea}^{TA} to be pressure insensitive.

The efficiency of energy transfer is graphically presented in Figure 19 as a function of pressure and acceptor concentration. These values were determined from the pressure dependence of the fluorescence quantum yield of DBA; the pressure dependence of the thermolysis of TMD [84]; the pressure and concentration dependencies of the intensity of chemiluminescence.

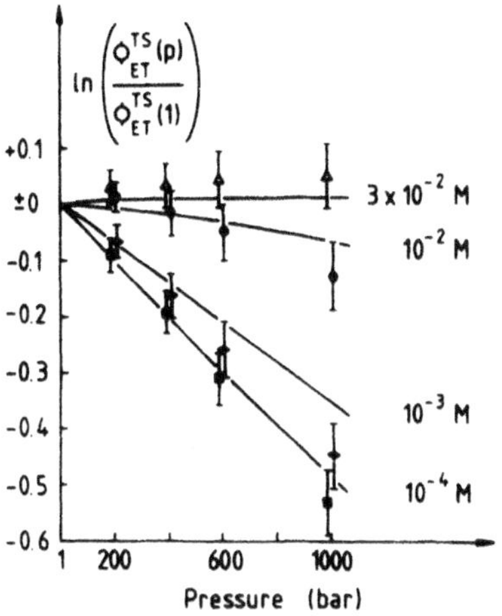

Figure 19. Pressure and concentration dependence of the energy transfer efficiency [81].

The curves drawn in Figure 19 were calculated according to Equation (16) using the given rate constants. Herein the pressure coefficient of k_{dif}^{DBA} was assumed to be equal to the reciprocal of the viscosity of the solvent toluene. Only the weak increase in ϕ_{isc}^{TS}, amounting to 5 % at 1000 bar, was adjusted to fit the data.

The good description of the results by the calculated curves establishes that the two-step, indirect energy transfer, in which triplet acetone is the donor and DBA the acceptor, operates according to the exchange mechanism.

In a more recent study we were able to demonstrate from high pressure measurements that the same mechanism is valid for energy transfer from triplet acetone to 9-bromoanthracene and 9,10-dichloroanthracene, respectively [85].

Recently, Yersin, Gliemann and Stock reported a strong pressure effect on the energy transfer in single crystals of

$Sm_2(Pt(CN)_4)_3 \cdot 18H_2O$ [86,87]. In these crystals $Pt(CN)_4^{2-}$ exists as single units arranged in a columnar structure and are separated by Sm^{3+} cations. By irradiating the anion bands not only was anion emission observed but also emission of the cations. This effect was attributed to an exchange energy transfer which was thought to be also possible in the crystalline form due to the small distance between $(Pt(CN)_4)^{2-}$ and Sm^{3+} of about 3 Å.

Upon exertion of pressure up to 25 kbar the donor emission bands are extensively red shifted by about 4000 cm^{-1}. At the same time, the acceptor emission disappears. The reason for the repression of the energy transfer can be explained in that this process becomes endothermic at high pressures and thereby looses its efficiency.

III.3. Photochemical Primary Processes

Photochemical reactions can be thought of as involving molecules in particular electronic states. Primary reactions originating from excited electronic states are generally simple chemical elementary processes e.g. bond formation or acid-base reactions. They occur in solution in a similar manner to energy transfer processes either from thermally equilibrated S_1- or T_1-states.

Photochemical primary processes originating from the first excited singlet state can be monitored by fluorescence spectroscopy providing light emission occurs. As the average lifetime of the S_1-state of organic molecules is ca. 10 ns, only such photochemical processes can be observed by fluorescence spectroscopy whose half lives are not much longer than 10 ns. Only then do these processes effectively compete with other radiationless deactivation processes and, in particular, fluorescence emission.

In the case of bimolecular reactions the half live τ_r is given by:

$$\tau_r = (k_b c_r)^{-1} \qquad (17)$$

Here k_b represents the bimolecular rate constant and c_r the concentration of the reaction partner. An upper limit of k_b is set for diffusion controlled reactions by k_{dif}. For typical organic solvents at 25°C, k_{dif} is in the order of 10^{10} M^{-1} s^{-1}.

Table 6 shows the concentration c_r of the reactant as a function of k_r for the case where τ_r = 10 ns. From these values it can be seen that for low concentrations of the reaction partner (c_r << 10^{-2}) photochemical reactions do not originate from the short lived singlet state. Generally, under these conditions, the time required for the encounter between the potential reacting partners is so long that the S_1-state has already been deactivated by other processes.

Table 6. Reactant concentration for a range of bimolecular rate constants where τ_r = 10 ns.

c_r [M]	10	1	10^{-1}	10^{-2}
k_b [M^{-1} s^{-1}]	10^7	10^8	10^9	10^{10}

However, processes with relatively small rate constants can still effectively originate from the S_1-state. They can be observed by fluorescence measurements if the reaction partner is the solvent as it has a high effective concentration. To such photochemical processes belong, among others, protolytic reactions of excited molecules in aqueous solutions.

III.3.1. Protolytic Reactions of Electronically Excited Molecules

Detailed kinetic information can also be obtained from the application of high pressure to protolytic reactions. In this contex we wish to present a brief description of such a high pressure investigation carried out by Leiber and Weller [88].

The dye acridine fluoresces in aqueous alkaline solutions. In

acridine acridinium cation

acid solution, where acridine is protonated, one can observe a red shift of the fluorescence spectrum corresponding to the acridine cation. In Figure 20 the fluorescence spectra of acridine at pH 13 and the acridinium cation at pH 3 are presented.

Whereas the cation fluorescence is pressure independent, the fluorescence spectrum of acridine at 5 kbar exhibits a small red shift which can be attributed to a slight formation of acridinium cations in their excited states.

The pressure effect on the fluorescence spectrum of an analogous solution at pH 8.7 is substantially greater, as shown in Figure 21.

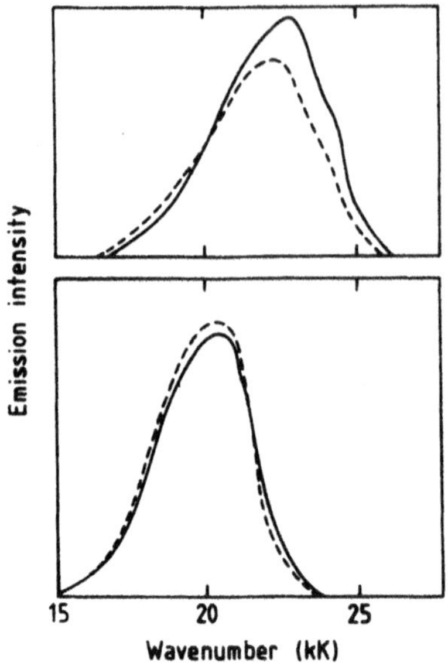

Figure 20. Fluorescence spectra at 1 bar (——) and at 5 kbar (---). Upper part acridine in water at pH 13, lower part acridinium cation in water at pH 3 [88].

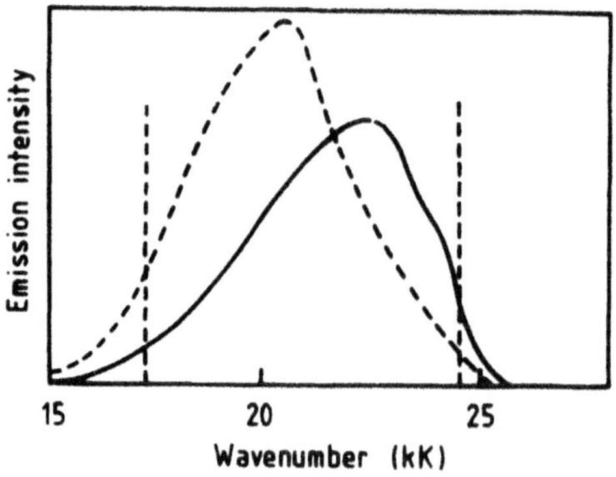

Figure 21. Fluorescence spectra of acridine in water (pH 8.7) at 1 bar (——) and at 5 kbar (---) [88].

At this pH the absorption spectrum of acridine can be seen, but not of the cation. In contrast, the fluorescence spectrum is a combination of both acridine and its cation, whereby an increase of 5 kbar causes the emission to shift further to favour cation fluorescence. This results from the addition of a proton to the excited acridine A^* from which an excited acridinium cation $(AH^+)^*$ is formed.

The most important condition for the occurance of the protolytic reaction of the excited acridine is the discontinuous change in the basic properties during excitation ($\sim 10^{-15}$ s). Whereas the equilibrium for the ground state reaction lies completely to the

$$A + H_2O \rightleftharpoons AH^+ + OH^- \qquad [A]/[AH^+] = 1.800$$

$$A^* + H_2O \rightleftharpoons (AH^+)^* + OH^- \qquad [A^*]/[(AH^+)^*] = 0.02$$

left at pH 8.7, for the reaction involving excited acridine it is shifted completely to the right. A^* is therefore a substantially stronger base than A. Thus the protonation reaction, which proceeds with the rate constant k_p, can be observed at this pH [89]. The equilibrium is not established because the $(AH^+)^*$ so formed is too rapidly deactivated.

k_p was calculated as a pseudo-first-order rate constant for 25°C at 1 bar to be 8.4×10^6 s^{-1}. Thus the half life of the reaction is τ_r = 80 ns. Thereby τ_r is in the order of magnitude of the lifetime of excited acridine, i.e. 14 ns, so that the increase in k_p with pressure can be followed by fluorescence spectroscopy. The essential bimolecular rate constant is then k_b = 1.5×10^5 M^{-1} s^{-1} due to the high concentration of water of 55 M.

The following equations apply to the fluorescence quantum yield of acridine ϕ_f^A and the acridinium cation ϕ_f^{AH}, as well as to the quantum yield for protonation ϕ_p:

$$\phi_f^A = k_f^A/(k_f^A + k_{dea}^A + k_p) \tag{18}$$

$$\phi_f^{AH} = k_f^{AH}/(k_f^{AH} + k_{dea}^{AH}) \tag{19}$$

$$\phi_p = k_p/(k_f^A + k_{dea}^A + k_p) \tag{20}$$

Here, k_f and k_{dea} are the rate constants for fluorescence and other deactivation processes, respectively. The respective indices A and AH represent acridine and the acridinium cation. The fluorescence intensities I and I', which are shown in Figure 21, represent acridine and its cation, respectively. The Equations (21) and (22) hold:

$$I = a\,\phi_f^A \quad (21) \qquad I' = b\,\phi_p\,\phi_f^{AH} \quad (22)$$

For the quotient $Q = I'/I$ one can derive:

$$Q = b\, k_p\, \phi_f^{AH}/(a\, k_f^{A}) \qquad (23)$$

As already mentioned ϕ_f^{AH} is pressure insensitive, b and a are proportionality constants and k_f may also be considered pressure independent. Thereby one can express Q as $Q = c\, k_p$, in which c is a pressure independent constant.

From measurements of the quotient Q at different pressures it is possible to calculate the activation volume $\Delta V^{\neq}$ from the following equation:

$$\Delta V^{\neq} = -RT\,(d \ln Q/dP)_T \qquad (24)$$

The data reported by Leiber and Weller are illustrated in Figure 22.

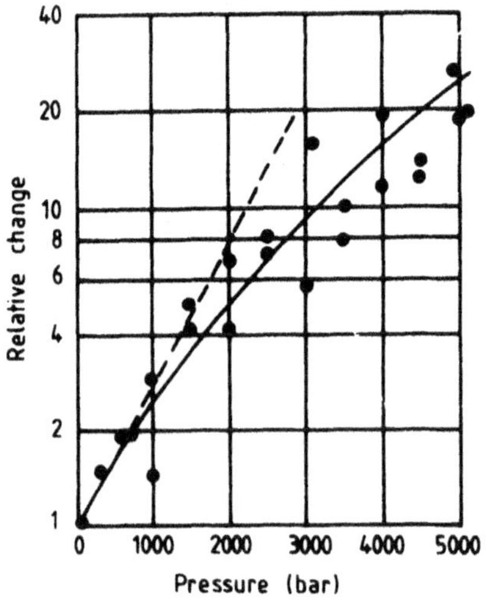

Figure 22. Pressure dependence of the relative rate constant Q [88].

Although the scatter is relatively high, one can recognize a significant increase in Q, and thereby in k_p with increasing pressure. As k_p at 25°C and 1 bar amounts to about $10^7\ s^{-1}$, k_p is then $2 \times 10^8\ s^{-1}$ at 5 kbar, which corresponds to a half life of the reaction at 5 kbar of only 5 ns.

From the initial slope an activation volume of - 25 cm^3/mole was calculated. This reduction in volume during the formation of the transition state can be essentially rationalized in terms of electrostriction resulting from the formation of an acridinium cation and a hydroxyl anion.

III.3.2. Formation of Excimers

The term "excimers" has become familiar as a name for atomic or molecular dimer aggregates, which are unstable in their ground state but stable in their electronically excited state. In 1954 Kasper and Förster [90] reported the behavior of pyrene which, in concentrated solutions, exhibits a red-shifted structureless band in addition to its normal fluorescence band, as shown in Figure 23. No corresponding feature in the absorption spectrum was discovered. Because of the supplementary concentration dependence of both components of the fluorescence spectrum, the red-shifted band was ascribed to a dimeric excited species D^* formed by the addition of an excited monomer M^* to a ground state monomer M according to the kinetic scheme:

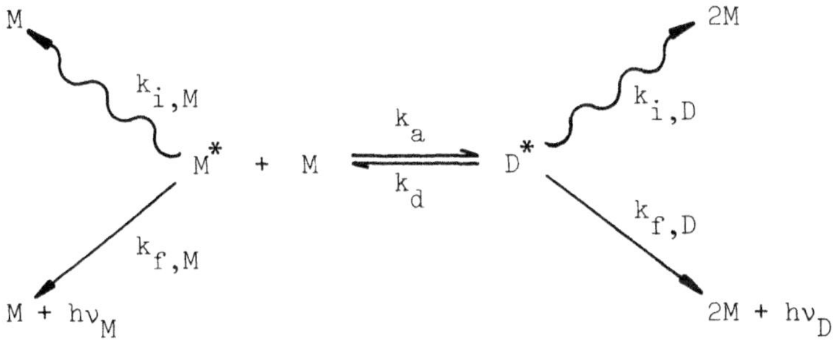

In this scheme $k_{f,M}$ and $k_{f,D}$ represent the radiative, and $k_{i,M}$ and $k_{i,D}$ the radiationless decay rates of the excited species M^* and D^*, respectively. Excimer formation is characterized by k_a, its dissociation by k_d. These processes compete with the total unimolecular decay processes $k_M = k_{f,M} + k_{i,M}$ and $k_D = k_{f,D} + k_{i,D}$, respectively.

The two rate constants k_d and k_D determine whether or not the excited state equilibrium is established. For $k_d \gg k_D$ equilibrium is reached whereas for the reverse situation, i.e. $k_d \ll k_D$, equilibrium is not established. Under which conditions these limiting cases may be reached can be determined with the help of fluorescence studies under high pressure.

Such investigations were first made by Förster and coworkers [91 - 93] using pyrene, 1,2-benzanthracene and various derivatives of naphthalene. For example, the influence of pressure on the excimer formation of pyrene at 296 K is shown in Figure 24.

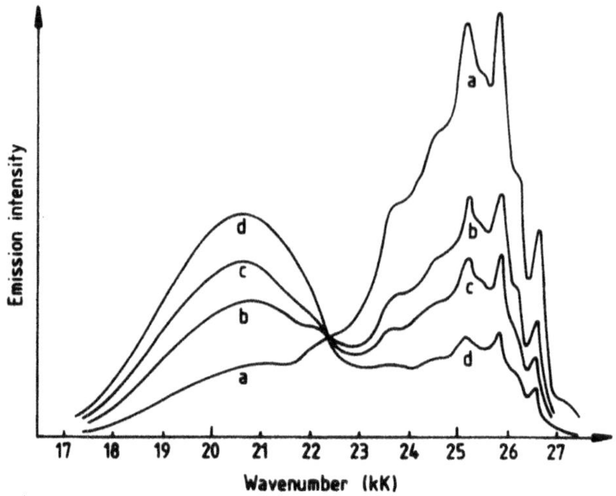

Figure 23. Fluorescence spectra of pyrene in n-heptane. T = 293 K, concentration (M): 5×10^{-5} (a), 1.8×10^{-4} (b), 3.1×10^{-4} (c) and 7.0×10^{-4} (d) [90].

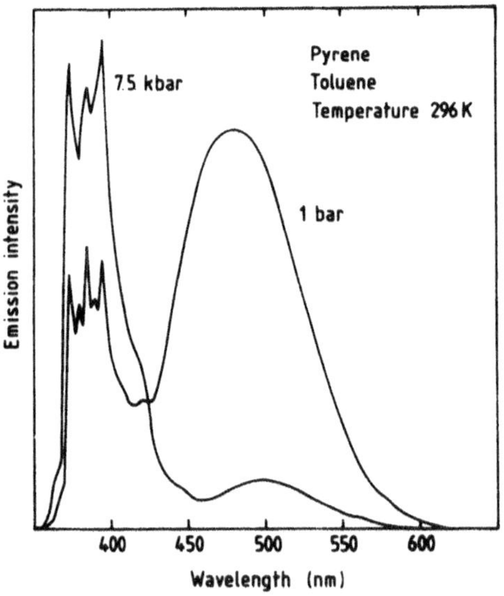

Figure 24. Fluorescence spectrum of 3×10^{-3} M pyrene in toluene at different pressures [94].

As can be observed, excimer formation dereases markedly at room temperature with increasing pressure. In order to evaluate the pressure experiments, Förster and coworkers determined the ratio of the maximum intensities of fluorescence of the dimer and monomer component, I_D/I_M, as a function of pressure. It could be shown that the pressure dependencies of these ratios (Q) are as follows:

$$\text{for } k_d \ll k_D: \quad Q = \frac{I_{D,P}\, I_{M,o}}{I_{D,o}\, I_{M,P}} = \frac{k_{a,P}}{k_{a,o}} \tag{25}$$

$$\text{for } k_d \gg k_D: \quad Q = \frac{I_{D,P}\, I_{M,o}}{I_{D,o}\, I_{M,P}} = \frac{k_{d,o}\, k_{a,P}}{k_{d,P}\, k_{a,o}} = \frac{K_P^*}{K_o^*} \tag{26}$$

The indices P and o characterize the data at pressure P and 1 bar, respectively.

In the first limiting case, Q equals the ratio of the rate constants of association, while in the second, Q corresponds to the ratio of the equilibrium constants of the excited states.

For the first case, where the excited state equilibrium is not reached, one can use high pressure studies to establish whether excimer formation follows a diffusion controlled process. Here the ratio Q should reflect the same pressure dependence as the ratio η_o/η_P, the viscosities of the solvent at 1 bar and pressure P.

In the situation where the excited state equilibrium is established, the ratio Q should increase with pressure. The volume change ΔV, associated with excimer formation, can be calculated according to:

$$d \ln Q/dP = d \ln K_P^*/dP - d \ln K_o^*/dP = -\Delta V/RT \tag{27}$$

The following results were obtained from Förster and coworkers:

For the pyrene system at room temperature the equilibrium is not reached. This can be seen qualitatively from Figure 24, where the dimer concentration decreases with increasing pressure. Figure 25 contains a plot of Q versus pressure, together with the pressure dependencies of the relative reciprocal solvent density and the relative reciprocal solvent viscosity. The strong decrease of Q with pressure reflects a reduction of the dimer fluorescence component compared to the monomer component. This can be explained by a diffusion controlled excimer formation, the latter being reduced at higher viscosities of the solvent. An inspection of Figure 25 demonstrates that the decrease in Q is almost identical with the change in the relative reciprocal viscosity. Similar results were obtained in methanol and isobutanol. For the pyrene system the equilibrium can only be reached at rather high temperatures.

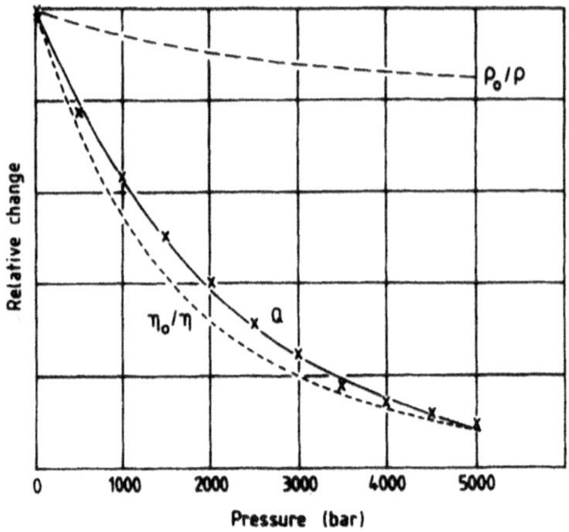

Figure 25. 5×10^{-3} M pyrene in toluene. Pressure dependence of: x-x-x Q, reduced excimer/monomer fluorescence yield; ---- ρ_o/ρ, reciprocal relative solvent density; ---- η_o/η, reciprocal relative solvent viscosity [91].

The pressure dependence of Q for a 0.1 molar solution of 1,2-benzanthracene in toluene, plotted in Figure 26, exhibits a different behavior. Despite the normal increase in the viscosity of the solvent with pressure the ratio Q increases up to ca. 2300 bar. In this system the equilibrium is established at 293 K and 1 bar. The value of Q is therefore not determined by the rate of association, but rather by the equilibrium constant K^*. An increase of this equilibrium constant with pressure is to be expected since the excimer formation is accompanied by a decrease in volume. At higher pressures Q decreases because of the increase in the viscosity of the solvent. Thus the rates of association and dissociation become so slow that the equilibrium is no longer established during the lifetimes of the excited species. From the initial slope of d ln Q/dP one obtaines a volume change of $- 6.2\ cm^3\ mol^{-1}$ for excimer formation at 1 bar.

For the naphthalene derivatives investigated, the excited state equilibria are also established at room temperature. The values of ΔV for these systems vary between -11 and $-17\ cm^3\ mol^{-1}$ and are of the expected order of magnitude. They indicate a molecular distance in the excimer of about 3 Å.

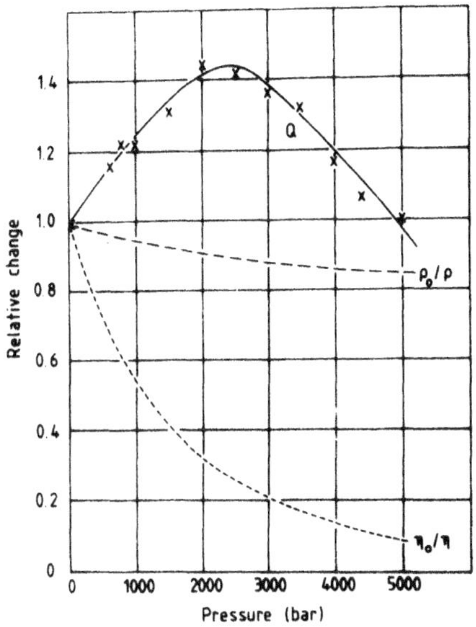

Figure 26. 0.1 M 1,2-benzanthracene in toluene [91].

The pyrene - toluene system has been studied in more detail by temperature and pressure dependent measurements of the emission lifetimes in order to separate the different kinetic and thermodynamic parameters.This approach enabled Johnson and Offen [94] to evaluate the enthalpies and volumes of activation for the association and dissociation processes. The values for the association process are related to the viscosity of the solvent, thus providing further evidence that excimer formation is diffusion controlled.

IV. ACKNOWLEDGEMENTS

The data in Figures 1-8 and 10-26, and in Tables 1-4 have been reproduced from: H. W. Offen, et al., Spectochimica Acta, 25A, 1023 (1969) with modifications by the Authors of the present paper, who would like to thank Pergamon Press Ltd. for their permission. They are especially indebted to Professor Drickamer for providing unpublished manuscripts. The results from our own laboratories produced here were supported by various grants from the Deutsche Forschungsgemeinschaft.

V. REFERENCES

1. O.E. Weigang, Jr., and W.W. Robertson, in High Pressure Physics and Chemistry (ed. R.S. Bradley), Academic Press, New York, 1963.
2. B. Oksengorn, H.Vu and B. Vodar, Physics of high Pressures and the Condensed Phase (ed. A. van Itterbeek), Wiley, London, 1965.
3. H.W. Offen, in Organic Molecular Photophysics (ed. J.B. Birks) Vol. 1, Wiley, London, 1973.
4. H.G. Drickamer and C.W. Frank, Electronic Transitions and the High Pressure Chemistry and Physics of Solids, Chapman and Hall, London, 1973.
5. R.A. Fitch, T.E. Slykhouse and H.G. Drickamer, J. Opt. Soc. Am. 47, 1015 (1957)
6. H.G. Drickamer and A.S. Balchan, in Modern Very High Pressure Techniques (ed. H.R. Wentorf, Jr.), Butterworths, London, 1962.
7. D.S. Hughes and W.W. Robertson, J. Opt. Soc. Am. 46, 557 (1956).
8. W.W. Robertson, Techniques Inorg. Chem. 1, 157 (1963).
9. H.W. Offen, R.L. Tanquary and K.F. Sterrett, J. Appl. Phys. 38, 5245 (1967).
10. H.W. Offen, Rev. Sci. Instr. 39, 1691 (1968).
11. N.J. Turro, Molecular Photochemistry, W.A. Benjamin, Inc., New York 1967.
12. Th. Förster, Fluoreszenz organischer Verbindungen, Vandenhoeck und Ruprecht, Göttingen, 1951.
13. H.W. Offen and R.A. Beardless, J. Chem. Phys. 48, 3584 (1968).
14. B. Vodar, Proc. Roy. Soc. A 255, 44 (1960).
15. B. Vodar, Spectrochim. Acta, 14, 213 (1959).
16. B. Oksengorn, J. Rech. C. N. R.S. 9, 339 (1958).
17. B. Oksengorn, Compt. Rend. Acad. Sci. 240, 2300 (1955); 242, 2324 (1956).
18. B. Oksengorn and B. Vodar, J. Phys. Radium 16, 893 (1955).
19. S.E. Babb, Jr., J.M. Robinson and W.W. Robertson, J. Chem. Phys. 30, 427 (1959)
20. W.W. Robertson and A.D. King, Jr., J. Chem. Phys. 31, 473 (1959).
21. A.D. King, Jr., and W.W. Robertson, J. Chem. Phys. 37, 1453 (1962).
22. A.D. King, Jr., and W.W. Robertson, J. Chem. Phys. 39, 3071 (1963).
23. W.W. Robertson, S.E. Babb, Jr., and F.A. Matsen, J. Chem. Phys. 26, 367 (1957).
24. W.W. Robertson and S.E. Babb, Jr., J. Chem. Phys. 28, 953 (1958).
25. W.W. Robertson, J.M. Robinson and S.E. Babb, Jr., J. Chem. Phys. 29, 243 (1958).

26. W.W. Robertson and F.A. Matsen, J. Chem. Phys. 23, 2468 (1955).
27. W.W. Robertson, O.E. Weigang, Jr., and F.A. Matsen, J. Mol. Spectroscopy 1, 1 (1957).
28. W.W. Robertson and A.D. King, Jr., J. Chem. Phys. 34, 1511 (1961).
29. B.Y. Okamoto and H.G. Drickamer, J. Chem. Phys. 61, 2870 (1974).
30. H.W. Offen, Ph. D. Thesis, University of California, Los Angeles, 1963; Diss. Abstracts 24, 1853 (1963).
31. M. Nicol, J. Opt. Soc. Am. 55, 1176 (1965).
32. H.W. Offen, J. Chem. Phys. 42, 2523 (1965).
33. H.W. Offen and E.H. Park, J. Chem. Phys. 43, 1848 (1965).
34. H.W. Offen and R.R. Eliason, J. Chem. Phys. 43, 4096 (1965).
35. E.H. Park, A.H. Kadhim and H.W. Offen, Photochem. Photobiol. 8, 261 (1968).
36. H.W. Offen, S.A. Balbo and R.L. Tanquary, Spectrochim. Acta 25A, 1023 (1969).
37. M. Nicol and J. Somekh, J. Opt. Soc. Am. 58, 233 (1968).
38. M. Nicol, W.D. Ellenson and R. Geffner, in Organic Scintillators and Liquid Scintillation Counting (eds. D.L. Horrocks and C.T. Peng), Academic Press, 1971.
39. T.T. Nakashima and H.W. Offen, J. Chem. Phys. 48, 4817 (1968).
40. W. Liptay, Angew. Chem. 81, 195 (1969).
41. W. Liptay, Z. Naturforsch. 20a, 1441 (1965).
42. J.R. Platt, J. Chem. Phys. 17, 484 (1949).
43. H.B. Klevens and J.R. Platt, J. Chem. Phys. 17, 470 (1949).
44. J.B. Birks, in Organic Molecular Photophysics (ed. by J.B. Birks), Wiley & Sons, New York, 1973.
45. A.H. Kadhim and H.W. Offen, J. Am. Chem. Soc. 89, 1805 (1967).
46. W.W. Robertson, A.D. King, Jr., and O.E. Weigang, Jr., J. Chem. Phys. 35, 464 (1961).
47. M. Nicol, J. Chem. Phys. 45, 4753 (1966).
48. I. Shirotani, T. Kajiwara and H. Inokuchi, Bull. Chem. Soc. Japan 42, 2387 (1969).
49. P.C. Johnson and H.W. Offen, J. Chem. Phys. 57, 336 (1972).
50. B.A. Baldwin and H.W. Offen, J. Chem. Phys. 46, 4509 (1967).
51. S. Arnold, W.B. Whitten and A.C. Damask, J. Chem. Phys. 61, 5162 (1974).
52. E.G. McRae, J. Phys. Chem. 61, 562 (1957).
53. Y. Oshika, J. Phys. Soc. Japan 9, 594 (1954).
54. N.S. Bayliss, J. Chem. Phys. 18, 292 (1950).
55. E. Lippert, Z. Elektrochem. 61, 962 (1957); Z. f. Naturforsch. 10a, 541 (1955).
56. L. Bilot and A. Kawski, Z. Naturforsch. 17a, 621 (1962).

57. A. Zipp and W. Kauzmann, J. Chem. Phys. 59, 4215 (1973).
58. H.G. Drickamer, C.W. Frank and C.P. Slichter, Proc. Natl. Acad. Sci. 69, 933 (1972).
59. B.Y. Okamoto and H.G. Drickamer, J. Chem. Phys. 61, 2878 (1974).
60. H.W. Offen and T.D. Phillips, J. Chem. Phys. 49, 3995 (1968).
61. J.J. Kim, R.A. Beardslee, D.T. Phillips and H.W. Offen, J. Chem. Phys. 51, 2761 (1969).
62. B.A. Baldwin and H.W. Offen, J. Chem. Phys. 48, 5358 (1968).
63. P.C. Johnson and H.W. Offen, Chem. Phys. Letters 6, 505 (1970).
64. S. Schoof, Ph. Thesis, Karlsruhe, KFK Nr. 1873 (1973).
65. T. Medinger and F. Wilkinson, Trans. Faraday Soc. 61, 620 (1965).
66. S. Schoof, H. Güsten and C. von Sonntag, Ber. Bunsenges. Phys. Chem. 81, 305 (1977).
67. W. Siebrand, J. Chem. Phys. 47, 2411 (1967).
68. R.W. Shaw and M. Nicol. Chem. Phys. Letters 39, 108 (1976).
69. F. Tanaka and J. Osugi, Rev. Phys. Chem. Japan 42, 85 (1972).
70. F. Tanaka and J. Osugi, Chem. Letters, 1279 (1973).
71. F. Tanaka, Rev. Phys. Chem. Japan 44, 65 (1974).
72. E.C. Lim, J.D. Laposa and Y.M.H. Yu, J. Mol. Spec. 19, 412 (1965).
73. R.G. Bennet and P.J. McCartin, J. Chem. Phys. 44, 1969 (1966).
74. W.R. Ware and B.A. Baldwin, J. Chem. Phys. 43, 1194 (1965).
75. D.J. Mitchell, G. Schuster and H.G. Drickamer, J. Amer. Chem. Soc. 99, 1145 (1977).
76. A.A. Lamola, Energy Transfer and Organic Photochemistry, Technique of Organic Chemistry XIV, Interscience Publishers, New York, 1969.
77. Th. Förster, Ann. Physik 2, 55 (1948).
78. Th. Förster, Discussions Faraday Soc. 27, 7 (1959).
79. D.L. Dexter, J. Chem. Phys. 21, 836 (1953).
80. K.W. Bieg and H.G. Drickamer, J. Chem. Phys. 66, 1437 (1977).
81. R. Schmidt, H. Kelm and H.-D. Brauer, Ber. Bunsenges. Phys. Chem. 81, 402 (1977).
82. N.J. Turro, P. Lechtken, N.E. Schore, G. Schuster, H.-C. Steinmetzer and A. Yekta, Acc. Chem. Res. 7, 97 (1974).
83. W. Adam, N. Duran and G.A. Simpson, J. Amer. Chem. Soc. 97, 5464 (1975).
84. R. Schmidt, H.-C. Steinmetzer, H.-D. Brauer and H. Kelm, J. Amer. Chem. Soc. 98, 8181 (1976).

85. R. Schmidt, H.-D. Brauer and H. Kelm, J. Photochem. 8, 217 (1978).
86. H. Yersin, Ber. Bunsenges. Phys. Chem. 80, 1237 (1976).
87. H. Yersin, M. Stock and G. Gliemann, private communication.
88. C.O. Leiber and A. Weller, Chem. Ing. Tech. 39, 563 (1967).
89. A. Weller, Ber. Bunsenges. Phys. Chem. 61, 956 (1957).
90. Th. Förster and K. Kasper, Z. physik. Chem. N. F. 1, 275 (1954).
91. Th. Förster, C.O. Leiber, H.-P. Seidel and A. Weller, Z. physik. Chem. N. F. 39, 265 (1963).
92. H.-P. Seidel and B.K. Selinger, Australian J. Chem. 18 977 (1965).
93. H. Braun and Th. Förster, Ber. Bunsenges. Phys. Chem. 70, 1091 (1966).
94. P.C. Johnson and H.W. Offen, J. Chem. Phys. 56, 1638 (1972).

LIST OF PARTICIPANTS

Askoy, A. I.T.Ü. Kimya, Fakültesi Sinai Kimya, Kürsüsü, Tesvikive-Istanbul.

Alder, B. Lawrence Livermore Laboratory, University of California, Livermore, Cal. 94550, USA.

Anastassopoulou, J. Dept. of Phys. Chem., National Technical University, 42 Patission Street, Athens 147, Greece.

Andersen, B. H. C. Ørsted Institute of Physical Chemistry, University of Copenhagen, Copenhagen, Denmark.

Arnold, I. Institute of Physical Chemistry, University of Frankfurt, Frankfurt am Main, F. R. Germany.

Bloch, D. CNRS, Laboratoire de Magnetisme, Grenoble, France.

Brauer, H.-D. Institute for Physical Chemistry, University of Frankfurt, Frankfurt am Main, F. R. Germany.

Cervone, E. Istituto Chimica Generale ed Inorganica, University of Rome, Rome, Italy.

Couterick, F. Department of Chemistry, Katholieke Universiteit Leuven, Leuven, Belgium.

Constantinides, D. Dept. of Chem., University of Athens, Athens, Greece.

Delaunois, D. Dept. of Chem., University of Brussels, Brussels, Belgium.

Deymann, H. Institut für Erdölforschung, Hannover, F. R. Germany.

Diomedi-Camasse, F. Istituto Chimica Generale ed Inorganica, University of Rome, Rome, Italy.

Drews, W. Institute for Physical Chemistry, University of Frankfurt, Frankfurt am Main, F. R. Germany.

van Eldik, R. State University of New York at Buffalo, Buffalo, N. Y., USA.

Elias, H. Institute for Inorganic Chemistry, Technical University Darmstadt, Darmstadt, F. R. Germany.

Evans, B. J. Institute of Mineralogy, University of Marburg, Marburg, F. R. Germany.

Forslund, B. Arrhenius Laboratory, University of Stockholm, Stockholm, Sweden.

Franck. E. U. Institute of Physical Chemistry, University of Karlsruhe, Karlsruhe, F. R. Germany.

Grivsky, E. Burroughs Wellcome Co., Research Park, N.C. 27709, USA.
Gronlund, F. H.C. Ørsted Institute of Physical Chemistry, University of Copenhagen, Copenhagen, Denmark.
Hammerich, B. Institute of Physical Chemistry, University of Frankfurt, Frankfurt am Main, F.R. Germany.
Hensel, F. Institute of Physical Chemistry, University of Marburg, Marburg, F.R. Germany.
Heremans, K. Department of Chemistry, Katholieke Universiteit Leuven, Leuven, Belgium.
Heydemann, P. National Bureau of Standards, Washington, D.C., USA.
Hippler, H. Institute of Physical Chemistry, University of Göttingen, Göttingen, F.R. Germany.
Höhne, K. Institute of Physical Chemistry, University of Frankfurt, Frankfurt am Main, F.R. Germany.
Hohmann, F. Institute of Physical Chemistry, University of Frankfurt, Frankfurt am Main, F.R. Germany.
Holzapfel, W. Max Planck Institut für Festkörperforschung, Stuttgart. F.R. Germany.
Hyde, K. Department of Chemistry, State University of New York at Oswego, Oswego, N.Y., USA.
Isaacs, N.S. Department of Chemistry, University of Reading, Reading, Berks., England.
Jenner, G. Institute de Chimie, Université de Strasbourg, Strasbourg, France.
Johansson, B. National Defense Research Institute, Stockholm, Sweden.
Jonas, J. Department of Chemistry, University of Illinois, Urbana, Illinois, USA.
Jouanne, J. Institute of Physical Chemistry, University of Frankfurt, Frankfurt am Main, F.R. Germany.
Katsaros, N. Department of Chemistry, Nuclear Research Center 'Democritos', Athens, Greece.
Kelm, H. Institute of Physical Chemistry, University of Frankfurt, Frankfurt am Main, F.R. Germany.
Koinis, S. Department of Chemistry, University of Athens, Athens, Greece.
Konstantatos, Y. Department of Chemistry, Nuclear Research Center, 'Democritos' Athens, Greece.
Kotowski, M. Institute of Physical Chemistry, University of Frankfurt, Frankfurt am Main, F.R. Germany.
Kotowski, S. Institute of Physical Chemistry, University of Frankfurt, Frankfurt am Main, F.R. Germany.
Lebert, K.-H. Institute of Physical Chemistry, University of Frankfurt, Frankfurt am Main, F.R. Germany.
Le Noble, W. Department of Chemistry, State University of New York at Stony Brook, Stony Brook, N.Y., USA.
Merbach, A. Institute de chimie minerale et analytique, Université Lausanne, Lausanne, Switzerland.
Meyer, F. Institute de chimie minerale et analytique, Université de Lausanne, Lausanne, Switzerland.

Mroczkowski, S. Department of Engineering and Applied Science, Yale University, New Haven, Connecticut, USA.
Ostoja-Starzewski, K.-H. Institute of Inorganic Chemistry, University of Frankfurt, Frankfurt am Main, F.R. Germany.
Palvra, A. Instituto Superior Technico, Centro de Quimica Estrulral das Universidades de Lisboa Complexo Interdisciplinar, Lisboa, Portugal.
Palmer, M.C. State University of New York at Buffalo, Buffalo, N.Y., USA.
Papaconstantinou, C. Department of Chemistry, Nuclear Research Center 'Democritos', Athens, Greece.
Papadopoulos, C. Department of Chemistry, Nuclear Research Center 'Democritos', Athens, Greece.
Paraskewas, S. Deparment of Chemistry, University of Athens, Athens, Greece.
Perry, A. Department of Chemistry, Nuclear Research Center 'Democritos', Athens, Greece.
Petrou, A. Department of Chemistry, Nuclear Research Center 'Democritos', Athens, Greece.
Pfeil, A. Institute of Chemistry, Fraunhofer Ges., Pfinstal-Berghausen, F.R. Germany.
da Ponte, Instituto Superior Technico, Centro Quimico Estrulral das Universidades de Lisboa Complexo Interdisciplinar, Lisboa, Portugal.
Rosengren, A. Nordish Institute for Theoretical Atomic Physics, Copenhagen, Denmanrk.
Roth, H. Nova Werke AG., Winterthur, Switzerland.
Ruijter, H.P. Koninlijke/Shell-Laboratorium, Amsterdam, Holland.
Schmidt, R. Institute of Physical Chemistry, University of Frankfurt, Frankfurt am Main, F.R. Germany.
Schrijver, J. Laboratory of Polymer Technology T/Tc CT-hal, University of Technology, Eindhoven, Holland.
Shidlovsky, I. David Sarnoff Research Center, Princeton, N.J., USA.
Seifert, K.-F. Mineralogy Institute, University of Bonn, Bonn, F.R. Germany.
Spitzer, M. Institute of Physical Chemistry, University of Frankfurt, Frankfurt am Main, F.R. Germany.
Stock, M. Institute of Solid State Physics, University of Regensburg, Regensburg, F.R. Germany.
Strohmeier, W. Institute of Physical Chemistry, University of Würzburg, F.R. Germany.
Sutton, J.F. Department of Chemistry, Stanford University, Stanford, Carolina, USA.
Troe, J. Institute of Physical Chemistry, University of Göttingen, Göttingen, F.R. Germany.
Tsatsas, A. Inorganic Chemistry Laboratory, University of Athens, Athens, Greece.
Vanni, H. Institute de chimie minerale et analytique, Université de Lausanne, Lausanne, Switzerland.

Vrachnou-Astra, E. Department of Chemistry, Nuclear Research Center 'Democritos', Athens, Greece.

Walker, D.D. Depatment of Chemistry, Stanford University, Stanford, Carolina, USA.

Walper, M. Institute of Physical Chemistry, University of Frankfurt, Frankfurt, F.R. Germany.

Wasgestian, F. Institute of Inorganic Chemistry, University of Köln, Köln, F.R. Germany.

Whalley, E. National Research Council of Canada, Division of Chemistry, Ottawa, Canada.

Yersin, H. Institute of Solid State Physics, University of Regensburg, Regensburg, F.R. Germany.

INDEX OF NAMES

Adam, W.,566
Adams, D. M.,135, 154
Adams, W. A., 155
Adler, D., 219
Alani, G. H., 255
Alben, R. 219
Alberts, H. L., 126, 220
Albritton, L. M., 196
Alder, B. J., 51, 63, 82, 109
Alekseev, V. A., 279
Alexeyev, K. A. 48
Alfrey, T., 398, 401
Allen, W. C., 109
Alley, W. E., 63
Amagat, E. H., 133, 154
d'Amour, H., 195, 197
Andeen, C., 21, 48
Andersen, H. C., 109
Anderson, P. W., 277, 433
Andreatch, P., 48
Andreev, A. A., 279
Andrussow, L., 433
Angel, W. T., 49
Applequist, J., 468, 485
Arndt, E., 109
Arnold, S., 565
Asai, H., 409
Asano, T., 343
Ashcroft, N. W., 279
Atanov, Y. A., 18, 19, 48
Averill, F. W., 196
Axe, J. D., 126

Babb, S. E., 18, 19, 48, 564
Bacon, G. E., 125, 126
Balbo, S. A., 565
Balchan, A. S., 134, 154, 564
Baldwin, B. A., 565, 566
Baldwin, R. L., 470, 485
Ballato, A. D., 48
Ballhausen, C. J., 174
Baloga, J. D., 219
Banus, M. D. 196
Barker, J. A., 63, 196
Barnett, J. D., 26, 48, 49, 135, 154, 157, 196
Barta, C., 150, 154
Bartlett, K. 397
Bartunik, H. D., 172, 174
Basalo, F., 465, 466
Basile, L. I., 256
Bassett, D. 414
Bassett, W. A., 48, 49, 196
Bastie, P., 219
Bastron, V. C., 172
Bates, F. E., 142, 154
Bayer, H., 107, 110
Bayliss, N. S., 533, 565
Bean, V. E., 18, 48, 49
Beardless, R. A., 564
Beardslee, R. A., 566
Bechmann, R., 23, 48
Behar, J. V., 309
Beille, J., 220
Bell, P. M., 49, 196, 197
Belluco, U., 466
Benedek, G. B., 66, 106, 108, 110, 173
Bennet, R. G., 566
Bennion, R. D., 126
Benson, S. W., 294, 295, 209
van der Bergh, H., 519
Berggren, K. F., 278
Bernasconi, C. F., 323

Berson, J. A., 294, 295, 309
Bertie, J. E., 137, 142, 154
Bertucci, S. T., 109
Besson, J. M., 157
Beyerlein, R. A., 126
Bieg, K. W., 548, 551, 566
Bilot, L., 565
Birch, 38
Birgeneau, R. J., 126
Birks, J. B., 565
Bjerrum, N., 434
Blander, M., 257
Bloch, D., 111, 125, 199, 219, 220
Block, R., 278
Block, S., 26, 48, 49, 154, 157, 196, 197
Bloembergen, N., 106, 109, 110
Bodanszky, A., 433
Boer, 141
Bonner, W. A., 126
Boolchand, P., 279
Boren, M. D., 48
Born, M., 154
Bosco, C. D., 49
Boussert, A. S., 487
Bowman, H. A., 48
Braddley, W. H., 466
Bradley, C. C., 126
Brafman, O., 134, 154
Bragg, J. K., 468, 485
Brandts, J. F., 469, 470, 485
Brasch, J. W., 146, 154
Brasche, 135
Brauer, H.-D., 309, 322, 521, 566, 567
Brauer, K. R., 434
Braun, H., 567
Bredig, M. A., 257
Bresciani, N., 466
Brevil, M., 279
Bridgman, P. W., 33, 38, 132, 135, 145, 154, 468
Brinkley, S. R.,
Brower, K. R., 311, 317, 322
Brugger, R. M., 126
Brush, S., 63
Buback, M., 256, 257
Buccaro, J. A., 150, 151, 155
Buchanan, J., 433
Buchenauer, C. J., 155
Buckley, F., 157
Bukshpan, S., 174
Bunker, D. L., 519
Bunting, E. N., 157
Buras, B., 197
Burke, W. J., 152, 153, 155
Burnham, C. W., 255
Burns, G., 519
Burova, L. L., 48
Busfield, W., 384
Bussey, R. J., 309

Cabane, B., 279
Cahay, R., 144, 155
Caldin, E. F., 317, 322, 323, 446, 465
Calligaris, M., 466
Campbell, J. H., 109, 151, 152, 154
Cantor, D. M., 109
Cardona, M., 155
Carey, L. R., 433, 465, 486
Carle, D. L. 465
Carnevale, E. H. 313, 319, 323
Cerdiera, F. 152, 153, 155
Ceuterick, F. 323, 473, 485, 486
Champion, A. R. 173
Chandler, D. 83, 85, 88, 89, 109
Chandraskhar, S. 519
Chapman, D. 470, 485
Charvel, M. 256
Chatterjee, A. 219
Chen, Y. C. 49
Chin, J. H. 486
Christian, S. D. 145, 155
Christoe, C. W. 172, 173, 196
Chu, C. W. 219
Chuang, T. J. 519
Claesson, S. 433
Clark, C. D. 135, 155
Clark, R. J. H. 433
Clauwaert, J. 323, 485, 487
Clegg, R. M. 322, 323
Clendenen, R. L. 196
Cohen, M. H. 268, 278
Cohen, R. L. 174
Colwell, J. H. 21, 48
Copley, J. R. D. 126
Coston, C. J. 172

Cox, B. G. 465
Cox, D. E. 219
Crawford, R. K. 154
Cross, J. L. 18, 19, 48
Crothers, D. M. 478, 486
Cubicciotti, D. 256
Cundill, M. A. 141, 155

Dadson, R. S. 10, 13, 48
Dainton, F. 383
Damask, A. C. 565
Damle, V. 485
Dammel, R. 519
Daniels, W. B. 126, 154, 219
Daud, H. G. 433
Davies, D. F. 434
Davies, N. F. 255
Davis, A. R. 144, 155
Davis, E. A. 277
Deb, P. 382
Debrunner, P. 173
Debye, P. 58, 312, 323
Decker, D. C. 23, 25, 26, 49
Decker, D. D. 23, 49
Decker, D. L. 1, 48, 126
Degrunner, P. 219
Denison, J. T. 434
Denner, W. 175, 195, 197
Desmedt, H. 323, 471
Deul, R. 256
Deventer, E. H. 257
Devillers, M. A. C. 278
Dexter, D. L. 546, 548, 551, 566
Dieterich, W. 195, 197
Dolling, G. 126
Dow, J. D. 264, 278
Drickamer, H. G. 48, 49, 86, 109, 134, 144, 145, 154, 155, 156, 172, 173, 196, 219, 324, 485, 525, 531, 532, 535, 537, 545, 547, 548, 551, 563, 564, 565, 566
Dubovka, G. T. 220
Dupuy, C. 519
Duran, N. 566
Dymond, J. 63, 85, 109
Dynes, R. C. 126

Eady, R. R. 486
Ebisuzaki, Y. 156
Eckert, C. A. 291, 297, 298, 307, 309, 519
Eckert, C. F. 433
Edge, C. K. 219
Egelstaff, P. A. 125, 279
Eggers, F. 485
Eigen, M. 311, 315, 322, 465
Eisenberg, H. 482, 487
Eisenthal, K. B. 519
Eliason, R. R. 565
Elleman, D. D. 109
Ellenson, W. D. 156, 565
Elliot, N. 174
Ellis, A. J. 433
Elson, E. L. 323
El'Yanov, B. S. 293, 309, 425, 433
Emrick, R. M. 219
Endo, S. 219, 279
Engelborghs, Y. 322, 482, 483, 487
England, J. 197
Enskog, 59
Ernst, G. 126
Evans, M. G. 309, 402
Even, U. 278
Everaert, J. 323
Eyring, H. 309, 485

Faber, T. E. 278
Farrar, T. C. 109
Fenner, J. 174
Ferraro, J. R. 154, 157, 256
Finger, L. W. 197
Finney, R. J. 109
Fischer, A. 433
Fischer, G. 275, 279
Fisher, D. C. 109, 172
Fisher, F. M. 434
Fishman, E. 134, 155
Fitch, R. A. 134, 155, 564
Flagler, E. A. 473, 486
Flaig, B. 487
Flygare, W. H. 109
Folland, R. 104, 109
Fontanella, J. 21, 48
Forman, R. A. 49, 157, 196
Forst, W. 520

Förster, Th. 547, 548, 551, 559, 561, 564, 566, 567
Fourme, R. 197
Francesconi, A. Z. 256
Franchesi, E. 219
Franck, E. U. 221, 255-257, 277, 433, 434
Frank, C. W. 144, 155, 173, 564, 566
Fravenfelder, M. 172, 219
Frazer, B. C. 126
Freud, P. J. 196
Freyland, W. 278
Friedel, J. 219, 279
Friedman, L. 278
De Fries, T. 69, 109
Fritz, I. J. 126
Fritzson, P. 278
Fuchs, R. 156
Fujioka, N. 219
Fuller, P. J. A. 48
Fung, S. C. 172
Fuoss, R. M. 315, 323, 434
Furumi, Y. 434, 465

Gaarz, V. 485
Gaedtke, H. 520
Gardiner, R. G. 155
Garland, C. W. 219
Gass, D. M. 63, 109
Gauper, F. P. 482, 487
Gay, D. L. 465
Geffner, R. 565
Gehrig, M. 255
Gennis, R. B. 486
Gerber, B. R. 487
Gernard, L. 197
Getting, I. C. 49
Giardini, A. A. 48
Gibb, T. C. 172
Gibson, Q. H. 486
Gill, S. J. 485
Glänzer, K. 520
Gläser, W. 278
Glatzel, G. 413
Gliemann, G. 553, 567
Glogovsky, R. L. 485
Goldanskij, V. I. 172
Goldmann, G. 256
Golinskin, H. S. 309
Gonikberg, M. G. 425, 433
Good, R. 465
Graffe, M. 487
Grant, M. W. 322, 323, 465
Gray, H. B. 174, 439, 465
Green, F. R. Jr. 196
Greenwood, N. N. 172
Greig, R. C. P. 48
Grenoble, D. C. 172
Greulich, K. O. 486
Grieger, R. A. 297, 309, 318, 322, 323
Grieger-Block, R. A. 323
Grimsditch, M. H. 155
Grison, E. 279
Grover, R. 63
Grunberg-Manago, M. 487
Grundes, J. 155
Gruverman, I. J. 172
Guastalla, G. 465
Gubbins, K. E. 255
Gurevich, I. I. 126
Güsten, H. 566
Gutfreund, H. 321, 323
Gutowsky, H. S. 102, 109, 486

Haas, J. L. Jr. 225, 255
Hague, D. N. 323
Hale, C. F. 434
Hall, H. T. 48, 49
Halleck, P. M. 196
Hamann, S. D. 146, 154-156, 257, 293, 309, 337, 343, 425, 433, 434
Hamilton, W. C. 156
Hammes, G. G. 486
El-Hanany, U. 278
Hanson, J. P. 63
Hara, K. 434
Harder, W. 256
Hardy, H. D. 155
Harrington, W. F. 487
Hartmann, H. 309
Hartridge, 311
Hasinoff, B. B. 322, 323, 434, 465
Hawke, S. 135, 156
Hawley, S. A. 468, 470, 485
Haygarth, J. C. 32, 49
Haynes, D. H. 470, 485

Hecht, C. E. 63
Hedwig, G. R. 465
Heger, K. 256
Heidberg, J. 77, 102, 109
Hellemans, L. 322, 323
Hemmes, P. 315, 323
Henderson, D. 63
Hensel, F. 252, 255, 257, 259, 277-379
Herber, R. H. 172, 173
Heremans, K. 311, 322, 323, 467, 485-487
Herrmann-Ronzaud, D. 219
Heydemann, P. L. M. 1, 15, 32, 48, 49
Hilbert, R. 255
Hill, J. D. 48
Hill, W. 256
Hindley, N. K. 267, 278
Hinze, W. 197
Hippler, H. 519, 520
Hoa, G. H. B. 487
Hochachka, P. W. 484, 486
Hoebeke, J. 487
Hoffman, G. W. 519
Høiland, H. 433
Holbrook, K. A. 520
Holloway, I. R. 255
Holzapfel, W. B. 156, 159, 173-175, 177, 196, 197, 257
Hook, J. W. III 485
Hord, B. L. 49
Horne, R. A. 465
Horton, R. F. 155
Hoshino, H. 277, 279
Houck, J. C. 15, 18, 30, 48, 49
Houston, L. C. 487
Hryckowian, E. 49
Huang, K. 141, 154
Hubbard, C. D. 323
Hubbard, C. R. 196
Hubbard, J. 270, 278
Huber, J. G. 220
Hughes, D. S. 564
Hunt, H. R. 465
Hwang, J. 108, 110
Hyde, K. 466
Hyne, J. B. 309
Ikkai, T. 486
Imoto, T. 309, 404
Inada, E. 434
Ingalls, R. L. 172, 174
Ingold, C. K. 437, 465
Inokuchi, H. 565
Ising, 468
Ishihara, J. 434
Ivanova, E. M. 48
Ivin, K. 383

Jacobson, B. S. 519
Jacucci, G. 109
Jaenicke, R. 486, 487
Jamieson, J. C. 126, 197
Jancovici, B. 63
Jayaraman, A. 15, 48, 174, 219
Jeffery, R. N. 32, 49
Jenner, G. 288, 309
Johnson, D. P. 8, 48
Johnson, F. H. 473, 485, 486
Johnson, I. W. 256
Johnson, P. C. 540, 542, 544, 563, 565-567
Jonas, A. 486
Jonas, J. 65, 69, 108, 109, 150, 151, 155
Jones, B. W. 141, 156
Jones, W. E. 433, 465
Jorgensen, J. D. 126
Jortner, J. 268, 278
Jost, A. 318, 322, 323, 433
von Jouanne, J. 77, 102, 109, 309
Jura, G. 173

Kadhim, A. M. 565
Kaindl, G. 174
Kajiwara, T. 565
Kalus, J. 126
Kalvius, G. M. 173, 174
Kamb, B. 140, 156
Kamitakahara, W. A. 126
Kankeleit, E. 173
Kaplyanskii, A. A. 154
Karim, A. 156
Kasper, K. 559, 567
Kaufman, R. D. 486
Kauzmann, W. 433, 469, 485, 533,

565
Kawai, N. 48, 219
Kawski, A. 565
Kay, R. L. 471, 485
Kearns, D. R. 487
Keeler, R. N. 49
Kegeles, G. 487
Keller, R. 174, 195-197
Kelm, H. 281, 309, 313, 325, 329, 330, 421, 433, 435, 465, 466, 521, 566, 567
Kennedy, G. C. 48, 49, 196, 255
Kestin, J. 255
Kechinashvili, N. N. 486
Kikoin, I. K. 277
Kim, J. J. 566
King, A. D. Jr. 564, 565
King, H. 197
Kitamura, J. 434
Kittel, C. 219
Kivelson, D. 110
Klaboe, P. 155
Klein, V. F. G. 172, 174
Klevens, H. B. 528, 565
Klostermeier, W. 257
Klotz, J. M. 433
Klug, D. D. 147, 156
Kluyev, Yu. A. 156
Knapp, G. S. 220
Knoche, W. 321, 323, 487
Knopoff, L. 219
Knowles, P. F. 485, 486
Knox, J. H. 434
Kohnstam, G. 309, 519
Kong, P. C. 433
Konyaev, Yu. S. 9, 48
Korte, F. 397
Kortman, P. J. 219
Körtüm, G. 433
Köster, H. 255
Kotowski, M. 466
Kozuba, T. 48
Kramers, H. A. 520
Krouse, H. R. 465
Kruse, N. 465
Kudchadker, A. P. 255
Kulakov, V. V. 154
Kushida, T. 106, 107, 108, 110
Kusubov, A. 49
Kut-Sav, A. R. 220
Lai, C. F. 109
Laidlaw, W. G. 309
Laidler, K. J. 475, 486
Lakaszek, T. J. 48
Lamola, A. A. 566
Lang, E. 323
Langford, C. H. 439, 465
Laposa, J. D. 566
Larsen, H. A. 144, 156
Larson, D. B. 49
Lartz, L. 197
Lavergne, A. 49, 133, 158
Lawn, D. A. 433
Leake, J. A. 126
Lechtken, P. 566
Lee, J. 433
Legrand, J. M. 220
Leiber, C. O. 567
Lekner, J. 279
Lentz, H. 255, 256
Lever, M. J. 473, 486
Levey, R. P. 173
Lewis, G. J. 135, 145, 146, 156
Lewis, G. K. Jr. 172
Li, T. M. 469, 470, 478, 485, 486
Libby, W. F. 486
Lim, E. C. 566
Lim, S. C. 433
Lin, F. L. 487
Lincoln, R. C. 49
Lindemann, F. A. 54, 56, 59, 63, 516, 517
Lindner, H. 134, 156
Linowski, J. W. 109
Linton, M. 155, 156, 257, 433
Lippert, E. 565
Lippincott, E. R. 154, 157
Liptay, W. 565
Litovitz, T. A. 313, 319, 323
Liu, L. 196
Liu, Nan-I, 109, 471, 485
Lo, S. T. D. 465
Lockyer, G. D. Jr. 486
Lovesey, S. W. 125
Ludwig, H. 475, 486
Lüedemann, H. D. 49, 256, 322, 323, 434, 485-486
Lundgreen, B. 434
Luther, K. 519, 520

Luu, S. H. 520
Lynch, R. W. 198
Lyon, R. K. 519

Mabrey, S. 486
MacDonald, A. G. 467, 473, 484, 486
MacLeod, R. M. 485
Madelung, E. 156
De Maeyer, L. 311, 315, 322, 323, 484, 487
Maier, M. 519
Maier, S. 255
Maines, R. G. 48
Malkin, B. Z. 154
Mann, B. R. 433
Mao, H. K. 49, 196
Maple, M. B. 220
March, N. H. 219
Mares, M. 465
Margrave, J. L. 196
Mariée, M. 152, 156
Marin, J. 486
Mark, M. 63
Markau, K. 487
Markov, Y. F. 154
Maroni, V. A. 157
Marquis, R. E. 484
Marsh, D. 485
Marshall, W. 125
Marshell, W. L. 256
Mathews, J. F. 255
Martino, F. 278
Masterton, W. L. 434
Mathieu, J. P. 152, 156
Matsen, F. A. 564, 565
Mattheis, L. F. 278
Mauer, F. A. 196
Maury, R. 219
Maxfield, B. W. 323
May, A. D. 157
McCann, D. R. 197
McCartin, P. J. 566
McConnell, H. M. 486
McDonald, I. R. 109
McGuire, M. J. 155
McQueen, R. G. 219
McRae, E. G. 533, 565
McSkimin, H. J. 23, 48
McWhan, D. B. 125, 126, 196, 197, 218, 219
Meadows, L. F. 519
Medinger, T. 566
Melveger, A. J. 154
Menten, 474
Merbach, A. E. 322, 323, 446, 465
Merrill, L. 48, 196
Meyer, J. 16, 197
Michaelis, 474
De Mies, W. M. 157
Miller, K. W. 473, 486
Millero, F. J. 433, 434
Minkiewicz, V. J. 126
Misawa, M. 279
Mitchell, A. C. 255
Mitchell, D. J. 566
Mitchell, R. M. 470, 485
Mitra, S. S. 141, 154, 156, 157
Miyosawa, Y. 485, 487
Möller, H. S. 173
Moon, T. W. 486
La Mori, P. N. 48, 196
Morild, E. 475, 486
Mössbauer, R. L. 172, 173, 174
Mott, N. L. 213, 260, 267, 270, 277, 278
Moyzis, J. 173
Mullen, J. G. 174
Murmann, R. K. 465
Murphy, R. B. 486
Murrell, J. N. 519, 520
Mustafa, T. 486

Naake, L. D. 256
Nagle, J. F. 471, 485
Nakahara, M. 434
Nakashima, T. T. 565
Nalepa, R. 465
Namjoshi, K. V. 141, 156
Nauwelaers, F. 323
Nees, F. W. 256
Le Neindre, B. 48, 255, 257
Nellis, W. L. 255
Neuman, R. C. Jr. 309, 475
Nicol, M. 135, 156, 173, 542, 565, 566
Nicholson, 470

Nieuwenhuysen, P. 323, 480, 485, 487
Le Noble, W. J. 309, 313, 325, 343, 345, 389, 519
Nochuch, H. J. 487
North, N. A. 427, 433
Noyes, R. M. 494, 519
Van Nuland, Y. 322, 478, 487
Nye, J. F. 126

O'Connor, T. M. 482, 487
Offen, H. W. 135, 156, 540, 542, 544, 563-567
Ogo, Y. 376, 377, 417
Ok, H. N. 174
Okamoto, B. Y. 525, 531, 532, 537, 565, 566
Okamoto, T. 324, 434
Oksengorn, B. 525, 564
Olbrechts, H. 485, 486
Olcese, G. L. 219
Oldenziel, J. G. 78, 101, 109
Oliveira, R. J. 485
Ollinger, B. 196
Ooi, T. 486
Oref, I. 520
Oosawa, F. 483, 487
Oshika, Y. 565
Osugi, J. 384, 434, 542, 566
Overhof, M. 278
Owen, N. B. 136, 156

Packard, J. R. 196
Palmer, D. A. 281, 309, 421, 433, 435, 465, 466
Panyushkin, V. N. 173
Parisot, G. 125, 126, 219
Park, E. H. 565
Parker, A. J. 465
Parkhurst, H. J. Jr. 109
Parsons, B. J. 135, 156
de Pasquali, G. 219
Paton, W. D. M. 486
Patterson, D. O. 173
Paul, W. 157, 219
Pauling, L. 174
Paureau, J. 111, 126
Payan, D. G. 486
Payen, T. A. J. 487
Payne, S. J. 135, 154
Pearce, P. J. 434
Pearson, R. G. 465
Pecht, I. 323
Peercy, P. S. 126, 219
Peeters, J. 486
Pequeux, A. 486
Perez-Albuerne, E. A. 25, 49, 196
Perron, J. C. 277
Persoons, A. 322, 323
Pfeifer, M. P. 272, 278
Phillips, T. D. 566
Piermarini, G. J. 26, 48, 49, 135, 154, 157, 195-197
Pipkorn, D. N. 219
Pistorius, C. W. T. F. 156
La Placa, S. J. 156
Plachy, W. Z. 106, 110
Platt, J. R. 528, 565
Pohl, F. M. 470, 485
Polanyi, M. 309
Pollack, F. H. 155
Pollock, E. L. 63
Ponyatovskii, E. G. 220
Poole, C. P. Jr. 110
Pörschke, D. 485
Postgate, J. R. 486
Postmus, C. 135, 154, 157
Poulter, T. C. 133, 157
Powles, J. G. 109
Prakash, A. 156
Pressley, R. J. 152, 155
Price, C. 398, 401
Price, J. H. 48
Privalov, P. L. 486
Prosser, H. 172, 174
Purcell, E. M. 66, 108

Quack, M. 520
Quist, A. S. 256

Rabinovitch, B. S. 520
Ramdas, A. K. 155
Ramsey, J. B. 434
Randaccio, L. 466
Rapoporte, 108, 110
Rauchschwalbe, R. 323
Redfield, D. 264,278
Reed, T. M. 255

Reese, R. A. 126
Reid, R. C. 519
Reisler, E. 482, 487
Reuter, K. 256
Revesz, A. 433
Rice, M. H. 219, 255
Ricevuto, V. 466
Richter, P. W. 149, 150, 157
Rifkind, J. M. 468, 485
Rigby, M. 63
Rigler, R. 323
Rijckenberg, J. 323, 486
Rinck, G. 309
Robb, W. L. 86, 109
Roberts, L. D. 173
Robertson, W. W. 525, 564, 565
Robinson, J. M. 564
Robinson, P. J. 520
Robinson, R. A. 434
Ross, R. G. 278
Ross, S. M. 104, 109
Van Rossen, A. 486
Rotter, C. A. 126
Roughton 311
Roult, G. 126
Rowland, F. S. 520
Rowlinson, I. S. 255
Runcorn, S. K. 135, 157
Ruoff, A. L. 49
Ruska, J. 275, 279
Russell, K. E. 309
Rynbrandt, J. D. 520

Sahlin, H. 63
Salomon, D. 174
Samara, G. A. 48, 219
Samson, F. 487
Sapse, A. M. 63
Sasaki, H. 219
Sayers, C. M. 219
Schaafsma, T. J. 106, 110
Schenk, G. 465
Schiferl, D. 195, 197
Schiff, D. 63
Schilling, J. S. 174
Schmid, G. 475, 486
Schmidt, R. 521, 566, 567
Schmutzler, R. W. 275, 277, 279
Schock, R. N. 133, 134, 157
Scholz, B. 256
Schönherr, G. 278
Schoof, S. 566
Schore, N. E. 566
Schröder, J. 519
Schuele, D. 21, 48
Schuetzle, 520
Schulz, E. 487
Schulz, H. 195, 197
Schuster, G. 566
Scott, G. J. 48, 373, 380
Sechenkov, A. R. 277
Seidel, H. P. 567
Seidell, C. W. 172
Selinger, B. K. 567
Senogles, E. 373, 380
Seright, R. S. 323
Shapiro, S. M. 126
Shaw, R. W. 542, 565
Shecter, H. 174
Shenoy, G. K. 172
Sherman, W. F. 135, 141, 147, 149, 155-157
Sherwood, T. K. 519
Shimizu, K. 434, 465
Shimshick, E. J. 486
Shirane, G. 126
Shirotani, I. 565
Siebrand, W. 542, 565
Simon, F. 413
Simpson, G. A. 566
Singer, 470
Singh, A. K. 196, 219
Sinha, S. K. 126
Skalyo, J. 126
Sleigh, M. A. 484
Slichter, C. P. 566
Slonczewski, J. C. 155
Slykhouse, T. E. 155, 564
De Smedt, H. 485, 486
Smith, F. A. 126
Smith, H. G. 126
Smith, J. A. S. 108, 110, 433
Smoluchowski, M. W. 312, 323
Smulovitch, P. P. 147, 149, 157
Snauwaert, J. 320, 322, 323, 486
Somekh, J. 565
Somero, G. N. 484
von Sonntag, C. 566
Spedding, F. H. 434

Spijkerman, J. 173
Spiro, T. G. 433
Stace, A. J. 519, 520
Stassis, C.126
Staun-Osen, J. 197
Steinmetzer, H. C. 566
Stejskal, K. F. 156, 564
Stevens, J. G. 172
Stevens, V. E. 172
Stock, M. 553, 567
Stokes, R. H. 434
Stover, B. J. 485
Strange, J. M. 76, 104, 109
Stranks, D. R. 306, 309, 323, 435, 437, 443, 451, 454, 459, 465
Strauss, H. L. 63
Strauss, W. 433, 434
Strehlow, H. 323
Stromberg, H. D. 133, 134, 157
Stryland, J. C. 157
Stuehr, J. 323
Sturtevant, J. M. 486
Suck, J. B. 278
Suito, K. 48
Sund, H. 487
Suromyi, P. 279
Sutherland, J. W. H. 481, 487
Suzuki, C. 487
Suzuki, K. 279, 434, 475, 485, 487
Swaddle, T. W. 309, 433, 435, 449, 465
Swenson, C. A. 196
Syassen, K. 156, 172-175, 195, 196,
Szafranski, A. W. 257
Szigetti, 141
Szwarc, M. 433

Taani, R. 255
Tait, P. G. 309
Takahashi, T. 49
Takaya, M. 157
Takenouchi, S. 255
Tamura, T. 309
Tanaka, F. 324, 542, 544-546, 566
Tang, Y. N. 520
Taniguchi, Y. 145, 157, 434, 485
Tanquary, R. L. 156, 564, 565
Tapavicza, S. V. 256
Tarasov, L. V. 126
Taube, H. 126, 465
Tauc, J. 265, 278
Teller, E. 63
Thirsk, H. R. 433
Thomson, R. N. F. 473, 486
Thorn, H. 275, 279
Thusius, D. 323, 482, 487
Thurston, R. N. 48
Tiltscher, H. 309
Tippelskritch, H. V. 255
Tödheide, K. 255-257
Tobe, M. L. 465
Tomizuka, C. T. 219
Tong, S. P. 465
Tourand, G. 279
Trappeniers, N. J. 78, 101, 109
Träuble, M. 470, 485
Traylor, J. G. 126
Treiber, G. 255, 257
Tregloan, P. A. 322
Trickey, S. B. 196
Trigueros, M. 63
Troe, J. 489, 519, 520
Troes, Dr. 312
Trudell, J. R. 472, 473, 486
Tsuchihashi, N. 434, 465
Tsuchyia, M. 487
Turgunov, T. 279
Turro, N. J. 564, 566
Twigg, M. V. 465

Uchtmann, H. 278
Uematsu, M. 256
Umebayashi, H. 126
Usha Devi, S. 219

Vaidya, S. N. 196
Van Valkenburg, A. 157
Vallade, M. 219
Vanderhoek, N. 443, 454, 465
Vandersypen, H. 320, 322, 323
Vanfleet, J. 433
Vanni, H. 323
Vaughan, J. 433
Vaughan, R. W. 77, 109, 173
Veleckis, E. 257

Vettier, C. 125, 126, 219
Vodar, B. 48, 255, 257, 525, 564
Vogel, K. 433
Voiron, J. 126
Vorob'ev, A. M. 156
Vu, H. 564

Wagner, F. E. 172, 174
Wainwright, T. E. 63, 109
Waisbluth, M. D. 322, 323
Walling, C. 384, 409
Walrafen, G. E. 134, 157
Walsh, J. M. 219, 255
Walsh, W. M. 108-110
Wang, Y. G. 309
Ware, W. R. 566
Warren, W. W. Jr. 277-279
Warschauer, D. M. 219
Watanabe, T. 434, 485
Wattiaux, R. 472, 486
Watts, A. 485
Watts, D. W. 465
Weale, K. E. 399, 519
Weaver, J. Scott, 25, 49
Weber, G. 322, 324, 485
Wedlake, R. J. 135, 155
Weeks, J. D. 109
Wegener, H. 172
Weigang, O. E. Jr. 564, 565
Weir, C. E. 157, 197
Weis, J. J. 63
Welch, B. E. 48
Weller, A. 567
Welles, H. 434
Welsch, H. 255
Wertheim, G. K. 172
Weser, G. 279
Westort, C. 485
Whalley, E. 43, 49, 127, 132-134, 136-138, 140, 142, 144-147, 150, 154-158, 302, 309, 345
Whitten, W. B. 565
Wick, F. G. 133, 158, 434
Wiese, G. 323
Wilbur, D. J. 109
Wilkins, R. G. 465
Wilkinson, F. 566
Will, G. 197
Williamson, D. L. 173, 174
Wilson, C. J. 323
Winstein, S. 388
Wishnia, A. A. 317, 323, 480, 487
Woessner, D. E. 109
Wohlfarth, E. P. 220
Wohlleben, D. 220
Wolfe, M. 109
Wong, P. T. T. 132, 137, 150, 155, 157, 158
Wong, W. H. 519
Worlton, T. G. 126
Wortmann, G. 172, 174, 175
Wu, C. 486
Wycoff, R. G. W. 144, 158
Wynne-Jones, The Lord 427, 433

Yamada, H. 77, 109, 132, 158
Yamamoto, T. 48, 49
Yates, M. G. 486
Yekta, A. 566
Yelon, W. B. 126, 219
Yersin, H. 553, 567
Yip, S. 63
Yomo, Shusuke 33
Yonezawa, F. 278
Young, D. A. 63
Yu, A. D. 322
Yu, Y. M. H. 566

Zahn, V. 174
Zeto, R. J.
Zeyen, C. M. E. 219
Ziman, J. M. 219, 270
Zimm, B. H. 468, 469, 485
Zimmerman, A. M. 485
Zipp, A. 485, 533, 566
Zussman, A. 108, 110
De Zwaan, J. 109
Zwolinski, B. I. 255

INDEX OF CHEMICAL COMPOUNDS

Acenaphthylene 383
acetic acid 335, 423
acetonitrile 221, 243, 244
acetyldiazene 357
N-acetyl-L-tryptophan ethyl ester 477
acrylamide 397
acrylonitrile 396, 397, 407
adamantane 102, 103
adenosine monophosphate 314
adipic acid 423
allenylidene 348
allyl acetate 383
alumina 116-119
aluminum 119
  alloys 116-118
  oxide 194
ammonia 221, 223, 243, 244, 332, 333, 426, 446
ammonium chloride 211, 249, 250
  ion 147, 149, 314, 334
aniline 426
anthracene 133, 134, 527-530
antimony iodosulfide 211
aquooxalatochromium (III) 453
aquopentaamminochromium (III) 445
  cobalt (III) 445
  iridium (III) 445
  rhenium (III) 445
argon 92, 93, 213, 223, 226, 232
arsenic 214
  sulfide 20, 22
azulene 530ff.

Barium 25, 30, 203
benzene 80-87, 223, 232, 351, 527-530
benzotrifluoride 99, 100
benzyne 348
Berylco 72
beryllium 189
betaine 304
Biebrich scarlet 317
biphenyl 341, 351
bismuth 17, 21, 25, 30, 32, 33, 121, 201
  germanate 20, 22
  trichloride 223, 248, 249
bisulfate ion 143
boron 179
bromide ion 446
bromophenol blue 314
brucite 133
n-butane 232
butyl acrylate 376
tert-butyl chloride 80, 339
butyl mercaptan 300
tert-butyl mercaptan 376
butyl methacrylate 376
butyl radical 508-511
butyraldehyde 384

Cadmium 201, 203
calcium carbonate 20, 22
  fluoride 21, 133
  ion 429
carbene 347
carbon 200
  dioxide 223, 231-235
  disulfide 70
carbonic acid 424

carbon monoxide 314
 tetrafluoride 80, 90-93
cerium 55, 214, 215
cesium 55, 200-204, 214, 215, 251, 260, 270
 bromide 141, 147, 149
 chloride 18, 23, 25
 chloroaurate 170, 171
 hydride 251
chloral 384
chloride ion 446
chlorine 503
o-chlorobenzotrifluoride 99, 100
p-chlorobenzotrifluoride 100
chloroform 151, 152
chromium 214
chrysene 530ff.
chymotrypsin 314, 317, 475-478
chymotrypsinogen 470
cobalt 204
 ion 429
 ion (2+) 446
 (II) oxide 165-167
copper 200, 201, 223
 alloys 116, 117, 118
coronene 530ff.
cycloheptatriene 507, 507
cyclohexane 105, 341
cyclopentadiene 292, 360
cytochrome C 320, 321

Deuterium oxide 69, 80, 94-98, 142
diamond 26, 27, 132-139, 162, 179, 185-191
diaquobis(ethylenediaminato) cobalt (III) 443-445, 454
diazene 357
dibromoethane 145
dichlorodipyridylplatinum (II) 457
dichloroethane 145
α, α-dideuterobenzyl cyanide 81
diethylenetetraminoplatinum (I) chloride 298
2,6-diisopropylpyridine 362
dimethylacetylenedicarboxylate 292
dimethylamine 426
dimethyl -2, 3-butadiene 388-391
dimethyl ether 446
dimethylethylphenylammonium iodide 340
dimethylformamide 341
2,6-dimethylpyridine 362
dimethyl selenide 446
dimethyl sulfide 446
dimethyl telluride 446
dioxane 80
dipalmitoyllecithin 472
diphenylpicrylhydrazyl radical 300
di-tert-butylproflavin 478
dysprosium 202

erbium 202
 nickel alloy 204
ethane 146, 223, 231, 232
ethanoic acid 423
ethanol, deuterated 118
ethidium bromide 478
ethylene 221, 237-240, 368, 384-387, 403
ethylenediaminopalladium (II) halide 460
 platinum (II) halide 458-460
eugenol 289
eugenoxide ion 289
europium 202
europium II compounds 163
europium III compounds 163
europium (II) oxide 164, 165
 (III) oxide 164
 selenide 165
 sulfate 164
 sulfide 164, 165
 telluride 164, 165
 titanate 164

ferric chloride 170
 ion 168, 169, 314
ferricyanide ion 320, 321
ferrous bromide 122
 chloride 122, 125

ion 165-167, 169
Fluorinert 118
fluorobenzene-d_5 80
perfluorobutane 80, 90
fluoronitrene 350
fluorotrichloromethane 80, 83, 89, 90
freon 244

Gadolinium 202
gallium 214
germanium 133, 147, 148, 152, 153, 214
glutamate dehydrogenase 479, 482
glutaric acid 423
glycine anion 335
gold 201, 203
cyanide 171
guanidine 333

Hafnium 203
helium 56, 57, 118, 124, 200, 213
heme 314
hemaglobin 314, 477
hexaaquochromium (III) 445
hexachlororhodate 456
hexa(dimethylformamido) chromium (III) 446
hexamethylethane 105
hexane 231
n-hexane 227, 229
hexa (sulfinylbis(methanato)) chromium (III) 446
hexyl mercaptan 300
holmium 202
cobaltide 204
hydrogen 56, 57
chloride 243
sulfide 424
ion 314

Ice 95, 140-142
indene 375, 376, 408, 409
iodine 491-503
iridium 203
iron 57-59, 201, 204, 216, 252
-rhodium alloys 217
silicide 59
sulfide 58
Irtrans 133
isoprene 295, 297, 377, 388-394, 408, 409

Krypton 124, 213

β-lactoglobulin 314
lanthanum 203
ion 429
lead 124, 201
lithium 202
hydride 251
lutetium 202
lysozyme 470

Magnesium ion 429
oxide 108
maleic anhydride 295, 297, 378-382, 405
malonic acid 423
manganese auride 122, 123
ion 429
oxide 210
sulfate 334
manganin 13, 18-21
mercurous bromide 149, 150
mercurous chloride 149
mercury 10, 21, 30, 31, 203, 221, 223, 226, 229, 230, 252, 253, 260-270
methane 223, 232
-^{13}C 101
methanol 80, 221, 223, 227-229, 243, 289
deuterated 118
methoxide ion 289
methylamine 335, 426
methyl iodide 67, 292, 301, 303, 307, 362
-d_3 68
α-methylstyrene 384, 401, 405, 407
molybdenum 203

Napthalene 526-530
neodymium 202
neon 92, 93, 124, 213
neopentane 80, 102, 104, 105

nickel 204
  chloride 242
  fluoride 120
  ion 429
  ion (2+) 446
  superalloys 252
niobium 203, 446
nitrene 350
nitrogen 137, 232
nitrogenase 473, 477
nitrogen dioxide 506, 507
p-nitrophenol 333
2-norbonyl ester 352
Nylon 74

tert-octyl mercaptan 300
octyl methacrylate 376
osmium 203
oxalic acid 423
oxygen 314

Palladium 203
pentaamminochromium (III)
  448-451
pentaamminocobalt (III)
  448-450
pentaamminonitratocobalt (II)
  ion 429
pentacyanatocobaltate 455
perylene 526-530
phenanthrene 527-530
phenol 333
phosphoric acid 424
phosphorus 214
  pentabromide 144
  pertachloride 144
picrate ion 314
pimelic acid 423
platinum 203
polonium 273
polyethylene 14-17, 221, 237-
  239, 412, 414
polytetrafluoroethylene 15,
  74
potassium 123, 124, 202, 223,
  270, 270
  aurochloride 171
  aurocyanide 171
  bromide 124, 141
  chloride 246, 247
  cyanide 121
  dihydrophosphate 211
  ferricyanide 144, 145
  ferrocyanide 144
praseodymium antimonide 123
proflavin 314, 476, 478
promethium 202
propane 137
propanoic acid 423
isopropanol 289
iso-propoxide ion 289
propylene 137, 383
pyridine 80, 292, 301, 303,
  362, 426
  -2-azodimethylaniline 446

Quartz 135

Rhenium 203
rhodium 124, 203
rubidium 202, 203, 260,
  270-272
  ion 429
ruby 5, 18, 26, 28, 186
ruthenium 203

Samarium 202
  sulfide 215
sapphire 133, 135
selenium 191, 192, 214, 260,
  273-276
silicon 133, 147, 148, 214
silver 183, 201, 203
sodium 124, 202, 251
  chloride 5, 18, 23, 25,
  26, 133, 134, 140, 180-
  183, 188, 223-226, 234,
  248
  fluorenyl 428
  hydride 251
  hydroxide 241
stainless steel 72
steel 8, 15
  maraging 116-118
strontium 203
  titanate 152, 153
styrene 368, 376, 378, 401,
  407
succinic acid 423
sulfur 260, 273, 276, 277

hexafluoride 80, 89
sulfuric acid 424

Tantalum 203, 446
technetium 203
tellurium 192, 214, 260, 273-276
ion 429
oxide 120, 124
terbium 202
cobaltide 204
tetrabutylammonium ion 314
picrate 335
tetracene 526-530
tetrahydrofuran 428
tetramethylsilane 80, 83, 85
thiocyanate ion 314
thorium 30
thulium 202
tin 214
titanium 214
alloy 72
toluene 151
$-d_8$ 68, 101
trimethylamine 426
trioxane 146
tripropylamine 307
tritriane 146
tropone 360
trypsin 314, 475, 477
tungsten 203, 223
carbide 8, 12, 13, 15, 179-181

Uranium 200
urea 340

Vanadium (III) oxide 213, 214
Vespel 74
vinyl acetate 376, 401

Water 80, 94, 95, 97, 101, 139, 140, 142, 221, 223, 224, 227, 229, 231-234, 241, 243, 424

Xenon 183, 185, 213, 232

Ytterbium 188, 202
yttrium 203
cobaltide 204

Zinc 201
ion 429
zirconium 203
zincide 217, 218

INDEX OF SUBJECTS

Absorption band broadening 525
Absorption coefficient spectral dependence of 264
absorption intensity 538
acid-base equilibrium 422
activation,
  chemical 512
  compressibility of 282
  pressure 282
  volume 281-309, 347, 435, 490, 514
    graphical determination of 295
    interpretation of 298
    of chain reactions 288
    of enzyme reactions 476, 477
    of first order reactions 283
    of first order reversible reactions 285
    of parallel reactions 287
    of reversible reactions 286
    of second order reactions 285
    of unimolecular reactions 515
    of unimolecular isomerization reactions 518
    pressure dependence of 306ff.
addition reaction 511
alkali metal
  chlorides, conductance of 248
  expanded 270ff.
A-mechanism 441
analysis by IR and Raman spectroscopy 143ff.
anesthesia 473ff.
anation reaction 456
  volume change in 454
Anderson transition 260, 268
anionic polymerization 388-392
antiferromagnet phase diagram of 205
aquation reaction 456
association 497ff.
azeotropic lines 407

Band crossing transition 260, 263
band gap closing of, in mercury 267
Benson-Berson equation 295
beta scission 387
binary systems 230
biochemical reactions 467-487
bond length 337
  rotation 342
Born-Oppenheimer approximation 200
Brillouin spectroscopy 150
bulk modulus 203

Cage effect 495, 507
  photolytic 490
cage recombination 496
carbene reactions 347-349
carbonium ions 345, 346
carboxylic acid reaction volume 423
ceiling temperature 373, 383, 384
charge delocalization 333
charge transfer 304
  complex 335, 404, 405
chemiluminescence 552
Claisen rearrangement 342
Clausius-Mosotti relation 62
cluster theory 91
collision efficiency 505
collision frequency 505
    clock 507
collisional quenching 505
  stabilization 508
complex formation, rapid 431ff.
complex, square-planar 449
compressibility 368, 410, 411
  coefficient 422
  correction 336
compressibility, isothermal 268, 272
  of alkali metals 202
  of liquids 330
  of solids 200
  of solvents 282
computer simulation 51-63
conductance 247
  of alkali chlorides 248
conductivity, electric, of mercury 261
  thermal 228
configuration coordinate model 535
conformational change 146
coordination complex, octahedral 442
coordination compounds 435-466

copolymerization 398-400, 404, 405
critical behaviour 207ff.
  density 53
  line for binary systems 230
    in gas mixtures 52
critical point 54, 55
  pressure of chalcogens 260
  temperature of chalcogens 260
  temperature of Jupiter's core 56
cryostat, for X-ray diffraction 184
crystallization 414
crystals, plastic 102
Curie point 206
cycloaddition reaction 360

Data processing, in high pressure kinetics 288
Debye equation 97
degradative transfer 374, 380, 383
density 271
  of alkali chlorides 248
depropagation effect 383
dielectric constant 243, 302
  permittivity 243
  properties 61, 242ff.
  solvent theory 533
Diels-Alder reaction 358
diffusion 228
  coefficient 59
  controlled process 552
dilote solutions 327
dipole moments 302
diradical intermediates 351
dissociation quenching 504
  volume 423
Drude-Nernst equation 332
D-mechanism 439
dye-nucleic acid interaction 478

Earth, interior of 58ff.
effective charge 140
electric conductivity 266, 270, 271, 274
  of mercury 261
electrical connections 43-47
  properties of polymers 415
  transport 269
electric field jump 319
  jump technique 316
  modulation in fast reaction study 314
electrolytic conductance 250
electron gas, in Jupiter's core 57

model of 53, 54, 57
electronic excitation 522
phase transition 212
phase transition, valence 214
structure, inner 202
of fluids 259-279
outer 202
electrostriction 304, 332, 333, 389, 424, 447, 453
solvent dependence of 334
emission spectra 530ff.
Enskog equation 229
theory 86
enzyme catalysis 474
kinetics 475
equation of state 272
calculation of 51
of alkali metal halides 23ff.
of fluids 223
of salt solutions 224
van der Waals 53
equilibria, rapid 421-434
equilibrium density 250
constant 421
ESR 106ff.
eutectics in earths core 58
exchange energy transfer 551
excimer, formation of 559
Eyring equation 328

Fast reaction 311-324
chemical control 312
physical control 312
Fermi resonance 146
ferromagnet, phase diagram of 205
flash photolysis 318
fluids 221-257
electronic structure of 259-279
hard sphere 82
semiconducting 259ff.
spectroscopy of 235
supercritical dense 68
fluorescence 526
lifetime 538, 539
lifetime study of fast reaction 314
spectra 530
Fokker-Planck equation 516
formation reaction 452
Förster-Dexter theory 548
Franck-Condon principle 524
Fuoss equation 315

Gages, *see* pressure gages
gas reaction kinetics 489-520
  self diffusion coefficient 66
gel effect 367, 378
Gibbs energy 234
glass transition 410

Hall coefficient 262, 270
  mobility 262
Hammett function 346
Hammond postulate 361-363
hard sphere fluids 82
  model 91, 272, 332
heat of compression 326
hot atom reaction 510
hydrogen bonding 95, 301, 424
hydrolysis-formation reaction 455-457
hydrolysis reaction 446-452
hyperfine interaction, in Mössbauer spectroscopy 161

I_a-mechanism 441
I_d-mechanism 440
intercalation 478
internal conversion 524
  transfer 386
intersystem crossing 524
ion formation 246ff.
ionic conductivity 246ff.
  dissociation of alkali chlorides 248
  polymerization 372, 387, 392
ionization 339
  auto- 331
  of gamma-haloamines 354ff.
  volume 331, 334
ion mobility 246
  nonclassical 352-354
  pair 388
    dissociation 334
    formation 427, 453
    loose 388, 389
    solvent separated 430
    tight 388-390
IR band intensity 142ff.
  band shape 150ff.
  spectroscopy 127-158, 305
    theory of 127-131
isomerization 341, 443
  quenching 504
isomer shifts 163-164, 171
isosbestic point 284

isotope effect 305, 363

Jablonski diagram 522, 523
Jahn-Teller effect 170
jump techniques 480
Jupiter, interior of 56ff.

kinetics of gas reactions 489-520
Kirkwood equation 307
  factors 244
Knight shift 263, 267, 269

Landau theory 207ff.
laser flash photolysis 314
  spectroscopy 493
Lennard-Jones potential 91
ligand substitution 435ff.
light scattering 61ff.
Lindemann-Hinshelwood model 516
  scheme 512
Lindemann's law 59
  rule 54
Liouville theorem 516
lipid bilayer 471
liquid compressibility 330
  molecular motion in 80
  NMR of 66
  self diffusion coefficient 66
Lorentz-Lorentz equation 533
low temperature neutron scattering 118
lubricants 36

Magnetic elastic scattering 114
  hyperfine splitting 165
  transition 216
magnon 124
many body complexes 498
Maxwell correlations 282
mechanical properties of polymers 416
mechanism of ligand substitution 437ff.
melting 54
  of cerium 55
  of polymers 367, 413
membranes, biological 472
Menschutkin reaction 291, 303, 307, 342
metal-insulator transition 212, 260
metal-nonmetal transition 252, 268
  in mercury 261
microtubuli 482, 483
miscibility of molten salts with liquid metals 251

molecular dynamics 51, 495
  model, configuration coordinate 535
    electron gas 54
    hard sphere 53, 54
    plasma 54
    random flight 494, 495, 516
  potential 51
    coulomb 52
    pseudo 52
monomer-polymer system 237
Morse function 305
Mössbauer effect 160
  spectroscopy 159-175
Mott theory 213, 267

Navier-Stokes equation 60
neutron monochromator 113
  scattering 111-126
  sources 111
  windows 117
nitrene reactions 350-351
nitriles, polymerization of 395
NMR 65
  apparatus, nonmagnetic materials for 72
  Fourier transform 99
  high resolution 77, 98ff.
  of liquids 66ff.
  of solids 76ff., 102
  pulse spectrometer 69
  relaxation 65, 80
  sample cell 72ff.
nonhydrostatic stress 151
nonmetal-metal transitions 252, 268
NQR 106ff.
nuclear elastic scattering 114
  inelastic scattering 115
nucleic acid-dye interaction 478
nucleic acid phase transition in 468, 469
nucleophilic substitution reaction 462

Ocean 478
optical absorption 276
  band gap 266
orbital interaction, secondary 359
order-disorder transitions 204
order of reaction, pressure dependance of 512
order parameter 206
organic model reactions 325-343
  problem reactions 345-363
oscillator strength 528

Pericyclic reactions 358
phase change 412, 414
  crystallographic 120
  magnetic 122
phase diagram 205
  computer simulation 52
  of binary systems 230
  of D_2O 69
  of ternary systems 234
phase, high pressure, recovery of at low temperature 136
  separation, in Jupiters core 56
    of monomer-polymer system 237
  transition 204ff.
    electronic 212
    in cesium 55
    in nucleic acids 468, 469
    in phospholipids 470, 471
    in proteins 468, 469
    of barium 25
    of bismuth 25, 32ff.
    valence electronic 214
phonon 124
phosphorescence lifetime 539
  spectra 531
  bilayers 473
  phase transition in 470, 471
photochemical primary processes 521-567
photolytic cage effect 490ff.
photophysical primary processes 521-567
  processes definition 522
planar complex, four coordinate 457
plasma, model of 54
polymerization, anionic 382-392
  mechanism 370, 374, 385, 388, 404
  of nitriles 395
  pseudocationic 393, 394
  radical co- 398, 407
  reaction 240, 370, 373, 384
  solid state 394-396
  ter- 305-307
polymorphism 414
pressure apparatus, for fast reaction study 317-320
  for mixing 328
  for phase equilibrium study 233
  for X-ray diffraction 180
pressure cell, anvil 138
  circular anvil 135
  for corrosive fluids 241
  for IR spectroscopy 236
  for neutron scattering experiments 116
  for X-ray diffraction 179ff.

for X-ray diffraction 179ff.
multianvil 135
pressure, definition of, 1
experimental measurement 3
gages, deadweight loaded piston 6
dielectric 21ff.
hydraulically loaded piston 15-18
mangamin 18ff.
piston 7-14
primary 5
ruby fluorescence 26ff.
sodium chloride 23ff.
transfer 5, 18ff.
ultrasonic 22-28
pressure, generation of 1-49
generators 37-40, 70, 71
hydrostatic 2, 70
jump technique 314, 316
movement of 1-49
of activation 282
rams 36
scale, fixed points on 1, 29ff.
primary 1, 3, 4
techniques, for Mössbauer spectroscopy 162
vessels 35
transparent 132ff.
protein assembly 480, 481
denaturation 470
folding 469, 470
phase transition in 468, 469
protolytic reaction of electronically excited molecules 555
proximity effect 495
pseudocationic polymerization 393, 394
pumps 35

Quadropole moments 302
splitting 171
quantum yield 490ff., 513, 514ff.
fluorescence 538

Racemization 341
radical copolymerization 398-407
decay-recombination 287
Raman band intensity 142ff.
shape 150
Raman spectroscopy 127-158
theory of 126-131
randomization problem 511
Rankine-Hugoniot equations 5

rare earth elements 56
rate constant 281
  unimolecular specific 504, 505, 512
reaction, diffusion-controlled 499
  volume 360, 447
reactivity ratio 398-405
recombination 494, 497ff.
  geminate 493
  of radicals 338
  nongeminate 493
redox reaction, biological 320
relaxation techniques 65, 80
  time 317
resonance energy transfer 547
ribosome subunit assembly 479, 480
Roedel mechanism 386
rotating sector method 372, 417

Seals 41ff.
self-diffusion coefficient, of liquids and gases 66
semiconducting elements 273
  fluids 259ff.
shear viscosity 86
shock waves 56, 500
  sonic 34ff.
Siebrands law 542
solid 199-120
  compressible 199ff.
  disordered 102
solidification 368, 413
solid state polymerization 394-396
  structure 149
solubility of polymers 368
solution equilibria 421-434
solvation 301
solvent dielectric theory 533
  effects on spectra 527
  shift theory 532
spectral shift 525, 528
spectroscopy of fluids 235
spin selection rule 522, 523
statistical method, Monte-Carlo 51
steady state 286
steric hindrance 460
Stern-Volmer plots 506, 508
Stokes-Einstein equation 60, 86, 97
Stokes law 517
stopped-flow techniques 311ff., 314, 315, 319, 476
stress, nonhydrostatic 151ff.
  tensor, deviatoric 2

symmetric 1
structure factor 271
substitution reaction 443, 446-457, 461
sun, interior of 57
superconductivity 213ff.
supramolecular structures 479
systems, ternary 234

Tacticity 409
Tait equation 427, 449, 553
functions 294
telomerization 408, 409
temperature control 326
jump technique 314, 316
terminal model 398, 399
ternary systems 234
terpolymerization 405-407
thermal conductivity 228, 411, 412
dissociation 513
reaction 511
thermal expansion coefficient 268
thermodynamic parameters, partial 328
properties of fluids 222
thermoelectric power 262, 270
Thomas Fermi approximation 201
trajectory calculation 495, 500-502, 518
transducers, capacitive 21ff.
transition state 307, 369, 372, 394
theory 281, 336, 515, 517
transport properties 59ff., 411, 412
of fluids 222
tricritical behaviour 209ff.

Ultrasonic study of fast reaction 314
unimolecular reaction 515, 516
Urbach rule 264

Valence transition 214
vapor pressure 250
vibrational frequency 137
viscosity 59, 228, 367, 373, 388, 412, 518
volume change 337
in anation reaction 454
of activation 281-309
profile diagram 459

Windows 43
Wöhler synthesis 340

X-ray diffraction 177-197

angular dispersive 179
energy dispersive 195ff.
single crystal 189ff.
theory 178
X-ray diffractometer, automatic four circle 192ff.

Zwitterionic intermediates 351

GPSR Compliance

The European Union's (EU) General Product Safety Regulation (GPSR) is a set of rules that requires consumer products to be safe and our obligations to ensure this.

If you have any concerns about our products, you can contact us on ProductSafety@springernature.com

In case Publisher is established outside the EU, the EU authorized representative is:

Springer Nature Customer Service Center GmbH
Europaplatz 3
69115 Heidelberg, Germany

Batch number: 10809436

Printed by Printforce, the Netherlands